Introduction to Magnetohydrodynamics

Magnetohydrodynamics (MHD) plays a crucial role in astrophysics, planetary magnetism, engineering and controlled nuclear fusion. This comprehensive textbook emphasises physical ideas, rather than mathematical detail, making it accessible to a broad audience. Starting from elementary chapters on fluid mechanics and electromagnetism, it takes the reader all the way through to the latest ideas in more advanced topics, including planetary dynamos, stellar magnetism, fusion plasmas and engineering applications.

With the new edition, readers will benefit from additional material on MHD instabilities, planetary dynamos and applications in astrophysics, as well as a whole new chapter on fusion plasma MHD. The development of the material from first principles and its pedagogical style makes this an ideal companion for both undergraduate students and postgraduate students in physics, applied mathematics and engineering. Elementary knowledge of vector calculus is the only prerequisite.

P. A. DAVIDSON is a professor in the Department of Engineering at the University of Cambridge. He has authored over 100 publications in the fields of magnetohydrodynamics and turbulence, including the books *Turbulence: An Introduction for Scientists and Engineers* and *Turbulence in Rotating, Stratified and Electrically Conducting Fluids*. He is also an associate editor of the *Journal of Fluid Mechanics*.

Cambridge Texts in Applied Mathematics

All titles listed below can be obtained from good booksellers or from Cambridge University Press. For a complete series listing, visit www.cambridge.org/mathematics.

Introduction to Magnetohydrodynamics

Second Edition

P. A. DAVIDSON
University of Cambridge

CAMBRIDGE
UNIVERSITY PRESS

University Printing House, Cambridge CB2 8BS, United Kingdom

One Liberty Plaza, 20th Floor, New York, NY 10006, USA

477 Williamstown Road, Port Melbourne, VIC 3207, Australia

4843/24, 2nd Floor, Ansari Road, Daryaganj, Delhi – 110002, India

79 Anson Road, #06-04/06, Singapore 079906

Cambridge University Press is part of the University of Cambridge.

It furthers the University's mission by disseminating knowledge in the pursuit of
education, learning, and research at the highest international levels of excellence.

www.cambridge.org
Information on this title: www.cambridge.org/9781107160163
10.1017/9781316672853

First edition © Cambridge University Press 2001
Second edition © P. A. Davidson 2017

First published 2001
Second edition 2017

A catalogue record for this publication is available from the British Library.

Library of Congress Cataloging-in-Publication Data
Names: Davidson, P. A. (Peter Alan), 1957– author.
Title: Introduction to magnetohydrodynamics / P.A. Davidson.
Other titles: Cambridge texts in applied mathematics.
Description: Second edition. | Cambridge, United Kingdom ; New York, NY : Cambridge University Press,
2017. | Series: Cambridge texts in applied mathematics | Includes bibliographical references and index.
Identifiers: LCCN 2016045802| ISBN 9781107160163 (hbk. ; alk. paper) | ISBN 1107160162 (hbk. ; alk.
paper) | ISBN 9781316613023 (pbk. ; alk. paper) | ISBN 131661302X (pbk. ; alk. paper)
Subjects: LCSH: Magnetohydrodynamics.
Classification: LCC QA920 .D38 2017 | DDC 538/.6–dc23
LC record available at https://lccn.loc.gov/2016045802

ISBN 978-1-107-16016-3 Hardback
ISBN 978-1-316-61302-3 Paperback

Dedicated to the memory of Henri

Contents

Preface to the Second Edition

Some 15 years have passed since the first edition of this book was published, and it seems natural to revisit the subject after so long a break, reacquainting oneself with an old friend, so to speak.

If an excuse were required to revisit MHD after such a prolonged absence, then the recent advances in geophysical and astrophysical applications provide ample motivation. Astrophysical MHD, for example, has made great progress, partially as a result of the extraordinary observational data gathered from spacecraft-based instruments. On the other hand the relentless rise in computing power has, for the first time, made it possible to compute certain (but certainly not all) aspects of planetary dynamos, heralding a new wave of dynamo theories. As a result, the geophysical and astrophysical applications of MHD are now more thought-provoking and inviting than ever before.

So how should one update a book in the light of these developments? Clearly there is a need to place more emphasis on the geophysical and astrophysical applications in a second edition, which in any event provides the perfect excuse for offering a more balanced presentation of MHD. So Chapter 14, on planetary dynamos, and Chapter 15, on astrophysical applications, are largely new. Another omission in the first edition was an absence of fusion plasma MHD, and it is hoped that this has been remedied by the addition of Chapter 16. Between them, Chapters 14 through 16 provide an introduction to many of the applications of MHD in physics, and the author thanks Felix Parra Diaz and Gordon Ogilvie for providing helpful comments on draft versions of Chapters 15 and 16. Perhaps the final major addition is an extended treatment of turbulence (in Chapter 8) and MHD turbulence (in Chapter 9), which reflect recent progress in theories of MHD turbulence. Despite this shift in emphasis, the engineering applications, which were a particular feature of the first edition, have been largely retained as they are sadly underrepresented elsewhere in textbooks on MHD.

Despite these changes, the ambition of the text remains largely the same: to provide a self-contained introduction to MHD for graduate and advanced undergraduate students, with background material on electromagnetism and fluid mechanics developed from first principles, and with the fundamental theory illustrated through a broad range of applications.

Preface to the First Edition

Prefaces are rarely inspiring and, one suspects, seldom read. They generally consist of a dry, factual account of the content of the book, its intended readership, and the names of those who assisted in its preparation. There are, of course, exceptions, of which Den Hartog's preface to a text on mechanics is amongst the wittiest. Musing whimsically on the futility of prefaces in general, and on the inevitable demise of those who, like Heaviside, use them to settle old scores, Den Hartog's preface contains barely a single relevant fact. Only in the final paragraph does he touch on more conventional matters with the observation that he has 'placed no deliberate errors in the book, but he has lived long enough to be quite familiar with his own imperfections'.

We, for our part, shall stay with a more conventional format. This work is more of a text than a monograph. Part I (the larger part of the book) is intended to serve as an introductory text for (advanced) undergraduates and post-graduate students in physics, applied mathematics and engineering. Part II, on the other hand, is more of a research monograph and we hope that it will serve as a useful reference for professional researchers in industry and academia. We have at all times attempted to use the appropriate level of mathematics required to expose the underlying phenomena. Too much mathematics can, in our opinion, obscure the interesting physics and needlessly frighten the student. Conversely, a studious avoidance of mathematics inevitably limits the degree to which the phenomena can be adequately explained.

It is our observation that physics graduates are often well versed in the use of Maxwell's equations, but have only a passing acquaintance with fluid mechanics. Engineering graduates often have the opposite background. Consequently, we have decided to develop, more or less from first principles, those aspects of electromagnetism and fluid mechanics which are most relevant to our subject, and which are often treated inadequately in elementary courses.

The material in the text is heavily weighted towards incompressible flows and to engineering (as distinct from astrophysical) applications. There are two reasons for this. The first is that there already exist several excellent texts on astrophysical, geophysical and plasma MHD, whereas texts oriented towards engineering applications are somewhat thinner on the ground. Second, in recent years we have witnessed a rapid growth in the application of MHD to metallurgical processes. This has spurred a great deal of fruitful research, much of which has yet to find its way into textbooks or monographs. It seems timely to summarise elements of this research. We have not tried to be exhaustive in our coverage of the metallurgical MHD, but we hope to have captured the key advances.

The author is indebted to S. Davidson for his careful perusal of the manuscript and his many incisive comments, to H. K. Moffatt and J. C. R. Hunt for their constant advice over the years, to K. Graham for typing the manuscript, and to C. Davidson for her patience.

Part I

From Maxwell's Equations to Magnetohydrodynamics

When I am dead, I hope it may be said:
'His sins were scarlet, but his books were read.'
Hilaire Belloc *(1923)*

Magnetohydrodynamics (MHD for short) is the study of the interaction between magnetic fields and moving, conducting fluids. In the following four chapters we establish the fundamental equations of MHD from first principles. The discussion is restricted to incompressible flows and we have given particular emphasis to the elucidation of physical principles rather than detailed mathematical solutions to particular problems.

We presuppose little or no background in fluid mechanics or electromagnetism, but rather develop these topics from first principles. We do, however, make extensive use of vector analysis and the reader is assumed to be fluent in vector calculus.

1

A Qualitative Overview of MHD

The neglected borderland between two branches of knowledge is often
that which best repays cultivation, or, to use a metaphor of Maxwell's, the
greatest benefits may be derived from a cross-fertilisation of the sciences.

Rayleigh, 1884

1.1 What Is MHD?

Magnetic fields influence many natural and man-made flows. They are routinely used
in industry to heat, pump, stir and levitate liquid metals. There is the terrestrial
magnetic field which is maintained by fluid motion in the earth's molten core, the
solar magnetic field, which generates sunspots and solar flares, and the interplanetary
magnetic field which spirals outward from the sun, carried by the solar wind.
The study of these flows is called magnetohydrodynamics (MHD, for short).
Formally, MHD is concerned with the mutual interaction of fluid flow and magnetic
fields. The fluids in question must be electrically conducting and non-magnetic[1],
which limits us to liquid metals, hot ionised gases (plasmas) and strong electrolytes.

The mutual interaction of a magnetic field, $\mathbf{B}$, and a velocity field, $\mathbf{u}$, arises
partially as a result of the laws of Faraday and Ampère, and partially because of the
Lorentz force experienced by a current-carrying body. The exact form of this
interaction is analysed in detail in the following chapters, but perhaps it is worth
stating now, without any form of proof, the nature of this coupling. It is convenient,
although somewhat artificial, to split the process into three parts.

(i) The relative movement of a conducting fluid and a magnetic field causes an
electromotive force (EMF) (of order $|\mathbf{u} \times \mathbf{B}|$) to develop in accordance with
Faraday's law of induction. In general, electrical currents will ensue, the
current density being of order $\sigma(\mathbf{u} \times \mathbf{B})$, σ being the electrical conductivity.

---

[1] The study of magnetically polarised fluids is called ferrohydrodynamics, and such fluids are referred to as
magnetic fluids.

3

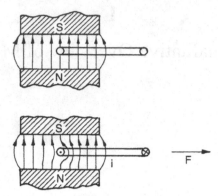

Figure 1.1 Interaction of a magnetic field and a moving wire loop.

(ii) These induced currents must, according to Ampère's law, give rise to a second, induced magnetic field. This adds to the original magnetic field and the change is usually such that the fluid appears to 'drag' the magnetic field lines along with it.

(iii) The combined magnetic field (imposed plus induced) interacts with the induced current density, $\mathbf{J}$, to give rise to a Lorentz force (per unit volume) of $\mathbf{J} \times \mathbf{B}$. This acts on the conductor and is generally directed so as to inhibit the relative movement of the magnetic field and the fluid.

Note that these last two effects have similar consequences. In both cases the relative movement of fluid and field tends to be reduced. Fluids can 'drag' magnetic field lines (effect (ii)) and magnetic fields can pull on conducting fluids (effect (iii)). It is this partial 'freezing together' of the medium and the magnetic field which is the hallmark of MHD.

These effects are, perhaps, more familiar in the context of conventional electro-dynamics. Consider a wire loop which is pulled through a magnetic field as shown in Figure 1.1. As the wire loop is pulled to the right, an EMF of order $|\mathbf{u} \times \mathbf{B}|$ is generated which drives a current as shown (effect (i)). The magnetic field associated with the induced current perturbs the original magnetic field and the net result is that the magnetic field lines seem to be dragged along by the wire (effect (ii)). The current also gives rise to a Lorentz force, $\mathbf{J} \times \mathbf{B}$, which acts on the wire in a direction opposite to that of the motion (effect (iii)). Thus it is necessary to provide a force, $\mathbf{F}$, to move the wire. In short, the wire appears to drag the field lines while the magnetic field reacts back on the wire, tending to oppose the relative movement of the two.

Let us consider effect (ii) in a little more detail. As we shall see later, the extent to which a velocity field influences an imposed magnetic field depends on the product of (i) the typical velocity of the motion, (ii) the conductivity of the fluid, and (iii) the characteristic length scale, ℓ, of the motion. Clearly, if the fluid is non-conducting or the velocity negligible there will be no significant induced magnetic field. (Consider the wire shown in Figure 1.1. If it is a poor conductor or moves very slowly, then the induced current and the associated magnetic field will be weak.) Conversely if σ or **u** is (in some sense) large, then the induced magnetic field may substantially alter the imposed field. The reason ℓ is important is a little less obvious but may be clarified by the following argument. The EMF generated by a relative movement of the imposed magnetic field and the conducting medium is of order $|\mathbf{u} \times \mathbf{B}|$ and so, by Ohm's law, the induced current density is of the order of $\sigma|\mathbf{u} \times \mathbf{B}|$. However, a modest current density spread over a large area can produce a strong magnetic field, whereas the same current density spread over a small area induces only a weak magnetic field. It is therefore the product $\sigma\mathbf{u}\ell$ which determines the ratio of the induced field to the applied magnetic field. In the limit $\sigma\mathbf{u}\ell \rightarrow \infty$ (typical of so-called ideal conductors), the induced and imposed magnetic fields are of the same order. In such cases it turns out that the combined magnetic field behaves as if it were locked into the fluid. Conversely, when $\sigma\mathbf{u}\ell \rightarrow 0$, the imposed magnetic field remains relatively unperturbed. Astrophysical MHD tends to be closer to the first situation, not so much because of the high conductivity of the plasmas involved, but because of the vast characteristic length scale. Liquid-metal MHD, on the other hand, usually lies closer to the second limit, with **u** leaving **B** relatively unperturbed. Nevertheless, it should be emphasised that effect (iii) is still strong in liquid metals, so that an imposed magnetic field can substantially alter the velocity field.

Perhaps it is worth taking a moment to consider the case of liquid metals in a little more detail. They have a reasonable conductivity ($\sim 10^6 \Omega^{-1}\mathrm{m}^{-1}$) but the velocity involved in a typical laboratory or industrial process is low (~ 1 m/s). As a consequence, the induced current densities are generally rather modest (a few Amps per cm^2). When this is combined with a small length scale (~ 0.1 m in the laboratory) the induced magnetic field is usually found to be negligible by comparison with the imposed field. There is very little 'freezing together' of the fluid and the magnetic field. However, the imposed magnetic field is often strong enough for the Lorentz force, $\mathbf{J} \times \mathbf{B}$, to dominate the motion of the fluid. We tend to think of the coupling as being one-way: **B** controls **u** through the Lorentz force, but **u** does not substantially alter the imposed field, **B**. There are, however, exceptions. Perhaps the most important of these is the earth's dynamo. Here, motion in the liquid-metal core of the earth twists, stretches and intensifies the terrestrial magnetic field, maintaining it against the natural processes of decay. It is the large length scales which are important here. While the induced current densities are weak, they are spread over a large area and as a result their combined effect is to induce a substantial magnetic field.

In summary, then, the freezing together of the magnetic field and the medium is usually strong in astrophysics, significant in geophysics, weak in metallurgical

MHD, and utterly negligible in electrolytes. However, the influence of **B** on **u** can be important in all four situations.

1.2 A Brief History of MHD

The laws of magnetism and fluid flow are hardly a twentieth-century innovation, yet MHD became a fully fledged subject only in the late 1930s or early 1940s. The reason, probably, is that there was little incentive for nineteenth-century engineers to capitalise on the possibilities offered by MHD. Thus, while there were a few isolated experiments by nineteenth-century physicists such as Faraday (he tried to measure the voltage across the Thames induced by its motion through the earth's magnetic field), the subject languished until the turn of the century. Things started to change, however, when astrophysicists realised just how ubiquitous magnetic fields and plasmas are throughout the universe. This culminated in 1942 with the discovery of the Alfvén wave, a phenomenon which is peculiar to MHD and important in astrophysics. (A magnetic field line can transmit transverse inertial waves, just like a plucked string.) Around the same time, geophysicists began to suspect that the earth's magnetic field was generated by dynamo action within its liquid-metal core, an hypothesis first put forward in 1919 by Larmor in the context of the sun's magnetic field. A period of intense research followed and continues to this day.

Plasma physicists, on the other hand, acquired an interest in MHD in the 1950s as the quest for controlled thermonuclear fusion gathered pace. They were particularly interested in the stability, or lack of stability, of plasmas confined by magnetic fields, and great advances in stability theory were made as a result. Indeed, stability techniques developed in the 1950s and 1960s by the plasma physicists have since found application in many other branches of fluid mechanics.

The development of MHD in engineering was slower and did not really get going until the 1960s. However, there was some early pioneering work by the engineer J. Hartmann, who invented the electromagnetic pump in 1918. Hartmann also undertook a systematic theoretical and experimental investigation of the flow of mercury in a homogeneous magnetic field. In the introduction to the 1937 paper describing his researches he observed:

The invention [his pump] is, as will be seen, no very ingenious one, the principle utilised being borrowed directly from a well-known apparatus for measuring strong magnetic fields. Neither does the device represent a particularly effective pump, the efficiency being extremely low due mainly to the large resistivity of mercury and still more to the contact resistance between the electrodes and the mercury. In spite hereof considerable interest was in the course of time bestowed on the apparatus, firstly because of a good many practical applications in cases where the efficiency is of small moment and then, during later years, owing to its inspiring nature. As a matter of fact, the study of the pump revealed to the author what he considered a new field of investigation, that of flow of a conducting liquid in a magnetic field, a field for which the name Hg-dynamics was suggested.

The name, of course, did not stick, but we may regard Hartmann as the father of liquid-metal MHD, and, indeed, the term 'Hartmann flow' is now used to describe duct flows in the presence of a magnetic field. Despite Hartmann's early researches, it was only in the early 1960s that MHD began to be exploited in engineering. The impetus for change came largely as a result of three technological innovations: (i) fast-breeder reactors use liquid sodium as a coolant, and this needs to be pumped; (ii) controlled thermonuclear fusion requires that the hot plasma be confined away from material surfaces by magnetic forces; and (iii) MHD power generation, in which ionised gas is propelled through a magnetic field, was thought to offer the prospect of improved power station efficiencies. This last innovation turned out to be quite impracticable, and its failure was rather widely publicised in the scientific community. However, as the interest in power generation declined, research into metallurgical MHD took off. Three decades later, magnetic fields are routinely used to heat, pump, stir and levitate liquid metals in the metallurgical industries. The key point is that the Lorentz force provides a non-intrusive means of controlling the flow of metals. With constant commercial pressure to produce cheaper, better and more consistent materials, MHD provides a unique means of exercising greater control over casting and refining processes.

So there now exist at least four overlapping communities who study MHD. Astrophysicists are concerned with the galactic magnetic field, the behaviour of (magnetically active) accretion discs, and the dynamics of stars. Planetary scientists study the generation of magnetic fields within the interior of planets, while plasma physicist are interested in the behaviour of magnetically confined plasmas. Finally, engineers study liquid-metal MHD, mostly in the context of the metallurgical industries. These communities have, of course, many common aims and problems, but they tend to use rather different vocabularies and occasionally have different ways of conceiving the same phenomena.

1.3 From Electrodynamics to MHD: A Simple Experiment

Now, the only difference between MHD and conventional electrodynamics lies in the fluidity of the conductor. This makes the interaction between **u** and **B** more subtle and difficult to quantify. Nevertheless, many of the important features of MHD are latent in electrodynamics and can be exposed by simple laboratory experiments. An elementary grasp of electromagnetism is then all that is required to understand the phenomena. Just such an experiment is described below. First, however, we shall discuss those features of MHD which the experiment is intended to illustrate.

1.3.1 Some Important Parameters in Electrodynamics and MHD

Let us introduce some notation. Let μ be the permeability of free space, σ and ρ denote the electrical conductivity and density of the conducting medium, and ℓ be a characteristic length scale. Three important parameters in MHD are

Magnetic Reynolds number, $R_m = \mu \sigma u \ell$
Alfvén velocity, $v_a = B/\sqrt{\rho \mu}$
Magnetic damping time, $\tau = [\sigma B^2 / \rho]^{-1}$

The first of these parameters may be considered as a dimensionless measure of the conductivity, while the second and third quantities have the dimensions of speed and time, respectively, as their names suggest.

Now we have already hinted that magnetic fields behave very differently depending on the conductivity of the medium. In fact, it turns out to be R_m, rather than σ, which is important. Where R_m is large, the magnetic field lines act rather like elastic bands frozen into the conducting medium. This has two consequences. First, the magnetic flux passing through any closed material loop (a loop always composed of the same material particles) tends to be conserved during the motion of the fluid. This is indicated in Figure 1.1. Second, as we shall see, small disturbances of the medium tend to result in near-elastic oscillations, with the magnetic field providing the restoring force for the vibration. In a fluid, this results in Alfvén waves, which turn out to have a frequency of $\varpi \sim v_a/\ell$.

When R_m is small, on the other hand, **u** has little influence on **B**, the induced field being negligible by comparison with the imposed field. The magnetic field then behaves quite differently. We shall see that it is dissipative in nature, rather than elastic, damping mechanical motion by converting kinetic energy into heat via Joule dissipation. The relevant time scale is now the damping time, τ, rather than ℓ/v_a.

All of this is dealt with more fully in Chapters 5, 6 and 7. The purpose of this section is to show how a familiar high school experiment is sufficient to expose these two very different types of behaviour, and to highlight the important roles played by R_m, v_a and τ.

1.3.2 Electromagnetism Remembered

Let us start with a reminder of the elementary laws of electromagnetism. (A more detailed discussion of these laws is given in Chapter 2.) The laws which concern us here are those of Ohm, Faraday and Ampère. We start with Ohm's law.

This is an empirical law which, for stationary conductors, takes the form $\mathbf{J} = \sigma \mathbf{E}$, where **E** is the electric field and **J** the current density. We interpret this as **J** being proportional to the Coulomb force $\mathbf{f} = q\mathbf{E}$ which acts on the free charge carriers,

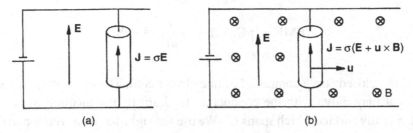

Figure 1.2 Ohm's law in stationary and moving conductors.

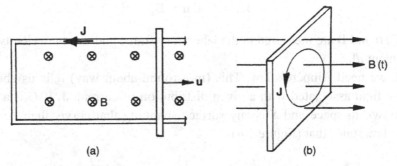

Figure 1.3 Faraday's law: (a) the EMF generated by movement of a conductor, (b) the EMF generated by a time-dependent magnetic field.

q being their charge. If, however, the conductor is moving in a magnetic field with velocity $\mathbf{u}$, the free charges will experience an additional force, $q\mathbf{u} \times \mathbf{B}$, and Ohm's law becomes (Figure 1.2)

$$\mathbf{J} = \sigma(\mathbf{E} + \mathbf{u} \times \mathbf{B}) \tag{1.1}$$

The quantity $\mathbf{E} + \mathbf{u} \times \mathbf{B}$, which is the total electromagnetic force per unit charge, arises frequently in electrodynamics and it is convenient to give it a label. We use

$$\mathbf{E}_r = \mathbf{E} + \mathbf{u} \times \mathbf{B} = \mathbf{f}/q. \tag{1.2}$$

Formally, $\mathbf{E}_r$ is the electric field measured in a frame of reference moving with velocity $\mathbf{u}$ relative to the laboratory frame (see Chapter 2). However, for our present purposes it is more useful to think of $\mathbf{E}_r$ as $\mathbf{f}/q$. Some authors refer to $\mathbf{E}_r$ as the *effective electric field*. In terms of $\mathbf{E}_r$, (1.1) becomes $\mathbf{J} = \sigma\mathbf{E}_r$.

Faraday's law tells us about the EMF which is generated in a conductor as a result of (i) a time-dependent magnetic field or (ii) the motion of a conductor within a magnetic field (Figure 1.3). In either case, Faraday's law may be written as

$$\text{EMF} = \oint_C \mathbf{E}_r \cdot d\mathbf{l} = -\frac{d}{dt}\int_S \mathbf{B} \cdot d\mathbf{S}. \qquad (1.3)$$

Here, C is a closed curve composed of line elements $d\mathbf{l}$. The curve may be fixed in space or it may move with the conducting medium (if the medium does indeed move). S is any surface which spans C. (We use the right-hand convention to define the positive directions of $d\mathbf{l}$ and $d\mathbf{S}$.) The subscript on $\mathbf{E}_r$ indicates that we must use the 'effective' electric field for each line element

$$\mathbf{E}_r = \mathbf{E} + \mathbf{u} \times \mathbf{B}, \qquad (1.4)$$

where $\mathbf{E}$, $\mathbf{u}$ and $\mathbf{B}$ are measured in the laboratory frame and $\mathbf{u}$ is the velocity of the line element $d\mathbf{l}$.

Next, we need Ampère's law. This (in a round-about way) tells us about the magnetic field associated with a given distribution of current, $\mathbf{J}$. If C is a closed curve drawn in space and S is any surface spanning that curve, then Ampère's circuital law states that (Figure 1.4)

$$\oint_C \mathbf{B} \cdot d\mathbf{l} = \mu \int_S \mathbf{J} \cdot d\mathbf{S}. \qquad (1.5)$$

Finally, there is the Lorentz force, $\mathbf{F}$. This acts on all conductors carrying a current in a magnetic field. It has its origins in the force acting on individual charge carriers, $\mathbf{f} = q(\mathbf{u} \times \mathbf{B})$, and it is easy to show that the force per unit volume of the conductor is given by

$$\mathbf{F} = \mathbf{J} \times \mathbf{B}. \qquad (1.6)$$

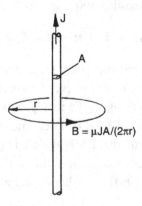

Figure 1.4 Ampère's law applied to a wire.

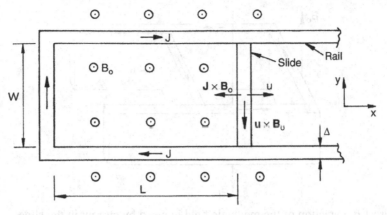

Figure 1.5 A simple experiment for illustrating MHD phenomena.

1.3.3 A Familiar High School Experiment

We now turn to the laboratory experiment. Consider the apparatus illustrated in Figure 1.5. This is frequently used to illustrate Faraday's law of induction. It consists of a horizontal, rectangular circuit sitting in a vertical magnetic field, B_0. The circuit is composed of a frictionless, conducting slide which is free to move horizontally between two rails. We take the rails and slide to have a common thickness Δ and to be made from the same material. To simplify matters, we shall also suppose that the depth of the apparatus is much greater than its lateral dimensions, L and W, so that we may treat the problem as two-dimensional. Also, we take Δ to be much smaller than L or W.

We now show that, if the slide is given a tap, and it has a high conductivity, it simply vibrates as if held in place by a (magnetic) spring. On the other hand, if the conductivity is low, it moves forward as if immersed in treacle, slowing down on a time scale of τ.

Suppose that, at $t = 0$, the slide is given a forward motion, $\mathbf{u}$. This movement of the slide will induce a current density, $\mathbf{J}$, as shown. This, in turn, produces an induced field $\mathbf{B}_i$ which is negligible outside the closed current path but is finite and uniform within the current loop. It may be shown, from Ampère's law, that $\mathbf{B}_i$ is directed downward (Figure 1.6) and has a magnitude and direction given by

$$\mathbf{B}_i = -(\mu \Delta J)\hat{\mathbf{e}}_z. \tag{1.7}$$

Note that the direction of $\mathbf{B}_i$ is such as to try to maintain a constant flux in the current loop (Lenz's law).

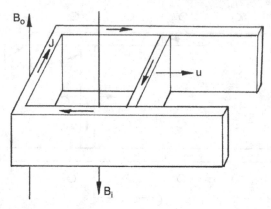

Figure 1.6 Direction of the magnetic field induced by current in the slide.

Next we combine (1.1) and (1.3) to give

$$\frac{1}{\sigma}\oint_C \mathbf{J} \cdot d\mathbf{l} = -\frac{d}{dt}\int_S \mathbf{B} \cdot d\mathbf{S}, \tag{1.8}$$

where C is the material circuit comprising the slide and the return path for $\mathbf{J}$. This yields

$$\frac{d\Phi}{dt} = \frac{d}{dt}[LW(B_0 - \mu\Delta J)] = 2(L+W)\frac{J}{\sigma}, \tag{1.9}$$

where $\Phi = (B_0 - \mu\Delta J)LW$ is the flux through the circuit. Finally, the Lorentz force (per unit depth) acting on the slide is

$$\mathbf{F} = -J(B_0 - \mu\Delta J/2)\Delta W\hat{\mathbf{e}}_x, \tag{1.10}$$

where the expression in brackets represents the average field within the slide (Figure 1.7). The equation of motion for the slide is therefore

$$\rho\frac{d^2 L}{dt^2} = \rho\frac{du}{dt} = -J(B_0 - \mu\Delta J/2), \tag{1.11}$$

where ρ is the density of the metal.

Equations (1.9) and (1.11) are sufficient to determine the two unknown functions $L(t)$ and $J(t)$. Let us introduce some simplifying notation: $B_i = \mu\Delta J$, $\ell = \Delta W/L$, $T = \mu\sigma\Delta W$ and $R_m = \mu\sigma u\ell = uT/L$. Evidently, B_i is the magnitude of the induced field, ℓ is a characteristic length scale, and T is a measure of the conductivity, σ, which happens to have the dimensions of time. Our two equations may be rewritten as

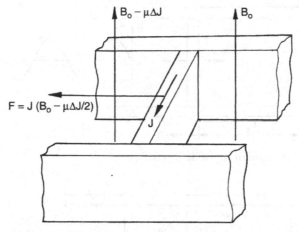

Figure 1.7 Lorentz force acting on the slide.

$$\frac{d}{dt}[L(B_0 - B_i)] = \frac{2(L + W)B_i}{T},\qquad(1.12)$$

and

$$\rho\Delta\frac{d^2L}{dt^2} = \rho\Delta\frac{du}{dt} = \frac{(B_0 - B_i)^2}{2\mu} - \frac{B_0^2}{2\mu}.\qquad(1.13)$$

Now we might anticipate that the solutions of (1.12) and (1.13) will depend on the conductivity of the apparatus, as represented by T, and so we consider two extreme cases:

(a) high conductivity limit, $\frac{u}{L} \gg \frac{1}{T}$, $(R_m = \mu\sigma u\ell \gg 1)$,
(b) low conductivity limit, $\frac{u}{L} \ll \frac{1}{T}$, $(R_m = \mu\sigma u\ell \ll 1)$.

In the high conductivity limit the right-hand side of (1.12) may be neglected and so the flux Φ linking the current path is conserved during the motion. In such cases we may look for solutions of (1.13) of the form $L = L_0 + \eta$, where η is an infinitesimal change of L and $L_0 = \Phi/B_0 W$. Noting that Φ is constant and equal to $L_0 B_0 W$, and retaining only leading-order terms in η, (1.13) yields

$$\frac{d^2\eta}{dt^2} + \frac{B_0^2}{\rho\mu\Delta L_0}\eta = 0.\qquad(1.14)$$

Thus, when the magnetic Reynolds number is high, the slide oscillates in an elastic manner, with an angular frequency of $\varpi \sim v_a/\sqrt{\Delta L_0}$, v_a being the Alfvén velocity.

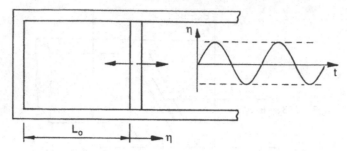

Figure 1.8 Oscillation of the slide when $R_m \gg 1$.

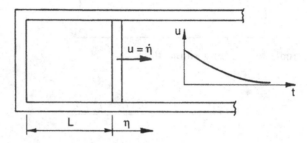

Figure 1.9 Motion of the slide when $R_m \ll 1$.

In short, if we tap the slide it will vibrate (Figure 1.8). It seems to be held in place by the magnetic field.

Now consider the low conductivity limit, $R_m \ll 1$. In this case the induction equation (1.12) tells us that $B_i \ll B_0$ and so the left-hand side of (1.12) reduces to uB_0. Substituting for B_i (in terms of u) in the equation of motion (1.13) then yields

$$\frac{du}{dt} + \frac{W}{2(L+W)} \left(\frac{\sigma B_0^2}{\rho} \right) u = 0. \tag{1.15}$$

Again we look for solutions of the form $L = L_0 + \eta$, with $\eta \ll L_0$ and $L_0 = L(t = 0)$. This time u declines exponentially on a time scale of $\tau = (\sigma B_0^2/\rho)^{-1}$, the magnetic damping time (Figure 1.9). The magnetic field now appears to play a dissipative role. Indeed, it is not difficult to show that

$$\frac{dE}{dt} = -\int (J^2/\sigma) dV, \tag{1.16}$$

where the volume integral is taken over the entire conductor and E is the kinetic energy of the slide. Thus the mechanical energy of the slide is lost to heat via Ohmic dissipation.

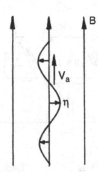

Figure 1.10 Alfvén waves. A magnetic field behaves like a plucked string, transmitting transverse inertial wave with a phase velocity of v_a.

Let us summarise our findings. When $R_m \gg 1$, and the slide is abruptly displaced from its equilibrium position, it oscillates in an elastic manner at a frequency proportional to the Alfvén velocity. During the oscillation the magnetic flux trapped between the slide and the rails remains constant. If $R_m \ll 1$, on the other hand, and the slide is given a push, it moves forward as if it were immersed in treacle. Its kinetic energy decays exponentially on a time scale of $\tau = (\sigma B_0^2/\rho)^{-1}$, the energy being lost to heat via Ohmic dissipation. Also, when R_m is small, the induced magnetic field is negligible.

We shall see that precisely the same behaviour occurs in fluids. The counterpart of the vibration is an Alfvén wave (Figure 1.10), which is a common feature of astrophysical MHD. In liquid-metal MHD, on the other hand, the primary role of **B** is to dissipate mechanical energy on a time scale of τ.

We have yet to explain these two types of behaviour. Consider first the high-R_m case. Here the key equation is Faraday's law in the form (1.8),

$$\frac{1}{\sigma}\oint_C \mathbf{J} \cdot d\mathbf{l} = -\frac{d}{dt}\int_S \mathbf{B} \cdot d\mathbf{S}. \qquad (1.17)$$

As $\sigma \to \infty$, the flux enclosed by the slide and rails must be conserved. If the slide is pushed forward, $J = B_i/\mu\Delta$ must rise to conserve Φ. The Lorentz force therefore increases until the slide is halted. At this point the Lorentz force $\mathbf{J} \times \mathbf{B}$ is finite but **u** is zero and so the slide starts to return. The induced field B_i, and hence J, now falls to maintain the magnetic flux. Eventually the slide returns to its equilibrium position and the Lorentz force falls to zero. However, the inertia of the slide carries it over its neutral point and the whole process now begins in reverse. This sequence of events is illustrated in Figure 1.11. It is the conservation of magnetic flux, combined with the inertia of the conductor, which leads to oscillations in this experiment, and to Alfvén waves in plasmas.

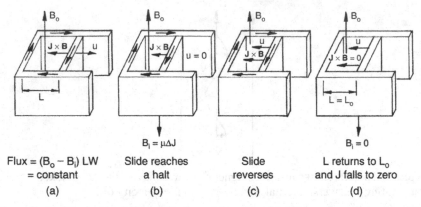

Figure 1.11 Mechanism for oscillation of the slide. (a) Slide moves forward while keeping the flux constant. (b) Slide reaches a halt. (c) Slide reverses. (d) L returns to L_0 and J falls to zero.

Now consider the case where $R_m \ll 1$. It is Ohm's law which plays the critical role here. The high resistivity of the circuit means that the currents, and hence induced field, are small. We may consider $\mathbf{B}$ to be approximately equal to the imposed field, $\mathbf{B}_0$. Since $\mathbf{B}$ is now almost constant the electric field must be irrotational:

$$\nabla \times \mathbf{E} = -\frac{\partial \mathbf{B}}{\partial t} \approx 0.$$

Ohm's law and the Lorentz force per unit volume now simplify to

$$\mathbf{J} = \sigma[-\nabla V + \mathbf{u} \times \mathbf{B}_0], \mathbf{F} = \mathbf{J} \times \mathbf{B}_0, \tag{1.18}$$

where V is the electrostatic potential. Integrating Ohm's law around the closed current loop eliminates V and yields a simple relationship between u and J,

$$2J(L + W) = \sigma W B_0 u.$$

The Lorentz force per unit mass becomes

$$\frac{\mathbf{F}}{\rho} = -\frac{W}{2(L + W)} \left(\frac{\sigma B_0^2}{\rho} \right) \mathbf{u} \sim -\frac{\mathbf{u}}{\tau}, \tag{1.19}$$

from which

$$\frac{d\mathbf{u}}{dt} \sim -\frac{\mathbf{u}}{\tau}. \tag{1.20}$$

Thus the slide slows down exponentially on a time scale of τ. The role of the induced current here is quite different from the high-R_m case. The fact that J creates an induced magnetic field is irrelevant. It is the contribution of J to the Lorentz force $\mathbf{J} \times \mathbf{B}_0$ which is important. This always acts to retard the motion. As we shall see, the two equations, $\mathbf{J} = \sigma[-\nabla V + \mathbf{u} \times \mathbf{B}_0]$ and $\mathbf{F} = \mathbf{J} \times \mathbf{B}_0$, are the hallmark of low-R_m MHD.

This familiar high school experiment encapsulates many of the phenomena which will be explored in the subsequent chapters. The main difference is that fluids have, of course, none of the rigidity of electrodynamic machines, and so they behave in more subtle and complex ways. Yet it is precisely this subtlety which makes MHD so intriguing. In summary, then, the implications of our experiment for MHD are as follows:

1. When the medium is highly conducting ($R_m \gg 1$), Faraday's law tells us that the flux through any closed material loop is conserved. When the material loop contracts or expands, currents flow so as to keep the flux constant. These currents lead to a Lorentz force which tends to oppose the contraction or expansion of the loop. The result is an elastic oscillation with a characteristic frequency of the order of v_a/ℓ, v_a being the Alfvén velocity.
2. When the medium is a poor conductor ($R_m \ll 1$), the magnetic field induced by motion is negligible by comparison with the imposed field, B_0. The Lorentz force and Ohm's law simplify to

$$\mathbf{F} = \mathbf{J} \times \mathbf{B}_0 \quad , \quad \mathbf{J} = \sigma[-\nabla V + \mathbf{u} \times \mathbf{B}_0].$$

The Lorentz force is now dissipative in nature, converting mechanical energy into heat on a time scale of the magnetic damping time, τ.

1.4 A Glimpse at the Astrophysical and Terrestrial Applications of MHD

We close this introductory chapter with a brief overview of the scope of MHD, and of this book. MHD operates on every scale, from the vast to the small. At the large scale, for example, magnetic fields play a central role in the dynamics of stellar accretion discs, generating the turbulence required to diffuse angular momentum across the disc, thus allowing the plasma to shed its angular momentum as it spirals radially inward towards the central star (Figure 1.12). Accretion discs are discussed in Chapter 15.

Closer to home, sunspots and solar flares are magnetic in origin. Sunspots are caused by buoyant magnetic flux tubes, perhaps 10^4km in diameter and 10^5km long, which are generated deep within the interior of the sun and occasionally erupt through the surface of the sun (the photosphere), as shown schematically in

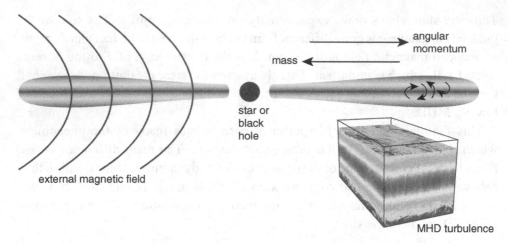

Figure 1.12 Schematic representation of an accretion disc. (Courtesy of Phil Armitage.)

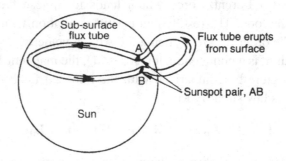

Figure 1.13 Schematic of sunspot formation.

Figure 1.13. Sunspots are the footprints of these flux tubes in the photosphere, where the local convective motions are suppressed by the intense magnetic field, leaving the plasma cooler (and hence darker) than it would otherwise be.

In fact, the solar atmosphere is threaded with a complex tangle of magnetic field lines whose origins lie deep below the surface (Figure 1.14). These are constantly being jostled by turbulent motion in the photosphere, which stretches and twists the field lines. Occasionally these magnetic field lines become so stretched that they snap and reconnect, releasing vast amounts of energy in the form of solar flares, eruptive prominences and coronal mass ejections, as shown in Figure 1.15. Sunspots, flares and coronal mass ejections are discussed in Chapter 15.

The mass and energy released by a solar flare enhances the solar wind which, even during quiescent periods (sunspot minimum), spirals radially outward from the sun, filling interplanetary space and carrying remnants of the solar magnetic

Figure 1.14 Magnetic flux loops arch up from the photosphere. (Courtesy of NASA/TRACE.)

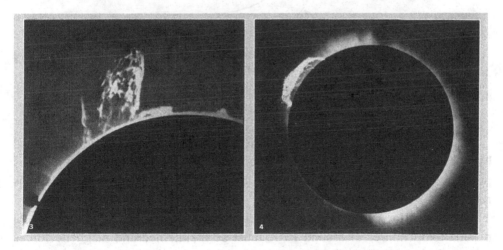

Figure 1.15 (a) An eruptive prominence. (b) A quiescent prominence.

field with it (Figure 1.16a). The solar wind sweeps past the earth with a typical velocity of around 400 km/s, but is deflected by the earth's magnetic field, which spares us from the worst excesses of the wind (Figure 1.16b). The solar wind is discussed in Chapter 15.

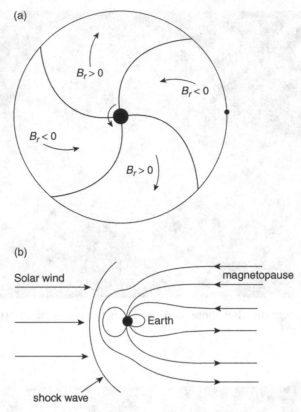

Figure 1.16 (a) The solar wind spirals outward from the sun, carrying remnants of the solar magnetic field with it. (b) The interaction of the earth's magnetic field with the solar wind.

Back on earth, the terrestrial magnetic field is now known to be maintained by fluid motion in the liquid-metal core of the earth (Figure 1.17). This process, called dynamo action, is still a source of much controversy and is reviewed in Chapter 14.

MHD is also an intrinsic part of controlled thermonuclear fusion. Here plasma temperatures of around 10^8K must be maintained and magnetic forces are used to confine the hot plasma away from the reactor walls. A simple example of a confinement scheme is shown in Figure 1.18. Unfortunately, such schemes are prone to hydrodynamic instabilities, the nature of which is discussed in Chapter 16.

In the metallurgical industries, magnetic fields are routinely used to heat, pump, stir and levitate liquid metals. Perhaps the earliest application of MHD is the electromagnetic pump (Figure 1.19). This simple device consists of mutually

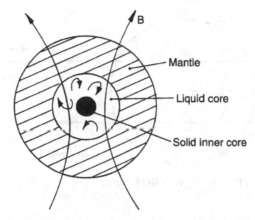

Figure 1.17 Motion in the earth's core maintains the terrestrial magnetic field.

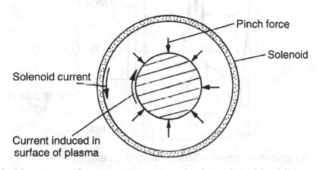

Figure 1.18 Plasma confinement. A current in the solenoid which surrounds the plasma induces opposite currents in the surface of plasma and the resulting Lorentz force pinches radially inward.

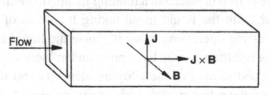

Figure 1.19 The electromagnetic pump.

perpendicular magnetic and electric fields arranged normal to the axis of a duct. Provided the duct is filled with a conducting liquid, so that currents can flow, the resulting Lorentz force provides the necessary pumping action. First proposed back in 1832, the electromagnetic pump has found its ideal application in fast-breeder nuclear reactors, where it is used to pump liquid sodium coolant through the reactor core.

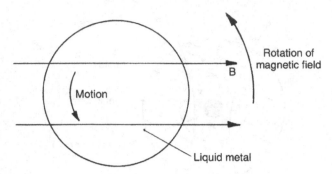

Figure 1.20 Magnetic stirring of a steel ingot.

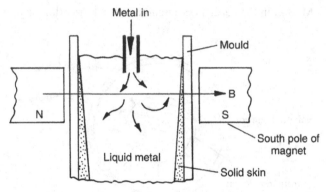

Figure 1.21 Magnetic damping of motion during casting.

Perhaps the most widespread application of MHD in engineering is the use of electromagnetic stirring. A simple example is shown in Figure 1.20. Here the liquid metal which is to be stirred is placed in a rotating magnetic field. In effect, we have an induction motor, with the liquid metal taking the place of the rotor. This is routinely used in casting operations to homogenise the liquid zone of a partially solidified ingot. The resulting motion has a profound influence on the solidification process, ensuring good mixing of the alloying elements and the continual fragmentation of the snowflake-like crystals which form in the melt. The result is a fine-structured, homogeneous ingot. This is discussed in detail in Chapter 11.

Perversely, in yet other casting operations, magnetic fields are used to dampen the motion of liquid metal. Here we take advantage of the ability of a static magnetic field to convert kinetic energy into heat via Joule dissipation (as discussed in the last section). A typical example is shown in Figure 1.21, in which an intense, static magnetic field is imposed on a casting mould. Such a device is used when the fluid motion within the mould has become so violent that the free surface of the liquid is disturbed, causing oxides and other pollutants to be entrained into the bulk.

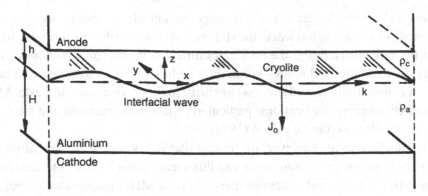

Figure 1.22 Instabilities in an aluminium reduction cell.

The use of magnetic damping promotes a more quiescent process, thus minimising contamination. The damping of jets and vortices is discussed in Chapters 6 and 11.

Another common application of MHD in metallurgy is magnetic levitation or confinement. This relies on the fact that a high-frequency induction coil repels conducting material by inducing opposing currents in any adjacent conductor. (Opposite currents repel each other.) Thus a 'basket' formed from a high-frequency induction coil can be used to levitate and melt highly reactive metals.

MHD is also important in electrolysis, particularly in those electrolysis cells used to reduce aluminium oxide to aluminium. These cells consist of broad but shallow layers of electrolyte and liquid aluminium, with the electrolyte lying on top. Oxide is continually dissolved in the electrolyte and a large current (perhaps 300kAmps) passes vertically downward through the two layers, which reduces the oxide to metal. The process is highly energy intensive, largely because of the high electrical resistance of the electrolyte. For example, in the USA, around 5 per cent of all generated electricity is used for aluminium production (Davidson, 1999). It has long been known that stray magnetic fields can destabilise the interface between the electrolyte and aluminium, in effect, through the generation of inter-facial gravity waves (Figure 1.22). In order to avoid this instability, the electrolyte layer must be maintained at a depth above some critical threshold, and this carries with it a severe energy penalty. This instability turns out to involve a rather subtle mechanism, in which interfacial oscillations absorb energy from the ambient magnetic field, converting it into kinetic energy. The stability of aluminium reduction cells is discussed in Chapter 13.

There are many other applications of MHD in engineering and metallurgy which, in the interests of brevity, we have not described here. This includes, for example, vacuum-arc melting as a means of purifying titanium and nickel ingots destined for use in aircraft parts (a process which resembles a gigantic electric welding rod – see Chapter 12); electromagnetic launchers, which have the same geometry as Figure 1.6, but where the slide is now a projectile and current is forced down the rails accelerating the slide; and finally a variety of induction melting processes. This latter technology is currently finding favour in the nuclear industry where it is used (with mixed success) to vitrify highly active nuclear waste. Many of these engineering applications, particularly those in aluminium and steel production, are discussed in Davidson (1999).

It would be wrong, however, to pretend that every engineering venture into MHD has been a success, and so we end this section on a lighter note, describing one of MHD's less notable developments: that of MHD propulsion for military submarines. Stealth is all-important in the military arena and so, in an attempt to eliminate the detectable (and therefore unwanted) cavitation noise associated with propellers, MHD pumps were once proposed as a propulsion mechanism for submarines. The idea is that sea water is drawn into ducts at the front of the submarine, passed through MHD pumps within the submarine hull, and then expelled at the rear of the vessel in the form of high-speed jets. It is an appealing idea, dating back to the 1960s, and in principle it works, as demonstrated in Japan by the surface ship YAMATO. Indeed, this idea has even found its way into popular fiction! The concept found renewed favour with the military authorities in the 1980s (the armaments race was at fever pitch) and serious design work commenced. Unfortunately, however, there is a catch. It turns out that the conductivity of sea water is so poor that the efficiency of such a device is, at best, a few per cent, nearly all of the energy going to heat the water. Worse still, the magnetic field required to produce a respectable thrust is massive, at the very limits of the most powerful superconducting magnets. So, while in principle it is possible to eliminate propeller cavitation, in the process a (highly detectable) magnetic signature is generated, to say nothing of the thermal and chemical signatures induced by electrolysis in the ducts. To track an MHD submarine, therefore, you simply have to borrow a Gauss meter, buy a thermometer, invest in litmus paper, or just follow the trail of dead fish!

Exercises

1.1 A bar of small but finite conductivity slides at a constant velocity u along conducting rails in a region of uniform magnetic field. The resistance in the circuit is R and the inductance is negligible. Calculate: (i) the current I flowing in the circuit; (ii) the power required to move the bar; and (iii) the Ohmic losses in the circuit.

1.2 A square metal bar of length ℓ and mass m slides without friction down parallel conducting rails of negligible resistance. The rails are connected to each other at the bottom by a resistanceless rail parallel to the bar, so that the bar and rails form a closed loop. The plane of the rails makes an angle θ with the horizontal, and a uniform vertical field, B, exists throughout the region. The bar has a small but finite conductivity and has a resistance of R. Show that the bar acquires a steady velocity of $u = mgR \sin \theta/(B\ell \cos \theta)^2$.

1.3 A steel rod is 0.5 m long and has a diameter of 1cm. It has a density and conductivity of $7 \times 10^3 \text{kg/m}^3$ and 10^6mho/m, respectively. It lies horizontally with its ends on two parallel rails, 0.5 m apart. The rails are perfectly conducting and are inclined at an angle of $15°$ to the horizontal. The rod slides up the rails with a coefficient of friction of 0.25, propelled by a battery which maintains a constant voltage difference of $2V$ between the rails. There is a uniform, unperturbed vertical magnetic field of 0.75T. Find the velocity of the bar when travelling steadily.

1.4 When Faraday's and Ohm's laws are combined, we obtain (1.8). Consider an isolated flux tube sitting in a perfectly conducting fluid and let C_m be a material curve (a curve always composed of the same material) which at some initial instant encircles the flux tube, lying on the surface of the tube. Show that the flux enclosed by C_m will remain constant as the flow evolves, and that this is true of each and every curve enclosing the tube at $t = 0$. This suggests that the tube itself moves with the fluid, as if frozen into the medium. Now suppose that the diameter of the flux tube is very small. What does this tell us about magnetic field lines in a perfectly conducting fluid?

1.5 Consider a two-dimensional flow consisting of an (initially) thin jet propagating in the x-direction and sitting in a uniform magnetic field which points in the y-direction. The magnetic Reynolds number is low. Show that the Lorentz force (per unit volume) acting on the fluid is $-\sigma u_x B^2 \hat{\mathbf{e}}_x$. Now consider a fluid particle sitting on the axis of the jet. It has an axial acceleration of $u_x(\partial u_x/\partial x)$. Show that the jet is annihilated within a finite distance of $L \sim u_0 \tau$, where u_0 is the initial value of u_x. (τ is the magnetic damping time, introduced in §1.3.1.)

1.6 Calculate the magnetic Reynolds number for motion in the core of the earth, using the radius of the core, $R_c = 3500 \text{km}$, as the characteristic length scale and $u \sim 2 \times 10^{-4} \text{m/s}$ as a typical velocity. Take the conductivity of iron as 10^6 mho/m. Now calculate the magnetic Reynolds number for motion in the outer regions of the sun, taking $\ell \sim 10^3 \text{km}$, $u \sim 1 \text{km/s}$ and $\sigma = 10^4 \text{mho/m}$. Explain why it is difficult (if not impossible) to model the dynamics of the solar dynamo and geodynamo using scaled laboratory experiments with liquid metals (although some people have tried – see Chapter 14).

1.7 Magnetic forces are sometimes used to levitate objects. For example, if a metal object is situated near a coil carrying an alternating current I, eddy currents will flow in the object and a repulsive force will result. Show that the force in the x-direction is $\frac{1}{2}I^2(\partial L/\partial x)$ if the object is allowed to move in the x-direction and L is the effective inductance of the coil.

2

The Governing Equations of Electrodynamics

> From a long view of history of mankind – seen from, say, ten thousand
> years from now – there can be little doubt that the most significant event
> of the nineteenth century will be judged as Maxwell's discovery of the
> laws of electrodynamics.
>
> *R.P. Feynman (1964)*

We are concerned here with electrically conducting, non-magnetic materials. For simplicity, we shall assume that all material properties, such as the electrical conductivity (σ), are spatially uniform, and that the medium is incompressible. The topics which concern us are the Lorentz force, Ohm's law, Ampère's law, the Biot–Savart law and Faraday's law. We shall examine each of these in turn, culminating in Maxwell's equations of electrodynamics. We shall then show that these equations may be greatly simplified in MHD, a simplification that arises from the observation that the density of free charges is invariably very small in electrically conducting fluids.

2.1 The Electric Field and the Lorentz Force

A particle moving with velocity **u** and carrying a charge q is, in general, subject to three electromagnetic forces,

$$\mathbf{f} = q\mathbf{E}_s + q\mathbf{E}_i + q\mathbf{u} \times \mathbf{B}. \tag{2.1}$$

The first is the electrostatic force, or Coulomb force, which arises from the mutual repulsion or attraction of electric charges. ($\mathbf{E}_s$ is the electrostatic field.) The second is the force which the charge experiences in the presence of a time-varying magnetic field, $\mathbf{E}_i$ being the electric field induced by the changing magnetic field. The third contribution is the Lorentz force, which arises from the movement of the charge through a magnetic field.

Now, Coulomb's law tells that $\mathbf{E}_s$ is irrotational and Gauss' law fixes the divergence of $\mathbf{E}_s$. Together these laws yield

$$\nabla \times \mathbf{E}_s = \mathbf{0}, \qquad \nabla \cdot \mathbf{E}_s = \rho_e/\varepsilon_0. \tag{2.2}$$

Here, ρ_e is the total charge density (free charges plus bound charges) and ε_0 is the permittivity of free space. In view of (2.2), we may introduce the electrostatic potential, V, defined by $\mathbf{E}_s = -\nabla V$. It follows that electrostatic fields are governed by $\nabla^2 V = -\rho_e/\varepsilon_0$. In infinite domains, this Poisson equation may be inverted to give

$$V(\mathbf{x}) = \frac{1}{4\pi\varepsilon_0} \int \frac{\rho_e(\mathbf{x}')}{|\mathbf{r}|} \, d\mathbf{x}', \qquad \mathbf{r} = \mathbf{x} - \mathbf{x}',$$

which is sometimes referred to as the Green's function solution of Poisson's equation.

The induced electric field, on the other hand, has zero divergence, while its curl is finite and governed by Faraday's law:

$$\nabla \cdot \mathbf{E}_i = 0, \qquad \nabla \times \mathbf{E}_i = -\frac{\partial \mathbf{B}}{\partial t}. \tag{2.3}$$

It is convenient to define the total electric field as $\mathbf{E} = \mathbf{E}_s + \mathbf{E}_i$, and so we have

$$\nabla \cdot \mathbf{E} = \rho_e/\varepsilon_0, \quad \nabla \times \mathbf{E} = -\frac{\partial \mathbf{B}}{\partial t}, \tag{2.4}$$

$$\text{(Gauss' law)} \qquad\qquad\qquad\qquad \text{(Faraday's law)}$$

$$\mathbf{f} = q(\mathbf{E} + \mathbf{u} \times \mathbf{B}). \tag{2.5}$$

$$\text{(Lorentz force)}$$

Equations (2.4) uniquely determine the electric field, since the requirements are that the divergence and curl of the field be known (and suitable boundary conditions be specified). It is customary to use Equation (2.5) to define the electric field $\mathbf{E}$ and the magnetic field $\mathbf{B}$. Thus, for example, the electric field $\mathbf{E}$ is the force per unit charge on a small test charge *at rest* in the observer's frame of reference.

Due attention must be given to frames of reference. Suppose that, in the laboratory frame, there is an electric field and a magnetic field. The electric field, $\mathbf{E}$, is defined by the force per unit charge on a charge at rest in that frame. If the charge is moving, the force due to the electric field is still given by $\mathbf{f} = q\mathbf{E}$ but the additional force $q\mathbf{u} \times \mathbf{B}$ appears, which is used to define $\mathbf{B}$. If, however, we use a frame of reference in which the charge is instantaneously *at rest* (but moving with velocity $\mathbf{u}$ relative to the laboratory frame), then the force on the charge can only be interpreted as due to an electric field, say $\mathbf{E}_r$. (The subscript r indicates 'relative to a moving frame'.) Newton's second law then gives, for the two frames, $\mathbf{f} = q(\mathbf{E} + \mathbf{u} \times \mathbf{B})$ and $\mathbf{f}_r = q\mathbf{E}_r$. However, Newtonian relativity (which is all that is required for MHD) tells us that $\mathbf{f} = \mathbf{f}_r$. It follows that the electric fields in the two frames are related by

$$\mathbf{E}_r = \mathbf{E} + \mathbf{u} \times \mathbf{B}. \qquad (2.6)$$

The magnetic fields $\mathbf{B}$ and $\mathbf{B}_r$ are equal (relativistic corrections apart).

We close this section by noting that $\mathbf{B}$ is a pseudo-vector and not a true vector. That is to say, while $\mathbf{B}$ at any point has an unambiguous magnitude and line of action, the sense (i.e. direction) of $\mathbf{B}$ along its line of action is somewhat arbitrary and a matter of convention; for example, $\mathbf{B}$ reverses direction if we move from a right-handed convention (the usual convention) to a left-handed one. This may be seen in one of several ways. For example, suppose we transform our coordinate system according to $\mathbf{x} \to \mathbf{x}' = -\mathbf{x}$. (This is referred to as an inversion of the coordinates, or else as a reflection about the origin.) We have moved from a right-handed coordinate system to a left-handed one in which, $x' = -x$, $y' = -y$, $z' = -z$, $\mathbf{i}' = -\mathbf{i}$, $\mathbf{j}' = -\mathbf{j}$ and $\mathbf{k}' = -\mathbf{k}$. Now the components of a true vector, such as force, $\mathbf{f}$, or velocity, $\mathbf{u}$, transform like the position vector, i.e. $f'_x = -f_x$ etc., which leaves the physical direction of the vector unchanged, since

$$\mathbf{f} = f_x\mathbf{i}_x + f_y\mathbf{i}_y + f_z\mathbf{i}_z = (-f'_x)(-\mathbf{i}'_x) + \left(-f'_y\right)(-\mathbf{i}'_y) + (-f'_z)(-\mathbf{i}'_z).$$

Thus, after an inversion of the coordinates, a true vector (such as $\mathbf{f}$ or $\mathbf{u}$) has the same magnitude and direction as before, although the numerical values of its components change sign. Now consider the definition of $\mathbf{B}$, $\mathbf{f} = q(\mathbf{u} \times \mathbf{B})$. Under an inversion of the coordinates, the components of $\mathbf{u}$ and $\mathbf{f}$ both change sign, and so those of $\mathbf{B}$ cannot. Thus the magnetic field transforms according to $B'_x = B_x$, etc. By implication, the physical direction of $\mathbf{B}$ reverses. (Such vectors are called pseudo-vectors.) So, if one morning we all agreed to change convention from a right-handed coordinate system to a left-handed one, all the magnetic field lines would reverse direction! The fact that $\mathbf{B}$ is a pseudo-vector is important in dynamo theory[1].

2.2 Ohm's Law and the Volumetric Lorentz Force

In a stationary conductor it is found that the current density, $\mathbf{J}$, is proportional to the force experienced by the free charges. This is reflected in the conventional form of

---

[1] An alternative way to distinguish true vectors from pseudo-vectors is to consider how a vector behaves when reflected into a 'mirror universe', as discussed in Chapter 52, Vol. I of Feynman, Leighton, and Sands, 1964.

Ohm's law, $\mathbf{J} = \sigma\,\mathbf{E}$. In a conducting fluid the same law applies, only now we must use the electric field measured in a frame moving with the local velocity of the conductor:

$$\mathbf{J} = \sigma\mathbf{E}_r = \sigma(\mathbf{E} + \mathbf{u} \times \mathbf{B}). \tag{2.7}$$

Note that $\mathbf{u}$ will, in general, vary with position.

Now the Lorentz force (2.5) is important not just because it lies behind Ohm's law, but also because the forces exerted on the free charges are ultimately transmitted to the conductor. In MHD we are less concerned with the forces on individual charges than with the bulk force acting on the medium. But this is readily found. If (2.5) is summed over a unit volume of the conductor, then Σq becomes the charge density, ρ_e, and $\Sigma q\mathbf{u}$ becomes the current density, $\mathbf{J}$. The volumetric version of (2.5) is therefore

$$\mathbf{F} = \rho_e\mathbf{E} + \mathbf{J} \times \mathbf{B}, \tag{2.8}$$

where $\mathbf{F}$ is the force per unit volume acting on the conductor. However, in conductors travelling at the sort of speeds we are interested in (much less than the speed of light), the first term in (2.8) is negligible. We may demonstrate this as follows. Conservation of charge requires that

$$\nabla \cdot \mathbf{J} = -\frac{\partial\rho_e}{\partial t}. \tag{2.9}$$

(This simply says that the rate at which charge is decreasing inside a small volume must equal the rate at which charge flows out across the surface of that volume.) By taking the divergence of both sides of (2.7), and using Gauss' law and (2.9), we find

$$\frac{\partial\rho_e}{\partial t} + \frac{\rho_e}{\tau_e} + \sigma\nabla \cdot (\mathbf{u} \times \mathbf{B}) = 0, \quad \tau_e = \varepsilon_0/\sigma.$$

The quantity τ_e is called the charge relaxation time, and for a typical metallic conductor it has a value of around 10^{-18} seconds. It is extremely small! To appreciate where its name comes from, consider the situation where $\mathbf{u} = \mathbf{0}$. In this case, $\partial\rho_e/\partial t + \rho_e/\tau_e = 0$, and so

$$\rho_e = \rho_e(0)\,\exp[-t/\tau_e].$$

Any net charge density which, at $t = 0$, lies in the interior of a conductor will move rapidly to the surface under the action of the electrostatic repulsion forces. It follows that ρ_e is always zero in stationary conductors, except during some minuscule period when a battery, say, is turned on. Now consider the case where **u** is non-zero. We are interested in events which take place on a time scale much longer than τ_e (we exclude events like batteries being turned on) and so we may neglect $\partial \rho_e/\partial t$ by comparison with ρ_e/τ_e. We are left with the pseudo-static equation

$$\rho_e = -\varepsilon_0 \nabla \cdot (\mathbf{u} \times \mathbf{B}). \tag{2.10}$$

Thus, when there is motion, we can sustain a finite charge density in the interior of the conductor. However, it turns out that ρ_e is very small, too small to produce any significant electric force, $\rho_e \mathbf{E}$. That is, from (2.10) we have $\rho_e \sim \varepsilon_0 u B/\ell$, while Ohm's law requires $\mathbf{E} \sim \mathbf{J}/\sigma$, and so

$$\rho_e|\mathbf{E}| \sim (\varepsilon_0 u B/\ell)(J/\sigma) \sim \frac{u\tau_e}{\ell} JB.$$

Here, ℓ is a typical length scale for the flow. Evidently, since $\tau_e \sim 10^{-18}$ seconds, $u\tau_e/\ell$ is tiny in all but the most extreme circumstances and so the magnetic force almost invariably dominates (2.8). Thus our force law simplifies to

$$\mathbf{F} = \mathbf{J} \times \mathbf{B}. \tag{2.11}$$

Note also that (2.10) is equivalent to ignoring $\partial \rho_e/\partial t$ in the charge conservation equation (2.9). That is to say, the charge density is so small that (2.9) simplifies to

$$\nabla \cdot \mathbf{J} = 0. \tag{2.12}$$

2.3 Ampère's Law and the Biot–Savart Law

The Ampère–Maxwell equation tells us something about the magnetic field generated by a given distribution of current. It is

$$\nabla \times \mathbf{B} = \mu \left[\mathbf{J} + \varepsilon_0 \frac{\partial \mathbf{E}}{\partial t} \right]. \tag{2.13}$$

The last term in (2.13) may be unfamiliar. It does not, for example, appear in Ampère's law of magnetostatics. This new term was introduced by Maxwell as a correction to Ampère's law and is called the displacement current. To see why it is necessary, we take the divergence of (2.13). Noting that $\nabla \cdot \nabla \times (\cdot) = 0$ and using Gauss' law, this yields

$$\nabla \cdot \mathbf{J} = -\varepsilon_0 \frac{\partial}{\partial t} \nabla \cdot \mathbf{E} = -\frac{\partial \rho_e}{\partial t}. \tag{2.14}$$

This is just the charge conservation equation, which, without the displacement current, would be violated. Put another way, Gauss' law plus charge conservation combine to yield

$$\nabla \cdot \left[\mathbf{J} + \varepsilon_0 \frac{\partial \mathbf{E}}{\partial t} \right] = 0, \tag{2.15}$$

and since the divergence of the left-hand side of (2.13) is zero, we must include the displacement current on the right of (2.13) to ensure that it, too, is solenoidal.

It turns out, however, that Maxwell's correction is not needed in MHD. That is, we have already noted that $\partial \rho_e / \partial t$ is negligible in conductors, and so we might anticipate that the contribution of $\varepsilon_0 \partial \mathbf{E} / \partial t$ to (2.13) is also small in MHD. This is readily confirmed:

$$\varepsilon_0 \frac{\partial \mathbf{E}}{\partial t} \sim \frac{\varepsilon_0}{\sigma} \frac{\partial \mathbf{J}}{\partial t} \sim \tau_e \frac{\partial \mathbf{J}}{\partial t} \ll \mathbf{J}.$$

We are therefore at liberty to use the pre-Maxwell form of (2.13), which is simply the differential form of Ampère's law,

$$\nabla \times \mathbf{B} = \mu \mathbf{J}. \tag{2.16}$$

This is consistent with approximation (2.12), since the divergence of (2.16) demands $\nabla \cdot \mathbf{J} = 0$.

Finally, we note that, in infinite domains, (2.16) may be inverted using the Biot–Savart law. That is, when the current density is a known, localised function of position, the magnetic field may be calculated directly from

$$\mathbf{B}(\mathbf{x}) = \frac{\mu}{4\pi} \int \frac{\mathbf{J}(\mathbf{x}') \times \mathbf{r}}{r^3} d\mathbf{x}', \quad \mathbf{r} = \mathbf{x} - \mathbf{x}', \tag{2.17}$$

where $r = |\mathbf{r}|$. This comes from the fact that a small volume element, $d\mathbf{x}'$, located at $\mathbf{x}'$ and carrying a current density of $\mathbf{J}(\mathbf{x}')$, induces a magnetic field at point $\mathbf{x}$ which is given by (see Figure 2.1)

$$d\mathbf{B}(\mathbf{x}) = \frac{\mu}{4\pi} \frac{\mathbf{J}(\mathbf{x}') \times \mathbf{r}}{r^3} d\mathbf{x}'. \tag{2.18}$$

Note that (2.17), which is equivalent to (2.16), reveals the true character of Ampère's law. It really tells about the structure of the magnetic field associated with a given current distribution. Perhaps it is worth taking a moment to explain where (2.17) comes from.

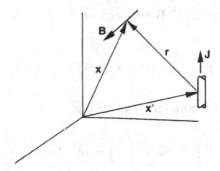

Figure 2.1 Coordinate system used in the Biot–Savart law.

Although (2.18) was first discovered by experiment, it is usual now to interpret it as a consequence of the Green's function solution of Poisson's equation. For example, we have already seen that the Poisson equation $\nabla^2 V = -\rho_e/\varepsilon_0$ may be inverted in an infinite domain to give the electrostatic potential:

$$V(\mathbf{x}) = \frac{1}{4\pi\varepsilon_0} \int \frac{\rho_e(\mathbf{x}')}{|\mathbf{r}|}\, d\mathbf{x}', \qquad \mathbf{r} = \mathbf{x} - \mathbf{x}'. \tag{2.19}$$

Now, as we shall see, $\mathbf{B}$ is solenoidal, $\nabla \cdot \mathbf{B} = 0$, and so we have $\nabla \times \nabla \times \mathbf{B} = -\nabla^2 \mathbf{B}$. Consequently, the curl of (2.16) yields the Poisson equation

$$\nabla^2 \mathbf{B} = -\mu \nabla \times \mathbf{J}, \tag{2.20}$$

whose individual components invert just like the electrostatic potential to give

$$\mathbf{B}(\mathbf{x}) = \frac{\mu}{4\pi} \int \frac{\nabla' \times \mathbf{J}'}{r}\, d\mathbf{x}', \qquad \mathbf{r} = \mathbf{x} - \mathbf{x}', \tag{2.21}$$

where $\mathbf{J}' = \mathbf{J}(\mathbf{x}')$. The next step is to show that (2.17) follows from (2.21). To that end we note that $\nabla' \times \mathbf{J}'/r = \nabla' \times (\mathbf{J}'/r) + \mathbf{J}' \times \nabla'(1/r)$, where the first term on the right integrates to zero since $\mathbf{J}$ is localised in space. Moreover, it is readily confirmed that $\nabla'(1/r) = -\nabla(1/r) = \mathbf{r}/r^3$, and so (2.21) becomes

$$\mathbf{B}(\mathbf{x}) = \frac{\mu}{4\pi} \int \nabla\left(\frac{1}{r}\right) \times \mathbf{J}'\, d\mathbf{x}' = \frac{\mu}{4\pi} \int \frac{\mathbf{J}' \times \mathbf{r}}{r^3}\, d\mathbf{x}'. \tag{2.22}$$

This brings us back to the Biot–Savart law (2.17), as required.

In fact, we can confirm by direct calculation that the Biot–Savart integral returns the correct values of both $\nabla \cdot \mathbf{B}$ and $\nabla \times \mathbf{B}$. For example, it is readily confirmed that the Biot–Savart law ensures $\nabla \cdot \mathbf{B} = 0$, because the divergence of the integrand in (2.22) is

$$\nabla \cdot \left(\nabla(r^{-1}) \times \mathbf{J'} \right) = \mathbf{J'} \cdot [\nabla \times \nabla(r^{-1})] - \nabla(r^{-1}) \cdot [\nabla \times \mathbf{J'}] = 0,$$

$\mathbf{J'}$ being independent of $\mathbf{x}$. Similarly, noting that

$$\nabla \times \left(\nabla\left(\frac{1}{r}\right) \times \mathbf{J'} \right) = -\mathbf{J'}\nabla^2\left(\frac{1}{r}\right) - (\mathbf{J'} \cdot \nabla')\nabla\left(\frac{1}{r}\right)$$

$$= -\mathbf{J'}\nabla^2\left(\frac{1}{r}\right) + \nabla\left(\frac{\nabla' \cdot \mathbf{J'}}{r}\right) - \nabla' \cdot \left[\mathbf{J'}\nabla\left(\frac{1}{r}\right)\right],$$

and that $-\nabla^2(1/4\pi r)$ is the three-dimensional Dirac delta function, we may evaluate $\nabla \times \mathbf{B}$ directly from the Biot–Savart integral. After a little work we find

$$\nabla \times \mathbf{B} = \mu \mathbf{J} + \frac{\mu}{4\pi} \nabla \left[\int \frac{\nabla' \cdot \mathbf{J'}}{r} \, d\mathbf{x'} \right]. \tag{2.23}$$

Now in general $\mathbf{J}$ is not solenoidal, since charge conservation requires $\nabla \cdot \mathbf{J} = -\partial \rho_e/\partial t$, but Ampère's law applies only in those situations where $\nabla \cdot \mathbf{J} = 0$, and in that case (2.23) reverts to (2.16), as required.

Example 2.1 Force-Free Fields

Magnetic fields of the form $\nabla \times \mathbf{B} = \alpha \mathbf{B}$, α = constant, are known as *force-free* fields, since $\mathbf{J} \times \mathbf{B} = \mathbf{0}$. (More generally, fields of the form $\nabla \times \mathbf{G} = \alpha \mathbf{G}$ are known as *Beltrami* fields.) They are important in plasma MHD where we frequently require the Lorentz force to vanish, or at least be very small. Show that, for a force-free field,

$$\left(\nabla^2 + \alpha^2\right)\mathbf{B} = 0. \tag{2.24}$$

Deduce that there are no force-free fields, other than $\mathbf{B} = \mathbf{0}$, for which $\mathbf{J}$ is localised in space and $\mathbf{B}$ is everywhere differentiable and $O(x^{-3})$ at infinity. ∎

2.4 Faraday's Law and the Vector Potential

Faraday's law is sometimes stated in integral form and sometimes in differential form. You have already met both. In §2.1 we stated it to be

$$\nabla \times \mathbf{E} = -\frac{\partial \mathbf{B}}{\partial t}. \tag{2.25}$$

This tells us about the electric field induced by a time-varying magnetic field. However, in Chapter 1 we gave the integral version [2],

$$\text{EMF} = \oint_C \mathbf{E}_r \cdot d\mathbf{l} = -\frac{d}{dt}\int_S \mathbf{B} \cdot d\mathbf{S}, \tag{2.26}$$

where $\mathbf{E}_r$ is the electric field measured in a frame of reference moving with line element $d\mathbf{l}$, i.e. $\mathbf{E}_r = \mathbf{E} + \mathbf{u} \times \mathbf{B}$, $\mathbf{u}$ being the velocity of the line element $d\mathbf{l}$. In fact, it is easily seen that (2.26) is a more powerful statement than the differential form of Faraday's law. In words, it states that the EMF around a closed loop is equal to the total rate of change of flux of $\mathbf{B}$ through that loop. In (2.26) the flux may change because $\mathbf{B}$ is changing with time, or because the loop is moving uniformly in an inhomogeneous field, or because the loop is changing shape. Whatever the cause, (2.26) gives the induced EMF. We shall return to the integral version of Faraday's law in §2.9, where we shall discuss its full significance. In the meantime, we shall show that the differential form of Faraday's law is a special case of (2.26).

Suppose that the loop is rigid and at rest in a laboratory frame. Then the EMF can arise only from a magnetic field which is time dependent. In this case (2.26) becomes

$$\oint_C \mathbf{E} \cdot d\mathbf{l} = \int_S (\nabla \times \mathbf{E}) \cdot d\mathbf{S} = -\int_S \frac{\partial \mathbf{B}}{\partial t} \cdot d\mathbf{S}.$$

Since this is true for any and all (fixed) surfaces S, we may equate the integrands in the two surface integrals. We then obtain the differential form of Faraday's law:

$$\nabla \times \mathbf{E} = -\frac{\partial \mathbf{B}}{\partial t}. \tag{2.27}$$

In this form, Faraday's law becomes one of Maxwell's equations (see §2.6). Note, however, that (2.27) is a weaker statement than (2.26). It only tells us about the electric field induced by a time-varying magnetic field.

Now (2.27) ensures that $\partial \mathbf{B}/\partial t$ is solenoidal, since $\nabla \cdot (\nabla \times \mathbf{E}) = 0$. In fact, it transpires that we can make an even stronger statement about $\mathbf{B}$. It turns out that $\mathbf{B}$ is itself solenoidal,

$$\nabla \cdot \mathbf{B} = 0. \tag{2.28}$$

---

[2] The EMF is defined as the force experienced by a unit charge, as measured by $\mathbf{E}_r$, integrated around a closed curve.

This allows us to introduce another field, $\mathbf{A}$, called the vector potential, defined (at least in part) by

$$\nabla \times \mathbf{A} = \mathbf{B}. \tag{2.29}$$

This definition automatically ensures that $\mathbf{B}$ is solenoidal, since $\nabla \cdot \nabla \times \mathbf{A} = 0$. If we substitute $\mathbf{A}$ for $\mathbf{B}$ in Faraday's equation we obtain

$$\nabla \times \mathbf{E} = -\nabla \times (\partial \mathbf{A}/\partial t),$$

which uncurls to yield

$$\mathbf{E} = -\frac{\partial \mathbf{A}}{\partial t} - \nabla V, \tag{2.30}$$

where V is an arbitrary scalar function. However we also have, from (2.2) and (2.3),

$$\mathbf{E} = \mathbf{E}_i + \mathbf{E}_s, \qquad \nabla \times \mathbf{E}_s = \mathbf{0}, \qquad \nabla \cdot \mathbf{E}_i = 0,$$

and so we might anticipate that $\mathbf{E}_i = -\partial \mathbf{A}/\partial t$ and $\mathbf{E}_s = -\nabla V$, where V is now interpreted as the electrostatic potential. For the particular case where $\nabla \cdot \mathbf{A} = 0$, this is readily confirmed by taking the divergence of (2.30) which, given $\nabla \cdot \mathbf{A} = 0$, shows that all of the divergence of $\mathbf{E}$ is captured by ∇V in (2.30), as required by (2.2) and (2.3).

We now need to say something about the divergence of $\mathbf{A}$, which is an issue of some subtlety. The vector potential is not uniquely defined by $\nabla \times \mathbf{A} = \mathbf{B}$ alone. Rather, to unambiguously define $\mathbf{A}$, we must specify both its curl and divergence, as well as appropriate boundary conditions (which are usually taken as $\mathbf{A} \to 0$ at infinity). So we need to decide on a definition of $\nabla \cdot \mathbf{A}$, which is known as *setting the gauge* for $\mathbf{A}$. It turns out that the most appropriate choice of $\nabla \cdot \mathbf{A}$ depends upon the circumstances. For cases where we may ignore the displacement current in the Ampère–Maxwell equation, such as magnetostatics and MHD, the most convenient choice is simply

$$\nabla \cdot \mathbf{A} = 0, \tag{2.31}$$

which is known as the *Coulomb gauge*. However, for those cases where we need to use the full set of Maxwell's equations (which is rarely the case in MHD), it turns out to be preferable to choose

$$\nabla \cdot \mathbf{A} = -\mu \varepsilon_0 \frac{\partial V}{\partial t} = -\frac{1}{c^2} \frac{\partial V}{\partial t}, \tag{2.32}$$

where c is the speed of light. This is called the *Lorentz gauge*. We shall return to (2.32) in §2.6 where we shall explain its significance.

Returning to the Coulomb gauge, $\nabla \cdot \mathbf{A} = 0$, which is appropriate for nearly all MHD, we see that Ampère's law (subject to $\nabla \cdot \mathbf{A} = 0$) yields a Poisson equation for $\mathbf{A}$,

$$\nabla^2 \mathbf{A} = -\mu \mathbf{J}. \tag{2.33}$$

Given the inversion formula (2.19), our Poisson equation inverts to give

$$\mathbf{A}(\mathbf{x}) = \frac{\mu}{4\pi} \int \frac{\mathbf{J}'}{r} \, d\mathbf{x}', \qquad \mathbf{r} = \mathbf{x} - \mathbf{x}', \tag{2.34}$$

where $\mathbf{J}' = \mathbf{J}(\mathbf{x}')$ and $r = |\mathbf{r}|$.

We can check that (2.34) yields the correct value of $\nabla \cdot \mathbf{A}$. Noting that $\mathbf{J}'$ is independent of $\mathbf{x}$, and that $\nabla(1/r) = -\nabla'(1/r)$, we have

$$\nabla \cdot (\mathbf{J}'/r) = \mathbf{J}' \cdot \nabla(1/r) = -\mathbf{J}' \cdot \nabla'(1/r) = -\nabla' \cdot (\mathbf{J}'/r) + (\nabla' \cdot \mathbf{J}')/r.$$

Since $\nabla' \cdot (\mathbf{J}'/r)$ integrates to zero for a localised distribution of currents, (2.34) yields

$$\nabla \cdot \mathbf{A} = \frac{\mu}{4\pi} \int \frac{\nabla' \cdot \mathbf{J}'}{r} \, d\mathbf{x}'.$$

However, $\nabla \cdot \mathbf{J} = 0$ in cases where Ampère's law holds (magnetostatics and MHD), and so we do indeed recover $\nabla \cdot \mathbf{A} = 0$.

In §2.6 we shall see how expressions (2.33) and (2.34) alter when displacement currents are taken into account, electromagnetic waves rear their head, and the Lorentz gauge is adopted.

Example 2.2 The Divergence of B
Faraday's law implies $(\partial/\partial t)(\nabla \cdot \mathbf{B}) = 0$. If this is also true relative to all sets of axes moving uniformly relative to one another, show that $\nabla \cdot \mathbf{B} = 0$. ∎

2.5 An Historical Aside: Faraday and the Concept of the Field

Faraday had a passionate interest in chemistry, electricity and magnetism, and especially the borderlands between those subjects. In chemistry, for example, he discovered benzene, established the laws of electrolysis, laid the foundations for the chemistry of plasmas, and pioneered the development of special alloy steels, presenting his friends in 1820 with gifts of novel non-rusting platinum steel razors! But his work on electricity and magnetism was to have a more profound impact, particularly his insistence that the concept of *fields* lay at the heart of the observed phenomena.

Even from the limited perspective of MHD, Faraday played an important role in the development of the subject. For example, he performed the first experiment in MHD when he tried to measure the voltage induced by the Thames flowing through the earth's magnetic field.[3] Moreover, his law of induction, discovered in 1831, demands that magnetic field lines in a perfectly conducting fluid must move with the fluid, as if frozen into the medium. This fundamental result is usually attributed to the twentieth-century astrophysicist and electrical engineer Hannes Alfvén, but really it follows directly from Faraday's law, as we shall see in §2.9. But undoubtedly the single most important contribution of Faraday was his development of the concept of the magnetic field.

Prior to the work of Faraday, the physics and mathematics communities were convinced that the laws of electromagnetism should be formulated in terms of action at a distance. The notion of a field did not exist. For example, Ampère had discovered that two current-carrying wires attract each other, and so, by analogy with Newton's law of gravitational attraction, it seemed natural to try and describe this force in terms of some kind of inverse square law. In this view, nothing of significance exists *between* the wires.

Faraday had a different vision, in which the medium between the wires plays a role. In his view, a wire which carries a current introduces a field into the medium surrounding it. This field (the magnetic field) exists whether or not a second wire is present. When the second wire is introduced it experiences a force by virtue of this field. Moreover, in Faraday's view, the field is not just some convenient mathematical intermediary. It has real physical significance, possessing energy, momentum and so on.

Of course, it is Faraday's view which now prevails, which is all the more remarkable because Faraday had no formal education and, as a consequence, little mathematical skill. James Clerk Maxwell was greatly impressed by Faraday, and in the preface to his classic treatise on Electricity and Magnetism he wrote,

Before I began the study of electricity I resolved to read no mathematics on the subject till I had first read through Faraday's Experimental Researches in Electricity. I was aware that there was supposed to be a difference between Faraday's way of conceiving phenomena and

---

[3] In Faraday's words: 'I made experiments therefore (by favour) at Waterloo bridge, extending a copper wire nine hundred and sixty feet in length upon the parapet of the bridge, and dropping from its extremities other wires with extensive plates of metal attached to them to complete contact with the water. Thus the wire and the water made one conducting circuit; and as the water ebbed and flowed with the tide, I hoped to obtain currents.' 1832

that of the mathematicians, so that neither he nor they were satisfied with each other's language. . .

As I proceeded with the study of Faraday, I perceived that his method of conceiving the phenomena was also a mathematical one, though not exhibited in the conventional form of mathematical symbols. . .

For instance, Faraday, in his mind's eye, saw lines of force traversing space where the mathematicians saw centres of force attracting at a distance: Faraday sought the seat of the phenomena in real actions going on in the medium, they were satisfied that they had found it in a power of action at a distance. . .

When I translated what I considered to be Faraday's ideas into mathematical form, I found that in general the results of the two methods coincided, so that the same phenomena were accounted for, and the same laws of action deduced by both methods, but that Faraday's methods resembled those in which we begin with the whole and arrive at the parts by analysis, while the ordinary mathematical methods were founded on the principle of beginning with the parts and building up the whole by synthesis. I also found that several of the most fertile methods of research discovered by the mathematicians could be expressed much better in terms of the ideas derived by Faraday than in their original form. . .

If by anything I have written I may assist any student in understanding Faraday's modes of thought and expression, I shall regard it as the accomplishment of one of my principle aims – to communicate to others the same delight which I have found myself in reading Faraday's 'Researches'. *(Maxwell, 1873)*

When Maxwell transcribed Faraday's ideas into mathematical form, correcting Ampère's law in the process, he arrived at the famous laws which now bear his name. Kelvin was similarly taken by Faraday's physical insight:

One of the most brilliant steps made in philosophical exposition of which any instance existed in the history of science was that in which Faraday stated, in three or four words, intensely full of meaning, the law of magnetic attraction or repulsion. . . .Mathematicians were content to investigate the general expression of the resultant force experienced by a globe of soft iron in all such cases; but Faraday, without any mathematics, divined the result of the mathematical investigations. Indeed, the whole language of the magnetic field and 'lines of force' is Faraday's. It must be said for the mathematicians that they greedily accepted it, and have ever since been most zealous in using it to the best advantage. *(Kelvin, 1872)*

The central role played by fields acquires special significance in relativistic mechanics where, because of the finite velocity of propagation of interactions, it is not meaningful to talk of direct interactions of particles (or currents) located at distant points. We can speak only of the field established by one particle and of the subsequent influence of this field on other particles. Of course, Faraday could not have foreseen this. Einstein explicitly noted the important role played by Faraday and Maxwell (Figure 2.2) in his popular introduction to Relativity:

During the second half of the nineteenth century, in conjunction with the researches of Faraday and Maxwell, it became more and more clear that the description of

Figure 2.2 Two leaders in the development of electromagnetism: (a) Faraday and (b) Maxwell.

electromagnetic processes in terms of fields was vastly superior to a treatment on the basis of the mechanical concepts of material points. . . . One psychological effect of this immense success was that the field concept, as opposed to the mechanistic framework of classical physics, gradually won greater independence.   *(Einstein, 1916)*

Of course, Faraday's contribution to magnetism did not stop with the introduction of fields. He also discovered electromagnetic induction. In fact, in 1831, in no more than ten full days of research, Faraday unravelled all of the essential features of electromagnet induction. We shall return to Faraday's law of induction in §2.9, where we shall see that in many ways it is a remarkable result, encompassing not just one physical law, but two.

2.6 Maxwell's Equations

We have mentioned Maxwell's equations several times. When combined with the force law (2.5) and the law of charge conservation (2.9), they embody all that we know about classical electrodynamics, and so it seems appropriate that, at some point, we should write them down and discuss their full significance. We offer such a discussion here, aimed at readers who are interested in the relationship between MHD, which uses only a reduced form of Maxwell's equations, and the broader range of phenomena associated with Maxwell's equations. Those readers impatient to pursue the story of MHD may choose to skip this section and move directly to §2.7, where we summarise the reduced form of Maxwell's equations required for MHD.

2.6.1 The Displacement Current and Electromagnetic Waves

For materials which are neither magnetic nor dielectric, Maxwell's equations state that

$$\nabla \cdot \mathbf{E} = \rho_e/\varepsilon_0, \qquad \text{(Gauss' law)} \qquad (2.35)$$

$$\nabla \cdot \mathbf{B} = 0, \qquad \text{(Solenoidal nature of } \mathbf{B}) \qquad (2.36)$$

$$\nabla \times \mathbf{E} = -\frac{\partial \mathbf{B}}{\partial t}, \qquad \text{(Faraday's law in differential form)} \qquad (2.37)$$

$$\nabla \times \mathbf{B} = \mu\left(\mathbf{J} + \varepsilon_0 \frac{\partial \mathbf{E}}{\partial t}\right). \qquad \text{(Ampere–Maxwell equation)} \qquad (2.38)$$

In addition, we have

$$\nabla \cdot \mathbf{J} = -\partial \rho_e/\partial t, \qquad \text{(charge conservation)} \qquad (2.39)$$

$$\mathbf{F} = q(\mathbf{E} + \mathbf{u} \times \mathbf{B}). \qquad \text{(force law)} \qquad (2.40)$$

Of course, not all of these laws are independent. For example, (2.36) follows (more or less) from the divergence of Faraday's law, while the divergence of the Ampère–Maxwell equation, combined with charge conservation, yields Gauss' law. Indeed, it was precisely the desire to make Ampère's law consistent with charge conservation and Gauss' law that led Maxwell to add the displacement current, $\varepsilon_0 \partial \mathbf{E}/\partial t$, to the right of (2.38).

Let us consider how the new term introduced by Maxwell works. To that end, consider the simple model problem of a long, straight, vertical wire which is charging a parallel-plate capacitor, as shown in Figure 2.3. Suppose the current in the wire is I, the instantaneous charge on the upper plate Q, the charge on the lower plate $-Q$, and the area of the capacitor plates A. Conservation of charge then demands that $I = dQ/dt$. Moreover, Gauss' law applied in integral form to a control volume that encloses the upper capacitor plate tells us that the electric field within

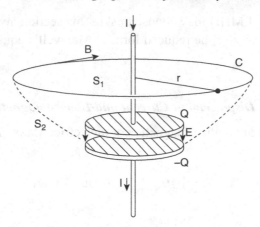

Figure 2.3 The role of Maxwell's displacement current in the charging of a parallel-plate capacitor.

the capacitor points downward from the upper to the lower plate, and has a magnitude of $E = Q/\varepsilon_0 A$. (We ignore edge effects here and take $\mathbf{E}$ to be uniform within the capacitor and zero outside.) Combining these simple results we have $I/A = \varepsilon_0 dE/dt$.

Now consider the integral version of the Ampère–Maxwell equation,

$$\oint_C \mathbf{B} \cdot d\mathbf{l} = \mu \int_S \mathbf{J} \cdot d\mathbf{S} + \mu \int_S \varepsilon_0 \frac{\partial \mathbf{E}}{\partial t} \cdot d\mathbf{S}, \qquad (2.41)$$

applied to a horizontal curve C which is circular and concentric with the wire. We let C have radius r and suppose that it lies well above the capacitor. The corresponding open surface S may be any surface that spans the curve C. We consider two particular examples of S: S_1 is horizontal and co-planar with the curve C, while S_2 is axially elongated, stretches all the way down to the capacitor, and passes between the capacitor plates. For the case of S_1 only the first integral on the right of (2.41) comes into play, and we simply obtain the usual result for a long, straight wire: $2\pi r B = \mu I$. For S_2 the first integral on the right is zero, since there is no net current passing though S_2. Yet we are obliged to obtain the same result as for S_1, i.e. $B = \mu I/2\pi r$. At this point Ampère's law fails us (there is no current passing through S_2 to support the field) and Maxwell's correction involving $\mathbf{E}$ comes to our rescue. That is, the second integral on the right of (2.41) has a value of $\varepsilon_0 \mu A dE/dt$, and when this is combined with $I = \varepsilon_0 A dE/dt$, the surface integral over S_2 becomes μI, exactly as required.

The key point is that Gauss' law plus charge conservation combine to yield

$$\nabla \cdot \left[\mathbf{J} + \varepsilon_0 \frac{\partial \mathbf{E}}{\partial t} \right] = 0. \tag{2.42}$$

By replacing $\mathbf{J}$ in Ampère's law by $\mathbf{J} + \varepsilon_0 \partial \mathbf{E}/\partial t$, we ensure that the right-hand side of the Ampère–Maxwell equation is solenoidal, and so integrates to give the same value for all possible surfaces S that span the curve C in the integral equation (2.41). So, as the current passing through the first surface integral in (2.41) falls away, the contribution from the second surface integral rises to compensate.

Note, however, that Maxwell's displacement current is important not just because it fixes up inconsistencies in Ampère's law. Rather, its primary importance lies in the fact that it leads to an entirely new phenomenon: electromagnetic waves. It is easy to see why this is so. The curl of the Ampère–Maxwell equation is

$$\nabla^2 \mathbf{B} = -\mu \nabla \times \mathbf{J} - \frac{1}{c^2} \frac{\partial}{\partial t} \nabla \times \mathbf{E},$$

which, combined with Faraday's law, yields

$$\nabla^2 \mathbf{B} - \frac{1}{c^2} \frac{\partial^2 \mathbf{B}}{\partial t^2} = -\mu \nabla \times \mathbf{J}. \tag{2.43}$$

We might compare this with the magnetostatic result (2.20), which excludes displacement currents:

$$\nabla^2 \mathbf{B} = -\mu \nabla \times \mathbf{J}.$$

Evidently, Maxwell's displacement current has transformed a Poisson equation into a wave equation, with the term $-\mu \nabla \times \mathbf{J}$ acting as a source of waves. In a vacuum, for example, we have

$$\nabla^2 \mathbf{B} - \frac{1}{c^2} \frac{\partial^2 \mathbf{B}}{\partial t^2} = 0, \tag{2.44}$$

which describes electromagnetic waves propagating at the speed of light.

2.6.2 Gauges, Retarded Potentials and the Biot–Savart Law Revisited

We now return to the issue of how to choose the optimal gauge (divergence) for the vector potential $\mathbf{A}$. As noted in §2.4, the simplest choice in magnetostatics and MHD, where we have $\nabla \cdot \mathbf{J} = 0$, is the *Coulomb gauge*, $\nabla \cdot \mathbf{A} = 0$. However, we have already suggested that for those cases in which $\nabla \cdot \mathbf{J} \neq 0$ and the displacement current is retained, it is preferable to choose a different definition for $\nabla \cdot \mathbf{A}$. In particular, when we require the full set of Maxwell's equations, it is preferable to choose the *Lorentz gauge*,

$$\nabla \cdot \mathbf{A} = -\frac{1}{c^2} \frac{\partial V}{\partial t}. \tag{2.45}$$

Let us see why this is so. Recall that the definition $\nabla \times \mathbf{A} = \mathbf{B}$ allows us to uncurl Faraday's law as

$$\mathbf{E} = -\frac{\partial \mathbf{A}}{\partial t} - \nabla V, \tag{2.46}$$

which effectively defines the scalar potential V. We now substitute for $\mathbf{E}$ in the Ampère–Maxwell equation, while making no assumption about $\nabla \cdot \mathbf{A}$:

$$\nabla \times \nabla \times \mathbf{A} = \mu \mathbf{J} - \frac{1}{c^2} \frac{\partial}{\partial t} \left[\frac{\partial \mathbf{A}}{\partial t} + \nabla V \right]. \tag{2.47}$$

Noting that $\nabla^2 \mathbf{A} = \nabla(\nabla \cdot \mathbf{A}) - \nabla \times \nabla \times \mathbf{A}$, this may be rewritten as

$$\nabla^2 \mathbf{A} - \nabla(\nabla \cdot \mathbf{A}) - \frac{1}{c^2} \frac{\partial^2 \mathbf{A}}{\partial t^2} = -\mu \mathbf{J} + \frac{1}{c^2} \frac{\partial}{\partial t} \nabla V. \tag{2.48}$$

Similarly, substituting for $\mathbf{E}$ in Gauss' law yields

$$\nabla^2 V + \frac{\partial}{\partial t} \nabla \cdot \mathbf{A} = -\frac{\rho_e}{\varepsilon_0}. \tag{2.49}$$

If we compare (2.48) and (2.49) we see that the two potentials are, in general, coupled. However, if we now adopt the Lorentz gauge (2.45), we obtain a great simplification:

$$\nabla^2 V - \frac{1}{c^2} \frac{\partial^2 V}{\partial t^2} = -\frac{\rho_e}{\varepsilon_0}, \tag{2.50}$$

$$\nabla^2 \mathbf{A} - \frac{1}{c^2} \frac{\partial^2 \mathbf{A}}{\partial t^2} = -\mu \mathbf{J}. \tag{2.51}$$

These equations are elegant in a number of respects. First, they are simple. Second, the governing equations for the vector and scalar potentials are now decoupled. Third, each potential is governed by a wave equation, with $-\rho_e/\varepsilon_0$ and $-\mu \mathbf{J}$ acting as source terms. This makes perfect sense since, when displacement currents are retained, information propagates at the speed of light in the form of electromagnetic waves. We might compare these equations with the equivalent magnetostatic results, in which we used the Coulomb gauge $\nabla \cdot \mathbf{A} = 0$:

$$\nabla^2 V = -\rho_e/\varepsilon_0, \quad \nabla^2 \mathbf{A} = -\mu \mathbf{J}, \tag{2.52}$$

Let us take things a little further. We have seen that the static equations may be inverted according to (2.19) and (2.34) as

$$V(\mathbf{x}, t) = \frac{1}{4\pi\varepsilon_0} \int \frac{\rho_e(\mathbf{x}', t)}{|\mathbf{r}|} \, d\mathbf{x}', \tag{2.53}$$

$$\mathbf{A}(\mathbf{x}, t) = \frac{\mu}{4\pi} \int \frac{\mathbf{J}(\mathbf{x}', t)}{|\mathbf{r}|} \, d\mathbf{x}', \tag{2.54}$$

where $\mathbf{r} = \mathbf{x} - \mathbf{x}'$ and we now explicitly note that V and ρ_e are evaluated at the same instant, as are $\mathbf{A}$ and $\mathbf{J}$. It is natural to ask if the wave equations (2.50) and (2.51) may be similarly inverted, and again the answer is rather elegant. The inversion formulae for our two wave equations turn out to be (see, for example, Jackson, 1999)

$$V(\mathbf{x}, t) = \frac{1}{4\pi\varepsilon_0} \int \frac{\rho_e(\mathbf{x}', t')}{|\mathbf{r}|} \, d\mathbf{x}', \tag{2.55}$$

$$\mathbf{A}(\mathbf{x}, t) = \frac{\mu}{4\pi} \int \frac{\mathbf{J}(\mathbf{x}', t')}{|\mathbf{r}|} \, d\mathbf{x}', \tag{2.56}$$

where $t' = t - r/c$. The quantity t' is known as the retarded time, and its appearance in (2.55) and (2.56) is entirely natural. That is, information can be transmitted no faster than it can be carried in the form of waves, and so a fluctuation in ρ_e or $\mathbf{J}$ at location $\mathbf{x}'$ is not felt instantaneously at the remote position $\mathbf{x}$, but rather a time r/c later. The potentials governed by (2.55) and (2.56) are known as the retarded potentials, and they are the usual starting point for exploring the properties of electromagnetic radiation.

Example 2.3 A Paradox

Consider a hollow plastic sphere which is mounted on a frictionless spindle and is free to rotate. Charged metal pellets are embedded in the surface of the sphere and a wire loop is placed near its centre, the axis of the loop being parallel to the rotation axis (Figure 2.4). The loop is connected to a battery, so that a current flows and

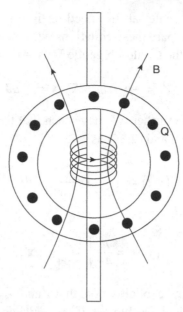

Figure 2.4 Charged metal pellets are embedded in the surface of a sphere and a wire loop carrying current is placed near its centre. (Courtesy of Avishek Ranjan.)

a dipole-like magnetic field is created. We now ensure that everything is stationary and (somehow) disconnect the battery. The magnetic field declines and so, by Faraday's law, we induce an electric field which is azimuthal, i.e. $\mathbf{E}$ takes the form of rings which are concentric with the axis of the wire loop. This electric field now acts on the charges to produce a torque on the sphere, causing it to spin up. At the end of the process we have gained some angular momentum in the sphere, but at the cost of the magnetic field. Apparently, we have contravened the principle of conservation of angular momentum! Can you unravel this paradox? (Hint: consult Feynman's 'Lectures on Physics', volume II.)

The earth has a negative charge on its surface, which gives rise to an average surface electric field of around 100V/m. It also has a dipole magnetic field, and rotates about an axis more-or-less aligned with the magnetic axis. Do you think the rotation rate of the earth changes when the earth's magnetic field reverses?　　■

Example 2.4 A Second Paradox
Combine (2.23), which follows from the Biot–Savart law of magnetostatics, with charge conservation, $\nabla \cdot \mathbf{J} = -\partial \rho_e / \partial t$, and the Green's function solution for the electrostatic potential, (2.19). Hence, show that, since $\mathbf{E} = -\nabla V$ in electrostatics, this yields

$$\nabla \times \mathbf{B} - \mu \mathbf{J} = \frac{\mu}{4\pi} \nabla \left[\int \frac{\nabla' \cdot \mathbf{J}'}{r} \, d\mathbf{x}' \right] = -\frac{1}{c^2} \frac{\partial}{\partial t} \nabla \left[\frac{1}{4\pi\varepsilon_0} \int \frac{\rho'_e}{r} \, d\mathbf{x}' \right]$$

$$= -\frac{1}{c^2} \frac{\partial}{\partial t} \nabla V = \frac{1}{c^2} \frac{\partial \mathbf{E}}{\partial t}.$$

This is the Ampère–Maxwell equation (2.38). However, the concept of the displacement current is a dynamic phenomenon, so how can we deduce it from the laws of magnetostatics and electrostatics? (A resolution of the paradox may be found in Figure 2.3, where $\mathbf{B}$ and $\mathbf{J}$ are steady, yet $\mathbf{E}$ is a linear function of time.) ∎

2.7 The Reduced Form of Maxwell's Equations for MHD

Let us now return to conducting fluids. For the purposes of MHD, Maxwell's equations may be simplified considerably. In §2.2 and §2.3 we showed that the charge density ρ_e is extremely small in a conducting fluid (see Equation (2.10)) and so plays no significant part in MHD. For example, we have seen that the electric force, $q\mathbf{E}$, is tiny by comparison with the Lorentz force, so that we may write $\mathbf{F} = \mathbf{J} \times \mathbf{B}$. Similarly, the contribution of $\partial \rho_e / \partial t$ to the charge conservation equation is also negligible, so that charge conservation reduces to the statement that $\nabla \cdot \mathbf{J} = 0$. Apparently ρ_e is significant only in Gauss' law, and so we simply drop Gauss' law and ignore ρ_e. Also, we have seen that in conducting fluids that the displacement currents are negligible by comparison with the current density, $\mathbf{J}$, and so the Ampère–Maxwell equation reduces to the differential form of Ampère's law.

We may now summarise the (pre-Maxwell) form of the electrodynamic equations used in nearly all MHD. They are:

- Ampère's law plus charge conservation,

$$\boxed{\nabla \times \mathbf{B} = \mu \mathbf{J}}, \quad \boxed{\nabla \cdot \mathbf{J} = 0}, \tag{2.57}$$

- Faraday's law plus the solenoidal constraint on $\mathbf{B}$,

$$\boxed{\nabla \times \mathbf{E} = -\frac{\partial \mathbf{B}}{\partial t}}, \quad \boxed{\nabla \cdot \mathbf{B} = 0}, \tag{2.58}$$

- Ohm's law plus the Lorentz force per unit volume,

$$\boxed{\mathbf{J} = \sigma(\mathbf{E} + \mathbf{u} \times \mathbf{B})}, \quad \boxed{\mathbf{F} = \mathbf{J} \times \mathbf{B}}. \tag{2.59}$$

These six equations constitute all the fundamental laws of electrodynamics required for MHD. To these we might add the Biot–Savart law,

$$\mathbf{B}(\mathbf{x}) = \frac{\mu}{4\pi}\int\frac{\mathbf{J}(\mathbf{x}')\times\mathbf{r}}{r^3}d\mathbf{x}', \quad \mathbf{r} = \mathbf{x} - \mathbf{x}', \tag{2.60}$$

which inverts Ampère's law and ensures that **B** is solenoidal.

If we choose to invoke the vector potential **A**, then we also have the auxiliary equations

$$\mathbf{B} = \nabla\times\mathbf{A}, \quad \nabla\cdot\mathbf{A} = 0, \tag{2.61}$$

where we have used the Coulomb gauge for **A**. Ampère's law now becomes

$$\nabla^2\mathbf{A} = -\mu\mathbf{J},$$

which inverts in an infinite domain to give

$$\mathbf{A}(\mathbf{x}, t) = \frac{\mu}{4\pi}\int\frac{\mathbf{J}(\mathbf{x}', t)}{|\mathbf{r}|}d\mathbf{x}', \quad \mathbf{r} = \mathbf{x} - \mathbf{x}'. \tag{2.62}$$

This is equivalent to the Biot–Savart law, (2.60). Finally, Faraday's law uncurls to give

$$\mathbf{E} = -\frac{\partial\mathbf{A}}{\partial t} - \nabla V. \tag{2.63}$$

Equations (2.57) through (2.63) encapsulate all that we need to know about electromagnetism for MHD.

Example 2.5 The Poynting Vector
Use Faraday's law and Ampère's law to show

$$\frac{d}{dt}\int_V(\mathbf{B}^2/2\mu)dV = -\int_V\mathbf{J}\cdot\mathbf{E}dV - \oint_S[(\mathbf{E}\times\mathbf{B})/\mu]\cdot d\mathbf{S}.$$

Now use Ohm's law to confirm that

$$\int_V\mathbf{J}\cdot\mathbf{E}dV = \frac{1}{\sigma}\int_V\mathbf{J}^2dV + \int_V(\mathbf{J}\times\mathbf{B})\cdot\mathbf{u}dV.$$

Combining the two we obtain the energy equation

$$\frac{d}{dt}\int(\mathbf{B}^2/2\mu)dV = -\frac{1}{\sigma}\int_V\mathbf{J}^2dV - \int_V(\mathbf{J}\times\mathbf{B})\cdot\mathbf{u}dV - \oint_S\mathbf{P}\cdot d\mathbf{S},$$

where $\mathbf{P} = (\mathbf{E} \times \mathbf{B})/\mu$ is called the Poynting vector. The first two integrals on the right represent Joule dissipation and the rate of loss of magnetic energy due to the rate of working of the Lorentz force on the medium. Conservation of energy then tells us that the third integral on the right must represent the rate at which electromagnetic energy flows out through the surface S. It follows that the Poynting vector is the electromagnetic energy flux density. ∎

2.8 A Transport Equation for the Magnetic Field

If we combine Ohm's law, Faraday's equation and Ampère's law we obtain an expression relating $\mathbf{B}$ to $\mathbf{u}$,

$$\frac{\partial \mathbf{B}}{\partial t} = -\nabla \times \mathbf{E} = -\nabla \times [(\mathbf{J}/\sigma) - \mathbf{u} \times \mathbf{B}] = \nabla \times [\mathbf{u} \times \mathbf{B} - \nabla \times \mathbf{B}/\mu\sigma].$$

Noting that $\nabla \times \nabla \times \mathbf{B} = -\nabla^2 \mathbf{B}$ (since $\mathbf{B}$ is solenoidal) this simplifies to

$$\boxed{\frac{\partial \mathbf{B}}{\partial t} = \nabla \times (\mathbf{u} \times \mathbf{B}) + \lambda \nabla^2 \mathbf{B}}, \quad \lambda = (\mu\sigma)^{-1}. \tag{2.64}$$

This is sometimes called the induction equation, although, as we shall see, a more descriptive name would be the advection-diffusion equation for $\mathbf{B}$. The quantity λ is called the *magnetic diffusivity*. Like all diffusivities, it has the units of m^2/s. Equation (2.64) is, in effect, a transport equation for $\mathbf{B}$, in the sense that if $\mathbf{u}$ is known then it dictates the spatial and temporal evolution of $\mathbf{B}$ from some specified initial condition. We shall spend much of Chapter 5 unpicking the physical implications of (2.64): it is one of the key equations in MHD. An important special case is that of a so-called perfect (or ideal) conductor in which $\lambda = 0$ and

$$\frac{\partial \mathbf{B}}{\partial t} = \nabla \times (\mathbf{u} \times \mathbf{B}).$$

Example 2.6 Decay of Force-Free Fields
Show that if, at $t = 0$, there exists a *force-free* field, $\nabla \times \mathbf{B} = \alpha\mathbf{B}$, in a stationary fluid, then that field will decay as $\mathbf{B} \sim \exp(-\lambda\alpha^2 t)$, remaining a force-free field. ∎

2.9 A Second Look at Faraday's Law

In this final section we return to Faraday's law, as it plays a particularly important role in MHD. In order to fully appreciate the significance of this law, we first need a simple but important kinematic result.

2.9.1 *An Important Kinematic Equation*

Suppose that $\mathbf{G}$ is a solenoidal field, $\nabla \cdot \mathbf{G} = 0$, and S_m is a surface which is embedded in a fluid medium, i.e. S_m is locked into the medium and moves as the fluid moves. (The subscript m indicates that it is a *material surface*.) Then it may be shown that

$$\frac{d}{dt} \int_{S_m} \mathbf{G} \cdot d\mathbf{S} = \int_{S_m} \left[\frac{\partial \mathbf{G}}{\partial t} - \nabla \times (\mathbf{u} \times \mathbf{G}) \right] \cdot d\mathbf{S}. \qquad (2.65)$$

A formal proof of (2.65) will be given in a moment. First, however, we might try to get a qualitative feel for its origins. The idea behind (2.65) is the following. The flux of $\mathbf{G}$ through S_m changes for two reasons. First, even if S_m were fixed in space there is a change in flux whenever $\mathbf{G}$ is time-dependent. This is the first term on the right of (2.65). Second, if the boundary of S_m moves it may expand at points to include additional flux, or perhaps contract at other points to exclude flux. It happens that, in a time δt, the surface adjacent to a line element $d\mathbf{l}$ increases by an amount $d\mathbf{S} = (\mathbf{u} \times d\mathbf{l})\delta t$, and so the increase in flux due to movement of the boundary C_m is

$$\delta \int_{S_m} \mathbf{G} \cdot d\mathbf{S} = \delta t \oint_{C_m} \mathbf{G} \cdot (\mathbf{u} \times d\mathbf{l}) = -\delta t \oint_{C_m} (\mathbf{u} \times \mathbf{G}) \cdot d\mathbf{l}.$$

Using Stokes' theorem, the final line integral above may be converted into a surface integral, which accounts for the second term on the right of (2.65). Of course, we have yet to show that $d\mathbf{S} = (\mathbf{u} \times d\mathbf{l})\delta t$.

A more formal proof of (2.65) proceeds as follows. The change in flux through S_m in a time δt is

$$\delta \int_{S_m} \mathbf{G} \cdot d\mathbf{S} = (\delta t) \int_{S_m} (\partial \mathbf{G}/\partial t) \cdot d\mathbf{S} + \oint_{C_m} \mathbf{G} \cdot \delta \mathbf{S},$$

where $\delta \mathbf{S}$ is the element of area swept out by the line element $d\mathbf{l}$ in time δt. However, $\delta \mathbf{S} = d\mathbf{l}' \times d\mathbf{l}$ where $d\mathbf{l}'$ is the infinitesimal displacement of the element $d\mathbf{l}$ in time δt (see Figure 2.5). Since $d\mathbf{l}' = \mathbf{u}\delta t$ we have $\delta \mathbf{S} = (\mathbf{u} \times d\mathbf{l})\delta t$, and so

$$\delta \int_{S_m} \mathbf{G} \cdot d\mathbf{S} = (\delta t) \int_{S_m} (\partial \mathbf{G}/\partial t) \cdot d\mathbf{S} - (\delta t) \oint_{C_m} \mathbf{u} \times \mathbf{G} \cdot d\mathbf{l}.$$

(We have used the cyclic properties of the scalar triple product to rearrange the terms in the line integral.) Finally, the application of Stokes' theorem to the line integral gets us back to (2.65), and this completes the proof.

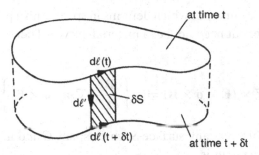

Figure 2.5 Movement of a material surface in a time δt.

Now, (2.65) should not be passed over lightly: it is a very useful result. The reason is that often in MHD (or conventional fluid mechanics) we find that certain vector fields obey a transport equation of the form

$$\frac{\partial \mathbf{G}}{\partial t} = \nabla \times [\mathbf{u} \times \mathbf{G}].$$

This is true of $\nabla \times \mathbf{u}$ in an unforced, inviscid flow (see Chapter 3) and of $\mathbf{B}$ in a perfect conductor (see Equation (2.64)). In such cases (2.65) tells us that the flux of $\mathbf{B}$ (or $\nabla \times \mathbf{u}$) through any material surface, S_m, is conserved as the flow evolves. We shall return to this idea time and again in subsequent chapters.

Note that it is not necessary to invoke the idea of a continuously moving medium and of material surfaces in order to arrive at (2.65). If we consider any curve, C, and any surface S that spans C and moves in space with a prescribed velocity $\mathbf{u}$, then

$$\frac{d}{dt} \int_S \mathbf{G} \cdot d\mathbf{S} = \int_S \left[\frac{\partial \mathbf{G}}{\partial t} - \nabla \times (\mathbf{u} \times \mathbf{G}) \right] \cdot d\mathbf{S}. \tag{2.66}$$

2.9.2 The Full Significance of Faraday's Law

We now return to electrodynamics. Recall that the differential form of Faraday's law is

$$\nabla \times \mathbf{E} = -\frac{\partial \mathbf{B}}{\partial t}. \tag{2.67}$$

As noted earlier, this is a weaker statement than the integral version (2.26), since it tells us only about the EMF induced by a time-dependent field. Let us now see if we can deduce the more general version of Faraday's law, (2.26), from (2.67).

Suppose we have a curve, C, which deforms in space with a prescribed velocity $\mathbf{u}$ $(\mathbf{x}, t)$. (This could be, but need not be, a material curve.) Then, at each point on the curve, (2.67) gives

$$\nabla \times (\mathbf{E} + \mathbf{u} \times \mathbf{B}) = -\left\{ \frac{\partial \mathbf{B}}{\partial t} - \nabla \times (\mathbf{u} \times \mathbf{B}) \right\}. \tag{2.68}$$

We now integrate this over any surface S which spans C and invoke the kinematic equation (2.66). The result is

$$\oint_C (\mathbf{E} + \mathbf{u} \times \mathbf{B}) \cdot d\mathbf{l} = -\frac{d}{dt} \int_S \mathbf{B} \cdot d\mathbf{S}. \tag{2.69}$$

So far we have used only Faraday's law in differential form. We now invoke the idea of the Lorentz force. This tells us that, in a frame of reference moving with velocity $\mathbf{u}$, the electric field is $\mathbf{E}_r = \mathbf{E} + \mathbf{u} \times \mathbf{B}$. Given that $\mathbf{E}$ transforms in this way we may rewrite our integral equation as

$$\oint_C \mathbf{E}_r \cdot d\mathbf{l} = -\frac{d}{dt} \int_S \mathbf{B} \cdot d\mathbf{S}. \tag{2.70}$$

Note that this applies to any curve C. For example, C may be fixed in space, move with the fluid, or execute some motion quite different to that of the fluid. It does not matter. The final step is to introduce the idea of an EMF, which is the force experienced by a unit charge, as measured by $\mathbf{E}_r$, integrated around a closed curve. Since the EMF is the closed-line integral of $\mathbf{E}_r$, (2.70) becomes

$$\text{EMF} = \oint_C \mathbf{E}_r \cdot d\mathbf{l} = -\frac{d}{dt} \int_S \mathbf{B} \cdot d\mathbf{S}. \tag{2.71}$$

We have arrived at the integral version of Faraday's law. Note, however, that to get from (2.67) to (2.71) we had to invoke the force law $\mathbf{F} = q(\mathbf{u} \times \mathbf{B})$, or equivalently the relationship $\mathbf{E}_r = \mathbf{E} + \mathbf{u} \times \mathbf{B}$. Note also that if C and S happen to be material curves and surfaces embedded in a fluid, then (2.71) becomes

$$\text{EMF} = \oint_{C_m} \mathbf{E}_r \cdot d\mathbf{l} = -\frac{d}{dt} \int_{S_m} \mathbf{B} \cdot d\mathbf{S}. \tag{2.72}$$

Now, it is intriguing that the integral version of Faraday's law describes the EMF generated in two very different situations, i.e. when $\mathbf{E}$ is induced by a time-dependent

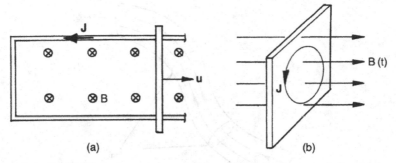

Figure 2.6 An EMF can be generated either (a) by movement of the conducting medium (motional EMF) or else (b) by variation of the magnetic field (transformer EMF).

magnetic field and when $\mathbf{E}_r$ is induced (at least in part) by motion of the circuit within a magnetic field. The two extremes are shown in Figure 2.6. If $\mathbf{B}$ is constant, and the EMF is due solely to movement of the circuit, then $\oint \mathbf{E}_r \cdot d\mathbf{l}$ is called a *motional* EMF. If the circuit is fixed and $\mathbf{B}$ is time-dependent then $\oint \mathbf{E} \cdot d\mathbf{l}$ is termed a *transformer* EMF. In either case, however, the EMF is equal to (minus) the rate of change of flux. Now, motional EMF is due essentially to the Lorentz force, $q\mathbf{u} \times \mathbf{B}$, while transformer EMF results from the Maxwell equation $\nabla \times \mathbf{E} = -\partial \mathbf{B}/\partial t$, which is usually regarded as a separate physical law. Yet both are described by the integral equation (2.71). Faraday's law is therefore an extraordinary result. It apparently embodies two quite different phenomena. It seems that it just so happens that motional EMF and transformer EMF can both be described by the same flux rule. (Actually, at a deeper level, both Maxwell's equations and the Lorentz force can, with some additional restrictions and assumptions, be deduced from Coulomb's law plus the Lorentz transformation of special relativity. This is discussed in, for example, Chapter 6 of Lorrain and Corson, 1970. So perhaps it is not just coincidence that Faraday's equation embraces two apparently distinct physical laws. Nevertheless, from a classical viewpoint, it represents a remarkably convenient equation.)

2.9.3 Faraday's Law in Ideal Conductors: Alfvén's Theorem

From Ohm's law, $\mathbf{J} = \sigma \mathbf{E}_r$, and (2.72) we have

$$\frac{1}{\sigma} \oint_{C_m} \mathbf{J} \cdot d\mathbf{l} = -\frac{d}{dt} \int_{S_m} \mathbf{B} \cdot d\mathbf{S} \tag{2.73}$$

for any material surface, S_m. Now suppose that $\sigma \to \infty$ (a perfect conductor). Then

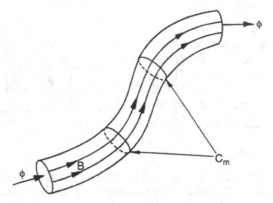

Figure 2.7 A magnetic flux tube.

$$\frac{d}{dt} \int_{S_m} \mathbf{B} \cdot d\mathbf{S} = 0, \ \sigma \to \infty. \tag{2.74}$$

We have arrived at a key result in MHD. That is to say, *in a perfect conductor, the flux through any material surface S_m is preserved as the flow evolves*. Now picture an individual flux tube sitting in a perfectly conducting fluid (Figure 2.7). Such a tube is, by analogy to a stream-tube in fluid mechanics, just an aggregate of magnetic field lines.

Since $\mathbf{B}$ is solenoidal ($\nabla \cdot \mathbf{B} = 0$), the flux of $\mathbf{B}$ along the tube, Φ, is constant. (This comes from applying Gauss' divergence theorem to a finite portion of the tube.) Now consider a material curve C_m which at some initial instant encircles the flux tube. From (2.74) the flux enclosed by C_m will remain constant as the flow evolves, and this is true of each and every material curve enclosing the tube at $t = 0$. This suggests (but does not prove) that the tube itself moves with the fluid, as if frozen into the medium. This, in turn, suggests that every field line moves with the fluid, since we could let the tube have a vanishingly small cross-section. We have arrived at Alfvén's theorem (Figure 2.8), which states that

magnetic field lines are frozen into a perfectly conducting fluid in the sense that they move with the fluid.

We shall give formal proof of Alfvén's theorem in Chapter 5.

Example 2.7 Magnetic Helicity
In general, the *magnetic helicity* of a magnetic field is defined as

$$h_m = \int \mathbf{A} \cdot \mathbf{B} dV.$$

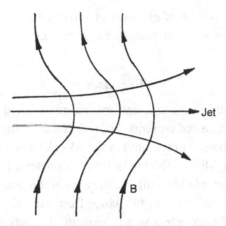

Figure 2.8 An example of Alfvén's theorem. Flow through a magnetic field causes the field lines to bow out.

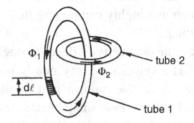

Figure 2.9 Two interlinked magnetic flux tubes.

Consider a magnetic field composed exclusively of two thin, interlinked magnetic flux tubes (Figure 2.9). Let the tubes have volumes V_1 and V_2, centrelines C_1 and C_2, and magnetic fluxes Φ_1 and Φ_2. Show that, for this particular case, the magnetic helicity is

$$h_m = \int_{V_1} \mathbf{A} \cdot \mathbf{B}dV + \int_{V_2} \mathbf{A} \cdot \mathbf{B}dV = \Phi_1 \oint_{C_1} \mathbf{A} \cdot d\mathbf{l} + \Phi_2 \oint_{C_2} \mathbf{A} \cdot d\mathbf{l}.$$

Hence, use Stokes' theorem to show that $h_m = \pm 2\Phi_1\Phi_2$, where the positive sign is applicable when the two flux tubes are connected in a right-handed fashion, and the negative sign when the connection is left-handed. Now use Alfvén's theorem to show that, if these two isolated flux tubes are embedded in, and swept around by, a perfectly conducting fluid, the magnetic helicity, h_m, is conserved. (No mathematics is required for this.) As we shall see in Chapter 5, this is a special case of a more general result: the magnetic helicity of a localised magnetic field distribution is always conserved in a perfectly conducting fluid, no matter how complicated the magnetic field or the motion of the fluid. ∎

This completes our survey of classical electrodynamics. Readers interested in more details could do worse than consult Feynman, Leighton, and Sands (1964).

Exercises

2.1 A conducting fluid flows in a uniform magnetic field which is negligibly perturbed by the induced currents. Show that the condition for there to be no net charge distribution in the fluid is that $\mathbf{B} \cdot (\nabla \times \mathbf{u}) = 0$.

2.2 A thin conducting disc of thickness h and diameter d is placed in a uniform alternating magnetic field parallel to the axis of the disc. What is the induced current density as a function of distance from the axis of the disc?

2.3 Show that a coil carrying a steady current, I, tends to orientate itself in a magnetic field in such a way that the total magnetic field linking the coil is a maximum. Also, show that the torque exerted on the coil is $\mathbf{m} \times \mathbf{B}$, where $\mathbf{m}$ is the dipole momentum of the coil. What do you think will happen to a small current loop in a highly conducting fluid which is permeated by a large-scale magnetic field?

2.4 A fluid of small but finite conductivity flows through a tube constructed of insulating material. The velocity is very nearly uniform and equal to u. To measure the velocity of the fluid, a part of the tube is subjected to a uniform magnetic field, B. Two small electrodes which are in contact with the fluid are installed through the tube walls. A voltmeter detects an induced EMF of V. What is the velocity of the fluid?

2.5 Show that it is impossible to construct a generator of electromotive force constant in time operating on the principle of electromagnetic induction.

3

A First Course in Fluid Dynamics

In his 1964 lectures on physics, R. P. Feynman noted that

The efforts of a child trying to dam a small stream flowing in the street, and his surprise at the strange way the water works its way out, has its analogue in our attempts over the years to understand the flow of fluids. We have tried to dam the water by getting the laws and equations ... but the water has broken through the dam and escaped our attempt to understand it.

In this chapter we build the dam and write down the equations. Later, particularly in Chapters 8 and 9 where we discuss turbulence, we shall see how the dam bursts open.

3.1 Different Categories of Fluid Flow

The beginner in fluid mechanics is often bewildered by the many diverse categories of fluid flow which appear in the textbooks. There are entire books dedicated to such topics as boundary layers, turbulence, vortex dynamics and so on. Yet the relationship between these different types of flow, and their relationship to 'real' flows, is often unclear. You might ask, if I want to understand natural convection in a room do I need a text on boundary layers, on turbulence or on vortex dynamics? The answer, probably, is all three. These subjects rarely exist in isolation but rather interact in some complex way. The purpose of §3.1 is to give some indication as to what expressions such as boundary layers, turbulence and vorticity mean, how these phenomena interact, and when they are likely to be important in practice. The discussion is essentially qualitative, and anticipates some of the results proved in the subsequent sections. So the reader will have to take certain facts at face value. Nevertheless, the intention is to provide a broad framework into which the many detailed calculations of the subsequent sections fit.

We shall describe why, for good physical reasons, fluid mechanics and fluid flows are often divided into different regimes. In particular, there are three very broad subdivisions in the subject. The first relates to the issue of when a fluid may be treated as inviscid, and when the finite viscosity possessed by all fluids (blood, air, liquid metals) must be taken into account. Here we shall see that, typically, viscosity and shear stresses are of great importance close to solid surfaces (within so-called *boundary layers*) but often of less importance at large distance from a surface. Next there is the subdivision between *laminar* (organised) flow and *turbulent* (chaotic) flow. In general, low-speed or very viscous flows are stable to small perturbations and so remain laminar, whereas most fluids operating at normal speeds are unstable to the slightest perturbation and rapidly develop a chaotic component of motion. The final, rather broad subdivision which occurs in fluid mechanics is between irrotational (sometimes called *potential*) flow and rotational flow. (By 'irrotational flow' we mean flows in which $\nabla \times \mathbf{u} = 0$.) Turbulent motion, internal flows, and boundary layers are always rotational. Sometimes, however, under very particular conditions, an external flow may be approximately irrotational, and indeed this kind of flow dominated the early literature in aerodynamics. In reality, though, such flows are rare in nature and the large space given over to potential flow theory in traditional texts probably owes more to the ease with which such flows are amenable to mathematical description than to their usefulness in interpreting real events.

3.1.1 Viscosity, the Reynolds Number and Boundary Layers

Let us now explore in a little more detail these three subdivisions. We need two elementary ideas as a starting point. We need to be able to quantify shear stresses and inertial forces in a fluid. Let us start with inertia. Suppose, for the sake of argument, that we have a steady flow. That is to say, the velocity field $\mathbf{u}$, which we normally write as $\mathbf{u}(\mathbf{x},t)$, is a function of $\mathbf{x}$ but not of t. It follows that the speed of the fluid at any one point in space is steady, the flow pattern does not change with time, and the streamlines (the analogue of $\mathbf{B}$-lines) represent particle trajectories for individual fluid 'lumps'. Now consider a particular streamline, C, as shown in Figure 3.1, and focus attention on a particular fluid blob as it moves along the streamline. Let s be a curvilinear coordinate measured along C, and $V(s)$ be the speed $|\mathbf{u}|$. Since the streamline represents a particle trajectory, we can apply the usual rules of mechanics and write

$$(\text{acceleration of lump}) = V\frac{dV}{ds}\hat{\mathbf{e}}_t - \frac{V^2}{R}\hat{\mathbf{e}}_n,$$

where R is the radius of curvature of the streamline and $\hat{\mathbf{e}}_t$, $\hat{\mathbf{e}}_n$ represent unit vectors tangential and normal to the streamline. In general, then, the acceleration of

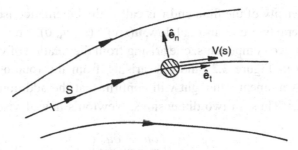

Figure 3.1 A fluid element in a steady flow.

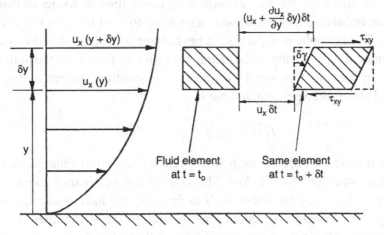

Figure 3.2 The distortion of a fluid element in a parallel shear flow.

a typical fluid element is of order $|\mathbf{u}|^2/\ell$, where ℓ is a characteristic length scale of the flow pattern.

Next we turn to shear stress in a fluid. In most fluids this is quantified using an empirical law known as Newton's law of viscosity. This is most simply understood in a one-dimensional flow, $u_x(y)$, as shown in Figure 3.2. Here fluid layers slide over each other due to the fact that u_x is a function of y. One measure of this rate of sliding is the angular distortion rate, $d\gamma/dt$, of an initially rectangular element. (See Figure 3.2 for the definition of γ.) Newton's law of viscosity says that a shear stress, τ, is required to cause the relative sliding of the fluid layers. Moreover, it states that τ is directly proportional to $d\gamma/dt$. However, it is clear from Figure 3.2 that $d\gamma/dt = \partial u_x/\partial y$, and so this law is usually written as

$$\tau = \rho \nu \frac{\partial u_x}{\partial y},$$

where ρ is the density of the fluid and ν is called the kinematic viscosity.

In a more general two-dimensional flow, $\mathbf{u}(x,y) = (u_x, u_y, 0)$, it turns out that γ, and hence $d\gamma/dt$, has two components, one arising from the rotation of vertical material lines, as shown in Figure 3.2, and one arising from the rotation of horizontal material lines. A moment's thought will confirm that the additional contribution to $d\gamma/dt$ is $\partial u_y/\partial x$. Thus, in two dimensions, Newton's law of viscosity becomes

$$\tau_{xy} = \rho\nu\left(\frac{\partial u_x}{\partial y} + \frac{\partial u_y}{\partial x}\right).$$

This generalises in an obvious way to three dimensions (see §3.2). Now, shear stresses are important not just because they cause fluid elements to distort, but because an imbalance in shear stress can give rise to a net force on individual fluid elements. For example, in Figure 3.2 a net horizontal force will be exerted on the fluid element if τ_{xy} at the top of the element is different from τ_{xy} at the bottom of the element. In fact, in this simple example it is readily confirmed that the net horizontal shear force per unit volume is

$$f_x = \partial\tau_{xy}/\partial y = \rho\nu\partial^2 u_x/\partial y^2.$$

We are now in a position to estimate the relative sizes of the inertial and viscous forces in a three-dimensional flow. The viscous forces per unit volume take the form of gradients in shear stress, such as $\partial\tau_{xy}/\partial y$, and have a size $f_\nu \sim \rho\nu|\mathbf{u}|/\ell_\perp^2$, where $\ell_\perp$ is a characteristic length scale normal to the streamlines. The inertial forces per unit volume, on the other hand, are of the order of $f_{in} \sim \rho \times$ (acceleration) $\sim \rho|\mathbf{u}|^2/\ell$, where ℓ is a typical geometric length scale. The ratio of the two is of order

$$\text{Re} = u\ell/\nu.$$

This is the *Reynolds number*. When Re is small, viscous forces outweigh inertial forces, and when Re is large viscous forces are relatively small. Now we come to the key point. When Re is calculated using some characteristic geometric length scale, it is almost always very large. This reflects the fact that the viscosity of nearly all common fluids, including liquid metal, is minute, of the order of $10^{-6}\text{m}^2/\text{s}$. Because of the large size of Re, it is tempting to dispense with viscosity altogether. However, this is extremely dangerous. For example, inviscid theory predicts that a sphere sitting in a uniform cross-flow experiences no drag (d'Alembert's paradox), and this is clearly not the case, even for 'thin' fluids like air.

Something seems to have gone wrong. The problem is that, no matter how small ν might be, there are always thin regions near surfaces where the shear stresses are

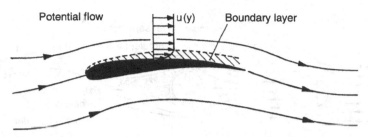

Figure 3.3 The boundary layer on an aerofoil.

significant, i.e. of the order of the inertial forces. These *boundary layers* give rise to the drag on, say, an aerofoil. Consider the flow over an aerofoil shown in Figure 3.3. Here we use a frame of reference moving with the foil. The value of Re for such a flow, based on the length of the aerofoil, will be very large, typically around 10^8. Consequently, away from the surface of the aerofoil, the fluid behaves as if it were inviscid. Close to the aerofoil, however, something else happens, and this is a direct result of a boundary condition called the *no-slip condition*. The no-slip condition says that all fluids are 'sticky', in the sense that there can be no relative motion between a fluid and a surface with which it comes into contact. The fluid 'sticks' to the surface. In the case of the aerofoil this means that there must be some transition region near the surface where the fluid velocity drops down from its *free-stream* value (the value it would have if the fluid were inviscid) to zero on the surface. This is the boundary layer. Boundary layers are usually very thin. We can estimate their thickness as follows. Within the boundary layer there must be some force acting on the fluid which pulls the velocity down from the free-stream value to zero at the surface. The force which does this is the viscous force and so within the boundary layer the inertial and viscous forces must be of similar order. Let δ be the boundary layer thickness and ℓ be the span of the aerofoil. The inertial forces are of order $\rho|\mathbf{u}|^2/\ell$, and the viscous forces are of order $\rho\nu|\mathbf{u}|/\delta^2$. Equating the two gives $\delta/\ell \sim \sqrt{\nu/u\ell} \ll 1$. Thus we see that, no matter how small we make ν, there is always some (thin) boundary layer in which shear stresses are important. This is why an aerofoil experiences drag even when ν is very small (Re is large).

We have reached the first of our three general subdivisions in fluid mechanics. That is to say, often (but not always) a high-Re flow may be divided into an external, inviscid flow plus one or more boundary layers. Viscous effects are then confined to the boundary layer. This idea was introduced by Prandtl in 1904 and works well for external flow over bodies, particularly streamlined bodies, but can lead to problems in confined flows. (It is true that boundary layers form at the boundaries in confined flows, and that shear stresses are usually large within the boundary layers and weak outside the boundary layers. However, the small but

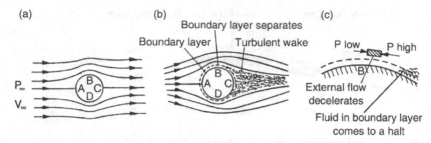

Figure 3.4 Flow over a sphere: (a) inviscid flow; (b) real flow at high Re; and (c) pressure forces which cause separation.

finite shear in the bulk of a confined fluid can, over long periods of time, have a profound influence on the overall flow pattern, as explained in §3.6.3.)

Boundary layers have another important characteristic, called *separation*. Suppose that, instead of an aerofoil, we consider flow over a sphere. If the fluid were inviscid (which no real fluid is!) we would get a symmetric flow pattern as shown in Figure 3.4(a). The pressure at the stagnation points A and C in front of and behind the sphere would be equal (by symmetry), and from Bernoulli's equation the pressure at these points would be high: $P_A = P_\infty + 1/2\,\rho V_\infty^2$, P_∞ and V_∞ being the upstream pressure and velocity. The real flow looks something like that shown in Figure 3.4(b). A boundary layer forms at the leading stagnation point and this remains thin as the fluid moves to the edges of the sphere. However, towards the rear of the sphere something unexpected happens. The boundary layer separates. That is to say, the fluid in the boundary layer is ejected into the external flow and a turbulent wake forms. This separation is caused by pressure forces. Outside the boundary layer the fluid, which tries to follow the inviscid flow pattern, starts to slow down as it passes over the outer edges of the sphere (points B and D) and heads towards the rear stagnation point. This deceleration is caused by pressure forces (pressure differences) which oppose the external flow. These same pressure forces are experienced by the fluid within the boundary layer and so this fluid also begins to decelerate (Figure 3.4(c)). However, the fluid in the boundary layer has less momentum than the corresponding external flow and very quickly it comes to a halt, reverses direction, and moves off into the external flow, thus forming a wake. Thus we see that the flow over a body at high Re can generally be divided into three regions: an inviscid external flow, boundary layers, and a turbulent wake.

3.1.2 Laminar versus Turbulent Flow

Now the fact that Re is invariably large has a second important consequence: most flows in nature are *turbulent*. This leads to a second classification in fluid mechanics. It is an empirical observation that at low values of Re flows are laminar (organised) while at high values of Re they are turbulent (chaotic). This was first demonstrated in 1883 by Reynolds, who studied flow in a pipe. In the case of a pipe, the transition from laminar to turbulent flow is rather sudden, and usually occurs at around Re ~2000, which typically constitutes a rather slow flow rate.

A turbulent flow is characterised by the fact that, superimposed on the mean (time-averaged) flow pattern, there is a random, chaotic motion. The velocity field is often decomposed into its time-averaged component, $\bar{\mathbf{u}}$, and random fluctuations about that mean: $\mathbf{u} = \bar{\mathbf{u}} + \mathbf{u}'$. The transition from laminar to turbulent flow occurs because, at a certain value of Re, instabilities develop in the laminar flow, usually driven by the inertial forces. At low values of Re these would-be inertial instabilities are damped out by viscosity, while at high values of Re the damping is inadequate. (Actually, some instabilities are triggered by viscous forces: see, for example, Acheson, 1990.)

There is a superficial analogy between turbulence and the kinetic theory of gases. The steady laminar flow of a gas has, at the macroscopic level, only a steady component of motion. However, at the molecular level, individual atoms not only possess the mean velocity of the flow, but also some random component of velocity which is related to their thermal energy. It is this random fluctuation in velocity which gives rise to the exchange of momentum between molecules and thus to the macroscopic property of viscosity. There is an analogy between individual atoms in a laminar flow and macroscopic blobs of fluid in a turbulent flow. Indeed, this (rather imperfect) analogy formed the basis of most early attempts to characterise turbulent flow. In particular, it was proposed that one should replace ν in Newton's law of viscosity, which for a gas arises from thermal agitation of the molecules, by an *eddy viscosity* ν_t, which arises from macroscopic fluctuations.

The transition from laminar to turbulent flow is rarely clear cut. For example, often some parts of a flow field are laminar while, at the same time, other parts are turbulent. The simplest example of this is the boundary layer on a flat plate. If the front of the plate is streamlined, and the turbulence level in the external flow is low, the boundary layer usually starts off as laminar. Of course, eventually it becomes unstable and turns turbulent (Figure 3.5).

Often periodic (non-turbulent) fluctuations in the laminar flow precede the onset of turbulence. This is illustrated in Figure 3.6, which shows flow over a cylinder at different values of Re. At low values of Re we get a symmetric flow pattern. This is called creeping flow. As Re rises above unity, steady vortices appear at the rear of the cylinder. By the time Re has reached ~100, these vortices start to peel off from the rear of the cylinder in a regular, periodic manner. (At this point the flow is still laminar.) This is called Kármán's vortex street. At yet higher values of Re the shed vortices become turbulent, but we still have a discernible vortex street. Finally, at a value of Re ~10^5, the flow at the rear of the cylinder loses its periodic structure

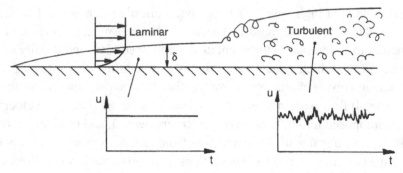

Figure 3.5 The development of a boundary layer on a flat plate.

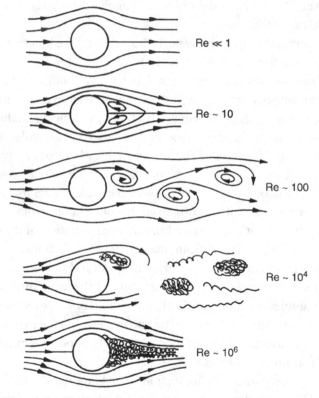

Figure 3.6 Flow behind a cylinder at various values of Re.

and becomes a turbulent wake. Notice that upstream of the cylinder, fluid blobs possess linear momentum but no angular momentum. In the Kármán street, however, certain fluid elements (those in the vortices) possess both linear and angular momentum. Moreover, the angular momentum seems to have come from the boundary layer on the cylinder. This leads us to our third and final subdivision in fluid mechanics. In some flows (potential flows) the fluid elements possess only linear momentum. In others (vortical flows) they possess both angular and linear momentum. In order to pursue this idea a little further we need some measure of the rotation or spin of individual fluid elements. This is called vorticity.

3.1.3 Rotational versus Irrotational Flow

So far we have discussed flow fields in terms of the velocity field $\mathbf{u}$. However, there is a closely related quantity, the vorticity, which is defined as $\boldsymbol{\omega} = \nabla \times \mathbf{u}$. From Stokes' theorem we have, for a small, disc-like fluid element, with cross-sectional area $d\mathbf{S}$ and circular boundary C,

$$\boldsymbol{\omega} \cdot d\mathbf{S} = \oint_C \mathbf{u} \cdot d\mathbf{l}.$$

We might anticipate, therefore, that $\boldsymbol{\omega}$ is a measure of the angular velocity of a fluid element about its centre, and this turns out to be true. In fact, it may be shown that the angular velocity, $\boldsymbol{\Omega}$, of a small fluid blob which is passing through point $\mathbf{x}_0$ at time t_0 is just $\boldsymbol{\omega}(\mathbf{x}_0, t_0)/2$. Thus, while $\mathbf{u}$ is related to the linear momentum of small fluid elements, $\boldsymbol{\omega}$ is related to their angular momentum, $\mathbf{H}$. (Note that we measure both $\boldsymbol{\Omega}$ and $\mathbf{H}$ relative to the centre of the fluid blob.) In short, $\boldsymbol{\omega}(\mathbf{x}, t)$ is twice the angular velocity of a small fluid element passing through point $\mathbf{x}$ at time t, that angular velocity being measured relative to the centre of the element. Now, $\boldsymbol{\omega}$ is a useful quantity because it turns out that, partially as a result of conservation of angular momentum, it cannot be created or destroyed within the interior of a fluid. At least that is the case in the absence of external forces such as buoyancy or the Lorentz force.

That is not to say that vorticity of a fluid particle is constant. Consider the vortices within a two-dimensional Kármán vortex street (Figure 3.6). It turns out that, as they are swept downstream, they grow in size in much the same way that a packet of hot fluid spreads heat by diffusion. Like heat, vorticity can diffuse. In particular, it diffuses between adjacent fluid particles as they sweep through the flow field. However, as with heat, this diffusion does not change the global amount of vorticity (heat) present in an isolated patch of fluid. Thus, as the vortices in the Kármán street spread, the intensity of $\boldsymbol{\omega}$ in each vortex falls, and it falls in such a way that $\int \boldsymbol{\omega} dA$ is conserved for each vortex.

There is a second way in which the vorticity in a given lump of fluid can change. Consider the ice skater who spins faster by pulling his or her arms inward. What is true for ice skaters is true for blobs of fluid. If a spinning fluid blob is stretched by the flow, say from a sphere to a cigar shape, it will spin faster, and the corresponding component of $\boldsymbol{\omega}$ increases to conserve angular momentum.

In summary, then, vorticity cannot be created within the interior of a fluid unless there are body forces present, but it spreads by diffusion and can be intensified by the stretching of fluid elements. The way in which we quantify this diffusion and intensification of vorticity will be discussed in §3.4. For the moment, however, the important point to grasp is that, like heat, vorticity cannot be created within the interior of a fluid unless there are rotational body forces.

So where does the vorticity evident in Figure 3.6 come from? Here the analogy to heat is useful. We shall see that, in the absence of stretching of fluid elements and of body forces, the governing equation for ω is identical to that for heat. It is transported by the mean flow (we say it is *advected*) and diffuses outward from regions of intense vorticity. Also, just like heat, it is in the boundaries that we find the sources of vorticity. In fact, boundary layers are filled with the vorticity which has diffused out from the adjacent surface. This gives us a new way of thinking about boundary layers: they are diffusion layers for the vorticity generated on a surface. Again there is an analogy to heat. Thermal boundary layers are diffusion layers for the heat which seeps into the fluid from a surface. In both cases the thickness of the boundary layer is fixed by the ratio of (i) the rate at which heat or vorticity diffuses across the streamlines from the surface and (ii) the rate at which heat or vorticity is swept downstream by the mean flow. Usually, when Re is large, the cross-stream diffusion is slow compared to the streamwise transport of vorticity, and this is why boundary layers are so thin.

We are now in a position to introduce our third and final classification in fluid mechanics. This is the distinction between so-called potential (vorticity-free) flows and vortical flows. Consider Figure 3.7(a). This represents classical aerodynamics. There is a boundary layer, which is filled with vorticity, and an external flow. The flow upstream of the aerofoil is (in classical aerodynamics) assumed to be irrotational (free of vorticity) and since the vorticity generated on the surface of the foil is confined to the boundary layer, the entire external flow is irrotational. This kind of external flow is called a potential flow.

The problem of computing the external motion is now reduced to solving two *kinematic equations*: $\nabla \cdot \mathbf{u} = 0$ (conservation of mass) and $\nabla \times \mathbf{u} = 0$ (zero vorticity). Indeed, Newton's second law seems to play almost no role, at least as far as mapping out the external velocity field is concerned. In effect, aerodynamics becomes aero-kinematics. However, potential flows (irrotational flows) are rare in nature. In fact, streamlined bodies moving through a still fluid (plus certain types of water waves) represent the only common examples. Almost all real flows are laden with vorticity; vorticity which has been generated somewhere in a boundary layer and then released into the bulk flow (see Figures 3.6 and 3.7(b)). The rustling

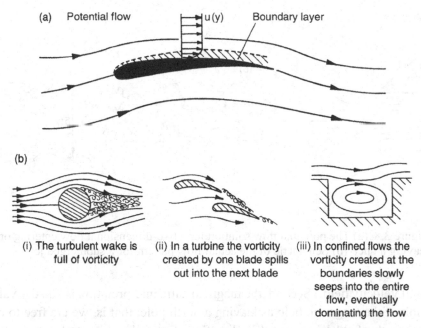

Figure 3.7 (a) Classical aerodynamics (potential flow). (b) Most real flows are laden with vorticity. For example, a turbulent wake, the vorticity shed from an upstream body (as in a cascade of blades), or vorticity slowly diffusing in from a boundary into a confined flow.

of leaves, the blood in our veins, the air in our lungs, natural convection in a room, the wind blowing down the street, the flows in the oceans and atmosphere, the solar wind, and the dramatic explosions on the surface of the sun are all examples of flows laden with vorticity.

The early mathematical texts on fluid mechanics tended to contain a great deal of potential flow theory, applied to all manner of complex geometries, but this probably owes more to the ease with which we can solve $\nabla \cdot \mathbf{u} = 0$ and $\nabla \times \mathbf{u} = 0$ than any link to reality. The dangers of potential flow theory are captured by Figure 3.8(a), which appears in many texts as the two-dimensional potential flow pattern at the entrance to a duct. This has a simple analytical solution, as noticed by Helmholtz in 1868. The same figure also appears on the cover of Volume II of Maxwell's classic *Treatise on Electricity and Magnetism,* and in that case it represents a magnetic field entering a south pole and governed by the magnetostatic equations $\nabla \cdot \mathbf{B} = 0$ and $\nabla \times \mathbf{B} = 0$. Two observations immediately come to mind. First, there is no concept of inertia associated with the static magnetic field, and this reminds us that the idea of inertia also plays no role when we map out a potential flow pattern. (Newton's second law, in the guise of Bernoulli's equation, enters the problem only if we wish to infer a pressure distribution from our prescribed

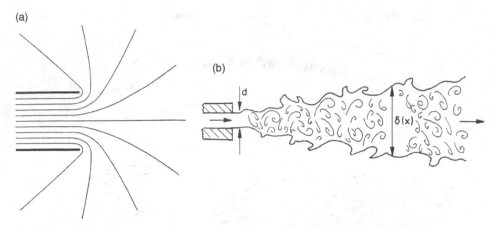

Figure 3.8 (a) The potential flow solution for a two-dimensional flow entering or leaving a duct. (b) Schematic of the actual flow pattern for a submerged jet.

potential velocity field.) Second, the magnetostatic interpretation is equally valid if we think of the magnetic field as leaving a north pole; that is, we are free to draw arrows on the field lines in either direction. Formally, this is also true of our potential flow, in that the same solution (with a change of sign) can be equally interpreted as fluid leaving a two-dimensional duct. But no self-respecting fluid dynamicist would ever pretend that Figure 3.8(a) actually represents such a flow, as in reality that would look very different, taking the form of a submerged jet with nearly parallel time-averaged streamlines (Figure 3.8(b)), a jet driven by the very inertia that potential flows lack!

So we have two types of flow: potential flows, which are easy to compute but infrequent in nature, and vortical flows, which are very common but much more difficult to understand. The art of quantifying this second category of flow consists of tracking the progress of the vorticity from the boundaries into the bulk of the fluid. Often this arises from wakes or from boundary-layer separation. Sometimes, as in the case of confined flows, it is due to a slow but finite diffusion of vorticity from the boundary into the interior of the flow. In either case, in the absence of body forces, it is the boundaries which generate the vorticity.

To these two classes of flow, potential flow and unforced vortical flow, we should add a third: that of MHD. Here the Lorentz force generates vorticity directly in the interior of the fluid. On the one hand this makes MHD more difficult to understand, but on the other it makes it more attractive. In MHD we have the opportunity to grab hold of the interior of a fluid and manipulate the flow. With this brief, qualitative overview of fluid mechanics we now set about quantifying the motion of a fluid. Our starting point is the equation of motion of a fluid blob.

3.2 The Navier–Stokes Equation

The Navier–Stokes equation is a statement about the changes in linear momentum of a small element of fluid as it progresses through a flow field. Let p be the pressure field, ρ the fluid density, and τ_{ij} the viscous stresses acting on the fluid. Then Newton's second law applied to a small blob of fluid of volume δV yields

$$(\rho \delta V) \frac{D\mathbf{u}}{Dt} = -(\nabla p)\delta V + [\partial \tau_{ij}/\partial x_j]\,\delta V. \tag{3.1}$$

That is to say, the mass of the element, $\rho \delta V$, times its acceleration, $D\mathbf{u}/Dt$, equals the net pressure force acting on the surface of the fluid blob,

$$\oint (-p)d\mathbf{S} = \int (-\nabla p)dV = -(\nabla p)\delta V,$$

plus the net surface force arising from the viscous stress τ_{ij},

$$\oint \tau_{ij}dS_j = \int (\partial \tau_{ij}/\partial x_j)dV = (\partial \tau_{ij}/\partial x_j)\delta V.$$

We shall take the fluid to be incompressible so that the conservation of mass, expressed as $\nabla \cdot (\rho \mathbf{u}) = -\partial \rho/\partial t$, reduces to the so-called continuity equation

$$\nabla \cdot \mathbf{u} = 0. \tag{3.2}$$

We also take the fluid to be Newtonian, so that the viscous stresses are given by the constitutive law,

$$\tau_{ij} = \rho \nu \left(\frac{\partial u_i}{\partial x_j} + \frac{\partial u_j}{\partial x_i} \right) = 2\rho\, \nu S_{ij}, \tag{3.3}$$

where ν is the kinematic viscosity of the fluid and S_{ij} is known as the rate-of-strain tensor. Substituting for τ_{ij} in (3.1) and dividing through by $\rho \delta V$ yields the conventional form of the Navier–Stokes equation,

$$\boxed{\frac{D\mathbf{u}}{Dt} = -\nabla (p/\rho) + \nu\, \nabla^2 \mathbf{u}.} \tag{3.4}$$

The boundary condition on $\mathbf{u}$ corresponding to (3.4) is that $\mathbf{u} = \mathbf{0}$ on any stationary, solid surface, i.e. the fluid 'sticks' to any solid surface. This is the *no-slip* condition.

The expression $D(\cdot)/Dt$ is called the convective derivative. It is the rate of change of a quantity measured in a frame of reference moving with a given element of fluid. This should not be confused with $\partial(\cdot)/\partial t$, which is, of course, the rate of change of a quantity at a fixed point in space. For example, DT/Dt is the rate of change of temperature of a given fluid lump as it slides down a streamline, whereas $\partial T/\partial t$ is the rate of change of temperature at a fixed point (through which a succession of fluid particles will pass). It follows that $D\mathbf{u}/Dt$ is the acceleration of a fluid element, which is why it appears on the left of (3.4).

An expression for $D\mathbf{u}/Dt$ may be obtained as follows. Consider a scalar function of position and time, $f(\mathbf{x}, t)$. We have $\delta f = (\partial f/\partial t)\delta t + (\partial f/\partial x)\delta x + \ldots$. If we are interested in the change in f following a fluid particle, then $\delta x = u_x \delta t$ etc., and so

$$\frac{Df}{Dt} = \frac{\partial f}{\partial t} + u_x \frac{\partial f}{\partial x} + \ldots = \frac{\partial f}{\partial t} + (\mathbf{u} \cdot \nabla)f.$$

The same expression may be applied to each of the components of the vector field, such as $\mathbf{u}$, which allows us to rewrite (3.4) in the form

$$\partial \mathbf{u}/\partial t + (\mathbf{u} \cdot \nabla)\mathbf{u} = -\nabla(p/\rho) + \nu\nabla^2 \mathbf{u}. \tag{3.5}$$

Note that, in steady flows, (i.e. flows in which $\partial \mathbf{u}/\partial t = 0$), the streamlines represent particle trajectories and the acceleration of a fluid element is $(\mathbf{u} \cdot \nabla)\mathbf{u}$. The physical origin of this expression becomes clearer when we rewrite $(\mathbf{u} \cdot \nabla)\mathbf{u}$ in terms of curvilinear coordinates attached to a streamline. As noted earlier,

$$(\mathbf{u} \cdot \nabla)\mathbf{u} = V\frac{\partial V}{\partial s}\hat{\mathbf{e}}_t - \frac{V^2}{R}\hat{\mathbf{e}}_n, \tag{3.6}$$

where $V = |\mathbf{u}|$, $\hat{\mathbf{e}}_t$ and $\hat{\mathbf{e}}_n$ are unit vectors in the tangential and principle normal directions, s is a streamwise coordinate, and R is the radius of curvature of the streamline. The first expression on the right is the rate of change of speed, DV/Dt, while the second is the centripetal acceleration, which is directed towards the centre of curvature of the streamline and is associated with the rate of change of direction of the velocity of a fluid particle.

3.3 Vorticity, Angular Momentum, and the Biot–Savart Law

So far we have concentrated on the velocity field, $\mathbf{u}$. However, in common with many other branches of fluid mechanics, in MHD it is often more fruitful to work with the vorticity field defined by

$$\boldsymbol{\omega} = \nabla \times \mathbf{u}. \tag{3.7}$$

The reason is two-fold. First, the rules governing the evolution of $\boldsymbol{\omega}$ are somewhat simpler than those governing $\mathbf{u}$. For example, pressure gradients appear as a source of linear momentum in (3.5), yet the pressure itself depends on the instantaneous (global) distribution of **u**. By focusing on vorticity, on the other hand, we may dispense with the pressure field entirely. (The reasons for this will become evident shortly.) The second reason for studying vorticity is that many flows are characterised by localised regions of intense rotation (i.e. vorticity). Dust whirls in the street, trailing vortices behind aircraft wings, whirlpools, tidal vortices, tornadoes, hurricanes, and the great red spot of Jupiter represent just a few examples.

Let us start by trying to endow ω with some physical meaning. Consider a small element of fluid in a two-dimensional flow, $\mathbf{u}(x,y) = (u_x, u_y, 0)$, $\boldsymbol{\omega} = (0,0,\omega_z)$, instantaneously centred at location $\mathbf{x}_0$. Suppose that, at some instant, the element is circular (a disc) with radius r. Let $\mathbf{u}_0$ be the linear velocity of the centre of the element and Ω be its mean angular velocity measured about $\mathbf{x}_0$, defined as the average rate of rotation of two mutually perpendicular material lines embedded within the element. From Stokes' theorem we have

$$\omega_z \pi \, r^2 = \int (\nabla \times \mathbf{u}) \cdot d\mathbf{S} = \oint \mathbf{u} \cdot d\mathbf{l}. \tag{3.8}$$

We might anticipate that the line integral on the right has a value of $(\Omega r)2\pi r$. If this were the case, then $\omega_z = 2\Omega$.

In fact exact analysis confirms that this is so. Consider Figure 3.9(a): the anti-clockwise rotation rate of a short line element, dx, orientated parallel to the x-axis is $\partial u_y/\partial x$, while the anti-clockwise rotation rate of a line element, dy, parallel to the y-axis is $-\partial u_x/\partial y$, giving $\Omega = (\partial u_y/\partial x - \partial u_x/\partial y)/2 = \omega_z/2$. It appears, therefore, that ω_z is twice the angular velocity of the fluid element measured about its centre. This result extends to three dimensions. The vorticity vector at a particular location and at a particular instant is twice the average angular velocity of a blob of fluid passing through that point at that instant, with Ω being measured about the blob's centre (Figure 3.9b).

It should be emphasised, however, that ω has nothing at all to do with the global rotation of a fluid. Rectilinear flows may possess vorticity while flows with circular streamlines need not. Consider, for example, the rectilinear shear flow $\mathbf{u}(y) = (\gamma y, 0, 0)$, where $\gamma = $ constant. The streamlines are straight and parallel yet the fluid elements rotate at a rate $\omega/2 = -\gamma/2$. This is because vertical line elements, dy, move faster at the top of the element than at the bottom, so they continually rotate towards the horizontal. Conversely, we can have global rotation of a flow without intrinsic rotation of the fluid elements. One example is the so-called free vortex $\mathbf{u}(r) = (0, k/r, 0)$ in (r, θ, z) coordinates, where k is a constant. It is readily confirmed that $\omega = \mathbf{0}$ in such a vortex, except for a singular point at the origin.

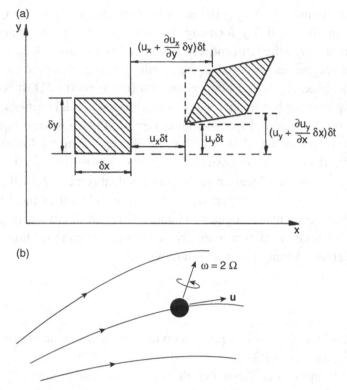

Figure 3.9 (a) In two dimensions the angular velocity of a small fluid element, defined as the average rate of rotation of two perpendicular material lines, is $\omega_z/2$. (b) In three dimensions the vorticity vector at a particular location and at a particular time is twice the average angular velocity of a small blob of fluid passing through that point at that instant.

So far we have been concerned only with kinematics. We now introduce some dynamics. Since we are interested in rotation, it is natural to focus on angular momentum rather than linear momentum. Consider the angular momentum, $\mathbf{H}$, of a small material element that is *instantaneously* spherical, with $\mathbf{H}$ measured about the centre of the element. Then $\mathbf{H} = I\boldsymbol{\omega}/2$ where I is the moment of inertia of the blob. This angular momentum will change at a rate determined by the tangential surface stresses alone. The pressure has no influence on $\mathbf{H}$ at the instant at which the element is spherical, since the pressure forces all point radially inward. Therefore, at one particular instant in time, we have

$$\frac{D\mathbf{H}}{Dt} = \mathbf{T}_\nu, \tag{3.9}$$

where $\mathbf{T}_\nu$ denotes the viscous torque acting on the sphere. Now the convective derivative satisfies the usual rules of differentiation and so we have

$$I\frac{D\omega}{Dt} = -\omega\frac{DI}{Dt} + 2\mathbf{T}_\nu. \tag{3.10}$$

Evidently, the terms on the right arise from the change in the moment of inertia of a fluid element and the viscous torque, respectively. In cases where viscous stresses are negligible (e.g. outside boundary layers) this simplifies to

$$\frac{D}{Dt}(I\omega) = 0. \tag{3.11}$$

Now (3.10) and (3.11) are not very useful (or even very meaningful) as they stand, since they apply only at the initial instant during which the fluid element is spherical. However, they suggest several results, all of which we shall confirm by exact analysis in the next section. First, there is no reference to pressure in (3.10) and (3.11) so that we might anticipate that ω evolves independently of p. Second, if ω is initially zero, and the flow is inviscid, then ω should remain zero in each fluid particle as it is swept around the flow field. This is the basis of potential flow theory in which we set $\omega = \mathbf{0}$ in the upstream fluid and so we can assume that ω is zero at all downstream points. Third, if I decreases in a fluid element (and $\nu = 0$), then the vorticity of that element should increase. For example, consider a blob of vorticity embedded in an otherwise potential flow field consisting of converging stream-lines, as shown in Figure 3.10. An initially spherical element will be stretched into an ellipsoid by the converging flow. The moment of inertia of the element about an axis parallel to ω decreases and consequently ω must rise to conserve $\mathbf{H}$. It is possible, therefore, to intensify vorticity by stretching fluid blobs. Intense rotation can result from this process, the familiar bathtub vortex being just one example. We shall see that something very similar happens to magnetic fields. They, too, can be intensified by stretching.

Finally we note that there is an analogy between the differential form of Ampère's law, $\nabla \times \mathbf{B} = \mu\mathbf{J}$, and the definition of vorticity, $\nabla \times \mathbf{u} = \omega$. Since both

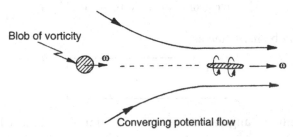

Figure 3.10 The stretching of fluid elements can intensify the vorticity.

B and **u** are solenoidal we can hijack the Biot–Savart law from electromagnetic theory to invert the relationship $\boldsymbol{\omega} = \nabla \times \mathbf{u}$. That is to say, in infinite domains,

$$\mathbf{u} = \frac{1}{4\pi} \int \frac{\boldsymbol{\omega}(\mathbf{x}') \times \mathbf{r}}{r^3} d^3\mathbf{x}', \qquad \mathbf{r} = \mathbf{x} - \mathbf{x}'. \tag{3.12}$$

Also, note that, like **u** and **B**, the vorticity field is solenoidal, $\nabla \cdot \boldsymbol{\omega} = 0$, since it is the curl of another vector. Consequently, we may invoke the idea of vortex tubes, which are analogous to magnetic flux tubes or streamtubes.

3.4 The Vorticity Equation and Vortex Line Stretching

We now formally derive the laws governing the evolution of vorticity. We start by writing (3.5) in the form

$$\frac{\partial \mathbf{u}}{\partial t} = \mathbf{u} \times \boldsymbol{\omega} - \nabla(p/\rho + u^2/2) + \nu \, \nabla^2 \mathbf{u}, \tag{3.13}$$

which follows from the identity

$$\nabla(u^2/2) = (\mathbf{u} \cdot \nabla)\mathbf{u} + \mathbf{u} \times \boldsymbol{\omega}.$$

Note, in passing, that steady, inviscid flows have the property that $\mathbf{u} \cdot \nabla(p/\rho + u^2/2) = 0$, so that $C = p/\rho + u^2/2$ is constant along a streamline. This is Bernoulli's theorem, C being Bernoulli's function.

We now take the curl of (3.13), noting that the gradient term disappears:

$$\frac{\partial \boldsymbol{\omega}}{\partial t} = \nabla \times [\mathbf{u} \times \boldsymbol{\omega}] + \nu \, \nabla^2 \boldsymbol{\omega}. \tag{3.14}$$

Compare this with (2.64). It appears that $\boldsymbol{\omega}$ and **B** obey precisely the same evolution equation! We shall exploit this analogy repeatedly in the subsequent chapters. Now, since **u** and $\boldsymbol{\omega}$ both are solenoidal, we have the vector relationship

$$\nabla \times (\mathbf{u} \times \boldsymbol{\omega}) = (\boldsymbol{\omega} \cdot \nabla)\mathbf{u} - (\mathbf{u} \cdot \nabla)\boldsymbol{\omega},$$

and so (3.14) may be rewritten as

$$\frac{D\boldsymbol{\omega}}{Dt} = (\boldsymbol{\omega} \cdot \nabla)\mathbf{u} + \nu \, \nabla^2 \boldsymbol{\omega}. \tag{3.15}$$

Compare this with our angular momentum equation for a blob which is instantaneously spherical:

$$\frac{D\omega}{Dt} = \omega \frac{D}{Dt} \ln(I^{-1}) + 2\mathbf{T}_\nu/I.$$

We might anticipate that the terms on the right of (3.15) represent (a) the change in the moment of inertia of a fluid element due to stretching of that element and (b) the viscous torque on an element. In other words, the rate of rotation of a fluid blob may increase or decrease due to changes in its moment of inertia, or change because it is spun up or spun down by viscous stresses. We shall see shortly that this interpretation of (3.15) is correct.

There is another way of looking at (3.15). It may be interpreted as an *advection-diffusion equation* for vorticity. The idea of an advection-diffusion equation is so fundamental that it is worth dwelling on its significance. Perhaps this is most readily understood in the context of two-dimensional flows, in which $\mathbf{u}(x,y) = (u_x, u_y, 0)$ and $\omega(x,y) = (0, 0, \omega_z)$. The first term on the right of (3.15) now vanishes to yield

$$\frac{D\omega_z}{Dt} = \nu \, \nabla^2 \omega_z. \tag{3.16}$$

Compare this with the equation governing the temperature, T, in a fluid,

$$\frac{DT}{Dt} = \alpha \nabla^2 T, \tag{3.17}$$

where α is the thermal diffusivity. This is the advection-diffusion equation for heat. In some ways (3.17) represents the prototype advection-diffusion equation and we shall take a moment to review its properties. When $\mathbf{u}$ is zero, we have, in effect, a solid, and the temperature field evolves according to

$$\frac{\partial T}{\partial t} = \alpha \nabla^2 T. \tag{3.18}$$

Heat soaks through material purely by virtue of thermal diffusion (conduction). At the other extreme, if $\mathbf{u}$ is non-zero but the fluid is thermally insulating ($\alpha = 0$) we have $DT/Dt = 0$. As each fluid lump moves around it conserves its heat, and hence its temperature. This is referred to as the advection of heat, i.e. the transport of heat by virtue of material movement. In general, though, we have both advection and diffusion of heat. To illustrate the combined effect of these processes, consider the unsteady, two-dimensional distribution of temperature in a uniform cross flow, $(u_x, 0, 0)$, as shown in Figure 3.11. From (3.17) we have

$$\frac{\partial T}{\partial t} + u_x \frac{\partial T}{\partial x} = \alpha \nabla^2 T. \tag{3.19}$$

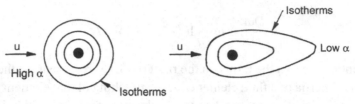

Figure 3.11 The advection and diffusion of heat from a hot wire.

Figure 3.12 Advection and diffusion of heat from a pulsed wire.

Suppose that heat is injected into the fluid from a hot wire. When the velocity is low and the conductivity high the isotherms around the wire will be almost circular. When the conductivity is low, however, each fluid element will tend to conserve its heat as it moves. The isotherms will then become elongated, as shown. The relative size of the advection to the diffusion term is given by the Peclet number, $Pe = u\ell/\alpha$. (Here ℓ is a characteristic length scale.) If the Peclet number is small, then the transfer of heat is dominated by diffusion; when it is large, advection dominates.

Now consider the case of a wire which is being pulsed with electric current to produce a sequence of hot fluid packets. These are swept downstream and grow by diffusion. In Figure 3.12 heat is restricted to the dotted volumes of fluid. Outside these volumes $T = 0$ (or equal to some reference temperature).

Note that advection and diffusion represent processes in which heat is redistributed. However, heat cannot be created nor destroyed by advection or diffusion. That is, the total amount of heat is conserved. This is most easily seen by integrating (3.17) over a fixed volume in space which encloses one of the hot blobs in Figure 3.12. If S is the surface of that volume then, using Gauss' theorem, we have

$$\frac{d}{dt}\int TdV = -\oint_S (\mathbf{u}T) \cdot d\mathbf{S} + \alpha\oint_S \nabla T \cdot d\mathbf{S} = 0, \qquad (3.20)$$

since $T = 0$ on S.

From (3.16) we see that the vorticity in a two-dimensional flow is advected and diffused just like heat. The analogue of the thermal diffusivity is ν and the Reynolds number, $Re = u\ell/\nu$, now plays the role of the Peclet number. In other words, vorticity is advected by $\mathbf{u}$ and diffused by the viscous stresses. Moreover, just like heat,

Figure 3.13 The Kármán vortices behind a cylinder.

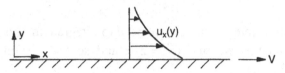

Figure 3.14 The diffusion of vorticity from a plate.

vorticity cannot be created or destroyed within the interior of the flow. The net vorticity within a fixed volume V can change only if vorticity is advected in or out of the volume, or else diffused across the boundary. In the absence of solid surfaces the global vorticity is conserved. Simple examples of this (analogous to the blobs of heat above) are the vortices in the Kármán vortex street behind a cylinder (see Figure 3.13). The vortices are advected by the velocity and spread by diffusion, but the total vorticity within each eddy remains constant as it moves downstream.

A simple illustration of the diffusion of vorticity is given by the following example. Suppose that a plate of infinite length is immersed in a still fluid. At time $t = 0$ it suddenly acquires a constant velocity V in its own plane. We want to find the subsequent motion, $\mathbf{u}(y, t)$.

Now, by the no-slip condition, the fluid adjacent to the plate sticks to it, and so moving the plate creates a gradient in velocity, which gives rise to vorticity. The plate becomes a source of vorticity, which subsequently diffuses into the fluid. Since u_x is not a function of x, the continuity equation gives $\partial u_y/\partial y = 0$, and since u_y is zero at the plate, $u_y = 0$ everywhere. The two-dimensional vorticity equation (3.16) then simplifies to

$$\frac{\partial \omega_z}{\partial t} = \nu \frac{\partial^2 \omega_z}{\partial y^2}, \quad \omega_z = -\frac{\partial u_x}{\partial y}. \tag{3.21}$$

This is identical to the equation describing the diffusion of heat through a semi-infinite solid whose surface temperature is suddenly raised. This sort of diffusion equation can be solved by looking for a similarity solution of the form

$$\omega = \frac{V}{\ell} f(y/\ell); \quad \ell = (2\nu t)^{1/2}. \tag{3.22}$$

Substituting this into (3.21) reduces our partial differential equation to

$$f'(\eta) + \eta f(\eta) = 0, \quad \eta = y/\ell, \tag{3.23}$$

which may be integrated to give

$$\omega_z = \frac{C_1 V}{\ell} \exp[-\eta^2/2]. \tag{3.24}$$

To fix the constant of integration, C_1, we need to integrate to find u_x on the surface of the plate. From this we find $C_1^2 = 2/\pi$ and so the vorticity distribution is

$$\omega_z = \frac{V}{(\pi \nu\, t)^{1/2}} \exp[-y^2/(4\nu t)]. \tag{3.25}$$

This may now be integrated once more to give the velocity field. However, the details of this solution are perhaps less important than the overall picture. That is, vorticity is created at the surface of the plate by the shear stresses acting on that surface. This vorticity then diffuses into the interior of the fluid in exactly the same way as heat diffuses in from a heated surface. There is no vorticity generation within the interior of the flow. The vorticity is merely redistributed (spread) by virtue of diffusion.

Let us now return to our general vorticity equation (3.15),

$$\frac{D\boldsymbol{\omega}}{Dt} = (\boldsymbol{\omega} \cdot \nabla)\mathbf{u} + \nu\, \nabla^2 \boldsymbol{\omega}. \tag{3.26}$$

In three-dimensional flows the first term on the right is non-zero and it is this additional effect which distinguishes three-dimensional flows from two-dimensional ones. It appears that the vorticity no longer behaves like a temperature field. We have already suggested, by comparing this with our angular momentum equation, that $(\boldsymbol{\omega}\cdot\nabla)\mathbf{u}$ represents intensification of vorticity by the stretching fluid elements. We shall now confirm that this is indeed the case.

Consider, by way of an example, the axisymmetric flow shown in Figure 3.15, which consists of converging streamlines (in the *r-z* plane) as well as a swirling component of velocity, u_θ. By writing $\nabla \times \mathbf{u}$ in terms of cylindrical coordinates, we find that, near the axis, the axial component of vorticity is

$$\omega_z = \frac{1}{r} \frac{\partial}{\partial r} (r u_\theta). \tag{3.27}$$

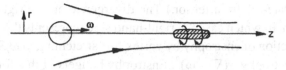

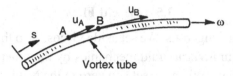

Figure 3.15 The stretching of a material element.

Figure 3.16 The stretching of a tube of vorticity.

Now consider the axial component of the vorticity equation (3.26) applied near $r = 0$. In addition to the usual advection and diffusion terms we have the expression

$$(\mathbf{\omega} \cdot \nabla)\mathbf{u} \sim \omega_z \frac{\partial u_z}{\partial z}. \tag{3.28}$$

This appears on the right of (3.26) and so acts like a source of axial vorticity. In particular, the vorticity, ω_z, intensifies if $\partial u_z / \partial z$ is positive, i.e. the streamlines converge. This is because fluid elements are stretched and elongated along the axis, as illustrated in Figure 3.15. This leads to a reduction in the axial moment of inertia of the element and so, by conservation of angular momentum, to an increase in ω_z.

More generally, consider a thin tube of vorticity, as shown in Figure 3.16. Let $u_{//}$ be the component of velocity parallel to the vortex tube and s be a coordinate measured along the tube. Then

$$|\mathbf{\omega}| \frac{du_{//}}{ds} = (\mathbf{\omega} \cdot \nabla)u_{//}. \tag{3.29}$$

Now the vortex line is being stretched if the velocity $u_{//}$ at point B is greater than $u_{//}$ at A. That is, the length of the material element AB increases if $du_{//}/ds > 0$. Thus the term $(\mathbf{\omega} \cdot \nabla)\mathbf{u}$ represents stretching of the vortex lines. This leads to an intensification of vorticity through conservation of angular momentum, confirming our interpretation of $(\mathbf{\omega} \cdot \nabla)\mathbf{u}$ in (3.26).

It is convenient at this point to introduce the *enstrophy*, defined as $\frac{1}{2}\omega^2$, whose evolution equation can be obtained by taking the dot product of $\mathbf{\omega}$ with (3.26):

$$\frac{D}{Dt}\frac{1}{2}\omega^2 = \omega_i\omega_j S_{ij} - \nu\left[(\nabla \times \mathbf{\omega})^2 + \nabla \cdot ((\nabla \times \mathbf{\omega}) \times \mathbf{\omega})\right]. \tag{3.30}$$

(Here S_{ij} is the rate-of-strain tensor.) The divergence on the right of (3.30) often integrates to zero, in which case the right-hand side represents the balance between the rate of production of enstrophy by vortex line stretching, $\omega_i \omega_j S_{ij}$, and the rate of destruction of enstrophy, $\nu(\nabla \times \boldsymbol{\omega})^2$. Enstrophy budgets of this form turn out to be a common theme in turbulence.

3.5 Inviscid Flow

We now do something dangerous. We set aside viscosity so that we can discuss the great advances made in inviscid fluid mechanics by the nineteenth-century physicists and mathematicians. When we set ν to zero in the Navier–Stokes equation we obtain the so called *Euler equation*. This is dangerous because, as we shall see, a fluid with no viscosity behaves very differently to a fluid with a small but finite viscosity. To emphasise this, John von Neumann referred to inviscid fluid mechanics as the study of *dry water* (Feynman et al., 1964).

3.5.1 Kelvin's Theorem

We now summarise the classical theorems of inviscid vortex dynamics. We start with the idea of a vortex tube. This is an aggregate of vortex lines, rather like a magnetic flux tube is composed of magnetic field lines. Since $\nabla \cdot \boldsymbol{\omega} = 0$, we have

$$\oint \boldsymbol{\omega} \cdot d\mathbf{S} = 0.$$

It follows that the flux of vorticity, $\Phi = \int \boldsymbol{\omega} \cdot d\mathbf{S}$, is constant along the length of a vortex tube since no flux crosses the side of the tube. A closely related quantity is the circulation, Γ. This is defined as the closed line integral of $\mathbf{u}$,

$$\Gamma = \oint_C \mathbf{u} \cdot d\mathbf{l}. \tag{3.31}$$

If the path C is taken as lying on the surface of a vortex tube (Figure 3.17), passing once around it, Stokes' theorem tells us that $\Gamma = \Phi$. Γ is sometimes called the strength of the vortex tube.

Kelvin's (1869) theorem is couched in terms of circulation. It says that, if $C_m(t)$ is a closed curve that always consists of the same fluid particles (a material curve), then the circulation,

$$\Gamma = \oint_{C_m(t)} \mathbf{u} \cdot d\mathbf{l}, \tag{3.32}$$

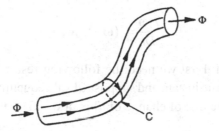

Figure 3.17 A vortex tube.

is independent of time. Note that this theorem does not hold true if C is fixed in space; C_m must be a material curve moving with the fluid. Nor does it apply if the fluid is subject to a rotational body force, such as $\mathbf{J} \times \mathbf{B}$, or for that matter if viscous forces are significant at any point on C_m.

The proof of Kelvin's theorem follows directly from the kinematic equation (2.65):

$$\frac{d}{dt} \int_{S_m} \mathbf{G} \cdot d\mathbf{S} = \int_{S_m} \left[\frac{\partial \mathbf{G}}{\partial t} - \nabla \times (\mathbf{u} \times \mathbf{G}) \right] \cdot d\mathbf{S}. \tag{3.33}$$

If we take $\mathbf{G} = \boldsymbol{\omega}$, invoke the vorticity equation (3.14), and use Stokes' theorem, we have

$$\frac{d\Gamma}{dt} = \frac{d}{dt} \int_{S_m} \boldsymbol{\omega} \cdot d\mathbf{S} = \nu \int_{S_m} \nabla^2 \boldsymbol{\omega} \cdot d\mathbf{S} = 0, \tag{3.34}$$

and Kelvin's theorem is proved.

3.5.2 Helmholtz's Laws

Helmholtz's laws are closely related to Kelvin's theorem. They were published before Kelvin's theorem, in 1858, and like Kelvin's theorem apply only to inviscid flows in the absence of rotational body forces. They state that:

(1) The fluid elements that lie on a vortex line at some initial instant continue to lie on that vortex line for all time, i.e. the vortex lines are frozen into the fluid.
(2) The flux of vorticity, $\Phi = \int \boldsymbol{\omega} \cdot d\mathbf{S}$, is the same at all cross-sections of a vortex tube and is independent of time.

Consider Helmholtz's first law. In two-dimensional flows it is a trivial consequence of $D\omega_z/Dt = 0$. In three dimensions, for which

$$\frac{D\boldsymbol{\omega}}{Dt} = (\boldsymbol{\omega} \cdot \nabla)\mathbf{u}, \tag{3.35}$$

more work is required. First we need the following result. Let $d\mathbf{l}$ be a short line drawn in the fluid at some instant and suppose $d\mathbf{l}$ subsequently moves with the fluid, like a dyeline. Then the rate of change of $d\mathbf{l}$ is $\mathbf{u}(\mathbf{x} + d\mathbf{l}) - \mathbf{u}(\mathbf{x})$ and so

$$\frac{D}{Dt}(d\mathbf{l}) = \mathbf{u}(\mathbf{x} + d\mathbf{l}) - \mathbf{u}(\mathbf{x}), \tag{3.36}$$

where $\mathbf{x}$ and $\mathbf{x} + d\mathbf{l}$ are the position vectors at the two ends of $d\mathbf{l}$. It follows that

$$\frac{D}{Dt}(d\mathbf{l}) = (d\mathbf{l} \cdot \nabla)\mathbf{u}. \tag{3.37}$$

Compare this with (3.35). Evidently, $\boldsymbol{\omega}$ and $d\mathbf{l}$ obey the same equation. Now suppose that at $t = 0$ we have $\boldsymbol{\omega} = \lambda d\mathbf{l}$, then from (3.35) and (3.37) we have $D\lambda/Dt = 0$ at $t = 0$ and so $\boldsymbol{\omega} = \lambda d\mathbf{l}$ for all subsequent times. That is to say, $\boldsymbol{\omega}$ and $d\mathbf{l}$ evolve in identical ways under the influence of $\mathbf{u}$ and so the vortex lines are frozen into the fluid, like dye lines.

Helmholtz's second law is now a trivial consequence of Kelvin's theorem and of $\nabla \cdot \boldsymbol{\omega} = 0$. The fact that the vorticity flux, Φ, is constant along a vortex tube follows from the solenoidal nature of $\boldsymbol{\omega}$, and the temporal invariance of Φ comes from the fact that a vortex tube moves with the fluid and so, from Kelvin's theorem, $\Gamma = \Phi =$ constant. (Here the curve C for Γ lies on the surface of the vortex tube, as in Figure 3.17.)

Helmholtz's first law, which states that vortex tubes are frozen into an inviscid fluid, has profound consequences for inviscid vortex dynamics. For example, if there exist two interlinked vortex tubes as shown in Figure 3.18, then as those tubes are swept around, they remain linked in the same manner, and the strength of each tube remains constant. Thus the tubes appear to be indestructible and their mutual topology is preserved forever. This state of permanence so impressed Kelvin that, in 1867, he developed an atomic theory of matter based on vortices. This rather bizarre theory of the vortex atom has not stood the test of time. However, when, in 1903, the Wright brothers first mastered powered flight, an entirely new incentive for studying vortex dynamics was born. Kelvin's theorem in particular plays a central role in aerodynamics[1].

---

[1] Ironically, Kelvin was not a great believer in powered flight. In 1890, on being invited to join the British Aeronautical Society, he is reputed to have said *'I have not the smallest molecule of faith in aerial navigation other than ballooning ... so you will understand that I would not care to be a member of the society.'*

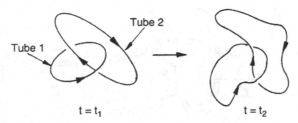

Figure 3.18 Interlinked vortex tubes preserve their topology as they are swept around in an inviscid fluid.

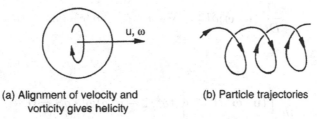

(a) Alignment of velocity and vorticity gives helicity

(b) Particle trajectories

Figure 3.19 A blob of fluid with helicity.

3.5.3 *Helicity Conservation*

The conservation of vortex-line topology implied by Helmholtz's laws is captured by an integral invariant called the helicity. This is defined as

$$h = \int_{V_\omega} \mathbf{u} \cdot \boldsymbol{\omega} \, dV, \tag{3.38}$$

where V_ω is a material volume (a volume composed always of the same fluid elements) for which $\boldsymbol{\omega} \cdot d\mathbf{S} = 0$. For example, the surface of V_ω may be composed of vortex lines. A blob of fluid has helicity if its velocity and vorticity are (at least partially) aligned, as indicated in Figure 3.19.

We may confirm that h is an inviscid invariant as follows. First we have

$$\frac{D}{Dt}(\mathbf{u} \cdot \boldsymbol{\omega}) = \frac{D\mathbf{u}}{Dt} \cdot \boldsymbol{\omega} + \frac{D\boldsymbol{\omega}}{Dt} \cdot \mathbf{u} = -\nabla(p/\rho) \cdot \boldsymbol{\omega} + (\boldsymbol{\omega} \cdot \nabla \mathbf{u}) \cdot \mathbf{u}. \tag{3.39}$$

Since $\boldsymbol{\omega}$ is solenoidal this may be written as,

$$\frac{D}{Dt}(\mathbf{u} \cdot \boldsymbol{\omega}) - \nabla \cdot [(u^2/2 - p/\rho)\boldsymbol{\omega}]. \tag{3.40}$$

Now consider an element of fluid of volume δV. The fluid is incompressible and so $D(\delta V)Dt = 0$. It follows that

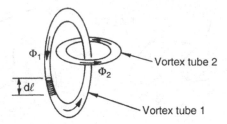

Figure 3.20 The physical interpretation of helicity in terms of flux.

$$\frac{D}{Dt}[(\mathbf{u} \cdot \boldsymbol{\omega})\delta V] = \nabla \cdot [(u^2/2 - p/\rho)\boldsymbol{\omega}]\delta V, \tag{3.41}$$

from which

$$\frac{d}{dt}\int_{V_\omega} (\mathbf{u} \cdot \boldsymbol{\omega})dV = \oint_{S_\omega}[(u^2/2 - p/\rho)\boldsymbol{\omega}] \cdot d\mathbf{S} = 0. \tag{3.42}$$

Thus the helicity, h, is indeed conserved. The connection to Helmholtz's laws and vortex-line topology may be established using the following simple example. Suppose that $\boldsymbol{\omega}$ is confined to two thin interlinked vortex tubes as shown in Figure 3.18, and that V_ω is taken as all space. Then h has two contributions, one from vortex tube 1, which has volume V_1 and flux Φ_1, and another from vortex tube 2. Let these be denoted by h_1 and h_2. Then,

$$h_1 = \int_{V_1} (\mathbf{u} \cdot \boldsymbol{\omega})dV = \oint_{C_1}\mathbf{u} \cdot (\Phi_1 d\mathbf{l}) = \Phi_1 \oint_{C_1}\mathbf{u} \cdot d\mathbf{l}, \tag{3.43}$$

since $\boldsymbol{\omega}dV = \Phi_1 d\mathbf{l}$ (see Figure 3.20). Here C_1 is the closed curve representing tube 1. However, $\oint_{C_1}\mathbf{u} \cdot d\mathbf{l}$ is, from Stokes' theorem, equal to Φ_2 provided that the loops are linked in a right-handed way. A similar calculation may be made for h_2, and we find

$$h = h_1 + h_2 = \Phi_1 \oint_{C_1}\mathbf{u} \cdot d\mathbf{l} + \Phi_2 \oint_{C_2}\mathbf{u} \cdot d\mathbf{l} = 2\Phi_1\Phi_2. \tag{3.44}$$

Note that, if the sense of direction of $\boldsymbol{\omega}$ in either tube were reversed, so that the linkage would become left-handed, then h would change sign. Moreover, if the tubes were not linked, then h would be zero. Thus the invariance of h in this simple example stems directly from the conservation of the vortex line topology and from Helmholtz's laws.

Finally, we note in passing that minimising kinetic energy subject to conservation of global helicity leads to a Beltrami field satisfying $\nabla \times \mathbf{u} = \alpha\mathbf{u}$, α being constant. We shall not pause to prove this result, but we shall make reference to it later.

This ends our discussion of inviscid vortex dynamics. From a mathematical perspective, inviscid fluid mechanics is attractive. The rules of the game are simple and straightforward. Unfortunately, the conclusions are often at odds with reality and so great care must be exercised in using such a theory. The dangers are nicely summarised by Rayleigh:

The general equations of (inviscid) fluid motion were laid down in quite early days by Euler and Lagrange . . . (unfortunately) some of the general propositions so arrived at were found to be in flagrant contradiction with observations, even in cases where at first sight it would not seem that viscosity was likely to be important. Thus a solid body, submerged to a sufficient depth, should experience no resistance to its motion through water. On this principle the screw of a submerged boat would be useless, but, on the other hand, its services would not be needed. It is little wonder that practical men should declare that theoretical hydrodynamics has nothing at all to do with real fluids.

Rayleigh (1914)

Rayleigh was, of course, referring to d'Alembert's paradox. With this warning, we now return to 'real' fluid dynamics.

3.6 Viscous Flow

There are three topics in particular which will be important in our exploration of MHD. The first is the need to quantify the viscous dissipation of energy. The second is an exact, non-trivial solution of the steady Navier–Stokes equation, called Burgers vortex. This is important because the smallest scales in a turbulent flow, which act as centres of dissipation, are thought to resemble such vortices. The third is the Prandtl–Batchelor theory which says, in effect, that a slow cross-stream diffusion of vorticity can be important even at high Re. The Prandtl–Batchelor theorem is important because it has its analogue in MHD (called 'flux expulsion').

3.6.1 The Dissipation of Energy

Consider a volume V of fluid whose surface is S. This surface exerts a viscous stress distribution, τ_{ij}, on the fluid, which manifests itself as the viscous surface forces $\tau_{ij}\,dS_j$. The rate of working of these surface forces on the fluid is then, using Gauss' theorem,

$$\dot{W} = \oint_S u_i \tau_{ij} dS_j = \int_V \frac{\partial}{\partial x_j}[u_i \tau_{ij}]dV. \tag{3.45}$$

Since $\tau_{ij} = \tau_{ji}$, the integrand on the right can be rewritten as

$$\frac{\partial}{\partial x_j}[u_i\tau_{ij}] = u_i\frac{\partial \tau_{ij}}{\partial x_j} + \tau_{ij}\frac{\partial u_i}{\partial x_j} = u_i\frac{\partial \tau_{ij}}{\partial x_j} + \tau_{ij}S_{ij}, \tag{3.46}$$

where

$$S_{ij} = \frac{1}{2}\left(\frac{\partial u_i}{\partial x_j} + \frac{\partial u_j}{\partial x_i}\right)$$

is the rate-of-strain tensor. However (3.1) tells us that $\partial \tau_{ij}/\partial x_j$ is the net viscous force per unit volume acting on the fluid, $\mathbf{f}_v$, while Newtonian fluids obey the constitutive law $\tau_{ij} = 2\rho v S_{ij}$. So (3.45) can be rewritten as

$$\dot{W} = 2\rho v \int_V S_{ij}S_{ij}dV + \int_V \mathbf{u}\cdot\mathbf{f}_v dV. \tag{3.47}$$

The energy created by $\dot{W}$ can appear as either mechanical energy or as internal energy, and our next task is to distinguish between the two. Now the force $\mathbf{f}_v$ appears as one of the forces in Newton's second law (the Navier–Stokes equation) applied to a lump of fluid, and so the second integral on the right of (3.47) is the rate of working of $\mathbf{f}_v$ on the fluid. It follows that this integral represents the rate of change in mechanical energy of the fluid by virtue of the action of $\mathbf{f}_v$ (Batchelor, 1967). The first integral must, therefore, be the rate of change of internal energy in the fluid, and we conclude that the rate of increase of internal energy per unit mass is

$$\varepsilon = 2v\ S_{ij}S_{ij}. \tag{3.48}$$

Often the surface stresses acting on S produce no net work, and in this situation ε represents the rate of transfer of mechanical energy to internal energy per unit mass of fluid, usually simply referred to as the rate of loss of mechanical energy per unit mass, or the dissipation rate per unit mass. Of course, when it comes to MHD, we will need to add to this the Joule dissipation.

3.6.2 The Burgers Vortex

The non-linearity of the Navier–Stokes equation ensures that it has very few non-trivial analytical solutions. One exception to this is Burgers vortex, which is axisymmetric and so usually expressed in cylindrical polar coordinates, (r, θ, z). It consists of a vortex tube centred on the z axis,

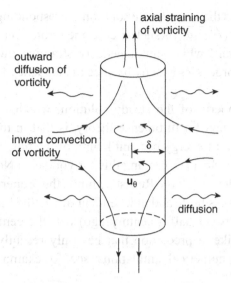

Figure 3.21 The structure of a Burgers vortex. (From Davidson, 2004, by permission of Oxford University Press.)

$$\boldsymbol{\omega} = \frac{\Phi_0}{\pi\delta^2} \exp[-r^2/\delta^2] \, \hat{\mathbf{e}}_z, \qquad \delta = \delta(t), \tag{3.49}$$

with the corresponding velocity field

$$\mathbf{u}^{(vor)} = \frac{\Phi_0}{2\pi r} [1 - \exp(-r^2/\delta^2)] \, \hat{\mathbf{e}}_\theta, \tag{3.50}$$

where Φ_0 is the net flux of vorticity along the vortex tube (or equivalently the circulation around the tube, Γ_0) and $\delta(t)$ is the characteristic radius of the vortex tube (Figure 3.21).

This vortex sits in an imposed, axisymmetric, irrotational strain field,

$$\mathbf{u}^{(irr)} = (u_r, 0, u_z) = \left(-\frac{1}{2}\alpha r, 0, \alpha z\right), \qquad \alpha = \alpha(t), \tag{3.51}$$

where $\alpha(t)$ is a measure of the imposed strain. It turns out that this combination of flows is an exact solution of the unsteady vorticity equation (3.26), provided that $\delta(t)$ and $\alpha(t)$ satisfy

$$\frac{d\delta^2}{dt} + \alpha(t)\delta^2 = 4\nu. \tag{3.52}$$

Of particular interest is the steady-state solution corresponding to a steady imposed strain, in which $\delta = \sqrt{4\nu/\alpha}$. In such a case the irrotational motion sweeps the vorticity radially inward while simultaneously straining the vortex in the axial direction. These two processes exactly balance the tendency for vorticity to diffuse radially outward.

One intriguing property of the steady solution, which particularly interested Burgers, and is important for turbulence theory, is that in the limit of $v \to 0$ the net viscous dissipation of energy per unit length of the tube is finite and independent of v, equal to $\alpha\Gamma_0^2/8\pi$. (Try proving this for yourself.) Now it is observed that fully developed turbulence at large Re also exhibits the property that the dissipation of energy per unit mass is finite and independent of v in the limit of $v \to 0$. This led Burgers to speculate (over half a century ago) that the centres of dissipation in turbulence are tube-like, a prediction that has only recently been confirmed by high-resolution direct numerical simulations (see, for example, Davidson, 2004).

3.6.3 The Prandtl–Batchelor Theorem

We now turn to the Prandtl–Batchelor theorem which, as we have said, has its analogue in MHD. This theorem has far-reaching consequences for internal flows. In effect, it states that a steady, laminar, two-dimensional motion with a high Reynolds number and closed streamlines must have uniform vorticity outside the boundary layers.

Consider a two-dimensional, closed-streamline flow which is steady and has a high Reynolds number. The velocity and vorticity are, in terms of the streamfunction ψ,

$$\mathbf{u} = (\partial\psi/\partial y, -\partial\psi/\partial x), \qquad \omega = -\nabla^2\psi. \tag{3.53}$$

In two dimensions the steady vorticity equation reduces to

$$(\mathbf{u} \cdot \nabla)\omega = \nu\,\nabla^2\omega, \tag{3.54}$$

and if we take the limit $v \to 0$ we obtain $(\mathbf{u}\cdot\nabla)\omega = 0$. The vorticity is then constant along the streamlines and so is a function only of ψ, $\omega = \omega(\psi)$. This is all the information which we may obtain from the inviscid equation of motion. Unfortunately, the problem appears to be underdetermined. There are infinite solutions to the equation

$$\nabla^2\psi = -\omega(\psi), \qquad \psi = 0 \text{ on the boundary,} \tag{3.55}$$

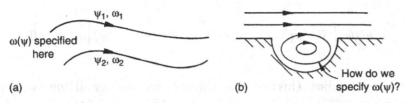

Figure 3.22 The specification of $\omega(\psi)$.

each solution corresponding to a different distribution of $\omega(\psi)$. (Note that $\mathbf{u} \cdot d\mathbf{S} = 0$ at the boundary requires $\psi = $ const. at the boundary and it is usual to take that constant as zero.) So what distribution of $\omega(\psi)$ does nature select? We appear to be missing some information. In cases where the streamlines are open the problem is readily resolved. We must specify the upstream distribution of ω and then track it downstream (see Figure 3.22(a)). However, when the streamlines are closed, such as in the cavity flow shown in Figure 3.22(b), we have no 'upstream point' at which we can specify $\omega(\psi)$.

Let us go back to the steady, viscous vorticity equation. If we integrate this over the area bounded by some closed streamline, we find, with the help of Gauss' theorem,

$$\nu \int \nabla^2 \omega dV = \int \nabla \cdot (\omega \mathbf{u}) dV = \oint \omega \, \mathbf{u} \cdot d\mathbf{S} = 0,$$

and in particular,

$$\nu \oint \nabla \omega \cdot d\mathbf{S} = 0. \tag{3.56}$$

This integral constraint must be satisfied for all finite values of ν, no matter how small ν may be. Now if Re is large then $\omega = \omega(\psi)$, plus some small correction due to the finite value of ν. Consequently, to within a small viscous correction, $\nabla \omega = \omega'(\psi) \nabla \psi$, where $\omega'(\psi)$ is proportional to the cross-stream gradient of vorticity. On substituting back into our integral equation, and noting that $\omega'(\psi)$ is constant along a streamline and so may be taken outside the integral, we have

$$\nu \left[\omega'(\psi) \oint \nabla \psi \cdot d\mathbf{S} + (\text{small correction}) \right] = 0. \tag{3.57}$$

We now assume ν is very small (Re is large) and throw away the small correction. Invoking Gauss' theorem once again, this time in reverse, and then Stokes' theorem, we have

$$\nu \; \omega'(\psi) \int \nabla^2 \psi dV = -\nu \; \omega'(\psi) \int \omega dA = -\nu \; \omega'(\psi) \oint_C \mathbf{u} \cdot d\mathbf{l} = 0,$$

where C is a streamline. This expression must be satisfied for all flows with a small but finite viscosity. Since ν is finite, and the circulation around C non-zero, the only possibility is that $\omega'(\psi) = 0$; that is, there can be no cross-stream gradient in ω.

We have proved the Prandtl–Batchelor theorem. It states that, for high Reynolds number flows with closed streamlines, the vorticity is uniform throughout the flow, $\omega = \omega_0$, except within the boundary layers. (We must exclude the boundary layers since our proof assumed that viscous effects are small and this is clearly not the case in the boundary layers). Thus, for example, in the cavity flow shown in Figure 3.22(b), ω will be constant within the region of closed streamlines (excluding the thin boundary layers). Of all possible vorticity distributions, $\omega(\psi)$, the Prandtl–Batchelor theorem tells us that nature will select the one where vorticity is not only constant along the streamlines, but is also constant *across* the streamlines.

Armed with the Prandtl–Batchelor theorem it is relatively straightforward to compute flow fields of the type sketched in Figure 3.22(b). One simply solves the equation $\nabla^2 \psi = -\omega_0$ with ψ zero on the boundary, where ω_0 is the (unknown but constant) vorticity in the flow. This yields $\mathbf{u}$ at all points outside the boundary layers. There is, however, still the unknown constant ω_0 to determine, and this is usually fixed by the boundary layer equations.

The physical interpretation of the Prandtl–Batchelor theorem is straightforward. Suppose a flow is initiated at, say, $t = 0$. Then, over a short timescale of the order of the eddy turnover time, the flow adopts a high Reynolds number form: i.e. $\omega = \omega$ (ψ). (Depending on how the flow is initiated, different distributions of $\omega(\psi)$ may appear.) There then begins a period of slow cross-stream diffusion of vorticity which continues until all the internal gradients in vorticity have been eliminated (except in the boundary layers). This takes a long time, but the flow does not become truly steady until the process is complete.

Example 3.1 The Thermal Analogue of the Prandtl–Batchelor Theorem
Starting with the energy equation

$$\frac{DT}{Dt} = \alpha \nabla^2 T,$$

show that for steady, laminar, high-Peclet-number, closed-streamline flows, the temperature outside the thermal boundary layers is uniform. This is the thermal equivalent of the Prandtl–Batchelor theorem. Give a physical interpretation of this result.   ∎

Example 3.2 The Prandtl–Batchelor Theorem in MHD

In two-dimensional MHD flows which have the form $\mathbf{u}(x,y) = (u_x, u_y, 0)$ and $\mathbf{B}(x, y) = (B_x, B_y, 0) = \nabla \times (A\hat{\mathbf{e}}_z)$, the induction equation (2.64) reduces to

$$\frac{DA}{Dt} = \lambda \nabla^2 A.$$

Where do you think the Prandtl–Batchelor arguments for closed streamline flows lead here? ∎

3.7 Boundary Layers, Reynolds Stresses and Elementary Turbulence Models

3.7.1 Boundary Layers

> During the last few years much work has been done in connection with artificial flight. We may hope that before long this may be coordinated and brought into closer relationship with theoretical hydrodynamics. In the meantime one can hardly deny that much of the latter science is out of touch with reality.
>
> *Rayleigh, 1916.*

We have already mentioned boundary layers, without really defining what we mean by this term, and so it seems appropriate to review briefly the key aspects of laminar boundary layers. (We leave turbulence to §3.7.2.) The concept of a boundary layer, and of boundary layer separation, was first conceived by the mechanical engineer L. Prandtl and it revolutionised fluid mechanics. It formed a bridge between the classical nineteenth-century mathematical studies of inviscid fluids and the subject of experimental fluid mechanics, and in doing so it resolved many traditional dilemmas such as d'Alembert's paradox. Prandtl first presented his ideas in 1904 in a short paper crammed with physical insight. Curiously though, it took many years for the full significance of his ideas to be generally appreciated.

Consider a high Reynolds number flow over, say, an aerofoil (Figure 3.23). By high Reynolds number, we mean that Re $= uL/\nu$ is large, where L is

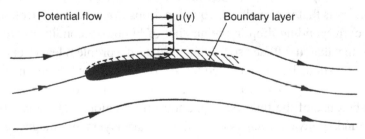

Figure 3.23 A boundary layer near a surface.

a characteristic geometric length scale, say the span of the aerofoil. Since Re is large we might be tempted to solve the inviscid equations of motion,

$$(\mathbf{u} \cdot \nabla)\mathbf{u} = -\nabla(p/\rho),$$

subject to the inviscid boundary condition $\mathbf{u} \cdot d\mathbf{S} = 0$ on all solid surfaces. This determines the so-called *external problem*.

Now in reality the fluid satisfies the no-slip boundary condition $\mathbf{u} = \mathbf{0}$ on $d\mathbf{S}$. (We take a frame of reference moving with the aerofoil.) Thus there must exist a region, surrounding the aerofoil, where the velocity given by the external problem adjusts to zero (see Figure 3.23). This region is called the boundary layer and it is easy to see that such a layer must be thin. The point is that the only mechanical forces available to cause a drop in velocity are viscous shear stresses. Thus the viscous term in the Navier–Stokes equation must be of the same order as the other terms within the boundary layer,

$$\nu \, \nabla^2 \mathbf{u} \sim (\mathbf{u} \cdot \nabla)\mathbf{u}.$$

This requires that the transverse length scale, δ, which appears in the Laplacian, be of order

$$\delta \sim \sqrt{\nu \, L/u} \sim \mathrm{Re}^{-1/2}L,$$

which fixes the boundary layer thickness. Since Re is large, this implies $\delta \ll L$. Note that, because the boundary layer is so thin, the pressure within a boundary layer is virtually the same as the pressure immediately outside the layer. (There can be no significant gradient in pressure across a boundary layer since this would imply a significant normal acceleration, which is not possible since the velocity is essentially parallel to the surface.)

Boundary layers occur in other branches of physics. It is not a phenomenon peculiar to velocity fields. In fact, it occurs whenever a small parameter, in this case ν, multiplies a term containing derivatives which are of higher order than the other derivatives appearing in the equation. In the case above, when we throw out $\nu\nabla^2\mathbf{u}$ on the basis that ν is small, our equation drops from second order to first order. There is a corresponding drop in the number of boundary conditions we can meet ($\mathbf{u} \cdot d\mathbf{S} = 0$ rather than $\mathbf{u} = \mathbf{0}$) and so solving the external problem leaves one boundary condition unsatisfied. This is corrected for in a thin transition region (in this case the velocity boundary layer) where the term we had thrown out, i.e. $\nu\nabla^2\mathbf{u}$, is now significant because of the thinness of the transition region. But we can have other types of boundary layers, such as thermal boundary layers and magnetic boundary layers. In the case of thermal boundary layers, the small parameter is α and it multiplies $\nabla^2 T$.

Note that the thickness of a boundary layer is not always $\sim \mathrm{Re}^{-1/2} L$. For example, we shall see later that the force balance within MHD boundary layers is more subtle than that indicated above, and so the estimate $\delta \sim \mathrm{Re}^{-1/2} L$ often needs modifying.

3.7.2 Turbulence and Simple Turbulence Models

We now consider the more elementary aspects of turbulent flow. A more detailed discussion is given in Chapter 8 where we consider the nature of turbulence itself. Here we restrict ourselves to the much simpler problem of characterising the influence of turbulence on the mean flow, and in particularly simple geometries, such as flat-plate boundary layers.

It is an empirical observation that if Re is made large enough (viscosity made small enough) a flow invariably becomes unstable and then turbulent. Suppose we have a turbulent flow in which $\mathbf{u}$ and p consist of a time-averaged component plus a fluctuating part (Figure 3.24), $\mathbf{u} = \bar{\mathbf{u}} + \mathbf{u}'$, $p = \bar{p} + p'$, where the overbar represents a time average.

When we time-average the Navier–Stokes equation new terms arise from the fluctuations in velocity. For example, the x-component of the time-averaged equation of motion is

$$(\bar{\mathbf{u}} \cdot \nabla)\bar{u}_x = -\frac{\partial}{\partial x}\left(\frac{\bar{p}}{\rho}\right) + \frac{\partial}{\partial x}\left[2\nu\frac{\partial \bar{u}_x}{\partial x}\right] + \frac{\partial}{\partial y}\left[\nu\left(\frac{\partial \bar{u}_x}{\partial y} + \frac{\partial \bar{u}_y}{\partial x}\right)\right]$$
$$+ \frac{\partial}{\partial z}\left[\nu\left(\frac{\partial \bar{u}_x}{\partial z} + \frac{\partial \bar{u}_z}{\partial x}\right)\right] + \frac{\partial}{\partial x}\left[-\overline{u'_x u'_x}\right]$$
$$+ \frac{\partial}{\partial y}\left[-\overline{u'_x u'_y}\right] + \frac{\partial}{\partial z}\left[-\overline{u'_x u'_z}\right].$$

Now the laminar stresses are, from Newton's law of viscosity, given by $\tau_{ij} = 2\rho\nu\, S_{ij}$, but the turbulence seems to have produced additional stresses. These are called Reynolds stresses in honour of Osborne Reynolds' pioneering work on turbulence (in the 1880s). These stresses are

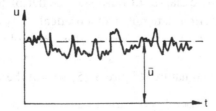

Figure 3.24 Velocity component in a turbulent flow.

$$\tau_{xx} = -\rho \overline{u'_x u'_x}, \quad \tau_{xy} = -\rho \overline{u'_x u'_y}, \quad \tau_{xz} = -\rho \overline{u'_x u'_z}. \tag{3.58}$$

We can now rewrite the x-component of the time-averaged equation in a more compact form:

$$(\overline{\mathbf{u}} \cdot \nabla)\overline{u}_x = -\frac{\partial}{\partial x}\left[\frac{\overline{p}}{\rho}\right] + \nu \, \nabla^2 \overline{u}_x + \frac{\partial}{\partial x_i}[-\rho \overline{u'_x u'_i}]. \tag{3.59}$$

Similar expressions may be written for the y and z components. If we wish to make predictions from this equation we need to be able to relate the Reynolds stresses, $-\rho \overline{u'_x u'_i}$, to some quantity which we know about, such as mean velocity gradients of the type $\partial \overline{u}_x / \partial y$. This is the purpose of so-called *one-point turbulence modelling*. Its aim is to recast the time-averaged equations in a form which may be solved, just like the Navier–Stokes equations. In effect, such a *turbulence model* provides a means of estimating the Reynolds stresses.

However, it should be emphasised from the outset that 'Reynolds stress modellers' live dangerously. The quest to find a *universal* turbulence model, which may be defended on theoretical grounds, is doomed to failure from the outset. There is no such thing as a universal turbulence model! The best that we can do is construct semi-empirical models, based on laboratory tests, and try to apply these models to flows not too different from the laboratory tests on which they were based. For example, Reynolds stress models developed from boundary layer experiments need not work well when applied to rapidly rotating flows.

The reason for this difficulty is the so-called 'closure problem' of turbulence. We can, in principle, derive rigorous equations for $\overline{u'_x u'_y}$, etc. (see Chapter 8). However, this involves quantities of the form $\overline{u'_x u'_y u'_z}$. When an equation for these new quantities is derived, we find that it involves yet more functions such as $\overline{u'_x u'_x u'_y u'_z}$, and so on. There are always more unknowns than equations, and it is impossible to close the set in a rigorous way. This is the price we pay in moving from the instantaneous equations of motion to a statistical (time-averaged) one[2]. This is all bad news since much of fluid mechanics centres around turbulent flows, and quantitative predictions of such flows requires a turbulence model. Fortunately, some of the simpler, semi-empirical turbulence models work reasonably well if restricted to the appropriate classes of flow, such as flat-plate boundary layers.

Historically, the first serious attempt at a theoretical study of turbulent flows was made by Boussinesq around 1877 (six years before Reynolds' famous pipe experiment). He proposed that the shear-stress strain-rate relationship for time-averaged flows of a one-dimensional nature (Figure 3.25) was of the form

---

[2] This is not to say that we cannot make rigorous and useful statements about turbulence; we can. We cannot, however, produce a rigorous Reynolds stress model, as discussed in Tennekes and Lumley (1972).

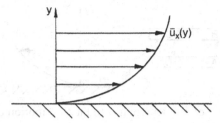

Figure 3.25 The time-averaged velocity in a turbulent flow.

$$\tau_{xy} = \rho(\nu + \nu_t)\frac{\partial \overline{u}_x}{\partial y}.$$

Boussinesq termed ν_t an *eddy viscosity*. While ν is a property of the fluid, ν_t will be a property of the turbulence. The first attempt to estimate ν_t was due to Prandtl in 1925. He invoked the idea of a *mixing length*, as we shall see shortly.

The idea of an eddy viscosity need not be restricted to simple shear flows of the type shown in Figure 3.25, i.e. $\overline{u}_x(y)$. It is common to introduce eddy viscosities into flows of arbitrary complexity. Then,

$$\tau_{xy} = \rho(\nu + \nu_t)\left[\frac{\partial \overline{u}_x}{\partial y} + \frac{\partial \overline{u}_y}{\partial x}\right], \quad \tau_{xz} = \rho(\nu + \nu_t)\left[\frac{\partial \overline{u}_x}{\partial z} + \frac{\partial \overline{u}_z}{\partial x}\right], \quad \text{etc.}$$

We have, in effect, accounted for the Reynolds stresses by replacing ν in Newton's law of viscosity by $\nu + \nu_t$. Now in most turbulent flows the eddy viscosity is much greater than ν (except very close to surfaces), so the viscous stresses may be dropped, giving

$$\tau_{xy} = -\rho\overline{u'_x u'_y} = \rho\nu_t\left[\frac{\partial \overline{u}_x}{\partial y} + \frac{\partial \overline{u}_y}{\partial x}\right], \quad \text{etc.} \tag{3.60}$$

We shall refer to these as Boussinesq's equations. The question now is, what is ν_t?

Prandtl was struck by the success of the kinetic theory of gases in predicting the properties of ordinary viscosity in which the 'mean free path length' plays a role. In fact, the simple kinetic theory of gases leads to the prediction $\nu = \ell V/3$, where V is the root-mean-square (rms) molecular velocity and ℓ is the mean free path length. Could the same thing be done for the eddy viscosity?

There is an analogy between Newton's law of viscosity and Reynolds stresses. In a laminar flow, layers of fluid which slide over each other experience a mutual shear stress, or drag, because thermally agitated molecules bounce around between the layers exchanging energy and momentum as they do so (Figure 3.26). For

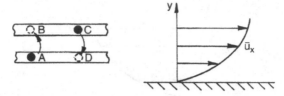

Figure 3.26 Exchange of momentum due to thermal motion of the molecules.

example, a molecule in the slow-moving layer at A may move up to B, slowing down the fast-moving layer. Conversely, a molecule in the fast-moving layer may drop down from C to D, speeding up the lower layer. This is the basic idea which lies behind the expression $v = \ell V/3$. But just the same sort of thing happens in a turbulent flow, albeit at the macroscopic, rather than molecular, level. Balls of fluid are exchanged between the layers due to turbulent fluctuations, and this causes a mixing of momentum across the layers.

This analogy between the transfer of momentum by molecules on the one hand and balls of fluid on the other led Prandtl to propose the relationship $v_t = \ell_m V_T$, where ℓ_m is called the *mixing length*. (Actually Boussinesq proposed something similar in 1870.) The length ℓ_m is a measure of the size of the large eddies in the flow, while V_T is a measure of u', and indicates the intensity of the turbulence. The more intense the fluctuations, the larger the cross-stream transfer of momentum, and so the larger the eddy viscosity.

To a large extent the equation above is simply a dimensional necessity, since v_t has the dimensions of m^2/s. Boussinesq's equations, however, are a little more worrying. At one level we may regard them as simply defining v_t, and so transferring the problem of estimating τ_{ij} to one of estimating v_t. But there is an important assumption here. We are assuming that the eddy viscosity in the x-y plane is the same as the eddy viscosity in the x-z plane and so on. This, in turn, requires that the turbulence be, in some sense, statistically isotropic. Still, let us see how far we can get with such an eddy viscosity model. We now need to find a way of estimating ℓ_m and V_T. For the particular case of simple shear flows (i.e. one-dimensional mean flows), Prandtl found a way of estimating V_T. This is known as Prandtl's *mixing length theory*.

Consider a mean flow, $\bar{u}_x$, which is purely a function of y (Figure 3.27). Then Prandtl's theory says that, in effect, the fluid at y, with mean velocity $\bar{u}_x(y)$, will, on average, have come from levels $y \pm \ell$, where ℓ is a measure of the mixing length. Suppose that, as a fluid lump is thrown from $y + \ell$ (or $y - \ell$) to y, it retains its forward momentum, which on average will be $\bar{u}_x(y \pm \ell)$. Then the mean velocities $\bar{u}_x(y \pm \ell)$ represent the spread of instantaneous velocities at position y. If ℓ is small (unfortunately it is not in turbulent flow!) then we have $\bar{u}_x(y \pm \ell) \approx \bar{u}_x(y) \pm \ell \partial \bar{u}_x/\partial y$, and it follows that

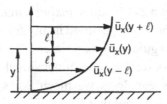

Figure 3.27 Prandtl's mixing length model.

$$\overline{(u'_x)^2} \sim \ell^2 \left[\frac{\partial \overline{u}_x}{\partial y} \right]^2 .$$

Next we note that there is a strong negative correlation between u'_x and u'_y, since a positive u'_x is consistent with fluid coming from $y + \ell$, requiring a negative u'_y. (If $\partial \overline{u}_x / \partial y$ is negative we expect a positive correlation.) Thus, $\overline{u'_x u'_y} = -c_1 \sqrt{\overline{(u'_x)^2}} \sqrt{\overline{(u'_y)^2}}$, where c_1 is some constant of order unity (called a correlation coefficient). If the rms values of u'_x and u'_y are of similar orders of magnitude we now have

$$\overline{u'_x u'_y} \sim \pm \overline{(u'_x)^2} = -c_2 \ell^2 \left| \frac{\partial \overline{u}_x}{\partial y} \right| \frac{\partial \overline{u}_x}{\partial y} ,$$

where c_2 is a second constant of order unity. Note the modulus on one of the $\partial \overline{u}_x / \partial y$ terms. This is needed to ensure that the correlation is negative when $\partial \overline{u}_x / \partial y > 0$ and positive when $\partial \overline{u}_x / \partial y < 0$. We now redefine our mixing length to absorb the unknown constant c_2 and the end result is

$$\tau_{xy} = -\rho \overline{u'_x u'_y} = \rho \ell_m^2 \left| \frac{\partial \overline{u}_x}{\partial y} \right| \frac{\partial \overline{u}_x}{\partial y} . \tag{3.61}$$

Compare this with Boussinesq's equation,

$$\tau_{xy} = \rho \nu_t \frac{\partial \overline{u}_x}{\partial y} = \rho \ell_m V_T \frac{\partial \overline{u}_x}{\partial y} .$$

Evidently, for this particular sub-class of one-dimensional shear flows, the eddy viscosity is

$$\nu_t = \ell_m^2 \left| \frac{\partial \overline{u}_x}{\partial y} \right| . \tag{3.62}$$

This represents what is known as Prandtl's *mixing length model*. Conceptually this is a tricky argument which cannot be justified in any formal way. Moreover, we still need to prescribe ℓ_m, perhaps guided by experiment. Nevertheless, Prandtl's mixing length model appears to work well for one-dimensional shear flows, provided ℓ_m is chosen appropriately. (By shear flows we mean flows like wakes and jets.) For turbulent boundary layers it is found that $\ell_m = \kappa y$ close to the wall, where κ, which has a value of $\kappa \approx 0.4$, is known as Kármán's constant.

Example 3.3 The α-Effect in Electrodynamics

The process of averaging chaotic or turbulent equations, in the spirit of Reynolds, is not restricted to the Navier–Stokes equation. For example, the heat equation or induction equation can be averaged in a similar way. Suppose we have a highly conducting, turbulent fluid in which $\mathbf{u} = \mathbf{u}_0 + \mathbf{v}$ and $\mathbf{B} = \mathbf{B}_0 + \mathbf{b}$ where $\mathbf{u}_0$ and $\mathbf{B}_0$ are steady or slowly varying and $\bar{\mathbf{v}} = 0$, $\bar{\mathbf{b}} = 0$. Show that the averaged induction equation is

$$\frac{\partial \mathbf{B}_0}{\partial t} = \nabla \times (\mathbf{u}_0 \times \mathbf{B}_0) + \lambda \nabla^2 \mathbf{B}_0 + \nabla \times \overline{(\mathbf{v} \times \mathbf{b})}.$$

The quantity $\overline{\mathbf{v} \times \mathbf{b}}$ is the electromagnetic analogue of the Reynolds stress. In some cases it is found that $\overline{\mathbf{v} \times \mathbf{b}} = \alpha \mathbf{B}_0$, where α is the analogue of Boussinesq's eddy diffusivity. This leads to the 'turbulent' induction equation

$$\frac{\partial \mathbf{B}_0}{\partial t} = \nabla \times (\mathbf{u}_0 \times \mathbf{B}_0) + \nabla \times (\alpha \mathbf{B}_0) + \lambda \nabla^2 \mathbf{B}_0.$$

In Chapter 14 you will see that the new term, called the α-effect, can give rise to the self-excited generation of a magnetic field.　■

3.8 Ekman Layers and Ekman Pumping in Rotating Fluids

We now consider the phenomenon of Ekman pumping which occurs whenever there is differential rotation between a viscous fluid and a solid surface. This turns out to be important in the magnetic stirring of partially solidified ingots (see Chapter 11) and, perhaps, in the geodynamo (Chapter 14). We start with Kármán's solution for laminar flow near the surface of a rotating disc.

Suppose we have an infinite disc rotating in an otherwise still liquid. A boundary layer will develop on the disc due to viscous coupling, and Kármán found an exact solution for this laminar boundary layer. Suppose that the disc rotates with angular velocity Ω. Then, on dimensional grounds, we might expect the boundary layer thickness to scale as $\delta \sim (\nu/\Omega)^{1/2}$. Kármán suggested looking for a solution in polar

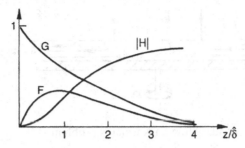

Figure 3.28 The solution of Kármán's problem (a rotating disc in a quiescent fluid).

coordinates in which z (which is normal to the disc) is normalised by δ. In fact, Kármán's solution is of the form

$$u_r = \Omega r F(\eta), \quad u_\theta = \Omega r G(\eta), \quad u_z = \Omega \hat{\delta} H(\eta),$$

where $\hat{\delta} = (\nu/\Omega)^{1/2}$ and $\eta = z/\hat{\delta}$.

If these expressions are substituted into the radial and azimuthal components of the Navier–Stokes equation and the equation of continuity, we obtain three ordinary differential equations for the three unknown functions F, G and H:

$$F^2 + F'H - G^2 = F'' \quad 2FG + HG' = G'', \quad 2F + H' = 0.$$

We take z to be measured from the surface of the disc and so we have the boundary conditions:

$$z = 0 : F = 0, G = 1, H = 0; \qquad z \to \infty : F = 0, G = 0.$$

Our three differential equations can be integrated numerically, subject to their boundary conditions, and the result is shown schematically in Figure 3.28.

This represents a flow which is radially outward within a boundary layer of thickness $\delta \approx 4\hat{\delta} = 4(\nu/\Omega)^{1/2}$. The flow pattern in the r-z plane is shown in Figure 3.29. Within the thin boundary layer the fluid is centrifuged radially outward so that each particle spirals outward to the edge of the disc. Outside the boundary layer u_r and u_θ are both zero, but surprisingly u_z is non-zero. Rather, there is a slow drift of fluid towards the disc, at a rate $|u_z| \approx 0.9\Omega\hat{\delta} \approx 0.2\Omega\delta$. Of course, this axial inflow is required to supply the radial outflow in the boundary layer. We have, in effect, a centrifugal fan.

Suppose now that the disc is stationary but that the fluid rotates like a rigid body ($u_\theta = \Omega r$) in the vicinity of the disc. Near the disc's surface this swirl is attenuated due to viscous drag and so, once again, a boundary layer forms. This problem was

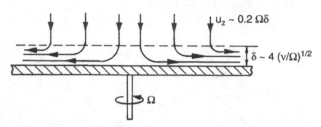

Figure 3.29 The secondary flow in Kármán's problem.

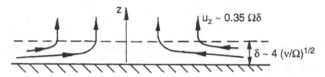

Figure 3.30 Bödewadt's problem (stationary disc, rotating fluid).

studied by Bödewadt, who showed that Kármán's procedure works as before. It is necessary only to change the boundary conditions. Once again the boundary layer thickness is constant and of the order of $4(v/\Omega)^{1/2}$. This time, however, the flow pattern in the *r-z* plane is reversed (Figure 3.30). Fluid particles spiral radially inward, eventually drifting out of the boundary layer. Outside the boundary layer we have rigid body rotation, $u_\theta = \Omega r$, plus a weak axial flow away from the surface, of magnitude $u_z \approx 1.4\Omega\hat{\delta} \approx 0.35\Omega\delta$.

The flow in the *r-z* plane is referred to as a secondary flow, inasmuch as the primary motion is a swirling flow. The reason for the secondary flow in Figure 3.30 is that, outside the boundary layer, we have the radial force balance

$$\frac{\partial p}{\partial r} = \rho \frac{u_\theta^2}{r}.$$

That is, the centrifugal force sets up a radial pressure gradient, with a relatively low pressure near the axis. This pressure gradient is imposed throughout the boundary layer on the plate. However, the swirl in this boundary layer is diminished through viscous drag, and so there is a local imbalance between the imposed pressure gradient and the centripetal acceleration. The result is a radial inflow with the fluid eventually drifting up and out of the boundary layer.

In general then, whenever we have a swirling fluid adjacent to a stationary surface we induce a secondary flow as sketched above. This is referred to as Ekman pumping, and the boundary layer is called an Ekman layer. The axial velocity induced by Ekman pumping is relatively small if the Reynolds number is large, $u_z \sim (v\Omega)^{1/2} \ll u_\theta$. Nevertheless, this weak secondary flow often has

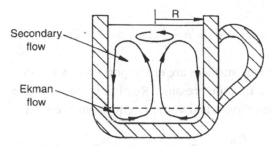

Figure 3.31 Spin-down of a stirred cup of tea.

profound consequences for the motion as a whole. Consider, for example, the problem of 'spin-down' of a stirred cup of tea (Figure 3.31). Suppose that, at $t = 0$, the tea is set into a state of (almost) inviscid rotation. Very quickly an Ekman layer will become established on the bottom of the cup, inducing a radial inflow at the base of the vessel. By continuity, this radial flow must eventually drift up and out of the boundary layer where it is recycled via side layers (called Stewartson layers) on the cylindrical walls of the cup. A secondary flow is established as shown in Figure 3.31. As each fluid particle passes through the Ekman layer it gives up a significant fraction of its kinetic energy. The tea finally comes to rest when all of the contents of the cup have been flushed through the Ekman boundary layer. The existence of the secondary flow is evidenced (in the days before tea-bags!) by the accumulation of tea-leaves at the centre of the cup.

The spin-down time, τ_{sd}, is therefore of the order of the turnover time of the secondary flow, $\tau_{sd} \sim R/u_z \sim R/\sqrt{\nu\,\Omega}$. By way of contrast, if there were no secondary flow the spin-down time would be controlled by the time taken for the core vorticity to diffuse to the walls, $\tau_{sd}^* \sim R^2/\nu$. Suppose that $R = 3\,\text{cm}$, $\nu = 10^{-6}\,\text{m}^2/\text{s}$ and $\Omega = 1\text{s}^{-1}$. Then $\tau_{sd} \sim 30$ seconds, which is about right, whereas $\tau_{sd}^* \sim 15$ minutes! Evidently, Ekman layers provide an efficient mechanism for destroying mechanical energy.

3.9 Waves and Columnar Vortices in Rotating Fluids

There are many cases in geophysics and astrophysics where a fluid is subject to a strong background rotation, say $\boldsymbol{\Omega} = \Omega\hat{\mathbf{e}}_z$. In such cases it often makes sense to transfer to a frame of reference rotating with the fluid. Of course, Newton's second law does not apply in such a non-inertial frame, but we can behave as if it does provided we add two fictitious forces to our equation of motion: the centrifugal force, which is irrotational and so merely augments the fluid pressure, and the Coriolis force, $2\mathbf{u} \times \boldsymbol{\Omega}$. The incompressible Navier–Stokes equation in the rotating frame then becomes

$$\frac{\partial \mathbf{u}}{\partial t} + (\mathbf{u} \cdot \nabla)\mathbf{u} = -\nabla\left[p/\rho - \frac{1}{2}(\mathbf{\Omega} \times \mathbf{x})^2\right] + 2\mathbf{u} \times \mathbf{\Omega} + \nu \nabla^2\mathbf{u}. \qquad (3.63)$$

In situations where free surfaces are not important it is usually sufficient to absorb the centrifugal force into the pressure. Retaining the symbol p for the augmented pressure, the Navier–Stokes and vorticity equations are then

$$\frac{D\mathbf{u}}{Dt} = -\nabla(p/\rho) + 2\mathbf{u} \times \mathbf{\Omega} + \nu \nabla^2\mathbf{u}, \qquad (3.64)$$

$$\frac{D\boldsymbol{\omega}}{Dt} = (\boldsymbol{\omega} \cdot \nabla)\mathbf{u} + 2(\mathbf{\Omega} \cdot \nabla)\,\mathbf{u} + \nu \nabla^2\boldsymbol{\omega}. \qquad (3.65)$$

The appearance of the Coriolis force in (3.64) and (3.65) has a number of curious consequences which, at first sight, seem counterintuitive. In particular, when the *Rossby number*,

$$\mathrm{Ro} = u/\Omega\ell, \qquad (3.66)$$

is small, so that departures from rigid-body rotation are weak, the fluid can sustain a form of internal wave motion, called *inertial waves*, the restoring force being the Coriolis force. It is also observed that, for $\mathrm{Ro} \ll 1$, the motion is often dominated by columnar vortices aligned with the rotation axis, usually referred to as *Taylor columns*. Indeed, we shall see that these two phenomena, waves and columnar vortices, are closely related, being different manifestations of the same underlying dynamics. Let us start, then, with Taylor columns.

3.9.1 The Taylor–Proudman Theorem

When $\mathrm{Ro} \ll 1$ and the influence of viscosity is small, the vorticity equation simplifies to

$$\frac{\partial \boldsymbol{\omega}}{\partial t} = 2(\mathbf{\Omega} \cdot \nabla)\mathbf{u}. \qquad (3.67)$$

If the motion is quasi-steady then it seems reasonable to neglect $\partial \boldsymbol{\omega}/\partial t$ and we obtain $(\mathbf{\Omega}\cdot\nabla)\mathbf{u} = 0$. We conclude that rapidly rotating, quasi-steady motion is subject to the surprising constraint that the velocity field, $\mathbf{u}$, is strictly two-dimensional, independent of the coordinate parallel to $\mathbf{\Omega}$. This is called the Taylor–Proudman theorem. To illustrate the power of this constraint, consider the experiment shown in Figure 3.32. An object is slowly towed across the base of a rotating tank which is

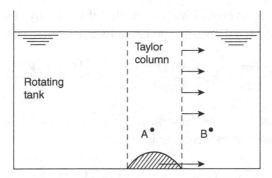

Figure 3.32 An object is slowly towed across the base of a rotating tank filled with water. It is observed that, as the object moves, it carries with it a Taylor column. (From Davidson, 2004, by permission of Oxford University Press.)

filled with water. As the object moves, the fluid ahead of it cannot flow up and over the object, as we might have expected, since that constitutes a three-dimensional flow pattern. Rather, the column of fluid located between the object and the surface of the fluid moves across the tank, as if rigidly attached to the object. This column is known as a *Taylor column*, and it is a direct consequence of the need to maintain two-dimensional motion. For example, a fluid particle initially at point A will move across the tank, always centred above the object below. It is natural to ask how the fluid lying above the object knows to move in tandem with it, and to answer this question we must introduce the idea of inertial waves.

3.9.2 Inertial Waves, Helicity Transport and the Formation of Taylor Columns

A rapidly rotating, incompressible fluid can support internal wave motion. To show that this is so, we note that application of the operator $\nabla \times (\partial/\partial t)$ to (3.67) yields the wave-like equation

$$\frac{\partial^2}{\partial t^2}\left(\nabla^2 \mathbf{u}\right) + 4(\mathbf{\Omega} \cdot \nabla)^2 \mathbf{u} = \mathbf{0}, \tag{3.68}$$

which allows for plane waves of the form $\mathbf{u} = \hat{\mathbf{u}}\exp[j(\mathbf{k} \cdot \mathbf{x} - \varpi t)]$. These are known as inertial waves and they have the dispersion relationship

$$\varpi = \pm 2(\mathbf{k} \cdot \mathbf{\Omega})/k, \quad k = |\mathbf{k}|. \tag{3.69}$$

Evidently the frequency of inertial waves is independent of k and is restricted to the range $0 \leq \varpi \leq 2\Omega$, with the lowest frequency corresponding to perpendicular wave vectors.

The group velocity of inertial waves, which is the velocity at which wave energy disperses from a localised disturbance in the form of wave packets, is then

$$\mathbf{c}_g = \frac{\partial \varpi}{\partial k_i} = \pm 2 \frac{\mathbf{k} \times (\mathbf{\Omega} \times \mathbf{k})}{|\mathbf{k}|^3} = \pm 2 \frac{k^2 \mathbf{\Omega} - (\mathbf{k} \cdot \mathbf{\Omega})\mathbf{k}}{|\mathbf{k}|^3}. \qquad (3.70)$$

(Readers unfamiliar with the concept of group velocity might want to consult Acheson, 1990.) Note in particular that

$$\mathbf{c_g} \cdot \mathbf{\Omega} = \pm 2k^{-3}\left[k^2\Omega^2 - (\mathbf{k} \cdot \mathbf{\Omega})^2 \right], \qquad (3.71)$$

which tells us that the positive sign in (3.69) corresponds to wave energy travelling upward, while the negative sign corresponds to energy propagating downward. Note also that the wave packets with the lowest frequency have the largest group velocity (for a given value of k), while those with the highest frequency have zero group velocity.

Inertial waves are invariably helical, with a non-zero helicity, $h = \mathbf{u} \cdot \boldsymbol{\omega}$. To see why this is so, consider Equation (3.67) combined with the dispersion relationship (3.69). This yields $\hat{\boldsymbol{\omega}} = \mp|\mathbf{k}|\hat{\mathbf{u}}$, where $\hat{\boldsymbol{\omega}}$ is the amplitude of the vorticity in the wave. Evidently the vorticity and velocity fields are parallel and in phase. We conclude that inertial waves have maximum possible helicity, with the positive sign in (3.71) corresponding to negative helicity, and the negative sign to positive helicity. In short, when wave packets disperse from a localised disturbance, those wave packets with negative helicity will propagate upward ($\mathbf{c}_g \cdot \mathbf{\Omega} > 0$), while wave packets with positive helicity travel downward ($\mathbf{c}_g \cdot \mathbf{\Omega} < 0$).

This phenomenon is illustrated by the numerical simulation shown in Figure 3.33, where a localised patch of turbulence spreads in a rotating fluid at Ro = 0.1 by emitting inertial wave packets. The left-hand image is the initial condition and the right-hand one corresponds to $\Omega t = 6$. Dark grey indicates negative helicity and light grey positive helicity, and it is clear that the upward (downward) propagating wave packets possess negative (positive) helicity. We shall return to this figure in §3.9.4.

Low-frequency inertial waves play a special role in many systems. They are characterised by $\mathbf{k} \cdot \mathbf{\Omega} \approx 0$ and a group velocity of $\mathbf{c}_g = \pm 2\mathbf{\Omega}/|\mathbf{k}|$, so that the wave crests move horizontally, yet wave energy propagates up and down the rotation axis with a speed of $\Omega\lambda/\pi$, where λ is the wavelength. This is shown schematically in Figure 3.34(a), where the waves are generated by a slowly oscillating disc.

To illustrate the importance of such waves, consider an unbounded initial-value problem in which we slowly move a disc of radius R along the rotation axis with a constant speed V, starting at $t = 0$. Evidently, low-frequency waves propagate in

(a) (b)

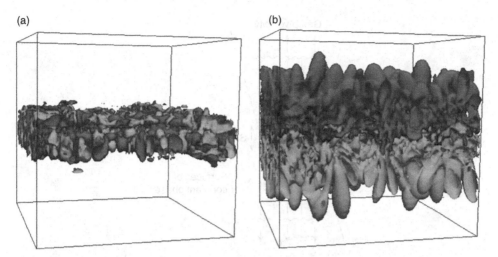

Figure 3.33 A localised patch of turbulence spreads in a rotating fluid at Ro = 0.1 by emitting inertial wave packets. Dark grey is negative helicity and light grey is positive. The left-hand image is the flow at $t = 0$ and the right-hand one corresponds to $\Omega t = 6$. (From Davidson, 2013.)

the $\pm\Omega$ directions, carrying energy away from the disc at a speed $c_g = 2\Omega/|\mathbf{k}|$. The smallest wave numbers (largest wavelengths) travel fastest, and these have a magnitude of $|\mathbf{k}| \approx \pi/R$. As a consequence, at time t we find wave fronts located a distance $(2/\pi)\Omega R t$ above and below the disc, as indicated in Figure 3.34(b).

We conclude that, after a time t, the inertial waves generated by the disc fill a column of radius R and length $2\ell \approx (4/\pi)\Omega R t$. Moreover, these waves carry with them the information that the disc is moving. An exact solution of this initial-value problem shows that the fluid which lies within this column has the same axial velocity as the disc, while that lying outside the column is unaware of the movement of the disc and so is quiescent in the rotating frame. In short, the inertial waves have created a transient Taylor column, whose length grows at the group velocity of low-frequency waves.

Returning to Figure 3.32, we can now understand how the Taylor column translates across the tank. When the object is towed slowly across the base of the tank it acts like a radio antenna, emitting low-frequency inertial waves which travel upward at their group velocity. These reach the fluid surface on a time scale which is very short by comparison with the slow timescale of the movement of the object. The waves then reflect at the surface to form a standing wave pattern below. Crucially, as the inertial waves travel upward they carry with them the information that the object is moving, and this provides the impetus for the fluid within the column to move horizontally, in line with the towed object below. In short, the moving Taylor column is maintained by a continual train of inertial wave packets radiated from the object, with the object itself acting rather like a radio antenna.

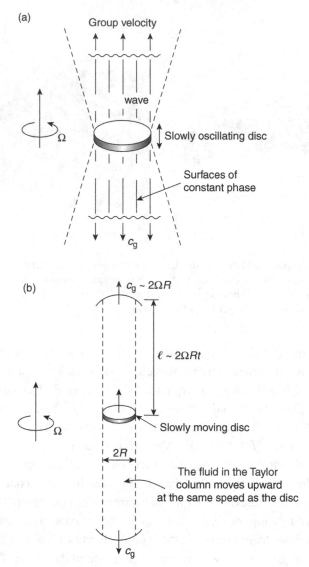

Figure 3.34 (a) Low-frequency inertial waves are characterised by $\mathbf{k}\cdot\mathbf{\Omega} \approx 0$ and $\mathbf{c}_g = \pm 2\mathbf{\Omega}/k$. (b) The formation of a transient Taylor column at Ro $\ll 1$. Wave fronts form at $z \approx \pm (2/\pi)\Omega R t$ and the fluid within the column moves with the same velocity as the disc.

3.9.3 Inertial Wave Packets, Columnar Vortices and Transient Taylor Columns

In Figures 3.32 and 3.34 we considered inertial waves driven by moving boundaries. In such cases there is an imposed time scale set by the speed at which the boundary moves. For example, slowly moving boundaries excite low-frequency waves and these propagate up and down the rotation axis. We now turn to

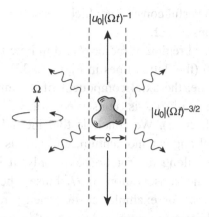

Figure 3.35 For localised disturbances the energy can, and indeed will, propagate in all directions (wiggly arrows). However, the axial component of the angular momentum can disperse along the rotation axis only (solid arrows). This biases the dispersion of energy onto the rotation axis.

boundary-free initial-value problems, which typifies Figure 3.33. Consider, for example, the initial-value problem consisting of a single eddy (a blob of vorticity) sitting in a rapidly rotating fluid (Figure 3.35). Here the situation is more complicated because there is no imposed time scale. In such a situation the direction of propagation of wave energy is determined by the distribution of energy amongst the various **k**-vectors in the initial velocity field. That is, the act of Fourier transforming the initial condition apportions a certain amount of energy to each wave vector, **k**. That energy then disperses in the form of wave packets in accordance with (3.70), with the direction of energy propagation being determined by the orientation of **k** relative to $\boldsymbol{\Omega}$. For an arbitrarily shaped eddy, then, we might expect the energy to disperse in all possible directions.

However, this is not the end of the story. Rather remarkably, even for an arbitrary disturbance, there is still a tendency for wave energy to propagate preferentially along the rotation axis to form columnar structures (Davidson, Staplehurst and Dalziel, 2006). That is, spatially compact eddies spontaneously evolve into columnar vortices in which $(\boldsymbol{\Omega}\cdot\nabla)\mathbf{u}$ is small, i.e. they evolve into a kind of transient Taylor column.

To illustrate the point, consider the initial value problem consisting of a localised blob of vorticity sitting in a rapidly rotating fluid (Figure 3.35). Let the characteristic scale of the blob be δ, a typical velocity be u, and suppose $\mathrm{Ro} = u/\Omega\delta \ll 1$. For $t > 0$ the motion consists of a spectrum of inertial waves whose energy disperses in all directions (as determined by the relative orientation of the various **k**-vectors), and with a typical speed of $|\mathbf{c}_g| \sim \Omega\delta$. However, as we shall see, this radiation of energy is subject to a powerful constraint which systematically favours dispersion of energy in the directions of $\pm\boldsymbol{\Omega}$.

Let V_R be the cylindrical region of radius R and infinite length that encloses the vorticity field $\boldsymbol{\omega}$ at $t = 0$ (the dotted lines in Figure 3.35). Then it may be shown that, in the rotating frame, the axial component of the angular momentum held within V_R, $H_z = \int_{V_R} [\mathbf{x} \times \mathbf{u}]_z dV$, is conserved for all time (see Davidson, Staplehurst and Dalziel, 2006). We conclude that inertial waves cannot sustain a net radial flux of axial angular momentum, and so this component of angular momentum can disperse along the rotation axis only. It is easy to see that the constraint imposed by the conservation of H_z biases the dispersion of energy towards the rotation axis. For example, as the energy radiates to fill a three-dimensional volume whose size grows as $V_{3D} \sim (c_g t)^3 \sim (\delta\Omega t)^3$, we would expect that conservation of energy requires a typical velocity outside V_R to decay as $|\mathbf{u}| \sim |\mathbf{u}_0|(\Omega t)^{-3/2}$. However, inside the cylinder V_R the axial component of angular momentum is confined to a cylindrical region whose volume grows as $V_{1D} \sim c_g t \delta^2 \sim \Omega t \delta^3$. Conservation of H_z then suggests that the characteristic velocity inside V_R falls more slowly, in fact as $|\mathbf{u}| \sim |\mathbf{u}_0|(\Omega t)^{-1}$. This is illustrated in Figure 3.35, and indeed these two temporal decay laws may be confirmed by exact asymptotic analysis (Davidson, Staplehurst and Dalziel, 2006). Thus the dominant influence of inertial wave radiation is to spread the energy of the initial vortex along the rotation axis.

An alternative way of understanding why the fluid within V_R exhibits a higher radiation density is to note that the group velocity of inertial waves is perpendicular to $\mathbf{k}$. Thus the energy associated with *all* horizontal wave vectors radiates along the rotation axis, and so all the energy contained within a thin horizontal disc in $\mathbf{k}$-space propagates along a narrow cylinder in real space. By way of contrast, only *one* orientation of $\mathbf{k}$ will transport energy to a particular location remote from the vertical axis passing through the disturbance. The process of channelling energy from a two-dimensional object in $\mathbf{k}$-space to a one-dimensional object (a narrow cylinder) in real space amplifies the radiation density on the rotation axis (Davidson, 2013).

Whichever explanation is preferred, we conclude that an eddy which is initially confined to the region $|\mathbf{x}| < \delta$ disperses energy in all directions, but that the energy density is inevitably highest on the rotation axis, creating a columnar vortex above and below the initial disturbance. This is illustrated in Figure 3.36, taken from Davidson (2013), which shows the axisymmetric dispersion of energy from a Gaussian eddy initially located at the origin. (Only the region $z > 0$ is shown.) These columnar structures are manifestly a form of transient Taylor column. They are wave packets that *propagate and spread* at a rate dictated by the low-frequency group velocity, and so their length grows as $\ell_z \sim \delta\Omega t$ while their centres propagate away from the initial disturbance at the rate $z \sim \pm \delta \Omega t$. Such columnar wave packets are clearly evident in Figure 3.33.

3.9.4 A Glimpse at Rapidly Rotating Turbulence

Let us return to Figure 3.33 and consider the implications of inertial wave dispersion for rotating turbulence. Consider the case where Re is large but Ro $= u/\Omega \delta$ is small, δ being the size of the large eddies. The evolution of the large eddies is then governed by (3.67). Perhaps the most important case is that of freely decaying turbulence, i.e. turbulence which has been stirred up and then left to itself. The large eddies will spontaneously emit inertial wave packets as discussed above, with a systematic bias towards radiating energy along the rotation axis. This, in turn, will elongate the turbulent eddies in the axial direction. In such cases we might tentatively neglect $\partial^2/\partial z^2$ in the Laplacian of (3.68), which then simplifies to

$$\frac{\partial^2}{\partial t^2} \left(\nabla_{\perp}^2 \mathbf{u} \right) + 4(\mathbf{\Omega} \cdot \nabla)^2 \mathbf{u} = 0, \tag{3.72}$$

where $\nabla_{\perp}^2$ represents the two-dimensional Laplacian defined in the transverse plane. We now perform a two-dimensional Fourier transform of this equation in the plane normal to $\mathbf{\Omega}$, to give

$$\frac{\partial^2 \hat{\mathbf{u}}}{\partial t^2} = \left(\frac{4\Omega^2}{k_{\perp}^2} \right) \frac{\partial^2 \hat{\mathbf{u}}}{\partial z^2}, \tag{3.73}$$

where $\hat{\mathbf{u}}$ is the Fourier transformed velocity. Evidently $\hat{\mathbf{u}}$, and hence wave energy, spreads along the Ω-axis by linear wave propagation, thus creating columnar structures.

This cartoon works well for the case of kinetic energy spreading from a localised patch of turbulence, as illustrated in Figure 3.33 and as discussed in Davidson (2013). Figure 3.33 shows the numerical simulation of a horizontal patch of turbulence spreading in a rapidly rotating fluid at Ro = 0.1. The two images correspond to $\Omega t = 0, 6$, with negative helicity marked dark grey and positive helicity light grey. We suggested earlier that the cloud grows by emitting low-frequency inertial waves which propagate along the rotation axis in the form of wave packets. Certainly the similarity to Figure 3.36 is apparent and the fact that the helicity is negative in the upper part of the flow, and positive in the lower regions, provides support for the suggestion that the columnar eddies are indeed wave packets. (The columnar structures seen in Figure 3.33 are unlikely to be the result of nonlinear dynamics because $\Omega t = 6$ combined with Ro $= 0.1$ gives $ut/\delta < 1$, so very little in the way of nonlinear dynamics has occurred.) The assertion that the

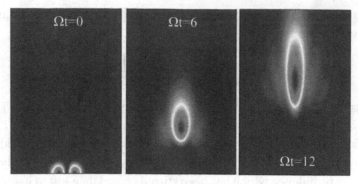

Figure 3.36 A Gaussian blob of vorticity located at the origin spontaneously evolves into a pair of columnar eddies through linear inertial wave propagation. Energy contours are shown in the half-space $z > 0$ at times $\Omega t = 0, 6, 12$. The columnar vortex elongates as $\ell_z \sim \delta \Omega t$ while its centre propagates away from the initial disturbance at the rate $z \sim \pm \delta \Omega t$. (From Davidson, 2013).

columnar eddies are simply linear wave packets can be confirmed beyond reasonable doubt by comparing the rate of spread of energy in the vertical direction with the group velocity of linear inertial waves, i.e. $\ell_z \sim \delta \Omega t$, and indeed the two match very closely in laboratory experiments (Davidson, Staplehurst and Dalziel, 2006), and also in related numerical simulations (Ranjan and Davidson, 2014).

Exercises

3.1 *The azimuthal-poloidal decomposition of axisymmetric flows.*

Consider an axisymmetric flow expressed in cylindrical polar coordinates, $(r, \theta, z,)$. This can be decomposed into its azimuthal, $\mathbf{u}_\theta$, and poloidal, $\mathbf{u}_p = (u_r, 0, u_z)$, components, both of which are solenoidal. Likewise, the vorticity field may be decomposed into $\boldsymbol{\omega}_\theta$ and $\boldsymbol{\omega}_p$. Show that $\boldsymbol{\omega}_\theta = \nabla \times \mathbf{u}_p$ and $\boldsymbol{\omega}_p = \nabla \times \mathbf{u}_\theta$, and that the instantaneous velocity distribution is uniquely determined by the two scalar fields $\Gamma = r u_\theta$ and ω_θ / r.

The governing equations of motion may also be decomposed into azimuthal and poloidal parts. Show that the azimuthal components of the inviscid momentum and vorticity equations yield the evolution equations

$$\frac{D\Gamma}{Dt} = 0, \quad \frac{D}{Dt}\left(\frac{\omega_\theta}{r}\right) = \frac{\partial}{\partial z}\left(\frac{\Gamma^2}{r^4}\right).$$

Explain the origin of the source term in the ω_θ equation, $\partial \Gamma^2 / \partial z$, in terms of the spiralling up of the poloidal vortex lines by their own velocity field, $\mathbf{u}_\theta$.

3.2 *The global conservation of azimuthal vorticity in steady, inviscid, axisymmetric flows.*

Consider now the special case of a steady flow and introduce the Stokes stream function for the poloidal velocity, $\mathbf{u}_p(r, z) = \nabla \times [(\psi/r)\hat{\mathbf{e}}_\theta]$, which satisfies $\mathbf{u} \cdot \nabla \psi = 0$. Show that the inviscid evolution equations of Exercise 3.1 can be rewritten as

$$\Gamma = \Gamma(\psi), \quad \mathbf{u} \cdot \nabla(\omega_\theta/r) = \mathbf{u} \cdot \nabla\left(\Gamma\Gamma'(\psi)/r^2\right),$$

from which

$$\omega_\theta/r = \Gamma\Gamma'(\psi)/r^2 - H'(\psi),$$

for some function $H(\psi)$. Consider an inviscid swirling flow between two parallel discs located at $z = 0$ and $z = H$. The fluid enters and leaves the domain $0 < r < R$ at radius $r = R$, with the fluid spiralling radially inward at low altitudes and then radially outward at higher altitudes. Use the expression above to show that, despite the source term $\partial\Gamma^2/\partial z$ in the azimuthal vorticity equation, the azimuthal vorticity on a given streamline enters and leaves the domain $0 < r < R$ with the same value.

3.3 *The Squire–Long equation for steady, inviscid, axisymmetric flows.*

Use the axial component of the inviscid momentum equation to confirm that the function H in the equations above is in fact Bernoulli's function, $H = \frac{1}{2}\mathbf{u}^2 + p/\rho$, which is constant along a streamline. Also show that, on substituting for ω_θ in terms of ψ,

$$\frac{\partial^2\psi}{\partial z^2} + r\frac{\partial}{\partial r}\frac{1}{r}\frac{\partial\psi}{\partial r} = r^2 H'(\psi) - \Gamma\Gamma'(\psi).$$

This is the Squire–Long equation. If the streamlines in a given flow are open, and the functions $\Gamma(\psi)$ and $H(\psi)$ can be specified upstream, then this equation may be integrated to give the entire flow field.

4

The Governing Equations of MHD

> The beauty of electricity or of any other force is not that the power is mysterious and unexpected, . . . but that it is under *law*, and that the taught intellect can even govern it.
>
> *Faraday (1858)*

While Faraday was at the forefront of harnessing the power of electromagnetism for the general benefit of mankind, he was also a great advocate of a unified theory of all natural forces, from the electromagnetic to the gravitational.

In this chapter we incorporate the Lorentz force into the Navier–Stokes equation, catalogue the governing equations of MHD, and consider some of their more elementary and immediate consequences.

4.1 The MHD Equations and Key Dimensionless Groups

Let us start by summarising the governing equations of electrodynamics required for MHD. They are the reduced form of Maxwell's equations in which the charge density and associated displacement currents are ignored,

$$\boxed{\nabla \times \mathbf{B} = \mu \mathbf{J}}, \qquad \boxed{\nabla \cdot \mathbf{J} = 0}, \tag{4.1}$$

$$\boxed{\nabla \times \mathbf{E} = -\frac{\partial \mathbf{B}}{\partial t}}, \qquad \boxed{\nabla \cdot \mathbf{B} = 0}. \tag{4.2}$$

These are supplemented by Ohm's law and the Lorentz force per unit volume:

$$\boxed{\mathbf{J} = \sigma(\mathbf{E} + \mathbf{u} \times \mathbf{B})}, \qquad \boxed{\mathbf{F} = \mathbf{J} \times \mathbf{B}}. \tag{4.3}$$

Equations (4.1) through (4.3) combine to give the induction equation,

$$\frac{\partial \mathbf{B}}{\partial t} = \nabla \times (\mathbf{u} \times \mathbf{B}) + \lambda \nabla^2 \mathbf{B}, \qquad \lambda = (\mu \sigma)^{-1},$$

or equivalently, given that **B** and **u** are both solenoidal,

$$\boxed{\frac{D\mathbf{B}}{Dt} = (\mathbf{B} \cdot \nabla)\mathbf{u} + \lambda \nabla^2 \mathbf{B}}. \tag{4.4}$$

On the other hand, Newton's second law gives us the Navier–Stokes equation which, if we incorporate the Lorentz force, takes the form

$$\boxed{\rho \frac{D\mathbf{u}}{Dt} = -\nabla p + \rho \nu \nabla^2 \mathbf{u} + \mathbf{J} \times \mathbf{B}}, \tag{4.5}$$

from which we obtain the vorticity equation

$$\boxed{\frac{D\boldsymbol{\omega}}{Dt} = (\boldsymbol{\omega} \cdot \nabla)\mathbf{u} + \nu \nabla^2 \boldsymbol{\omega} + \rho^{-1} \nabla \times (\mathbf{J} \times \mathbf{B})}. \tag{4.6}$$

Since we are restricting ourselves to incompressible fluids, we also have the continuity equation

$$\boxed{\nabla \cdot \mathbf{u} = 0}. \tag{4.7}$$

Of course, the vorticity field is also solenoidal, although this has nothing to do with incompressibility, but rather follows from the fact that the divergence of the curl of any vector field is zero.

In Chapter 1 we saw that three important parameters in MHD are

- magnetic Reynolds number, $R_m = \mu \sigma u \ell = u\ell/\lambda$,
- Alfvén velocity, $v_a = B/\sqrt{\rho \mu}$,
- magnetic damping time, $\tau = [\sigma B^2/\rho]^{-1}$,

where ℓ is a characteristic length scale of the motion. The first of these is dimensionless and characterises the ratio of advection to diffusion of a magnetic field governed by (4.4). When R_m is large, as in highly conducting fluids, diffusion is weak. The second parameter is important when the magnetic Reynolds number is large, so that perturbed magnetic field lines can oscillate in an elastic manner, rather like plucked violin strings. In particular, we showed in Chapter 1 that the frequency of these elastic oscillations is of the order of $\varpi \sim v_a/\ell$. Finally, the magnetic damping time, τ, is important when R_m is small, being the characteristic timescale over which the Lorentz force appreciably influences a flow at low R_m (see §1.3.3).

Since both ρ and μ are constants in an incompressible flow, some authors find it convenient to work with the scaled magnetic field

$$\mathbf{h} = \mathbf{B}/\sqrt{\rho\mu} \tag{4.8}$$

which, like v_a, has the dimensions of a velocity. Invoking Ampère's law we see that the Lorentz force per unit mass then becomes

$$\mathbf{F} = \mathbf{J} \times \mathbf{B}/\rho = (\nabla \times \mathbf{h}) \times \mathbf{h}. \tag{4.9}$$

There are a number of dimensionless groups which regularly appear in the MHD literature, and we shall discuss some of these here. Let us start by asking how many independent dimensionless groups one might expect when the fluid is incompressible. If we consider the Navier–Stokes equation with the Lorentz force rewritten as (4.9), the physical variables that concern us are u, $|\mathbf{h}|$, ν and ℓ, where u is a typical velocity and $|\mathbf{h}|$ is a typical scaled magnetic field. When we include the induction equation in our analysis we must add λ to this list. This gives us a total of five parameters which contain between them only two dimensions: length and time. The Buckingham pi theorem then tells us there are only three independent dimensionless groups. Clearly one is $R_m = u\ell/\lambda$ and another $Re = u\ell/\nu$, both of which represent the relative magnitudes of advection to diffusion; in one case the diffusion of a magnetic field and in the other case the diffusion of momentum or vorticity. Of course, we know from Chapter 3 that Re can also be interpreted as the ratio of inertial to viscous forces.

This gives us a clue as to the third independent dimensionless group. Since $\mathbf{h}$ has the dimensions of a velocity, the grouping $|\mathbf{h}|\ell/(diffusivity)$ is also dimensionless, and this third group is conventionally taken to be $|\mathbf{h}|\ell/\sqrt{\nu\lambda}$. In terms of the magnetic field this becomes

$$\mathrm{Ha} = B\ell(\sigma/\rho\nu)^{1/2}, \tag{4.10}$$

which is known as the *Hartmann number.* It turns out that, when $R_m \ll 1$, the square of the Hartmann number represents the ratio of Lorentz to viscous forces, as discussed below.

From these three (independent) dimensionless groups we can construct many others. For example, a commonly used dimensionless group is the obscurely named *interaction parameter*,

$$N = \frac{\sigma B^2 \ell}{\rho u} = \frac{\ell}{u\tau}, \tag{4.11}$$

Table 4.1 *Some dimensionless groups in MHD.*

Name	Symbol	Definition	Significance
Reynolds number	Re	$u\ell/\nu$	Ratio of inertial to viscous forces
Interaction parameter	N	$\sigma B^2 \ell/\rho u$	Ratio of Lorentz to inertial forces at low R_m
Hartman number	Ha	$B\ell(\sigma/\rho\nu)^{1/2}$	Ratio of Lorentz to viscous forces at low R_m
Magnetic Reynolds number	R_m	$u\ell/\lambda$	Ratio of advection to diffusion of **B**

where τ is the magnetic damping time, $\tau = [\sigma B^2/\rho]^{-1}$. This is particularly relevant in those situations where **J** is primarily driven by $\mathbf{u} \times \mathbf{B}$ in Ohm's law, and so $|\mathbf{J}| \sim \sigma u B$, as often occurs when $R_m \ll 1$. In such a case N represents the ratio of the Lorentz force, $\mathbf{J} \times \mathbf{B}$, to inertia, $\rho(\mathbf{u}\cdot\nabla)\mathbf{u}$. We can write the Hartmann number as a hybrid of Re and N:

$$\text{Ha} = B\ell(\sigma/\rho\nu)^{1/2} = (N\text{Re})^{1/2}. \tag{4.12}$$

Since N is the ratio of Lorentz forces to inertia (when $R_m \ll 1$), and Re the ratio of inertial to viscous forces, we conclude that $(\text{Ha})^2 = N\,\text{Re}$ represents the ratio of the Lorentz to viscous forces at low R_m, as noted above. These various groups are tabulated above.

From these four dimensionless groups we can construct many others. For example, the ratio of the magnetic Reynolds number to Re is

$$\text{Pr}_m = \frac{\nu}{\lambda},$$

which is called the *magnetic Prandtl number.* Another common group is $|\mathbf{h}|\ell/\lambda$, or in terms of the magnetic field, $B\ell\sigma\sqrt{\mu/\rho}$. This is called the *Lundquist number,* and it provides a convenient measure of the degree to which Alfvén waves can be considered to be free from magnetic dissipation.

Note that in the discussion above the characteristic length scale of the flow, ℓ, need not be known in advance, but rather it may emerge from some internal force balance. The obvious example is a laminar boundary layer (in the absence of a magnetic field) where often we find that $\delta \sim \sqrt{\nu\, L/u}$ is the boundary layer thickness when L is the characteristic length scale of the external flow (see §3.7.1). In such cases a Reynolds number based on L or δ is large, yet one based on the length scale $\ell = \delta^2/L$ is of the order of unity. That is the whole point about boundary layers; viscous and inertial forces are always of the same order, no matter how small the viscosity may be, and so a Reynolds number based on an appropriate length scale must be of order unity. One must always be careful in the choice of length scale when constructing meaningful dimensionless groups. In general, each case must be treated on its merits. Nevertheless, dimensionless groups are extremely useful. Often, when they are very large or very small, they allow us to throw out certain terms in the governing equations, thereby greatly simplifying the problem.

4.2 Energy Considerations

The Lorentz force is important because it allows energy to transfer back and forth between the magnetic field and the fluid, as we now show. The scalar product of **B** with Faraday's law yields, with some help from Ampère's law,

$$\frac{\partial}{\partial t}\left[\frac{\mathbf{B}^2}{2\mu}\right] = -\mathbf{J}\cdot\mathbf{E} - \nabla\cdot[(\mathbf{E}\times\mathbf{B})/\mu]. \tag{4.13}$$

Substituting for **E** using Ohm's law we obtain an energy equation for the magnetic field:

$$\frac{\partial}{\partial t}\left[\frac{\mathbf{B}^2}{2\mu}\right] = -\mathbf{J}^2/\sigma - (\mathbf{J}\times\mathbf{B})\cdot\mathbf{u} - \nabla\cdot[(\mathbf{E}\times\mathbf{B})/\mu]. \tag{4.14}$$

The left-hand side is the rate of change of magnetic energy per unit volume, and so the terms on the right must represent either a conversion of magnetic energy into some other form of energy, or else a redistribution of magnetic energy in space. $\mathbf{J}^2/\sigma$ is the Joule dissipation, and so represents the conversion of magnetic energy into heat, while $(\mathbf{J}\times\mathbf{B})\cdot\mathbf{u}$ is the rate of working of the Lorentz force, representing the conversion of magnetic energy into mechanical energy. On integrating (4.14) over a fixed volume V we obtain

$$\frac{d}{dt}\int_V (\mathbf{B}^2/2\mu)dV = -\int_V (\mathbf{J}^2/\sigma)dV - \int_V [(\mathbf{J}\times\mathbf{B})\cdot\mathbf{u}]dV - \oint_S [(\mathbf{E}\times\mathbf{B})/\mu]\cdot d\mathbf{S}.$$

$$\tag{4.15}$$

The surface integral in (4.15) represents the flux of magnetic energy out through the surface of the control volume. As noted in Example 2.5, the quantity $\mathbf{P} = (\mathbf{E}\times\mathbf{B})/\mu$ is known as the *Poynting vector,* or Poynting's electromagnetic energy flux density. It acts to redistribute electromagnetic energy in space.

When integrated over all space, the surface integral in (4.15) vanishes and we are left with

$$\frac{d}{dt}\int_{V_\infty} (\mathbf{B}^2/2\mu)dV = -\int_{V_\infty} [(\mathbf{J} \times \mathbf{B}) \cdot \mathbf{u}]dV - \int_{V_\infty} (\mathbf{J}^2/\sigma)dV. \qquad (4.16)$$

On the other hand, integrating the dot product of **u** with (4.5) over all space, invoking (3.46) to rewrite the rate of working of the viscous force in terms of viscous dissipation, and dispensing with all surface integrals, we find

$$\frac{d}{dt}\int_{V_\infty} (\rho\mathbf{u}^2/2)dV = \int_{V_\infty} [(\mathbf{J} \times \mathbf{B}) \cdot \mathbf{u}]dV - 2\rho\nu\int_{V_\infty} S_{ij}S_{ij}dV. \qquad (4.17)$$

Note that the rate of working of $\mathbf{J} \times \mathbf{B}$ appears with opposite signs in these two equations, confirming its role as a means of exchanging energy between the magnetic and velocity fields. If we add (4.16) and (4.17) we find that the total energy, magnetic plus kinetic, declines as a result of Joule and viscous dissipation.

4.3 Maxwell's Stresses and Faraday's Tension

It is instructive to consider different forms of the Lorentz force. From Ampère's law we may rewrite the Lorentz force in terms of **B** alone. Starting with the vector identity,

$$\nabla(\mathbf{B}^2/2) = (\mathbf{B} \cdot \nabla)\mathbf{B} + \mathbf{B} \times \nabla \times \mathbf{B},$$

and using $\nabla \times \mathbf{B} = \mu\mathbf{J}$, yields

$$\mathbf{J} \times \mathbf{B} = (\mathbf{B} \cdot \nabla)(\mathbf{B}/\mu) - \nabla(\mathbf{B}^2/2\mu). \qquad (4.18)$$

The second term on the right of (4.18) acts on the fluid in exactly the same way as the pressure force $-\nabla p$. It is irrotational and so makes no contribution to the vorticity equation. In incompressible flows without a free surface its role is simply to augment the fluid pressure. (Its absence from the vorticity equation implies that it cannot influence the flow field.) For this reason, $\mathbf{B}^2/2\mu$ is called the magnetic pressure and in many, if not most, incompressible flows it is of no dynamical significance. This brings us to the first term on the right. We can rewrite the ith component of this force as

$$\mathbf{B} \cdot \nabla(B_i/\mu) = \frac{\partial}{\partial x_j}\left[\frac{B_iB_j}{\mu}\right], \qquad (4.19)$$

where there is an implied summation over the index j. From this we may show that the effect of the body force in (4.19) is exactly equivalent to a distributed set of fictitious stresses, $B_i B_j / \mu$, acting on the surface of fluid elements. One approach is simply to compare the right-hand side of (4.19) with the viscous force in (3.1). Alternatively, this can be established by integrating (4.19) over an arbitrary volume V and invoking Gauss' theorem. Since $\nabla \cdot (B_i \mathbf{B}) = (\mathbf{B} \cdot \nabla) B_i$, we find

$$\int [\mathbf{B} \cdot \nabla (B_i / \mu)] dV = \oint (B_i / \mu) \mathbf{B} \cdot d\mathbf{S}. \qquad (4.20)$$

The surface integral on the right of (4.20) is equal to the cumulative effect of the distributed stress system $B_i B_j / \mu$ acting over the surface of V. However, Equation (4.20) tells us that this surface stress distribution is, in turn, equivalent to the integrated effect of the volume force $(\mathbf{B} \cdot \nabla)(\mathbf{B}/\mu)$. Since this is true for any volume V it follows that the body force $(\mathbf{B} \cdot \nabla)(\mathbf{B}/\mu)$ and the stress system $B_i B_j / \mu$ are entirely equivalent in their mechanical action.

In summary then, we may replace the Lorentz force, $\mathbf{J} \times \mathbf{B}$, by an imaginary set of stresses

$$\tau_{ij}^M = \left(B_i B_j / \mu \right) - \left(\mathbf{B}^2 / 2\mu \right) \delta_{ij}, \qquad (4.21)$$

where the second term on the right is the magnetic pressure. These are called *Maxwell's stresses,* and their primary utility lies in the fact that the integrated effect of a distributed body force can be represented by surface stresses alone.

Now there is another, perhaps more useful, representation of $\mathbf{J} \times \mathbf{B}$. This comes from replacing $\mathbf{u}$ with $\mathbf{B}$ in (3.6):

$$(\mathbf{B} \cdot \nabla) \mathbf{B} = B \frac{\partial B}{\partial s} \hat{\mathbf{e}}_t - \frac{B^2}{R} \hat{\mathbf{e}}_n. \qquad (4.22)$$

Here, s is now a coordinate measured along a magnetic field line, $\hat{\mathbf{e}}_t$ and $\hat{\mathbf{e}}_n$ are unit vectors in the tangential and normal direction, $B = |\mathbf{B}|$, and R is the local radius of curvature of the field line. It follows that the Lorentz force may be written as

$$\mathbf{J} \times \mathbf{B} = \frac{\partial}{\partial s} \left[\frac{B^2}{2\mu} \right] \hat{\mathbf{e}}_t - \frac{B^2}{\mu R} \hat{\mathbf{e}}_n - \nabla \left(\mathbf{B}^2 / 2\mu \right). \qquad (4.23)$$

We now have two alternative representations of $\mathbf{J} \times \mathbf{B}$. In cases where the magnetic pressure is unimportant, which is usually the case in an incompressible flow without free surfaces, we are concerned only with $\mathbf{B} \cdot \nabla(\mathbf{B}/\mu)$. In such situations $\mathbf{J} \times \mathbf{B}$ may be pictured either as the result of the Maxwell stress system $B_i B_j / \mu$, or else it may be written in the form

$$(\mathbf{B} \cdot \nabla)(\mathbf{B}/\mu) = \frac{\partial}{\partial s} \left[\frac{B^2}{2\mu} \right] \hat{\mathbf{e}}_t - \frac{B^2}{\mu R} \hat{\mathbf{e}}_n. \qquad (4.24)$$

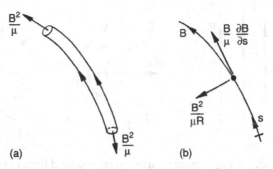

Figure 4.1 The tangential and normal components of $\mu^{-1}(\mathbf{B} \cdot \nabla)\mathbf{B}$ acting on a flux tube can be interpreted as a consequence of Faraday's tension, B^2/μ, acting along the axis of the flux tube.

Crucially, it would seem that there is a Lorentz force normal to the field lines which is associated with the *curvature* of those lines.

To illustrate the physical nature of (4.24), consider a thin, isolated flux tube carrying a magnetic flux $\Phi = B(s)A(s)$, where A is the cross-sectional area of the flux tube (Figure 4.1). If we think in terms of Maxwell stresses, then the force $\mu^{-1}(\mathbf{B} \cdot \nabla)\mathbf{B}$ can be represented by the action of the stress $B_i B_j/\mu$, and for this particularly simple geometry this reduces to a tensile stress of B^2/μ acting along the axis of the isolated flux tube. This tensile stress is known as *Faraday's tension* and it leads to a tensile force in the flux tube of $T(s) = (B^2/\mu)A$. Now a string carrying a tensile force $T(s)$ experiences a force per unit length of $\delta\mathbf{F} = (dT/ds)\hat{\mathbf{e}}_t - (T/R)\hat{\mathbf{e}}_n$, where R is the local radius of curvature of the string. It follows that the curvature of the thin flux tube, and the variation with s of Faraday's tensile force, $T = \Phi B/\mu$, leads to a force per unit length acting on the flux tube of

$$\delta\mathbf{F} = \frac{dT}{ds}\hat{\mathbf{e}}_t - \frac{T}{R}\hat{\mathbf{e}}_n = \frac{1}{\mu}\frac{d(\Phi B)}{ds}\hat{\mathbf{e}}_t - \frac{B^2 A}{\mu R}\hat{\mathbf{e}}_n. \tag{4.25}$$

Noting that Φ is independent of s and dividing through by A brings us back to (4.24). Thus the tangential and normal components of $\mu^{-1}(\mathbf{B} \cdot \nabla)\mathbf{B}$ in (4.24) arise simply from the action of Faraday's tension, B^2/μ, acting along the axis of the flux tube, as shown in Figure 4.1.

The idea of Faraday's tension provides a useful qualitative means picturing the effect of the Lorentz force. We may think of magnetic field lines as being in tension, exerting a force on the fluid, rather like that of a stretched elastic band. In particular, if the field lines are curved, then there is a Lorentz force normal to the field lines which is associated with that curvature. This underlies the phenomenon of Alfvén waves.

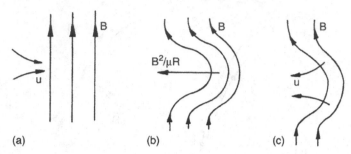

Figure 4.2 In a highly conducting fluid the magnetic field lines behave like elastic bands glued to the fluid. One consequence of this is Alfvén wave propagation.

4.4 A Glimpse at Alfvén Waves

We have already seen in §2.9.3 that, when R_m is large, the magnetic field lines tend to be frozen into a fluid. Now we see that the field lines also behave as if they are in tension. Consider what happens then if, at $t = 0$, we apply an impulsive force that pushes a region of highly conducting fluid past a magnetic field. The field lines will be swept along with the flow and the resulting curvature of the lines will create a back reaction, $B^2/\mu R$, on the fluid (Figure 4.2). The larger the distortion of the field lines, the larger the Lorentz force, and so eventually the fluid will come to rest and the Faraday tensions will then reverse the flow, pushing the field lines back towards their equilibrium position. However, the finite inertia of the fluid will carry it, and hence also the magnetic field lines, past the equilibrium position and so the entire process now begins in reverse. Evidently, oscillations (Alfvén waves) result. This is effectively what happens in the experiment described in §1.3.3.

Let us see if we can estimate the frequency of oscillation of Alfvén waves. A plucked violin string has a wave speed of $c = \sqrt{T/\rho_s}$, where T is the tension in the string and ρ_s is the mass per unit length of the string. Now we have just seen that a thin magnetic flux tube of cross-sectional area A behaves as if it carries a tension of $T = (B^2/\mu)A$, and of course it has a mass per unit length of ρA. This suggests that Alfvén waves have a wave speed of $c = \sqrt{B^2/\rho\mu}$, and this turns out to be correct: Alfvén waves are non-dispersive transverse waves which travel along magnetic field lines, just like waves on a string, and they have a wave speed of $v_a = B/\sqrt{\rho\mu}$. The frequency of oscillation of such waves is then $\varpi = v_a k = (B/\sqrt{\rho\mu})k$, where k is the wave number.

We shall return to Alfvén waves in Chapter 7, where we shall formally derive the dispersion relationship $\varpi = (B/\sqrt{\rho\mu})k$ and catalogue some of the properties of these waves.

Part II

The Fundamentals of Incompressible MHD

Nothing can be more fatal to progress than a too confident reliance on
mathematical symbols; for the student is only too apt to take the easier
course, and consider the formula and not the fact as the physical reality.
Kelvin, 1879.

We now discuss the fundamental theorems and phenomena encountered in MHD.
In order to minimise the algebra, we continue to restrict the discussion to incompressible fluids. This is entirely appropriate for liquid metal MHD, but somewhat
artificial in the case of astrophysical plasmas.

Perhaps one point is worth emphasising from the outset. The governing equations of MHD consist simply of Newton's laws of motion and the pre-Maxwell
form of the laws of electrodynamics. The reader is likely to be familiar with
elements of both sets of laws and many of the phenomena associated with them.
Thus, while the mathematical formulation of MHD may often seem daunting, the
underlying physical phenomena are usually fairly straightforward. It pays, therefore, when confronted with a welter of mathematical detail, to follow the advice of
Kelvin and keep asking the question: 'What is really going on?'

5

Kinematics: Advection, Diffusion and Intensification of Magnetic Fields

We adopt the suggestion of Ampère, and use the term *Kinematics* for the purely geometrical science of motion in the abstract. Keeping in view the properties of language, and following the example of most logical writers, we employ the term 'dynamics' in its true sense as the science which treats the action of force.

Kelvin (1879), preface to Natural Philosophy.

We now consider one-half of the coupling between **B** and **u**. Specifically, we look at the influence of **u** on **B** without worrying about the origin of the velocity field or the back reaction of the Lorentz forces on the fluid. In effect, we take **u** to be prescribed, forget about the Navier–Stokes equation, and focus on the role of **u** in Maxwell's equations. This is referred to as the *kinematics of MHD*.

5.1 The Analogy to Vorticity

In Chapter 2 we showed that Maxwell's equations lead to the induction equation

$$\frac{\partial \mathbf{B}}{\partial t} = \nabla \times (\mathbf{u} \times \mathbf{B}) + \lambda \nabla^2 \mathbf{B}, \tag{5.1}$$

where $\lambda = (\mu\sigma)^{-1}$. Compare this with the transport equation for vorticity,

$$\frac{\partial \boldsymbol{\omega}}{\partial t} = \nabla \times (\mathbf{u} \times \boldsymbol{\omega}) + \nu \nabla^2 \boldsymbol{\omega}. \tag{5.2}$$

There appears to be an exact analogy. In fact, the analogy is not perfect because $\boldsymbol{\omega}$ is functionally related to **u** in a way that **B** is not. Nevertheless, this does not stop us from borrowing many of the theorems of classical vortex dynamics and re-interpreting them in terms of MHD, with **B** playing the role of $\boldsymbol{\omega}$. For example, **B** is advected by **u** and diffused by λ, and in the limit $\lambda \to 0$, the counterpart of Helmholtz's first law of vortex dynamics is that **B** is frozen into the fluid.

5.2 Diffusion of a Magnetic Field

When $\mathbf{u} = 0$ we have

$$\frac{\partial \mathbf{B}}{\partial t} = \lambda \nabla^2 \mathbf{B}, \tag{5.3}$$

which may be compared with the diffusion equation for heat,

$$\frac{\partial T}{\partial t} = \alpha \nabla^2 T. \tag{5.4}$$

It appears that, like heat, magnetic fields will diffuse through a conducting medium at a finite rate. We cannot suddenly 'impose' a distribution of $\mathbf{B}$ throughout a conductor. All we can do is specify values at the boundaries and wait for it to diffuse inward. For example, suppose we have a semi-infinite region of conducting material occupying $y > 0$, and at $t = 0$ we apply a magnetic field $B_0 \hat{\mathbf{e}}_x$ at the surface $y = 0$. Then $\mathbf{B}$ will diffuse into the conductor in precisely the same way as heat or vorticity diffuses. In fact, to find the distribution of $\mathbf{B}$ at any instant we may simply lift the solution directly from the equivalent thermal problem. Such diffusion problems were discussed in §3.4, where we found that T (or $\boldsymbol{\omega}$) diffuses a distance $l \sim \sqrt{\alpha t}$, (or $\sqrt{\nu t}$) in a time t. By implication, $\mathbf{B}$ diffuses a distance of order $\sqrt{\lambda t}$ in the same time.

Example 5.1 Extinction of a Magnetic Field
Consider a long conducting cylinder of radius R which, at $t = 0$, contains an axial magnetic field $B_0(r)$, where r is a radial coordinate measured from the centre of the cylinder. The field outside the cylinder is zero. The axial field inside the cylinder will decay according to the diffusion equation

$$\frac{\partial B}{\partial t} = \lambda \frac{1}{r} \frac{\partial}{\partial r} \left(r \frac{\partial B}{\partial r} \right),$$

subject to $B = 0$ at $r = R$ and $B = B_0(r)$ at $t = 0$. Show that a Fourier–Bessel series of the form

$$B = \sum_{n=1}^{\infty} A_n \mathrm{J}_0(\gamma_n r/R) \exp\left(-\gamma_n^2 \lambda t/R^2\right)$$

is one possible representation of the solution, where J_0 is the usual Bessel function, γ_n are the zeros of J_0, and A_n represents a set of amplitudes. Deduce that the field decays on a time scale of $R^2/5.75\lambda$.   ∎

5.3 Advection in Ideal Conductors: Alfvén's Theorem

5.3.1 Alfvén's Theorem

We now consider the other extreme, where there is no diffusion ($\lambda = 0$) but $\mathbf{u}$ is finite. This applies to conducting fluids with a very high conductivity (ideal conductors). Consider the similarity between

$$\frac{\partial \mathbf{B}}{\partial t} = \nabla \times (\mathbf{u} \times \mathbf{B}) \qquad (5.5)$$

and the vorticity equation for an inviscid fluid,

$$\frac{\partial \boldsymbol{\omega}}{\partial t} = \nabla \times (\mathbf{u} \times \boldsymbol{\omega}).$$

We might anticipate, correctly as it turns out, that Helmholtz's first law and Kelvin's theorem (which is, in effect, Helmholtz's second law) have their analogues in MHD. The equivalent theorems are:

Theorem I: (analogue of Helmholtz's first law)	The fluid elements that lie on a magnetic field line at some initial instant continue to lie on that field line for all time, i.e. the field lines are frozen into the fluid.
Theorem II: (analogue of Kelvin's theorem)	The magnetic flux linking any material loop moving with the fluid is conserved.

These two results are collectively known as Alfvén's theorem. In fact, theorem II is a direct consequence of the generalised version of Faraday's law, as explained in §2.9.3. Moreover, the first theorem may be proved in precisely the same manner as Helmholtz's first law, the proof relying on the analogy between (5.5) in the form

$$\frac{D\mathbf{B}}{Dt} = (\mathbf{B} \cdot \nabla)\mathbf{u}, \qquad (5.6)$$

and Equation (3.37) for a material line element $d\mathbf{l}$,

$$\frac{D}{Dt}(d\mathbf{l}) = (d\mathbf{l} \cdot \nabla)\mathbf{u}.$$

The 'frozen-in' nature of magnetic fields is of crucial importance in astrophysics, where R_m is usually very high. For example, one might ask, 'Why do many stars possess dipole fields of the order of 10 or 100 Gauss?' The answer, possibly, is that there exists a weak galactic field of $\sim 10^{-5}$ Gauss. As a star starts to form due to the

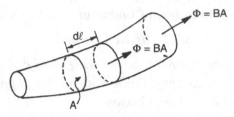

Figure 5.1 The stretching of a flux tube intensifies B.

gravitational collapse of an interstellar cloud, the galactic field, which is trapped in the plasma, becomes concentrated by the inward radial movement. A simple estimate of the increase in **B** due to this mechanism can be obtained if we assume the cloud remains spherical, of radius r, during the collapse. Two invariants of the cloud are its mass, $M \sim \rho r^3$, and the flux of the galactic field which traverses the cloud, $\Phi \sim Br^2$. It follows that during the collapse, $B \propto \rho^{2/3}$ which suggests $(B_{star}/B_{gal}) \sim (\rho_{star}/\rho_{gal})^{2/3}$. Actually, this overestimates B_{star} somewhat, possibly because the collapse is not spherical but rather disc-like (see Chapter 15), and possibly because there is some turbulent diffusion of **B**, despite the high value of R_m.

Now the analogy between **B** and $\boldsymbol{\omega}$ can be pushed even further. For example, our experience with vorticity suggests that, in three dimensions, we can stretch the magnetic field lines (or flux tubes) leading to an intensification of **B**. That is, the left of (5.6) represents the material advection of the magnetic field, so that when $(\mathbf{B}\cdot\nabla)\mathbf{u} = 0$ (as would be the case in certain two-dimensional flows), the magnetic field is passively advected. However, in three-dimensional flows $(\mathbf{B}\cdot\nabla)\mathbf{u}$ need not be zero and, because of the analogy with vortex tubes, we would expect this to lead to a rise in **B** whenever the flux tubes are stretched by the flow (see §3.4). In fact, this turns out to be true, as it must because the mathematics in the two cases are formally identical. However, the physical interpretation of this process of intensification is different in the two situations. In vortex dynamics it is a direct consequence of the conservation of angular momentum. In MHD, however, it follows from a combination of the conservation of mass, $\rho\delta V = \rho(d\ell\delta A)$, and flux, $\Phi = B\delta A$, applied to a short portion of a thin flux tube of cross-sectional area δA, as shown in Figure 5.1. If the flux tube is stretched, δA decreases and so B rises to conserve flux. This is the basis of dynamo theory in MHD, whereby magnetic fields are intensified by the continual stretching of the flux tubes.

5.3.2 An Aside: Sunspots

As an illustration of the 'frozen-in' behaviour of magnetic fields and of flux-tube stretching we shall describe here the phenomenon of sunspots. We give only

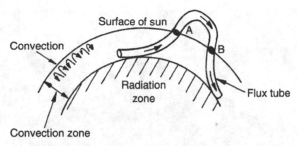

Figure 5.2 Schematic representation of the formation of sunspots. A buoyant flux tube erupts through the surface of the sun. Sunspots form at A and B where the magnetic field suppresses the turbulence, cooling the surface.

a qualitative description, but the interested reader will find more details in the references for this chapter (Moffatt, 1978, Biskamp, 1993, Priest, 2014).

If you look at the sun through darkened glass it is possible to discern small dark spots on its surface. These come and go, with a typical lifetime of several days. These spots (sunspots) typically appear in pairs and are concentrated around the equatorial plane. The spots have a diameter of $\sim 10^4$ km, which is around the same size of the earth! To understand how they arise, you must first know a little bit about the structure of the sun.

The surface of the sun is not uniformly bright but rather has a granular appearance. This is because the outer layer of the sun is in a state of convective turbulence. This *convection zone* has a thickness of 2×10^5km (the radius of the sun is around 7×10^5km) and consists of a continually evolving pattern of convection cells, rather like those seen in Bénard convection (Figure 5.2). The cells nearest the surface are about 10^3km across. Where hot fluid rises to the surface, the sun appears bright, while the cooler fluid, which falls at the junction of adjacent cells, appears dark. A typical convective velocity is around 1km/s and estimates of Re and R_m are Re $\sim 10^{11}$, $R_m \sim 10^8$, i.e. very large!

Now the sun has an average dipole magnetic field of a few Gauss, rather like that of the earth. Because R_m is large, this dipole field tends to be frozen into the fluid in the convection zone. Large-scale differential rotation at the base of this zone stretches and intensifies the dipole field until large field strengths (perhaps 10^4 Gauss, or greater) build up at the base of the convection zone in the form of azimuthal (east-west) flux tubes. The pressure inside these azimuthal flux tubes is significantly less than the ambient pressure in the convective zone, essentially because the Lorentz forces in a flux tube point radially outward. The density inside the tubes is correspondingly less and so the tubes experience a buoyancy force which tends to propel them towards the photosphere (the surface of the sun). This force is strongest in the thickest flux tubes, parts of which become convectively unstable and drift upward, with a rise time

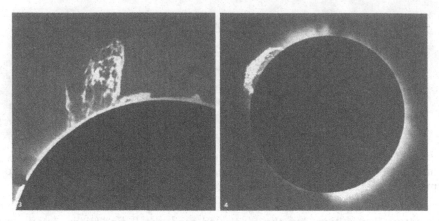

Figure 5.3 Magnetic activity in the solar atmosphere (Encyclopaedia Britannica): (a) an eruptive prominence; (b) a quiescent prominence.

of perhaps a month. Periodically, then, flux tubes of diameter $\sim 10^4$ km burst through the surface into the sun's atmosphere (Figure 5.3b). Sunspots are the foot-points where the tubes pierce the photosphere (A and B in Figure 5.2). These foot-points appear dark because the intense magnetic field in the flux tubes (of the order of 3000 Gauss) locally suppresses the fluid motion and convective heat transfer, thus cooling the surface.

This entire phenomenon relies on the magnetic field being (partially) frozen into the plasma. It is this which allows intense east-west fields to form in the first place, and which ensures that the buoyant fluid at the core of a flux tube carries the tube with it as it floats upward. We shall return to this topic in Chapter 15, where we shall see that sunspots are often accompanied by other magnetic phenomena, such as solar flares. A detailed discussion of sunspots and solar flares may be found in Biskamp (1993) and Priest (2014).

5.4 Helicity Invariants in Ideal MHD

5.4.1 Magnetic Helicity

We can take the analogy between $\boldsymbol{\omega}$ and $\mathbf{B}$ yet further. In §3.5.3 we saw that the helicity,

$$h = \int_{V_\omega} \mathbf{u} \cdot \boldsymbol{\omega} dV,$$

is conserved in an inviscid flow. Moreover, this is a direct consequence of the conservation of vortex-line topology which is enforced by Helmholtz's laws. We would expect, therefore, that the magnetic helicity,

$$h_m = \int_{V_B} \mathbf{A} \cdot \mathbf{B} \, dV, \tag{5.7}$$

will be conserved in an ideal conductor as a consequence of Alfvén's theorem. ($\mathbf{A}$ is the vector potential defined via (2.29).) This is readily confirmed. First we uncurl (5.5) to give

$$\frac{\partial \mathbf{A}}{\partial t} = \mathbf{u} \times \mathbf{B} + \nabla \varphi, \tag{5.8}$$

where φ is a scalar defined by the divergence of (5.8). From (5.8) and (5.5) we have

$$\frac{\partial}{\partial t}(\mathbf{A} \cdot \mathbf{B}) = \nabla \cdot (\varphi \mathbf{B}) + \mathbf{A} \cdot [\nabla \times (\mathbf{u} \times \mathbf{B})],$$

which, with the help of the vector relationship,

$$\nabla \cdot [\mathbf{A} \times (\mathbf{B} \times \mathbf{u})] = \nabla \cdot [(\mathbf{A} \cdot \mathbf{u})\mathbf{B} - (\mathbf{A} \cdot \mathbf{B})\mathbf{u}] = \mathbf{A} \cdot \nabla \times [\mathbf{u} \times \mathbf{B}],$$

becomes

$$\frac{D}{Dt}(\mathbf{A} \cdot \mathbf{B}) = \nabla \cdot [(\varphi + \mathbf{A} \cdot \mathbf{u})\mathbf{B}]. \tag{5.9}$$

We now integrate (5.9) over a material volume V_B which always consists of the same fluid particles (each of volume δV) and for which $\mathbf{B} \cdot d\mathbf{S} = 0$ on the bounding surface. Remembering that $D(\delta V)Dt = 0$ in an incompressible fluid, we obtain,

$$\frac{d}{dt} \int_{V_B} (\mathbf{A} \cdot \mathbf{B}) dV = 0,$$

as required.

As with the helicity of a vorticity field, this conservation law is topological in nature. It stems from the fact that interlinked flux tubes in an ideal conductor remain linked for all time (Figure 5.4), conserving their relative topology as well as their individual fluxes. (See §3.5.3 for the topological interpretation of kinetic helicity.)

For example, consider two thin, interlinked flux tubes with centre lines C_1 and C_2, carrying fluxes Φ_1 and Φ_2. Since $\mathbf{B}dV = \Phi d\mathbf{l}$ for a short portion of a flux tube, we have

$$h_m = \oint_{C_1} \mathbf{A} \cdot (\Phi_1 d\mathbf{l}) + \oint_{C_2} \mathbf{A} \cdot (\Phi_2 d\mathbf{l}) = \Phi_1 \oint_{C_1} \mathbf{A} \cdot d\mathbf{l} + \Phi_2 \oint_{C_2} \mathbf{A} \cdot d\mathbf{l} = \pm 2\Phi_1 \Phi_2,$$

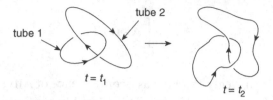

Figure 5.4 Interlinked flux tubes.

with the $+\,(-)$ sign corresponding to a right-handed (left-handed) linkage of the flux tubes. Since Φ_1, Φ_2 and the linkage of the tubes are all conserved, h_m is an invariant.

5.4.2 Minimum Energy States

We now show that minimising magnetic energy subject to the conservation of magnetic helicity leads to the *force-free* field $\nabla \times \mathbf{B} = \alpha \mathbf{B}$. Consider the ratio

$$g = \int \mathbf{B}^2 dV \Big/ \int \mathbf{A} \cdot \mathbf{B} dV,$$

which we wish to minimise. If we consider infinitesimal changes to $\mathbf{B}$ about $g = g_0$, then we have, to leading order,

$$2\int \mathbf{B} \cdot \delta \mathbf{B} dV = \delta g \int \mathbf{A} \cdot \mathbf{B} dV + g_0 \int [\mathbf{A} \cdot \delta \mathbf{B} + \mathbf{B} \cdot \delta \mathbf{A}] dV.$$

Thus stationary values of g correspond to

$$2\int \mathbf{B} \cdot \delta \mathbf{B} dV = g_0 \int [\mathbf{A} \cdot \delta \mathbf{B} + \mathbf{B} \cdot \delta \mathbf{A}] dV = 2g_0 \int \mathbf{A} \cdot \delta \mathbf{B} dV,$$

where we have taken the surface integral of $\mathbf{A} \times \delta \mathbf{A}$ to be zero. We conclude that

$$\int (\mathbf{B} - g_0 \mathbf{A}) \cdot \delta \mathbf{B} dV = 0,$$

and since this holds for all possible choices of $\delta \mathbf{B}$, stationary values of g must correspond to $\mathbf{B} - g_0 \mathbf{A} = 0$. This, in turn, corresponds to the force-free field $\nabla \times \mathbf{B} = \alpha \mathbf{B}$. A little more effort is required to show that this is a minimum in $|g|$, and hence a minimum energy state. In Chapter 9 we shall see that a dissipative relaxation to a force-free magnetic field is a common process in MHD turbulence.

5.4.3 Cross Helicity

There is one other topological invariant of ideal (i.e. diffusionless) MHD:

$$h_c = \int \mathbf{u} \cdot \mathbf{B} dV = \text{constant}. \tag{5.10}$$

Like h_m, this is an invariant whenever $\mathbf{B}$ is localised in space and the integral encloses the magnetic field. The proof of (5.10) proceeds as follows. From (5.1) and the Euler equation we have

$$\frac{D\mathbf{B}}{Dt} = \mathbf{B} \cdot \nabla \mathbf{u},$$

$$\rho \frac{D\mathbf{u}}{Dt} = -\nabla p + \mathbf{J} \times \mathbf{B},$$

from which

$$\frac{D}{Dt}(\mathbf{u} \cdot \mathbf{B}) = \nabla \cdot [(u^2/2 - p/\rho)\mathbf{B}].$$

If we now integrate over any volume which encloses $\mathbf{B}$, we recover (5.10). The invariant h_c is called the *cross helicity* of $\mathbf{u}$ and $\mathbf{B}$, and it plays an important role in astrophysical turbulence, as discussed in Chapter 9.

The physical significance of cross helicity can be exposed by considering a thin isolated vortex tube which interlinks with (but does not overlap) a thin isolated flux tube, as shown in Figure 5.4. Just as h tells us about the degree of linkage of two vortex tubes (see §3.5.3) and h_m tells us about the degree of linkage of two magnetic flux tubes (see §5.4.1), so h_c is a measure of the linkage of the vortex and flux tubes, which is conserved in an ideal fluid. (The proof of this statement is left as an exercise for the reader.)

5.5 Advection Plus Diffusion

We now consider the combined effects of diffusion and advection. For simplicity we focus on two-dimensional flows in which there is no flux-tube stretching. In such cases it is convenient to work with the vector potential $\mathbf{A}$, rather than $\mathbf{B}$. Suppose that $\mathbf{u} = (\partial\psi/\partial y, -\partial\psi/\partial x, 0)$ and $\mathbf{B} = (\partial A/\partial y, -\partial A/\partial x, 0)$ where ψ is the stream function for $\mathbf{u}$, $\mathbf{u} = \nabla \times (\psi\hat{\mathbf{e}}_z)$, and A is the analogous *flux function* for $\mathbf{B}$, $\mathbf{B} = \nabla \times (A\hat{\mathbf{e}}_z)$. Then the induction equation (5.1) uncurls to give $\partial\mathbf{A}/\partial t = \mathbf{u} \times \mathbf{B} + \lambda\nabla^2\mathbf{A}$, from which

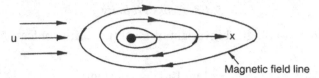

Figure 5.5 The magnetic field induced by a current-carrying wire in a cross-flow.

$$\frac{DA}{Dt} = \lambda \nabla^2 A. \qquad (5.11)$$

Note that the contours of constant A represent magnetic field lines. Also, as noted in §4.1, $R_m = \mu \sigma u \ell = u\ell/\lambda$ is a measure of the relative strengths of advection and diffusion.

5.5.1 Field Sweeping

Now A is transported just like heat, c.f. (5.4). Let us start, therefore, with a problem which is analogous to a heated wire in a cross flow, as this example was discussed at some length in §3.4. The equivalent MHD problem is sketched in Figure 5.5. We have a thin wire carrying a current I (directed into the page) which sits in a uniform cross flow, **u**. The magnetic field lines surrounding the wire are swept downstream by **u**, just like the isotherms in Figure 3.11. In the steady state (5.11) can be written as

$$u\frac{\partial A}{\partial x} = \lambda \nabla^2 A, \qquad (5.12)$$

where the origin is taken as the centre of the wire and x is in the stream-wise direction.

Crucially, there is no natural length scale for this problem. (The wire is considered to be vanishingly thin.) The only way of constructing a magnetic Reynolds number is to use position, say the radial coordinate r, as the characteristic length scale. Thus we have $R_m = \mu \sigma u r = ur/\lambda$. Near the wire, therefore, we will have a diffusion-dominated regime ($R_m \ll 1$) while at large distances from the wire $R_m \gg 1$) advection of **B** will dominate. It turns out that Equation (5.12) may be solved by looking for solutions of the form $A = f(x, y)\exp(ux/2\lambda)$. This yields $\nabla^2 f = (u/2\lambda)^2 f$, and so the solution for A is

$$A = CK_0(ur/2\lambda)\exp(ux/2\lambda),$$

where K_0 is the zero-order modified Bessel function normally denoted by K. The constant C may be determined by matching this expression to the diffusive solution $A = (\mu I/2\pi)\ln(1/r) + $ constant, which dominates for $r \to 0$. This gives $C = \mu I/2\pi$. (The details are spelt out in Shercliff, 1965.) The shape of the field lines is as shown in Figure 5.5. They are identical to the isotherms in the analogous thermal problem.

5.5.2 Flux Expulsion

We now consider another example of combined advection and diffusion. This is a phenomenon called *flux expulsion* which, from the mathematical point of view, is nothing more than the Prandtl–Batchelor theorem applied to **B** rather than **ω**. Suppose that we have a steady, two-dimensional flow consisting of a region of closed streamlines of size ℓ, and that $R_m = u\ell/\lambda$ is large. Then we may show that any magnetic field which lies within that region is gradually expelled (see Figure 5.6). The proof is essentially the same as that for the Prandtl–Batchelor theorem. In brief, the argument goes as follows. We have seen that A satisfies an advection diffusion equation, just like vorticity. When R_m is large we find A is almost constant along the streamlines. However, a small but finite diffusion slowly eradicates cross-stream gradients in A until it is perfectly uniform, giving **B** = **0**. We now work through the details, starting with the high-R_m equation $DA/Dt \approx 0$.

In the steady-state $DA/Dt = 0$ simplifies to $\mathbf{u} \cdot \nabla A = 0$ which in turn implies $A = A(\psi)$. That is, A is constant along the streamlines so that **B** and **u** are co-linear. Now suppose that λ is extremely small, but nevertheless finite. The steady version of (5.11), $\mathbf{u} \cdot \nabla A = \lambda \nabla^2 A$, yields the integral equation

$$I = \lambda \int_{V_\psi} \nabla^2 A \ dV = 0, \tag{5.13}$$

where V_ψ is the volume enclosed by a closed streamline. Now (5.13) must hold true for any finite value of λ, and in particular it remains valid when λ is very small, so that $A \approx A(\psi)$. Let us now explore the consequences of the integral constraint (5.13) for our high-R_m flow. We have, using Gauss' theorem,

$$I = \lambda \int_{V_\psi} \nabla^2 A \ dV = \lambda \int_{S_\psi} \nabla A \cdot d\mathbf{S} = \lambda A'(\psi) \int_{S_\psi} \nabla \psi \cdot d\mathbf{S},$$

where $A'(\psi)$ is the cross-stream gradient of A, which is constant on the surface S_ψ. However, the integral on the right is readily evaluated. We use Gauss' and Stokes' theorems as follows,

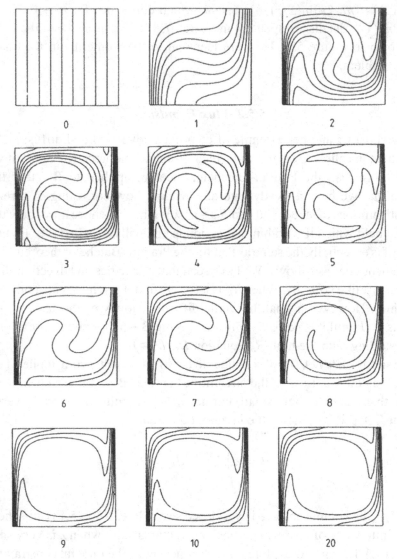

Figure 5.6 An example of flux expulsion in a square at $R_m = 100$, based on computations by N.O. Weiss. The figures show the distortion of an initially uniform field by a clockwise eddy. (From Galloway and Weiss, 1981 © AAAS. Reproduced with permission.)

$$\int_{S_\psi} \nabla\psi \cdot d\mathbf{S} = \int_{V_\psi} \nabla^2\psi dV = -\int_{V_\psi} \omega dV = -\oint_{C_\psi} \mathbf{u} \cdot d\mathbf{l},$$

where C_ψ is the streamline which defines S_ψ. Evidently our integral constraint demands

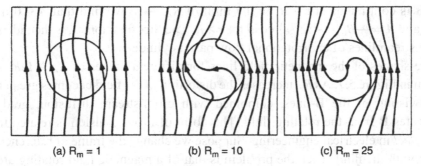

(a) $R_m = 1$ (b) $R_m = 10$ (c) $R_m = 25$

Figure 5.7 Distortion and expulsion of a magnetic field by differential rotation. (From Moffatt, 1978, with permission.)

$$I = -\lambda A'(\psi) \oint_{C_\psi} \mathbf{u} \cdot d\mathbf{l} = 0. \tag{5.14}$$

Again it is emphasised that this holds true no matter how small we make λ. It is only necessary that λ be finite. Now it follows from (5.14) that $A'(\psi) = 0$, since the line integral cannot be zero. We conclude, therefore, that in a region of closed streamlines with a high value of R_m, the flux function is constant. It follows that $\mathbf{B} - \mathbf{0}$. This phenomenon is known as flux expulsion.

An example of this process is shown in Figure 5.7. A magnetic field $\mathbf{B} = B_0 \hat{\mathbf{e}}_y$ pervades a conducting fluid and a region of this fluid, $r < R$, is in a state of rigid body rotation, the remainder being quiescent. This rotation distorts $\mathbf{B}$ and the distortion is readily calculated. Let Ω be the angular velocity of the fluid. In the steady state we have

$$\frac{\Omega}{\lambda} \frac{\partial A}{\partial \theta} = \nabla^2 A, \quad 0 < r < R,$$

with $\nabla^2 A = 0$ for $r > R$. It is natural to look for solutions of the form $A = f(r)\exp(j\theta)$, where we extract only the real part of A. This yields

$$\frac{j\Omega}{\lambda} f(r) = \left(\frac{1}{r} \frac{d}{dr} r \frac{d}{dr} - \frac{1}{r^2} \right) f, \quad 0 < r < R.$$

The solution for f is then,

$$f = -B_0 r + C/r, \quad r > R,$$
$$f = D J_1(pr), \quad 0 < r < R,$$

where C and D are constants, J_1 is the usual first-order Bessel function, and $p = (1 - j)(\Omega/2\lambda)^{1/2}$. The unknown constants can be evaluated from the condition that $\mathbf{B}$ is continuous at $r = R$. In the limit of $R_m \to \infty$ we find that $A = 0$ inside $r = R$ and $A = -B_0 \left(r - R^2/r \right) \cos\theta$ for $r > R$. The flux function, A, is then identical to the streamlines of an irrotational flow past a cylinder.

The shapes of the magnetic field lines for different values of $R_m = \Omega R^2/\lambda$ are shown in Figure 5.7. As R_m increases the distortion of the field becomes greater and this twisting of the $\mathbf{B}$-lines, combined with cross-stream diffusion, gradually eradicates $\mathbf{B}$ from the rotating fluid. This form of flux expulsion is related to the *skin effect* in electrical engineering. Suppose we change the frame of reference and rotate with the fluid. Then the problem is that of a magnetic field rotating around a stationary conductor. In such a case it is well known that the field will penetrate only a finite distance, $\delta = \sqrt{2\lambda/\Omega}$, into the conductor. This distance is known as the skin depth. As $\Omega \to \infty$, the field is excluded from the interior of the conductor.

5.5.3 Azimuthal Field Generation by Differential Rotation: the Ω-Effect

Our next example of combined advection and diffusion is axisymmetric rather than planar. It is mainly of interest in astrophysics and concerns a differentially rotating fluid permeated by a magnetic field. It happens that the interior of planets or stars do not always rotate as a rigid body. Our own sun, for example, exhibits a variation of surface rotation rate with latitude. It turns out that this differential rotation has a profound influence on the structure of the planetary or stellar magnetic field, as we now show.

Consider a non-uniformly rotating star which possesses an axisymmetric, poloidal magnetic field, i.e. a field of the form $\mathbf{B}_p(r, z) = (B_r, 0, B_z)$ in (r, θ, z) coordinates. Suppose the star rotates faster at the equator than at its poles, with velocity $\mathbf{u} = (0, \Omega(z) r, 0)$. By Alfvén's theorem, the poloidal field lines will tend to be advected as shown in Figure 5.8. The field lines will bow out until such time as the diffusion created by the distortion is large enough to counter the effects of field sweeping.

This is readily seen from the azimuthal component of the steady induction equation

$$\frac{\partial \mathbf{B}_\theta}{\partial t} = \nabla \times (\mathbf{u} \times \mathbf{B}_p) + \lambda \nabla^2 \mathbf{B}_\theta = 0,$$

where $\mathbf{B}_p$ is the poloidal magnetic field. The source of azimuthal field is the term $\nabla \times (\mathbf{u} \times \mathbf{B}_p)$, which may be rewritten as $[r(\mathbf{B}_p \cdot \nabla)\Omega]\hat{\mathbf{e}}_\theta$, making explicit the role played by $\Omega(z)$ in generating the azimuthal field. Note that, if λ is very small, then extremely large azimuthal fields may be generated by this mechanism, of order $R_m|\mathbf{B}_p|$. This is a key process in many theories relating to solar MHD, such as the origin of sunspots and of the solar dynamo.

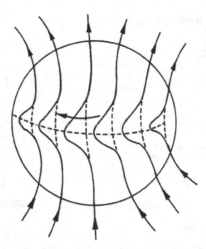

Figure 5.8 Distortion of the magnetic field lines by differential rotation.

5.5.4 Stretched Flux Tubes and Current Sheets

We now consider the magnetic analogue of the Burgers vortex (as described in §3.6.2). Consider the magnetic flux tube

$$\mathbf{B} = \frac{\Phi}{\pi \delta^2} \exp[-r^2/\delta^2] \, \hat{\mathbf{e}}_z, \quad \delta = \delta(t), \tag{5.15}$$

where Φ is the net magnetic flux in the tube (a constant), δ is the characteristic radius of the tube, and we are using cylindrical polar coordinates, (r, θ, z). Suppose this flux tube sits in an axisymmetric, irrotational strain field, $u_r = -\frac{1}{2}\alpha r$ and $u_z = \alpha z$, as shown in Figure 5.9(a), where $\alpha(t)$ is an imposed, time-dependent strain. This represents an exact solution of (5.1), provided that δ satisfies

$$\frac{d\delta^2}{dt} + \alpha(t)\delta^2 = 4\lambda. \tag{5.16}$$

(Compare this with Equation (3.52) for a Burgers vortex.) For the particular case where the imposed strain α is steady, the solution of (5.16) is

$$\delta^2 = \delta_0^2 e^{-\alpha t} + \delta_\infty^2 [1 - e^{-\alpha t}],$$

where $\delta_0 = \delta(0)$ and $\delta_\infty = \sqrt{4\lambda/\alpha}$. Evidently, the characteristic tube radius eventually settles down to a steady value of $\sqrt{4\lambda/\alpha}$, irrespective of its initial value, and once $\delta = \delta_\infty$ we have an exact balance between inward convection, outward diffusion, and axial stretching of the flux tube.

There is a Cartesian analogue of (5.15) in the form of the flux sheet

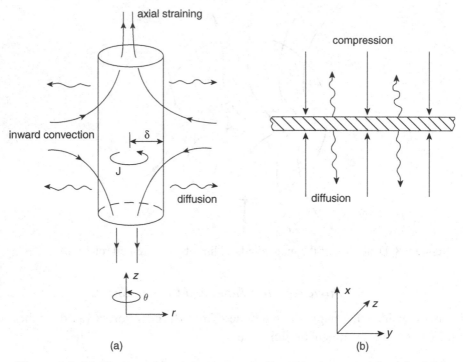

Figure 5.9 (a) The stretching of a magnetic flux tube by an axial strain field. (b) The compression of a current sheet by a biaxial strain field.

$$\mathbf{B} = \frac{\Phi}{\sqrt{\pi}\delta} \exp[-x^2/\delta^2]\,\hat{\mathbf{e}}_y, \quad \delta = \delta(t), \tag{5.17}$$

subject to the irrotational biaxial strain field $(u_x, u_y, 0) = (-\alpha x, \alpha y, 0)$, as shown in Figure 5.9(b). This is also an exact solution of (5.1), in this case governed by

$$\frac{d\delta^2}{dt} + 2\alpha\delta^2 = 4\lambda. \tag{5.18}$$

When α is steady, the flux sheet invariably reaches the equilibrium thickness $\delta = \sqrt{2\lambda/\alpha}$. Note that, while we have labelled this as a flux sheet, we could equally think of it as a current sheet.

With one eye to the topic of the next section, it is of some interest to temporarily set aside kinematics and consider the force balance within such a current sheet. The current density is in the z direction and given by $\mu J_z = \partial B_y/\partial x$, and consequently the Lorentz force per unit mass is simply

$$\mathbf{F} = -\left(B_y J_z / \rho\right)\hat{\mathbf{e}}_x = -\nabla\left(B_y^2 / 2\rho\mu\right).$$

Now $\mathbf{u}$ is irrotational and so we may write $\mathbf{u} = \nabla\varphi$, and the Navier–Stokes equation reduces to

$$\frac{\partial}{\partial t}\nabla\varphi = -\nabla\left[\frac{1}{2}\mathbf{u}^2 + \frac{p}{\rho} + \frac{B_y^2}{2\rho\mu}\right].$$

Evidently, in such a sheet we have

$$\frac{\partial\varphi}{\partial t} + \frac{1}{2}\mathbf{u}^2 + \frac{p}{\rho} + \frac{B_y^2}{2\rho\mu} = 0, \tag{5.19}$$

which is just Bernoulli's equation with the fluid pressure supplemented by the magnetic pressure. We shall return to this energy equation shortly.

5.5.5 Magnetic Reconnection

Finally, we consider the role played by a small but finite diffusivity in the reconnection of magnetic flux tubes. Such reconnections occur through the transient formation of thin current sheets which facilitate rapid diffusion in small, localised regions of space. Fast magnetic reconnections are particularly important in the triggering of solar flares, where the nominal value of R_m is very large, yet the snapping and reconnection of magnetic field lines are crucial events. Such reconnections allow the magnetic field to relax to a lower energy state and so liberate vast amounts of stored magnetic energy on a short time scale.

By way of an example of reconnection, consider a flux tube in the form of a ring which, at $t = 0$, sits in a differentially rotating fluid, as shown in Figure 5.10. When the two branches of the tube come into contact, the field lines locally compress and the gradients in $\mathbf{B}$ become large, leading to a local current sheet. Eventually the gradients become so large that, despite the smallness of λ, significant diffusion sets in. The result is that the magnetic field lines reconnect, forming two smaller flux tubes.

Models of magnetic reconnection tend to focus on current sheets, which are usually assumed to take a form similar to that shown in Figure 5.9(b), in which the field lines are pushed together by an external strain field, say the biaxial strain $(u_x, u_y, 0) = (-\alpha x, \alpha y, 0)$ if the flow is two-dimensional. The main difference is that, in order to get reconnection, the direction of the magnetic field is taken to be antisymmetric about the mid-plane of the current sheet (see Figure 5.11), say

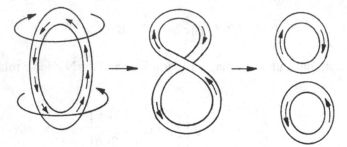

Figure 5.10 The severing of a flux tube.

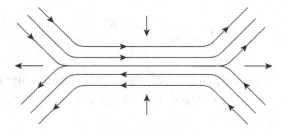

Figure 5.11 The reconnection of magnetic field lines through a planar current sheet.

$$\mathbf{B} = B_0 \frac{x}{\ell} \exp[-x^2/\ell^2]\, \hat{\mathbf{e}}_y.$$

The simplest model of such a current sheet is known as the Sweet–Parker model, and assumes that the sheet has a finite horizontal width, $2W$, with $W \gg \ell$, and that the fluid pressure near the top and bottom of the sheet, say at $x = \pm\ell$, is the same of as that at the edges of the sheet, $y = \pm W$. We then apply (5.19) to a streamline that enters the sheet near the centreline $y = 0$ and leaves near the mid-plane $x = 0$, say from the point $x = \ell$, $y = 0$ to $x = 0$, $y = W$. Since we have assumed negligible change in fluid pressure, we have $(\alpha\ell)^2 + B_0^2/\rho\mu \sim (\alpha W)^2$, or else $B_0^2/\rho\mu \sim (\alpha W)^2$, since $W \gg \ell$. If we label the inlet and exit velocities as u_i and u_e, then we have

$$\alpha \sim u_i/\ell \sim u_e/W, \quad u_e \sim v_a, \quad \ell \sim \sqrt{\lambda/\alpha},$$

where v_a is the Alfvén velocity based on B_0. These combine to yield

$$\frac{u_i}{v_a} \sim \frac{\ell}{W} \sim \sqrt{\frac{\lambda}{v_a W}} = (\mathrm{Lu})^{-1/2}, \qquad (5.20)$$

where Lu is the Lundquist number, first introduced in §4.1.

Of course this is a very crude analysis, but it does capture the essential features of a planar current sheet and it has provided the starting point for more realistic models. Actually, (5.20) substantially overestimates the reconnection time observed in solar flares. In the solar corona Lu may be as large as 10^{10}, so (5.20) estimates $u_i \sim 10^{-5} v_a$, which is much slower than the observed reconnection rates. More realistic models of reconnection are described in Biskamp (1993) and Priest (2014).

5.6 Field Generation by Flux-Tube Stretching: a Glimpse at Dynamo Theory

We close with a brief introduction to a topic which dominates Chapter 14: that of dynamo theory. This concerns the spontaneous amplification and long-term maintenance of a large-scale magnetic field within a conducting fluid through the continual stretching and twisting of the flux tubes. That is, the fluid converts mechanical energy into magnetic energy by continually stretching the magnetic field lines. In kinematic dynamo theory one takes the fluid velocity as specified and then asks if the prescribed flow will amplify a seed magnetic field, an approach which fits well within the framework of this chapter. The altogether more difficult question of the back reaction of the magnetic field on the fluid, which causes dynamos to saturate, is left until Chapter 14.

Of course, dynamo action is of crucial importance in astrophysics, where it is responsible for the maintenance of stellar and planetary magnetic fields. Consequently, most astrophysical dynamo theory is developed within the context of closed spherical fluid domains. However, perhaps the simplest examples of laminar fluid dynamos occur in other geometries, particularly those cases in which the conducting fluid is allowed to extend to infinity in one or more directions. So in this section we consider the simple case of helical flow in a long, straight tube, leaving the more difficult spherical geometry for Chapter 14.

Perhaps the simplest known example of a kinematic dynamo is that of Ponomarenko, which consists of a swirling pipe flow of diameter d embedded in an otherwise stationary conducting medium of the same electrical conductivity. The velocity field is helical and takes the form $\mathbf{u} = (0, \Omega r, V)$ in (r, θ, z) coordinates, Ω and V being constants. For such a simple flow the induction equation admits separable solutions of the type $\mathbf{B} \sim \exp[j(kz + m\theta + \varpi t)]$, and the helical pitch most favourable to dynamo action turns out to be $V = 0.657\Omega d$. The resulting eigenvalue problem yields growing solutions whenever $R_m = u_{\max} d/\lambda$ exceeds a value of $R_m = 35.4$, where $u_{max} = (V^2 + \Omega^2 d^2/4)^{1/2}$. The growing solutions are asymmetric, despite the symmetry of the base flow, with the marginally unstable mode corresponding to $m = 1$ and an axial wavenumber of $kd = -0.775$. The growing field takes the form of spiralled flux tubes concentrated near $r = d/2$ and propagating along the z axis.

This simple example of a laminar dynamo was first discovered in 1973 and subsequently generalised to include velocity profiles which are smooth functions of radius and satisfy the no-slip boundary condition at $r = d/2$. (Such profiles avoid the discontinuity in velocity at $r = d/2$.) Although there were earlier examples of laminar kinematic dynamos, this simple flow is particularly easy to analyse and so remains popular. For example, it was the basis for the first successful laboratory demonstration of dynamo action, which occurred at the Latvian Institute of Physics in Riga some 27 years after Ponomarenko first published his simple analytical dynamo.

This model problem exhibits a number of features which turn out to be common in most natural dynamos. First, note that the resulting magnetic field is asymmetric, despite the axial symmetry of the base flow. It turns out that strictly axisymmetric dynamos are not possible, a result known as Cowling's theorem. Second, note that the flow is helical, which is a recurring theme of Chapter 14. While helicity is not strictly necessary for dynamo action, nearly all working dynamos are helical. Finally, we observe that a dynamo occurs only when R_m exceeds a few multiples of 10. Similar limits on R_m exist in other geometries, including spherical fluid domains.

Exercises

5.1 A semi-infinite region of conducting material is subject to mutually perpendicular electric and magnetic fields of frequency ω at, and parallel to, its plane boundary. There are no fields deep inside the stationary conductor. Derive expressions for the variation of amplitude and phase of the magnetic field as a function of distance from the surface.

5.2 A perfectly conducting fluid undergoes an axisymmetric motion and contains an azimuthal magnetic field $\mathbf{B}_\theta$. Show that B_θ/r is conserved by each fluid element.

5.3 An electromagnetic flow meter consists of a circular pipe under a uniform transverse magnetic field. The voltage induced by the fluid motion between electrodes, placed at the ends of a diameter of the pipe perpendicular to the field, is used to indicate the flow rate. The pipe walls are insulated and the flow axisymmetric. Show that the induced voltage depends only on the total flow rate and not on the velocity profile.

5.4 A perfectly conducting, incompressible fluid is deforming in such a way that the magnetic field lines are being stretched with a rate of strain S. Show that the magnetic energy rises at a rate SB^2/μ per unit volume.

5.5 Fluid flows with uniform velocity passed an insulated, thin, flat plate containing a steady current sheet orientated perpendicular to the velocity. The intensity of the current sheet varies sinusoidally with the streamwise coordinate and the electric field in the fluid is zero. Find the form of the magnetic field and show that it is confined to boundary layers when R_m is large.

6

Dynamics at Low Magnetic Reynolds Numbers

It was perhaps for the advantage of science that Faraday, though thoroughly conscious of the forms of space, time and force, was not a professed mathematician. He was not tempted to enter into the many interesting researches in pure mathematics... and he did not feel called upon either to force his results into a shape acceptable to the mathematical taste of the time, or to express them in a form which the mathematicians might attack. He was thus left to his proper work, to coordinate his ideas with his facts, and to express them in natural, uncomplicated language.

Maxwell, 1873

In Chapter 5 we looked at the effect of a prescribed fluid motion on a magnetic field, without worrying about the back reaction of **B** on **u**. We now consider the reverse problem, in which **B** influences **u** (via the Lorentz force) but **u** does not significantly perturb **B**. In short, we look at the effect of a prescribed magnetic field on the flow. To ensure that **B** remains unaffected by **u** we must restrict ourselves to low magnetic Reynolds numbers:

$$R_m = u\ell/\lambda = \mu\sigma u\ell \ll 1. \tag{6.1}$$

However, this is not overly restrictive, at least not in the case of liquid-metal MHD. For example, in most laboratory experiments, or industrial processes, we have $\lambda \sim 1 \text{ m}^2/\text{s}$, $\ell \sim 0.1$ m and internal friction keeps **u** to a level of around 0.01 m/s $\rightarrow$ 1 m/s. This gives $R_m \sim 0.001 \rightarrow 0.1$. The primary exception to this is the geodynamo, where motion in the earth's core spans the range of scales $1 < R_m < 600$, as discussed in Chapter 14. We also note in passing that the kinematic viscosity, v, of liquid metal is similar to that of water and somewhat less than that of air, so the Reynolds numbers of most liquid-metal flows is very high.

Now a magnetic field can alter **u** in one of three ways. It can suppress bulk motion, excite bulk motion, or alter the structure of the boundary layers in some way. We look at the first of these possibilities in §6.2, where we discuss the damping of flows using a static magnetic field. We tackle the second possibility in §6.4, where the effect of a rotating magnetic field is investigated. Finally, we examine MHD boundary layers (so-called Hartmann layers) in §6.5.

We shall see that, because of Joule dissipation, an imposed, static magnetic field tends to dampen out fluid motion, while simultaneously creating a form of aniso-tropy in which the gradients in **u** parallel to **B** are preferentially destroyed. Thus turbulence in the presence of a strong magnetic field becomes quasi-two-dimensional as the eddies elongate in the direction of **B**. Travelling or rotating magnetic fields, on the other hand, tend to induce a motion which reduces the relative speed of the field and fluid. The magnitude of the induced velocity is controlled by a balance between friction and the Lorentz force, the second of which tends to drag the fluid along with the translating or rotating magnetic field. Finally, we shall see that magnetic fields alter the structure of boundary layers, which are now controlled by a competition between the Lorentz force and shear. All in all, it seems that magnetic fields provide a versatile, non-intrusive means of controlling liquid-metal and plasma flows. Let us start, however, with the governing equations of low-R_m MHD.

6.1 The Low-Magnetic Reynolds Number Approximation

The essence of the low-R_m approximation is that the magnetic field associated with the induced current, $\mathbf{J} \sim \sigma\, \mathbf{u} \times \mathbf{B}$, is negligible by comparison with the imposed magnetic field. That is, Ampère's law tells us that the magnetic field associated with the induced current, $\mathbf{J} \sim \sigma\, \mathbf{u} \times \mathbf{B}$, has a magnitude of $|\mathbf{b}| \sim \mu \ell (\sigma u B) \sim R_m B$, and so $|\mathbf{b}| \ll B$ when R_m is small. There are three distinct cases which commonly arise:

(i) The imposed magnetic field is static, the flow is induced by some external agency and friction keeps **u** to a modest level in the sense that $|\mathbf{u}| \ll \lambda/\ell$.

(ii) The imposed magnetic field travels or rotates uniformly and slowly such that $\mathbf{u}_{field} \ll \lambda/\ell$. This induces a flow **u** which, due to friction in the fluid, is somewhat slower than the speed of the field.

(iii) The imposed magnetic field oscillates extremely rapidly, in the sense that the skin-depth $\delta = (2/\mu\sigma\varpi)^{1/2}$ is much less than ℓ, ϖ being the field frequency. The magnetic field is then excluded from the interior of the conductor and inertia or friction in the fluid ensures that $|\mathbf{u}| \ll \varpi\ell$.

Categories (i) through (iii) cover the majority of flows in engineering applications. Typical examples are the magnetic damping of jets, vortices or turbulence (case (i)), magnetic stirring using a rotating magnetic field (case (ii)) and magnetic levitation (case (iii)). In this chapter we focus on cases (i) and (ii). Of course, if the imposed magnetic field travels or rotates in a uniform manner then a suitable change of frame of reference will convert problems of type (ii) into those of type (i). Thus, without loss of generality, in this section we may take **B** to be steady.

We now discuss the simplifications which result in the governing equations when R_m is low and the imposed magnetic field is steady. Let $\mathbf{E}_0$, $\mathbf{J}_0$ and $\mathbf{B}_0$ represent the fields which would exist in a given situation if $\mathbf{u} = 0$, and let **e**, **j** and **b** be the infinitesimal perturbations in **E**, **J** and **B** which occur due to the presence of a weak velocity field. These quantities are governed by

$$\nabla \times \mathbf{E}_0 = 0, \qquad \mathbf{J}_0 = \sigma \mathbf{E}_0, \tag{6.2}$$

$$\nabla \times \mathbf{e} = -\partial \mathbf{b}/\partial t, \qquad \mathbf{j} = \sigma(\mathbf{e} + \mathbf{u} \times \mathbf{B}_0), \tag{6.3}$$

where we have neglected the second-order term, $\mathbf{u} \times \mathbf{b}$, in Ohm's law. Now Faraday's equation tells us that the rotational part of **e** has magnitude $|\mathbf{e}_{rot}| \sim u|\mathbf{b}| \sim R_m u B_0$, while the divergence of Ohm's law tells us that the solenoidal part of **e** is of order $|\mathbf{e}_{sol}| \sim u B_0$. So, to leading order in R_m, the total electric field, $\mathbf{E}_0 + \mathbf{e}$, is irrotational and may be written as $-\nabla V$, where V is an electrostatic potential. Ohm's law now takes the particularly simple form

$$\boxed{\mathbf{J} = \sigma(-\nabla V + \mathbf{u} \times \mathbf{B}_0)}, \tag{6.4}$$

where $\mathbf{J} = \mathbf{J}_0 + \mathbf{j}$. The leading-order term in the Lorentz force (per unit volume) is evidently

$$\boxed{\mathbf{F} = \mathbf{J} \times \mathbf{B}_0}. \tag{6.5}$$

Equations (6.4) and (6.5) are all that we require to evaluate the Lorentz force in low-R_m MHD. There is no need to calculate **b** since it does not appear in the Lorentz force. Moreover, **J** is uniquely determined by (6.4) since

$$\nabla \cdot \mathbf{J} = 0, \qquad \nabla \times \mathbf{J} = \sigma \nabla \times (\mathbf{u} \times \mathbf{B}_0), \tag{6.6}$$

and a vector field is unambiguously determined if its divergence and curl are known (and some suitable boundary conditions are specified). From now on we shall drop the subscript on $\mathbf{B}_0$, on the understanding that **B** represents the imposed magnetic field.

6.2 The Suppression of Motion

6.2.1 Magnetic Damping

There are many industrial and laboratory processes in which an intense, static magnetic field is used to suppress unwanted motion in a liquid metal. For example, in the continuous casting of large steel slabs, an intense DC magnetic field ($\sim 10^4$ Gauss) is commonly used to suppress motion within the mould. Sometimes the motion takes the form of a submerged jet which feeds the mould from above, at other times it takes the form of large vortices. In both cases the aim is to keep the free surface of the liquid quiescent, thus avoiding the entrainment of surface debris. Magnetic damping is also used in the laboratory measurements of chemical and thermal diffusivities, particularly where thermal or solutal buoyancy can disrupt the measurement technique. These examples are discussed in more detail in Chapter 11. Here we present just a glimpse of the possibilities offered by magnetic damping. We shall consider the fluid to be infinite in extent, or else bounded by an electrically insulating surface, S. For simplicity, we neglect the viscous forces and take the imposed magnetic field to be uniform.

Let us start by considering the destruction of mechanical energy via Joule dissipation. To some extent, the mechanism of magnetic damping is clear. Motion across magnetic field lines induces a current. This leads to Joule dissipation and the resulting rise in thermal energy is accompanied by a corresponding fall in kinetic energy. This is evident from (6.4) and (6.5), which give the rate of working of the Lorentz force as

$$(\mathbf{J} \times \mathbf{B}) \cdot \mathbf{u} = -\mathbf{J} \cdot (\mathbf{u} \times \mathbf{B}) = -(\mathbf{J}^2/\sigma) - \nabla \cdot [V\mathbf{J}],$$

while the product of the inviscid equation of motion with $\mathbf{u}$ yields

$$\frac{D}{Dt} \frac{\rho \mathbf{u}^2}{2} = (\mathbf{J} \times \mathbf{B}) \cdot \mathbf{u} - \nabla \cdot [p\mathbf{u}].$$

Combining the two furnishes the energy equation,

$$\frac{d}{dt} \int \frac{1}{2} \rho \mathbf{u}^2 dV = -\frac{1}{\sigma} \int \mathbf{J}^2 dV. \tag{6.7}$$

Thus, as anticipated, Joule dissipation leads to a fall in kinetic energy. However, there are other, more subtle effects associated with magnetic damping. Specifically, the action of a magnetic field is anisotropic. It opposes motion normal to the field lines but leaves motion parallel to $\mathbf{B}$ unopposed. Moreover, as we shall see, vorticity and linear momentum tend to propagate along the field lines by a pseudo-diffusion process. These anisotropic effects can be understood in terms of field sweeping and the resulting Faraday tension in the $\mathbf{B}$-lines, as discussed in §4.3.

For example, consider a jet which is directed at right angles to a uniform magnetic field. Motion across the field lines induces a second, weak, magnetic field. The combined field is then bowed slightly in the direction of $\mathbf{u}$ and the

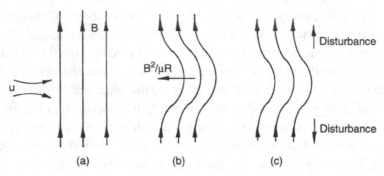

Figure 6.1 Motion across the field lines distorts those lines and the resulting curvature gives rise to a force $B^2/\mu R$ opposing the motion. The disturbance also propagates laterally along the magnetic field lines.

resulting curvature gives rise to a Lorentz force $B^2/\mu R$ which opposes the motion. The tension in the field lines then causes the disturbance to spread laterally along the **B**-lines.

Now all of this is, to say the least, a little heuristic. However, a couple of simple examples will help establish the general ideas. We start with the jet shown in Figure 6.1.

6.2.2 The Damping of a Two-Dimensional Jet

The Lorentz force per unit mass acting on the jet shown in Figure 6.1 is, from (6.4),

$$\mathbf{F} = (\mathbf{J} \times \mathbf{B})/\rho = -\mathbf{u}_\perp/\tau - \nabla \times (\sigma V \mathbf{B}/\rho). \tag{6.8}$$

Here $\mathbf{u}_\perp$ represents the velocity components normal to **B** and τ is the magnetic damping time, $\tau = \left(\sigma B^2/\rho\right)^{-1}$. Note the anisotropic nature of this force. Pressure forces and the effect of V apart, each fluid particle decelerates on a time scale of τ, according to

$$\frac{D\mathbf{u}_\perp}{Dt} \sim -\frac{\mathbf{u}_\perp}{\tau}, \qquad \frac{D\mathbf{u}_{//}}{Dt} \sim 0.$$

It is as if each element of fluid which tries to cross a magnetic field line experiences a frictional drag. As a simple example, consider a thin, steady, two-dimensional jet, $\mathbf{u}(x,y) = (u_x, u_y, 0)$, directed along the x-axis and passing through a uniform field imposed in the y-direction. This geometry is particularly easy to handle since both the pressure p and potential V are zero (or constant), as we now show. The divergence of Ohm's law gives

$$\nabla^2 V = \nabla \cdot (\mathbf{u} \times \mathbf{B}) = \mathbf{B} \cdot \boldsymbol{\omega} = 0, \tag{6.9}$$

and so V is zero, provided there is no electrostatic field imposed from the boundaries. (We exclude such cases.) The induced current, $\mathbf{J} = \sigma\mathbf{u} \times \mathbf{B}$, is then directed along the z-axis and the Lorentz force, $\mathbf{J} \times \mathbf{B} = -\sigma u_x B^2 \hat{\mathbf{e}}_x$, acts to retard the flow. Moreover, the fluid surrounding the jet is quiescent and so $\nabla p = 0$ outside the jet. Provided the jet is thin, in the sense that its characteristic thickness, δ, is much less than a characteristic axial length scale, ℓ, then ∇p is also negligible within the jet. (If the streamlines are virtually straight and parallel, then there can be no significant pressure gradients normal to the streamlines.) In this simple example, then, both the pressure forces and $\nabla V \times \mathbf{B}$ are zero. It follows that

$$\mathbf{u} \cdot \nabla u_x = -u_x/\tau. \tag{6.10}$$

Equation (6.10) is readily solved. We look for a similarity solution of the form $u_x = u_0(x)f(y/\delta(x))$, $u_0\delta^2 = $ constant, where u_0 is the velocity on the x-axis (the centreline of the jet). Then (6.10) applied to the axis gives $u'_0(x) = -1/\tau$. Next we find u_y using continuity, evaluate $\mathbf{u} \cdot \nabla u_x$, and substitute for this term in (6.10). This yields,

$$f^2 - \frac{1}{2}f'(\eta)\int_0^\eta f d\eta = f, \qquad \eta = y/\delta,$$

which has solution $f = \text{sech}^2(\eta)$. Thus the velocity distribution in the jet is

$$u_x = [U - x/\tau]\text{sech}^2(y/\delta), \tag{6.11}$$

where U is the initial centreline velocity. The most striking feature of this solution is that the jet is destroyed within a finite distance, $L = U\tau$. The situation is as shown in Figure 6.2(a). Note that our solution ceases to be valid as we approach $x = U\tau$, since δ is not small there.

We shall return to the topic of MHD jets in Chapter 11, where we look at more complex flows. Interestingly, it turns out that the picture above is quite misleading when it comes to three-dimensional jets. In fact, a three-dimensional jet maintains its linear momentum and so cannot come to a halt. It has the shape shown in Figure 6.2(b).

6.2.3 The Damping of a Vortex

Let us now consider a second example, designed to bring out the tendency for vorticity to diffuse along the magnetic field lines. As before, we take $\mathbf{B}$ to be uniform. This time, however, we consider the initial velocity field to be an

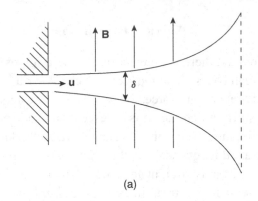

(a)

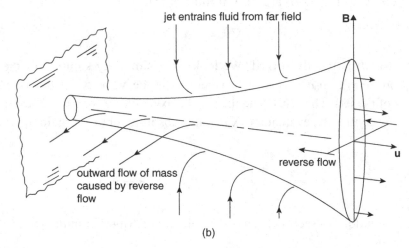

(b)

Figure 6.2 (a) A two-dimensional jet is destroyed by a magnetic field within a distance $U\tau$. (b) By contrast, a three-dimensional MHD jet maintains its linear momentum flux.

axisymmetric, swirling vortex, $\mathbf{u} = (0, \Gamma/r, 0)$ in (r, θ, z) coordinates. $\mathbf{B}$ is taken to be parallel to the z-axis. At $t = 0$ the angular momentum per unit mass, $\Gamma(r, z)$, is assumed to be confined to a spherical region of characteristic scale δ, as shown in Figure 6.3.

Now the axial gradients in Γ will, via the centrifugal force, tend to induce a poloidal component of motion, $\mathbf{u}_p = (u_r, 0, u_z)$. That is, if Γ is a function of z then the centripetal force, $(\Gamma^2/r^3)\hat{\mathbf{e}}_r$, is rotational and cannot be balanced by a radial pressure gradient. A secondary, poloidal motion then results, which complicates the problem. However, in the interests of simplicity, we shall take $\mathbf{J} \times \mathbf{B} \gg \mathbf{u} \cdot \nabla\mathbf{u}$, which is equivalent to specifying that the magnetic damping time, τ, is much less than the inertial timescale δ/u_θ. Since poloidal motion grows on a timescale of δ/u_θ, we may then neglect $\mathbf{u}_p$ for times of order τ.

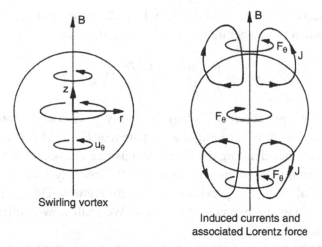

Swirling vortex

Induced currents and
associated Lorentz force

Figure 6.3 An initially spherical vortex is damped by a magnetic field.

Let us now determine the induced current, $\mathbf{J}$, and hence the Lorentz force which acts on the (initially) spherical vortex. The term $\mathbf{u} \times \mathbf{B}$ in Ohm's law gives rise to a radial component of current, J_r. However, the current lines must form closed paths and so an electrostatic potential, $V(r, z)$, is established whose primary function is to ensure that the $\mathbf{J}$-lines close. The distribution of V is determined by $\nabla^2 V = \nabla \cdot (\mathbf{u} \times \mathbf{B}) = \mathbf{B} \cdot \boldsymbol{\omega}$ and it drives an axial component of current, thus allowing $\mathbf{J}$ to form closed current paths in the r–z plane, as shown in Figure 6.3. Since $\mathbf{J}$ is solenoidal we can introduce a vector potential defined by

$$\mathbf{J} = \nabla \times [(\varphi/r)\hat{\mathbf{e}}_\theta] = \left(-\frac{1}{r}\frac{\partial \varphi}{\partial z}, \quad 0, \quad \frac{1}{r}\frac{\partial \varphi}{\partial r} \right).$$

In the fluid mechanics literature φ would be called the Stokes streamfunction for $\mathbf{J}$. For reasons which will become apparent shortly, it is convenient to take the curl of this,

$$\nabla \times \mathbf{J} = -r^{-1}\left(\nabla_*^2 \varphi\right)\hat{\mathbf{e}}_\theta,$$

where ∇_*^2 is the Laplacian-like operator,

$$\nabla_*^2 \equiv \frac{\partial^2}{\partial z^2} + r\frac{\partial}{\partial r}\left(\frac{1}{r}\frac{\partial}{\partial r}\right).$$

However, Ohm's law (6.6) gives us is $\nabla \times \mathbf{J} = \sigma B \partial \mathbf{u}/\partial z$, and so φ satisfies

$$\nabla_*^2 \varphi = -\sigma B \partial \Gamma / \partial z. \tag{6.12}$$

We have managed to relate φ, and hence $\mathbf{J}$, to the flow field. This allows us to evaluate the Lorentz force per unit mass, $\mathbf{F} = -(J_r B/\rho)\,\hat{\mathbf{e}}_\theta = F_\theta\,\hat{\mathbf{e}}_\theta$, in terms of Γ:

$$rF_\theta = \frac{B}{\rho}\frac{\partial\varphi}{\partial z} = -\frac{1}{\tau}\frac{\partial^2}{\partial z^2}\left(\nabla_*^{-2}\Gamma\right).$$

Here the inverse operator $f = \nabla_*^{-2}(g)$ is simply a symbolic representation of $\nabla_*^2 f = g$. From Figure 6.3 we might anticipate that $F_\theta \sim -J_r B$ is negative in the core of the vortex, decelerating the fluid, and positive above and below the vortex where the current returns to the z-axis and $J_r < 0$. Crucially, this returning current induces azimuthal motion in previously quiescent regions. This, in turn, suggests that Γ spreads along the magnetic field lines. We shall now confirm that this is indeed the case.

The azimuthal equation of motion is

$$\frac{D\Gamma}{Dt} = rF_\theta.$$

Note that, in the absence of the Lorentz force, angular momentum is materially conserved (i.e. preserved by each fluid particle), there being no azimuthal pressure gradient in an axisymmetric flow. Since we are neglecting the poloidal motion on a timescale of τ, our equation of motion becomes

$$\frac{\partial\Gamma}{\partial t} = \frac{B}{\rho}\frac{\partial\varphi}{\partial z} = -\frac{1}{\tau}\frac{\partial^2}{\partial z^2}\left(\nabla_*^{-2}\Gamma\right). \tag{6.13}$$

The first thing to note from (6.13) is that the global angular momentum, H, of the vortex is conserved,

$$\frac{dH}{dt} = \frac{d}{dt}\int\Gamma dV = \frac{B}{\rho}\int\nabla\cdot[\varphi\hat{\mathbf{e}}_z]dV = 0, \tag{6.14}$$

yet energy is continually dissipated in accordance with (6.7):

$$\frac{dE}{dt} = -\frac{1}{\rho\sigma}\int J^2 dV, \qquad E = \frac{1}{2}\int u^2 dV.$$

(For simplicity, we omit the constant ρ in our definitions of angular momentum and energy.) How can the vortex preserve its angular momentum in the face of continual Joule dissipation? We shall see that the answer to this question holds the key to the evolution of the vortex. Let ℓ_r and ℓ_z be characteristic radial and axial length scales for the vortex. At $t = 0$ we have $\ell_r = \ell_z = \delta$, and we shall suppose that ℓ_r

remains of order δ throughout the life of the vortex, there being no reason to suppose otherwise. (We shall confirm this shortly.) Then (6.6), i.e. $\nabla \times \mathbf{J} = \sigma \nabla \times (\mathbf{u} \times \mathbf{B}_0)$, allows us to estimate the magnitude of $\nabla \times \mathbf{J}$, and hence $\mathbf{J}$, from which

$$\frac{dE}{dt} \sim -\frac{1}{\tau}\left(\frac{\delta}{\ell_z}\right)^2 E, \qquad E \sim \Gamma^2 \ell_\perp. \tag{6.15}$$

However we also have

$$H \sim \Gamma \delta^2 \ell_z = \text{constant}. \tag{6.16}$$

It is evident that ℓ_z must increase with time since otherwise E would decay exponentially on a timescale of τ, which contradicts (6.16). In fact the only way of satisfying both (6.15) and (6.16) is if Γ and ℓ_z scale as

$$\Gamma \sim \Gamma_0(1 + t/\tau)^{-1/2}, \qquad \ell_z \sim \delta(1 + t/\tau)^{1/2}, \tag{6.17}$$

which, in turn, suggests that the kinetic energy of the vortex declines as $(t/\tau)^{-1/2}$. It seems that the vortex evolves from a sphere to an elongated cigar-like shape on a timescale of τ (see Figure 6.4). This is the first hint of the pseudo-diffusion process mentioned earlier. In fact, we might have anticipated (6.17) from the two-dimensional Fourier transform of (6.13),

$$\frac{\partial \hat{\Gamma}}{\partial t} \sim \frac{1}{\tau}\frac{\partial^2}{\partial z^2}\left(k^{-2}\hat{\Gamma}\right) \sim \frac{\delta^2}{\tau}\frac{\partial^2 \hat{\Gamma}}{\partial z^2}, \tag{6.18}$$

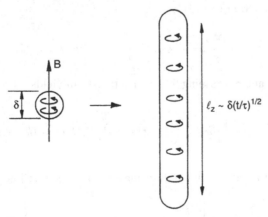

Figure 6.4 Diffusion of angular momentum along the magnetic field lines causes an initially spherical vortex to elongate into a cigar-like shape.

where $\hat{\Gamma}$ is the transform of Γ in the x–y plane. This suggests diffusion along the magnetic field lines with a diffusivity of $\alpha_B \sim \delta^2/\tau$. Recalling that the diffusion rate in a typical thermal problem is $\ell \sim \sqrt{\alpha t}$, we have $\ell_z \sim \delta(t/\tau)^{1/2}$, as in (6.17).

In the spirit of field sweeping and Faraday tensions we might picture this diffusion process as a spiralling up of the magnetic field lines, which then slowly unwind, propagating angular momentum along the z-axis. We shall return to this idea in §7.1, where we show that this pseudo-diffusion is the last vestige of Alfvén wave propagation.

Example 6.1 An Exercise for the Enthusiastic or the Sceptical

The estimates (6.17) may be confirmed by detailed analysis. The most direct method of solving (6.13) is to use Fourier transforms. In axisymmetric problems the three-dimensional Fourier transform reduces to the so-called cosine-Hankel transform, defined by the transform pair,

$$\hat{u}_\theta(k_r, k_z) = 4\pi \int_0^\infty \int_0^\infty u_\theta J_1(k_r r)\cos(k_z z) r\,dr\,dz,$$

$$u_\theta(r, z) = \frac{1}{2\pi^2} \int_0^\infty \int_0^\infty \hat{u}_\theta J_1(k_r r)\cos(k_z z) k_r\,dk_r\,dk_z.$$

Here $\hat{u}_\theta$ is the transform of u_θ, $\hat{u}_\theta = F(u_\theta)$. This transform has the convenient properties,

$$F(\partial^2 u_\theta/\partial z^2) = -k_z^2 F(u_\theta), \quad F\left(r^{-1}\nabla_*^2(ru_\theta)\right) = -\left(k_r^2 + k_z^2\right)F(u_\theta) = -k^2 F(u_\theta),$$

and so the transform of (6.13) is

$$\frac{\partial \hat{u}_\theta}{\partial t} = -[\cos^2\hat{\theta}]\frac{\hat{u}_\theta}{\tau}, \qquad \cos\hat{\theta} = k_z/k.$$

Solving for $\hat{u}_\theta$ and performing the inverse transform yields

$$\Gamma = \frac{r}{2\pi^2} \int_0^\infty \int_0^\infty [\hat{u}_0 \exp\left(-(t/\tau)\cos^2\hat{\theta}\right)] J_1(k_r r)\cos(k_z z) k_r\,dk_r\,dk_z,$$

where $\hat{u}_0 = F(u_\theta)$ at $t = 0$. Confirm that, for $t \gg \tau$, this integral takes the form

$$\Gamma(\hat{t} \to \infty) = \frac{r\hat{t}^{-1/2}}{2\pi^2} \int_0^\infty \int_0^\infty k\hat{u}_0 e^{-q^2} J_1(kr) \cos\left(kqz/\hat{t}^{1/2}\right) k\,dk\,dq, \qquad \hat{t} = t/\tau.$$

Thus confirm that

$$\Gamma \sim (t/\tau)^{-1/2}, \qquad \ell_z \sim (t/\tau)^{1/2},$$

as suggested earlier on the basis of qualitative arguments. Evidently, the vortex distorts from a sphere to a column, growing axially at a rate $\ell_z \sim \delta(t/\tau)^{1/2}$. This axial elongation is essential to preserving the angular momentum of the vortex. ∎

6.2.4 The Damping of Turbulence at Low R_m

The example above suggests that a turbulent flow evolving in a magnetic field will behave very differently from conventional turbulence, and this is indeed the case. We shall examine this issue in detail in Chapter 9; here we just give a flavour of some of the underlying ideas. A classic problem in conventional turbulence theory is the so-called free decay of a turbulent flow, and perhaps it is worth considering this purely hydrodynamic problem first.

Suppose that the fluid in a large vessel is stirred vigorously and then left to itself. Suppose also that the eddies created by the stirring are randomly oriented and distributed throughout the vessel, so that the initial turbulence is approximately statistically homogeneous and isotropic. Let the vessel have size L and a typical eddy have size ℓ and velocity u. We take $\mathbf{B}$ to be zero and $L \gg \ell$ so that the boundaries have little influence on the bulk of the motion. The first thing which happens is that some of the eddies which are set up at $t = 0$ break up through inertially driven instabilities, creating a whole spectrum of eddy sizes, from ℓ down to $\ell_{min} \sim (u\ell/\nu)^{-3/4}\ell$, the latter length scale being the smallest eddy size which may exist in a turbulent flow without being eradicated by viscosity. (Eddies of size ℓ_{min} are characterised by $\nu\nabla^2\mathbf{u} \sim \mathbf{u} \cdot \nabla\mathbf{u}$ – see Chapter 8.) There then follows a period of decay in which energy is extracted from the turbulence via the destruction of small-scale eddies by viscous stresses, kinetic energy being continually passed down from the large scales to the small scales through the stretching of the smaller vortices by larger eddies. This free-decay process is characterised by the facts that (i) the turbulence remains approximately homogeneous and isotropic during the decay and (ii) the energy (per unit mass) declines according to Kolmogorov's decay law $E \sim E_0 (u_0t/\ell_0)^{-10/7}$, or else Saffman's decay law $E \sim E_0 (u_0t/\ell_0)^{-6/5}$. (Here u_0 and ℓ_0 are the initial values of u and ℓ.) Again, the details are spelled out in Chapter 8.

Now suppose that we repeat this process but in the presence of a uniform magnetic field $\mathbf{B} = B\hat{\mathbf{e}}_z$. For simplicity we take the fluid to be inviscid and to be housed in a large electrically insulated sphere of radius R, with $R \gg \ell$ (see Figure 6.5). From (6.6) and (6.7) we have

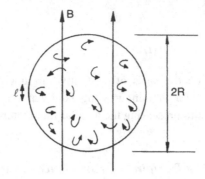

Figure 6.5 Initial condition for decaying turbulence in a magnetic field.

$$\frac{dE}{dt} = -\frac{1}{\rho\sigma}\int J^2 dV, \qquad E = \frac{1}{2}\int u^2 dV, \tag{6.19}$$

$$\nabla \times \mathbf{J} = \sigma B \frac{\partial \mathbf{u}}{\partial z}, \qquad \nabla \cdot \mathbf{J} = 0. \tag{6.20}$$

Clearly the kinetic energy of the flow falls monotonically and this process ceases if, and only if, $\mathbf{u}$ is independent of z, i.e. $\mathbf{J} = 0$. However, one component of angular momentum is conserved during this decay. Formally, this may be seen by transforming the expression for the component of torque parallel to $\mathbf{B}$ as follows,

$$[\mathbf{x} \times (\mathbf{J} \times \mathbf{B})] \cdot \mathbf{B} = [(\mathbf{x} \cdot \mathbf{B})\mathbf{J} - (\mathbf{x} \cdot \mathbf{J})\mathbf{B}] \cdot \mathbf{B} = -(B^2/2)\nabla \cdot [(x^2 + y^2)\mathbf{J}]. \tag{6.21}$$

This integrates to zero over the sphere (remember that $\mathbf{J} \cdot d\mathbf{S} = 0$). Thus the global Lorentz torque parallel to $\mathbf{B}$ is zero and so, since there are no viscous forces, one component of angular momentum,

$$\mathbf{H}_{//} = \int (\mathbf{x} \times \mathbf{u})_{//} dV,$$

is conserved as the flow evolves. (We take the origin of coordinates to lie at the centre of the sphere and use // and $\perp$ to indicate components of a vector parallel and normal to $\mathbf{B}$.)

The physical interpretation of (6.21) is straightforward. The current density, $\mathbf{J}$, may be considered to be composed of many current tubes, and each of these may, in turn, be considered to be the sum of many infinitesimal current loops, as in the proof of Stokes' theorem. However, the torque on each elementary current loop is $d\mathbf{m} \times \mathbf{B}$, where $d\mathbf{m}$ is its dipole moment, and this is perpendicular to $\mathbf{B}$.

Consequently, the global torque, which is the sum of many such terms, can have no component parallel to **B**. Conservation of $\mathbf{H}_{//}$ then follows.

As we shall see, this conservation law is fundamental to the evolution of a turbulent flow. In fact, we may show that, as in the example discussed in §6.2.3, the conservation of $\mathbf{H}_{//}$, combined with continual Joule dissipation, leads to an elongation of the eddies. Let us pursue this idea a little further. Since $\mathbf{H}_{//}$ is conserved, the energy of the flow cannot fall to zero. Yet (6.20) tells us that **J** is non-zero, and the Joule dissipation finite, as long as **u** is a function of z. It follows that, eventually, the flow must settle down to a two-dimensional one, in which **u** exhibits no variation along the field lines. We may determine how quickly this happens as follows. Noting that the ith component of the Lorentz torque may be written as

$$2[\mathbf{x} \times (\mathbf{J} \times \mathbf{B})]_i = [(\mathbf{x} \times \mathbf{J}) \times \mathbf{B}]_i + \nabla \cdot [(\mathbf{x} \times (\mathbf{x} \times \mathbf{B}))_i \mathbf{J}], \qquad (6.22)$$

we may rewrite the global Lorentz torque in terms of the dipole moment, **m**, of **J**:

$$\mathbf{T} = \int \mathbf{x} \times (\mathbf{J} \times \mathbf{B}) dV = \frac{1}{2} \left\{ \int \mathbf{x} \times \mathbf{J} dV \right\} \times \mathbf{B} = \mathbf{m} \times \mathbf{B}.$$

Also, from Ohm's law (6.4), we have

$$(\mathbf{x} \times \mathbf{J}) = \sigma[\mathbf{x} \times (\mathbf{u} \times \mathbf{B}) + \nabla \times (\mathbf{x}V)].$$

(Here V now stands for the electrostatic potential, rather than volume.) On integrating this expression over the spherical volume, the second term on the right converts to a surface integral which is zero, since $\mathbf{x} \times d\mathbf{S} = 0$. The first contribution on the right may be rewritten (using a version of (6.22) in which **u** replaces **J**) as $\frac{1}{2}(\mathbf{x} \times \mathbf{u}) \times \mathbf{B}$ plus a divergence, which also integrates to zero. It follows that

$$\mathbf{m} = \frac{1}{2} \int \mathbf{x} \times \mathbf{J} dV = (\sigma/4)\, \mathbf{H} \times \mathbf{B}, \qquad (6.23)$$

and so the global Lorentz torque becomes

$$\mathbf{T} = -\frac{\sigma B^2}{4} \mathbf{H}_\perp.$$

The global angular momentum equation,

$$\rho \frac{\partial \mathbf{H}}{\partial t} = \mathbf{T} = -\frac{\sigma B^2}{4} \mathbf{H}_\perp, \qquad (6.24)$$

then yields (Davidson, 1997)

$$\mathbf{H}_{//} = \text{constant}, \qquad \mathbf{H}_{\perp} = \mathbf{H}_{\perp}(0)\exp(-t/4\tau). \qquad (6.25)$$

As expected, $\mathbf{H}_{//}$ is conserved while $\mathbf{H}_{\perp}$ decays exponentially on a time scale of τ. The simplicity of this inviscid result is surprising, partially because of its generality (the initial conditions may be quite random), and partially because the local momentum equation,

$$\rho\left(\frac{\partial\mathbf{u}}{\partial t} + \mathbf{u}\cdot\nabla\mathbf{u}\right) = -\nabla p + \mathbf{J}\times\mathbf{B},$$

is quadratic in $\mathbf{u}$ and so possesses analytical solutions only for the most trivial of flows.

Equation (6.25) is highly suggestive. The preferential destruction of $\mathbf{H}_{\perp}$ suggests that vortices whose axes are perpendicular to $\mathbf{B}$ are annihilated, leading to a quasi-two-dimensional flow. We may quantify this as follows. First we need the Schwarz integral inequality. In its simplest form this states that any two functions, f and g, satisfy the inequality

$$\left[\int fg\,dV\right]^2 \leq \int f^2\,dV \int g^2\,dV. \qquad (6.26)$$

The analogous results for arbitrary vector fields $\mathbf{A}$ and $\mathbf{B}$ are

$$\left[\int \mathbf{A}\cdot\mathbf{B}\,dV\right]^2 \leq \int \mathbf{A}^2\,dV \int \mathbf{B}^2\,dV \qquad (6.27)$$

and

$$\left[\int \mathbf{A}\times\mathbf{B}\,dV\right]^2 \leq \int \mathbf{A}^2\,dV \int \mathbf{B}^2\,dV. \qquad (6.28)$$

In the present context, this yields

$$\mathbf{H}_{//}^2 \leq \int \mathbf{x}_{\perp}^2\,dV \int \mathbf{u}_{\perp}^2\,dV,$$

which, in turn, furnishes a lower bound on the kinetic energy, E:

$$E \geq \mathbf{H}_{//}^2\left[2\int \mathbf{x}_{\perp}^2\,dV\right]^{-1}. \qquad (6.29)$$

Thus, provided $\mathbf{H}_{//}$ is non-zero, the flow cannot come to rest. Yet (6.20) tells us that, as long as there is some variation of velocity along the $\mathbf{B}$-lines, the Joule dissipation remains finite, and E falls. Consequently, whatever the initial condition, the flow must evolve to a steady state that is strictly two-dimensional, exhibiting no variation of $\mathbf{u}$ along the field lines. In short, the flow adopts the form of one or more columnar vortices, each aligned with the $\mathbf{B}$-field, all other components of angular momentum being destroyed on a time scale of 4τ. The simplicity of this result is intriguing, particularly since it is valid for any value of $u\tau/\ell$. That is, unlike the example discussed in §6.2.3, this is valid for any ratio of $|\mathbf{J} \times \mathbf{B}|$ to inertia.

It appears, therefore, that magnetic fields tend to induce a strong anisotropy in a turbulent flow, elongating the eddies in the direction of $\mathbf{B}$. Of course, any real fluid is viscous and so this elongation will be accompanied by viscous dissipation, just as in conventional turbulence. The eddies become elongated only if they survive long enough. This, in turn, requires that $\mathbf{J} \times \mathbf{B}$ be at least of order $(\mathbf{u} \cdot \nabla)\mathbf{u}$ and so we would expect strong anisotropy in a real flow only if the interaction parameter, $N = \ell/u\tau$, is greater than unity. We return to this topic in §6.3 and again in Chapter 9.

6.2.5 Natural Convection in a Magnetic Field: Rayleigh–Bénard Convection

As a final example of the dissipative effect of a static magnetic field we consider the influence of a uniform, imposed field on natural convection. We start by considering the case of Rayleigh–Bénard convection; that is, convective motion between two flat parallel plates, where the lower plate is heated.

Let us first consider Rayleigh–Bénard convection in the absence of a magnetic field. It is a common experience that a fluid pool heated from below exhibits natural convection. Hot, buoyant fluid rises from the base of the pool. When this fluid reaches the surface, it cools and sinks back down to the base. Such a flow is characterised by the continual conversion of gravitational energy into kinetic energy, the potential energy being released as light fluid rises and heavy fluid falls. However, this motion is opposed by viscous dissipation, and if the heating is uniform across the base of the pool, and the viscosity high enough, no motion takes place. Rather, the fluid remains in a state of hydrostatic equilibrium and heat diffuses upwards by conduction alone. The transition between the static, diffusive state and that of natural convection occurs at a critical value of

$$\mathrm{Ra} = g\beta\Delta T d^3 / \nu\alpha, \qquad (6.30)$$

called the Rayleigh number. Here ΔT is the imposed temperature difference between the top and the bottom of the pool, d the depth of the pool, β the expansion coefficient of the fluid (in units of K^{-1}) and α is the thermal diffusivity. The sudden transition from one state to another is called the Rayleigh–Bénard instability, in recognition of Bénard's experimental work in 1900 and the subsequent analytical investigation by Rayleigh in 1916. Rayleigh described Bénard's experiment thus:

Bénard worked with very thin layers, only about 1 mm deep, standing on a levelled metallic plate which was maintained at a uniform temperature... The layer rapidly resolves itself into a number of 'cells', the motion being an ascension in the middle of a cell and a descension at the common boundary between a cell and its neighbours.

Inspired by these experiments, Rayleigh developed the theory of convective instability for a thin layer of fluid between horizontal planes. He found that the destabilising effect of buoyancy (heavy fluid sitting over light fluid) wins out over the stabilising influence of viscosity only when Ra exceeds a critical value, $(\mathrm{Ra})_c$. For fluid bounded by two solid planes the critical value is 1708, while an open pool with a free upper surface has $(\mathrm{Ra})_c = 1101$. In principle, one can also do the calculation where both the bottom and top surfaces are free (although the physical significance of such a geometry is unclear) and this yields $(\mathrm{Ra})_c = 658$ (see Chandrasekhar, 1961). Ironically, many years later, it was discovered that the motions observed by Bénard were driven, for the most part, by surface tension, and not by buoyancy. (This is because Bénard used very thin layers.) Nevertheless Rayleigh's analysis of convective instability remains valid. We now extend this analysis to incorporate the stabilising (dissipative) effect of a vertical magnetic field, as shown in Figure 6.6.

When dealing with natural convection in a liquid it is conventional and convenient to use the Boussinesq approximation. In effect, this says that density variations are so small that we may continue to treat the fluid as incompressible and having uniform density, ρ, except to the extent that it introduces a buoyancy force per unit volume, $\delta\rho\mathbf{g}$, into the Navier–Stokes equation. This buoyancy force is usually rewritten as $-\rho\beta T\mathbf{g}$, where β is the expansion coefficient, $-\rho^{-1}(\partial\rho/\partial T)$, and T is the temperature (relative to some datum). The governing equations in the presence of an imposed, vertical field, $\mathbf{B}_0$, are then

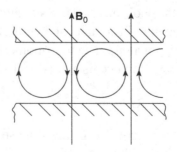

Figure 6.6 Bénard convection cells in the presence of an imposed magnetic field.

$$\frac{D\mathbf{u}}{Dt} = -\nabla\left(\frac{p}{\rho}\right) + \frac{1}{\rho}(\mathbf{J} \times \mathbf{B}_0) + \nu\nabla^2\mathbf{u} - \beta T\mathbf{g}, \qquad (6.31)$$

$$\nabla \cdot \mathbf{u} = 0 , \qquad \mathbf{J} = \sigma(-\nabla V + \mathbf{u} \times \mathbf{B}_0), \qquad (6.32)$$

$$\frac{DT}{Dt} = \alpha\nabla^2 T. \qquad (6.33)$$

Here we have ignored the internal heating due to viscous and Joule dissipation by comparison with the heat transfer from the lower boundary. The stationary configuration whose stability is in question is

$$\mathbf{u}_0 = 0, \qquad T_0 = \Delta T(1 - z/d), \qquad \mathbf{J}_0 = 0.$$

Here we take z to point vertically upward, the top and bottom surfaces to lie at $z = 0$, d, and ΔT to be the imposed temperature difference $T(z = 0) - T(z = d)$. Now the formal method of determining the stability of such a base state is straightforward. One looks for slightly perturbed solutions of the form $T = T_0 + \delta T$, $\mathbf{u} = \mathbf{u}_0 + \delta\mathbf{u}$ and $\mathbf{J} = \mathbf{J}_0 + \delta\mathbf{J} = \delta\mathbf{J}$, substitute these into the governing equations, discard terms which are quadratic in the disturbance, and look for separable solutions of the linearised equations of the form $\mathbf{u} = \hat{\mathbf{u}}(\mathbf{x})\exp(jst)$. If all goes well, this results in an eigenvalue problem, the eigenvalues of which determine the growth (or decay) rate of some initial disturbance. This process is long and tedious, resulting in an eighth-order differential system, and we do not intend to do it. Rather, we shall give an heuristic description of the instability which captures the key physics of the process and yields a surprisingly accurate estimate of $(\text{Ra})_c$. It is based on an approximate energy budget.

If we take the product of the Navier–Stokes equation with $\mathbf{u}$, we obtain

$$\frac{\partial}{\partial t}\left(\frac{u^2}{2}\right) = -\nabla \cdot \left[\left(\frac{p}{\rho} + \frac{u^2}{2}\right)\mathbf{u}\right] + \frac{1}{\rho}(\mathbf{J} \times \mathbf{B}_0) \cdot \mathbf{u} - \nu\left(\mathbf{u} \cdot (\nabla \times \boldsymbol{\omega})\right) + g\beta T u_z.$$

$$(6.34)$$

The rate of working of the Lorentz, viscous and buoyancy forces may be rewritten as

$$\frac{1}{\rho}(\mathbf{J} \times \mathbf{B}_0) \cdot \mathbf{u} = -\frac{1}{\rho}(\mathbf{u} \times \mathbf{B}_0) \cdot \mathbf{J} = -\frac{J^2}{\sigma\rho} - \nabla \cdot (V\mathbf{J}/\rho), \qquad (6.35)$$

$$v[\mathbf{u} \cdot (\nabla \times \boldsymbol{\omega})] = v[\omega^2 + \nabla \cdot (\boldsymbol{\omega} \times \mathbf{u})], \tag{6.36}$$

$$g\beta(Tu_z) = g\beta\left(\delta Tu_z + \Delta T(1 - z/d)u_z\right) = g\beta\left[\delta Tu_z + \nabla \cdot \left((z - z^2/2d)\Delta T\mathbf{u}\right)\right], \tag{6.37}$$

where δT is the (small) departure of T from the static, linear distribution. We now gather all the divergence terms together and rewrite our energy equation as

$$\frac{\partial}{\partial t}\left(\frac{u^2}{2}\right) = \nabla \cdot (\sim) - \frac{\mathbf{J}^2}{\sigma\rho} - v\omega^2 + g\beta u_z \delta T. \tag{6.38}$$

Now the divergence term vanishes when this equation is integrated over the entire pool (or a single convection cell) and we obtain

$$\frac{d}{dt}\int\left(\frac{u^2}{2}\right)dV = -\frac{1}{\rho\sigma}\int J^2 dV - v\int \omega^2 dV + g\beta\int u_z \delta T dV. \tag{6.39}$$

The dissipative roles of the viscous and Lorentz forces are now apparent, as is the source of potential energy, $g\beta\delta Tu_z$. We would expect that the equilibrium is unstable wherever the fluid can arrange for

$$g\beta\int u_z \delta T dV \geq \frac{1}{\rho\sigma}\int J^2 dV + v\int \omega^2 dV, \tag{6.40}$$

with marginal stability corresponding to the equality sign. Let us now try to estimate the various integrals above. Suppose that the convection cells are two-dimensional, taking the form of rolls, with $\mathbf{u}$ confined to the x–z plane. We might approximate the shape of these by the streamfunction $\psi(\mathbf{x}, t) = \hat{\psi}(t)\sin(\pi z/d)\sin(\pi x/\ell)$, where 2ℓ is the wavelength of the instability. Also, we suppose the onset of the instability to be non-oscillatory, so that $s = 0$ at Ra = (Ra)$_c$. (All of the experimental and analytical evidence suggests that this is the case, except perhaps in certain hot plasmas in which $v > \lambda$.) Since the electrostatic potential, V, is zero for two-dimensional flow, (6.4) gives, for one cell,

$$\frac{1}{\rho\sigma}\int \mathbf{J}^2 dV = \frac{\sigma B_0^2}{\rho}\int u_x^2 dV = \frac{\sigma B_0^2}{\rho}\left(\frac{\pi}{2d}\right)^2 \hat{\psi}^2 \ell d. \tag{6.41}$$

The viscous dissipation, on the other hand, takes the form

$$\nu \int \omega^2 dV = (\nu/4)[(\pi/\ell)^2 + (\pi/d)^2]^2 \hat{\psi}^2 \ell d. \qquad (6.42)$$

Finally, the buoyancy integral can be estimated with the aid of $(\mathbf{u} \cdot \nabla) T_0 = \alpha \nabla^2 (\delta T)$, which yields

$$\alpha \nabla^2 (\delta T) = -\alpha[(\pi/\ell)^2 + (\pi/d)^2]\delta T = -u_z(\Delta T/d). \qquad (6.43)$$

This gives the estimate,

$$g\beta \int \delta T u_z dV = g\beta \alpha^{-1}[(\pi/\ell)^2 + (\pi/d)^2]^{-1}(\Delta T/d)(\pi/2\ell)^2 \hat{\psi}^2 \ell d. \qquad (6.44)$$

Thus the transition to instability occurs when

$$\underbrace{\left[\frac{g\beta\Delta T d^3}{\alpha}\right]\frac{(\pi/\ell)^2}{[(\pi/\ell)^2 + (\pi/d)^2]}}_{\text{(driving force)}} = \pi^2\underbrace{\left[\frac{\sigma B_0^2 d^2}{\rho}\right]}_{\text{(Joule dissipation)}} + \nu\underbrace{\left[\pi^2 + \left(\frac{\pi d}{\ell}\right)^2\right]^2}_{\text{(viscous dissipation)}}. \qquad (6.45)$$

Introducing the cell aspect ratio, $a = (\pi d /\ell)$, this simplifies to

$$(\text{Ra})_c = a^{-2}(a^2 + \pi^2)[(\pi^2 + a^2)^2 + \pi^2(\text{Ha})^2], \qquad (6.46)$$

where $\text{Ha} = B_0 d\sqrt{\sigma/\rho\nu}$ is the Hartmann number introduced in Chapter 4. It remains to estimate a. We now suppose that the cell shape is chosen so as to maximise the rate of working of the buoyancy force and minimise the dissipation. That is to say, we choose ℓ/d such that $(\text{Ra})_c$ is a minimum. This yields

$$(2a_c^2 - \pi^2)(a_c^2 + \pi^2)^2 = \pi^4(\text{Ha})^2, \qquad (6.47)$$

from which we find

$$\text{Ha} \to 0: \qquad a_c = \pi/\sqrt{2}, \qquad (\text{Ra})_c = 658,$$
$$\text{Ha} \to \infty: \qquad a_c = (\pi^4/2)^{1/6}(\text{Ha})^{1/3}, \qquad (\text{Ra})_c = \pi^2(\text{Ha})^2.$$

Note that, for $\text{Ha} = 0$, the convection rolls are predicted to have an aspect ratio d/ℓ of the order of unity, while the cells are narrow and deep at high Ha. We have made many assumptions in deriving these criteria and so we must now turn to the exact analysis to see how our guesses have faired. Fortuitously, it turns out that our estimate of $(\text{Ra})_c$ at large Ha is exactly correct! Our estimate of $(\text{Ra})_c = 658$

at Ha = 0 is less good though. Depending on the boundary conditions at $z = 0$ and d, an exact analysis gives $(\text{Ra})_c = 658$ (two free surfaces), 1101 (one free, one solid), 1708 (two solid surfaces). (The various exact solutions are given in Chandrasekhar, 1961.) Still, our energy analysis seems to have caught the essence of the process, and it is satisfying that its predictions are exact at high Ha. (The errors at low Ha are due to the assumed distribution of ψ.) It would seem that the cell size automatically adjusts to give the best possibility of an instability, minimising dissipation while maximising the rate of working of the buoyancy flow.

We close with an aside about natural convection in other configurations. Perhaps the first point to note is that Rayleigh–Bénard convection represents a singular geometry in the sense that it admits a static solution of the governing equations, i.e. **u** = **0**. Convection appears only because this solution is unstable at high values of ΔT or low values of v. In most geometries (for example a heated plate whose flat faces are vertical) motion develops irrespective of the size of v and ΔT. There is no static solution of the governing equations. In such cases the smallest temperature difference will drive motion. The influence of an imposed magnetic field is then different. It does not delay the onset of convection, as in the Rayleigh–Bénard geometry, but rather moderates the motion which inevitably occurs.

Consider the case of a vertical plate held at a temperature ΔT above the ambient fluid temperature. Here the motion is confined to a thermal boundary layer, δ, which grows from the base of the plate as the fluid passes upward. When there is no imposed magnetic field, we can estimate u_z and δ from the equations

$$u_z(\partial u_z/\partial z) \sim g\beta\Delta T, \quad \text{(vertical equation of motion)}$$

$$u_z(\partial T/\partial z) \sim \alpha\Delta T/\delta^2. \quad \text{(heat balance)}$$

This yields

$$u_z \sim (g\beta\Delta Tz)^{1/2}, \quad \delta \sim (\alpha^2 z/g\beta\Delta T)^{1/4}, \tag{6.48}$$

where z is measured from the base of the plate. (Actually, these estimates are accurate only for low Prandtl number fluids, $v/\alpha \ll 1$, such as liquid metals. When the Prandtl number is of order unity or greater, the viscous term vu/δ^2 must be included in the vertical force balance, leading to a modification in the estimate of δ. However, we shall stay with liquid metals for the moment.) Let us see how magnetic damping alters the situation. The imposition of a horizontal magnetic field, B_x, normal to the plate, modifies the first of these equations to

$$u_z(\partial u_z/\partial z) \sim g\beta\Delta T - u_z/\tau \, , \qquad \tau^{-1} = \sigma B_x^2/\rho. \tag{6.49}$$

Evidently, the fluid ceases to accelerate when u_z reaches a value of u_z^* given by $u_z^* \sim (g\beta\Delta T)\tau$. For a plate of length ℓ, the ratio of u_{max} with and without a magnetic field is therefore

$$\frac{u_z^*}{u_z} \sim \frac{(g\beta\Delta T\ell)^{1/2}}{\sigma B_x^2\ell/\rho} \sim \frac{(\mathrm{Ra})^{1/2}(\alpha/\nu)^{1/2}}{\mathrm{Ha}^2}, \tag{6.50}$$

so that the damping effect goes as $\sim B^2$. (The expression above assumes τ is small enough, or ℓ large enough, for u_z to saturate before it leaves the plate.) For efficient damping, therefore, we require

$$\mathrm{Ha}^2 \geq (\mathrm{Ra})^{1/2}(\alpha/\nu)^{1/2},$$

or equivalently,

$$\ell/\tau \gg \sqrt{g\beta\Delta T\ell}.$$

Note that, despite the appearance of diffusivities in the definitions of Ra and Ha, in fact ν and α play no role here.

The use of magnetic fields to curtail unwanted natural convection is quite common. For example, in the casting of aluminium, the natural convection currents in a partially solidified ingot are significant (a few cm/s) and are thought to be detrimental to the ingot structure, causing a non-uniformity of the alloying elements through the transport of crystal fragments. Static magnetic fields have been used to minimise this natural convection. In the laboratory, on the other hand, the standard method of measuring the thermal diffusivity of liquid metals relies on injecting heat into the metal and measuring the rate of spread of heat by conduction. However, natural convection disrupts this procedure, and since it is difficult to design an apparatus free from convection, magnetic damping is employed to minimise the flow. Convection in a magnetic field is also important in geophysics. The terrestrial magnetic field, for example, is maintained by motion in the liquid core of the earth and this is driven, in part, by solutal and thermal convection. However, this convection is damped by the terrestrial field (see Chapter 14).

6.3 An Aside: a Glimpse at the Damping of Turbulence at Arbitrary R_m

Let us now return to the topic of §6.2.4, that of MHD turbulence, only this time we lift the restriction that R_m be small.

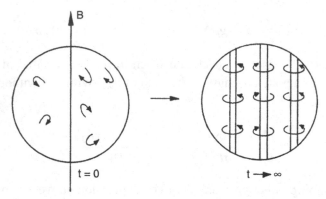

Figure 6.7 Turbulence evolves towards a two-dimensional state under the influence of an imposed magnetic field. This holds for any value of R_m.

The discussion in §6.2.3 illustrates the important role played by the conservation of angular momentum in the magnetic damping of a low-R_m vortex. That is, such a vortex cannot be destroyed by $\mathbf{J} \times \mathbf{B}_0$ since it must conserve $\mathbf{H}_{//}$, the component of angular momentum parallel to $\mathbf{B}_0$. (As before, we use subscripts $\perp$ and $//$ to indicate components perpendicular and parallel to $\mathbf{B}_0$.) Since the kinetic energy of the vortex, E, falls off as $\left(\ell_\perp/\ell_{//}\right)^2 E/\tau$, as described by Equation (6.15), we conclude that $\ell_{//}/\ell_\perp$ must grow with time. If it did not, E would decay exponentially on a time-scale of τ, which is forbidden by angular momentum conservation. Thus the conservation of angular momentum forces the vortex to become anisotropic, with the eddy elongating along the $\mathbf{B}_0$-lines. It is natural to ask if this phenomenon carries over to a field of turbulence evolving in an imposed magnetic field, irrespective of the value of R_m. We shall see that it does.

Consider the following idealised problem. An inviscid fluid is contained within an electrically insulated sphere of radius R and volume V_R. The fluid is set into turbulent motion with a typical eddy size much less that R, as shown in Figure 6.7. The fluid is permeated by a uniform magnetic field, $\mathbf{B}_0$, and we wish to determine the influence of this field on the subsequent motion. We place no restriction on the initial value of R_m, nor any restriction on the value of the interaction parameter, $N = \ell/u\tau$, so that inertia may be large or small.

As in §6.2.4, our starting point is to consider the global angular momentum of the fluid, $\mathbf{H} = \int \mathbf{x} \times \mathbf{u} dV$. Since the fluid is inviscid, $\mathbf{H}$ can change as a result of the Lorentz torque only,

$$\rho \frac{d\mathbf{H}}{dt} = \int \mathbf{x} \times (\mathbf{J} \times \mathbf{B}_0)dV + \int \mathbf{x} \times (\mathbf{J} \times \mathbf{b})dV,$$

where **b** is the induced magnetic field associated with **J**. However, a closed system of currents produces zero net torque when it interacts with its self-field, **b**, and so the second integral on the left is zero. The first integral may be transformed using (6.22) to give

$$\rho \frac{d\mathbf{H}}{dt} = \left\{ \frac{1}{2} \int \mathbf{x} \times \mathbf{J} dV \right\} \times \mathbf{B}_0 = \mathbf{m} \times \mathbf{B}_0, \tag{6.51}$$

where **m** is the net dipole moment of the induced current distribution. Once again we find that $\mathbf{H}_{//}$ is conserved, and this is true for both high- and low-R_m turbulence. This, in turn, places a lower bound on the kinetic energy, E_u, of the turbulence, since the Schwarz inequality applied to $\mathbf{H}_{//} = \int \mathbf{x}_\perp \times \mathbf{u}_\perp dV$ demands

$$E_u = \int \frac{1}{2} \rho \mathbf{u}^2 dV \geq \int \frac{1}{2} \rho \mathbf{u}_\perp^2 dV \geq \rho \mathbf{H}_{//}^2 / 2 \int \mathbf{x}_\perp^2 dV. \tag{6.52}$$

However, the total energy (kinetic plus magnetic) declines due to Joule dissipation in accordance with

$$\frac{dE}{dt} = \frac{d}{dt} \int_{V_R} \left(\frac{1}{2} \rho \mathbf{u}^2 \right) dV + \frac{d}{dt} \int_{V_\infty} (\mathbf{b}^2 / 2\mu) dV = -\frac{1}{\sigma} \int_{V_R} \mathbf{J}^2 dV. \tag{6.53}$$

Moreover, this decay of energy ceases only when **J** is everywhere zero. Evidently, in order to conserve $\mathbf{H}_{//}$, the system must evolve towards a state in which $\mathbf{J} = 0$ yet E_u is non-zero. However, if $\mathbf{J} = 0$ then $\mathbf{E} = -\mathbf{u} \times \mathbf{B}_0$, while Faraday's law requires $\nabla \times \mathbf{E} = 0$. It follows that the end state is one in which $\nabla \times (\mathbf{u} \times \mathbf{B}_0) = \mathbf{B}_0 \cdot \nabla \mathbf{u} = 0$, i.e. the flow is strictly two-dimensional. Once again we see that the destruction of energy subject to the constraint of conservation of $\mathbf{H}_{//}$ leads to anisotropy, with the flow structures elongated in the direction of the imposed field (Figure 6.7).

If we restrict ourselves to the case of low R_m, then we can determine how quickly this state is approached, as shown in §6.2.4. That is, the low-R_m form of Ohm's law, combined with (6.22), yields

$$\mathbf{m} = \frac{1}{2} \int \mathbf{x} \times \mathbf{J} dV = (\sigma/4) \mathbf{H} \times \mathbf{B}_0.$$

When combined with (6.51), this gives

$$\frac{d\mathbf{H}}{dt} = \rho^{-1} \mathbf{m} \times \mathbf{B}_0 = -\frac{\mathbf{H}_\perp}{4\tau}, \qquad \tau = \left(\sigma B_0^2 / \rho \right)^{-1}, \tag{6.54}$$

which is valid for any value of N. Clearly $\mathbf{H}_\perp$ decays exponentially on a time scale of 4τ, while $\mathbf{H}_{//}$ is conserved.

6.4 The Generation of Motion

6.4.1 Rotating Fields and Swirling Motion

Let us now consider a problem which frequently arises in engineering; that of stirring of a long column of liquid metal. This illustrates the capacity for magnetic fields to induce motion as well as suppress it. Suppose that fluid is held in a long cylinder of radius R and that a uniform magnetic field $\mathbf{B}_0$ rotates about the cylinder with angular velocity Ω, as shown in Figure 6.8. In effect, we have a simple induction motor, with the fluid playing the role of rotor. The rotating magnetic field therefore induces an azimuthal velocity, $u_\theta(r)$, in the fluid, stirring the contents of the cylinder.

The use of magnetic stirring is very common in the continuous casting of steel. Here, alloying elements tend to segregate out of the host metal during solidification, giving rise to inhomogeneity in the final ingot. Moreover, small cavities can form in the ingot either because of trapped gas or because of the shrinkage of the metal during freezing. All of these defects can be alleviated by stirring the liquid pool.

We now try to estimate the magnitude of the induced velocity. Let us start by evaluating the Lorentz force. For simplicity, we suppose that the field rotation rate is low (in the sense that $\Omega R \ll \lambda/R$). Next we change frames of reference so that the field is stationary and pointing in the x-direction. The fluid then appears to rotate in a clockwise direction at a rate $\hat{u} = (\Omega r - u_\theta)$. We now satisfy the low-R_m conditions of §6.1 and so we may assume that the imposed magnetic field is unperturbed and the low-R_m form of Ohm's law applies:

$$\mathbf{J} = \sigma(-\nabla V + \hat{\mathbf{u}} \times \mathbf{B}_0), \qquad \mathbf{F} = \mathbf{J} \times \mathbf{B}_0. \qquad (6.55)$$

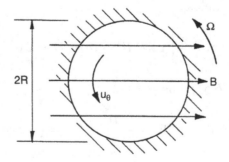

Figure 6.8 Magnetic stirring using a rotating magnetic field.

Now the divergence of (6.55) gives us $\nabla^2 V = 0$, and so we may take $V = 0$, provided that no electrostatic field is applied at the boundaries. It follows that

$$\mathbf{F} = \mathbf{J} \times \mathbf{B}_0 = \sigma(\hat{\mathbf{u}} \times \mathbf{B}_0) \times \mathbf{B}_0 = -\sigma B_0^2 \hat{\mathbf{u}}_\perp, \tag{6.56}$$

where $\hat{\mathbf{u}}_\perp$ is the component of $\hat{\mathbf{u}}$ perpendicular to $\mathbf{B}_0$. In (r, θ) coordinates this becomes

$$\mathbf{F} = \sigma B_0^2 (\Omega r - u_\theta) \cos\theta [\sin\theta \ \hat{\mathbf{e}}_r + \cos\theta \ \hat{\mathbf{e}}_\theta],$$

which may be conveniently split into two parts:

$$\mathbf{F} = \frac{1}{2} \sigma B^2 (\Omega r - u_\theta) \hat{\mathbf{e}}_\theta + \frac{1}{4} \sigma B^2 (\Omega - u_\theta / r) \nabla (r^2 \sin 2\theta). \tag{6.57}$$

Now, although we have assumed that $\Omega R \ll \lambda/R$ (i.e. $\mu\sigma\Omega R^2 \ll 1$), it may be shown that expression (6.57) is a good approximation up to values of $\mu\sigma\Omega R^2 \sim 1$, with a maximum error of ~4 per cent. (There is some hint of this in Figure 5.7 which shows very little magnetic field distortion at $R_m = 1$.) It turns out that this is useful since most engineering applications are characterised by the double inequality

$$u_\theta \ll \Omega R \leq \lambda/R,$$

and so, for most practical purposes, we may take

$$\mathbf{F} = \frac{1}{2} \sigma B_0^2 \Omega r \hat{\mathbf{e}}_\theta + \nabla\varphi, \qquad \varphi = \frac{1}{4} \sigma B_0^2 \Omega r^2 \sin 2\theta.$$

The second term may now be dropped since φ simply augments the pressure distribution in the fluid and plays no role in the dynamics of the flow. Finally we end up with

$$\mathbf{F} = \frac{1}{2} \sigma B^2 \Omega r \hat{\mathbf{e}}_\theta, \tag{6.58}$$

where we have dropped the subscript on B, on the understanding that it is the unperturbed field.

We now return to the laboratory frame of reference and consider the equation of motion for the fluid. The radial component of the Navier–Stokes equation simply expresses the balance between $\partial p/\partial r$ and the centripetal acceleration. The azimuthal component gives, in the steady state,

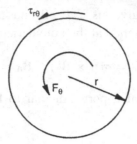

Figure 6.9 Torque balance between the Lorentz force and viscous stresses.

$$\tau_{r\theta}r^2 = -\int_0^r r^2 F_\theta dr.$$

This represents the torque balance on a cylinder of fluid of radius r (Figure 6.9).

Assuming laminar flow and substituting for $\tau_{r\theta}$ using Newton's law of viscosity yields

$$\tau_{r\theta} = \rho\nu r\frac{d}{dr}\left(\frac{u_\theta}{r}\right) = -\frac{1}{8}\sigma B^2\Omega r^2, \tag{6.59}$$

which may be integrated to give

$$u_\theta/r = \frac{\sigma B^2\Omega}{16\rho\nu}\left(R^2 - r^2\right). \tag{6.60}$$

Unfortunately (6.60) is of little practical value since very few flows of this type are laminar. The viscosity, ν, of most liquid metals is similar to that of water, and so the Reynolds' number in practical applications is invariably high, implying a turbulent motion. In such cases we must return to (6.59) and replace the laminar shear stress by the turbulent *Reynolds stress* which appears in the time-averaged equations of motion for a turbulent flow (see §3.7). This gives

$$\tau_{r\theta} = -\rho\,\overline{u'_r u'_\theta} = -\frac{1}{8}\sigma B^2\Omega r^2, \tag{6.61}$$

where $\mathbf{u}'$ represents the fluctuating component of velocity and the overbar denotes a time average. We now need some means of estimating the Reynolds stress. As noted in §3.7, a commonly used, though ultimately empirical, model for turbulent shear flows is Prandtl's *mixing length* model. The essence of this model is that we approximate the Reynolds stress in a planar flow adjacent to a wall by

$$\tau_{xy} = \rho \ell_m^2 \left| \frac{\partial \bar{u}_x}{\partial y} \right| \frac{\partial \bar{u}_x}{\partial y}, \qquad \ell_m = \kappa y,$$

where $\bar{u}_x$ is the time-averaged velocity, ℓ_m is called the mixing length, and y is the distance from the wall. The empirical constant κ is usually taken to be 0.4. The rotational equivalent of this is

$$\tau_{r\theta} = \rho \ell_m^2 r^2 \left| \frac{\partial}{\partial r} \left(\frac{\bar{u}_\theta}{r} \right) \right| \frac{\partial}{\partial r} \left(\frac{\bar{u}_\theta}{r} \right), \qquad \ell_m = \kappa (R - r).$$

Substituting into (6.61) and integrating yields (see Exercise 6.3 at the end of this chapter)

$$(\bar{u}_\theta / r)_{r=0} = \Omega_f \left\{ \frac{1}{2\sqrt{2}\kappa} \ln \left(\frac{\Omega_f R^2}{\nu} \right) + 1.0 \right\}, \tag{6.62}$$

where $\Omega_f^2 = \sigma \Omega B^2 / \rho$. Note that in a turbulent flow $\bar{u}_\theta$ scales linearly with B (with a logarithmic correction) whereas in a laminar flow u_θ scales as B^2.

Equation (6.62) gives values of $\bar{u}_\theta$ which compare favourably with estimates obtained using more complicated turbulence models. However, its main limitation is the fact that few engineering applications are strictly one-dimensional. The problem is immediately obvious if we refer back to Figure 3.31, showing spin-down of a stirred cup of tea. In a confined domain, rotation invariably induces a secondary flow via Ekman pumping. The mean inertial forces in the bulk of the fluid are then no longer zero and, in fact, at high values of Re, these forces greatly exceed the shear stresses, even the Reynolds stresses. However (6.62) is based entirely on a balance between $\mathbf{J} \times \mathbf{B}$ and shear, the mean inertia associated with secondary flow being ignored. Evidently, such a balance is rarely satisfied in practice, and so estimate (6.62) must be regarded with caution.

We shall examine the practical consequences of Ekman pumping in some detail in Chapter 11, where we shall see that (6.62) is often quite misleading. In the meantime, we can gain some hint as to the difficulties involved by considering a second, related example.

6.4.2 Swirling Flow Induced between Two Parallel Plates

We can gain some insight into the role of Ekman pumping by considering a second model problem. Suppose we have two infinite, parallel discs located at $z = 0$, and $z = 2w$, and that the gap is filled with liquid metal. The body force $F_\theta = \frac{1}{2} \sigma \Omega B^2 r$ is applied to the fluid, inducing a steady, laminar swirling flow. We choose F_θ so that Re is high and look for a steady solution of the Kármán type (see §3.8):

$$u_r = \Omega_c r F(z/\ell), \quad u_\theta = \Omega_c r G(z/\ell), \quad u_z = \Omega_c \ell H(z/\ell), \qquad (6.63)$$

$$p = \frac{1}{2}\rho\Omega_c^2 r^2 + \rho\Omega_c^2 \ell^2 P(z/\ell). \qquad (6.64)$$

Here ℓ is some characteristic length scale, yet to be determined, and Ω_c is a characteristic rotation rate in the core of the flow. We might anticipate that the flow divides into thin Bödewadt layers on the disc surfaces, between which lies an inviscid core. In fact, this is precisely what happens, as we now show.

Consider the lower half of the flow $0 \leq z \leq w$. Away from the disc we take $\ell = \ell_c = w$. (The subscript on ℓ denotes the core flow.) In the boundary layer, on the other hand, we try the scaling $\ell = \ell_b = (\nu/\Omega_c)^{1/2}$, which is Bödewadt boundary layer scaling. The square of the ratio of these length scales is the Ekman number, $\nu/\Omega_c w^2$, a sort of inverse Reynolds number. Let

$$\varepsilon = (\nu/\Omega_c w^2)^{1/2} = \ell_b/\ell_c. \qquad (6.65)$$

We shall take ε to be small and match the velocity profiles in the two regions using the method of *matched asymptotic expansions*. Substituting our proposed velocity distributions into the Navier–Stokes and continuity equations yields, for both the core and the boundary layer,

$$F^2 + HF' - G^2 + 1 = (\nu/\Omega_c\ell^2)F'', \quad \text{(radial equation)}$$

$$2FG + G'H = (\nu/\Omega_c\ell^2)G'' + \frac{1}{2}\Omega_f^2/\Omega_c^2, \quad (\theta \text{ equation})$$

$$H' + 2F = 0, \quad \text{(continuity)}$$

where the prime indicates differentiation with respect to z/ℓ and, as before, $\Omega_f^2 = \sigma\Omega B^2/\rho$. In the core of the flow, where $\ell = \ell_c = w$, these yield, for $\varepsilon \to 0$,

$$F_c^2 + H_c F'_c - G_c^2 = -1, \qquad (6.66)$$

$$2F_c G_c + G'_c H_c = \frac{1}{2}\Omega_f^2/\Omega_c^2, \qquad (6.67)$$

$$H'_c + 2F_c = 0, \tag{6.68}$$

where again the prime represents differentiation with respect to z/ℓ_c. The boundary conditions for F_c, G_c and H_c arise from the fact that the flow must be symmetric about $z = w$ and that u_z in the core and boundary layer must match at the interface between the two, in the sense that

$$\lim_{z/\ell_c \to 0} u_z = \lim_{z/\ell_b \to \infty} u_z.$$

This gives the boundary conditions

$$z/\ell_c = 1 \ : \ H_c = 0, \quad F'_c = 0, \quad G'_c = 0, \quad \text{(symmetry)}$$

$$z/\ell_c \to 0 \ : \ \ell_c H_c = \ell_b H_b(\infty), \quad \text{or} \quad H_c = \varepsilon H_b(\infty), \quad \text{(matching condition)}$$

$$z/\ell_c \to 0 \ : \ G_c = 1. \quad \text{(definition of } \Omega_c)$$

Here $H_b(\infty)$ is the far-field value of H furnished by the boundary-layer solution, while the condition $G_c = 1$ effectively defines Ω_c.

We now expand F_c, G_c, and H_c in polynomials of ε and substitute these into the governing core equations. To leading order in ε we find

$$F_c = \frac{1}{2}\varepsilon H_b(\infty), \quad G_c = 1, \quad H_c = \varepsilon H_b(\infty)[1 - z/w],$$

so the core velocity is

$$\mathbf{u}_c = \left(\frac{1}{2}\varepsilon H_b(\infty)\Omega_c r, \quad \Omega_c r, \quad \varepsilon H_b(\infty)\Omega_c[w - z] \right). \tag{6.69}$$

Note that we have rigid-body rotation in the core and that u_r and u_z are of order $\sim\varepsilon u_\theta$. It appears that, for small ε, the core velocity is determined by only three parameters: ε, Ω_c and $H_b(\infty)$. The second of these, Ω_c, is fixed by the azimuthal component of the core equation of motion, which may be rearranged to give

$$\Omega_c = \frac{\Omega_f}{[2H_b(\infty)]^{2/3}} \left[\frac{\Omega_f w^2}{\nu} \right]^{1/3}. \tag{6.70}$$

It now remains to find $H_b(\infty)$ and this is furnished by the boundary-layer equations. Immediately adjacent to the disc, the azimuthal equation of motion becomes

$$2F_bG_b + G'_bH_b = G''_b + \frac{1}{2}\Omega_f^2/\Omega_c^2,$$

where the prime now represents differentiation with respect to z/ℓ_b. However, we have already shown that the last term on the right of this equation is of order ε. Consequently, magnetic forcing is negligible in the boundary layer (by comparison with viscous and inertial forces) and the equations locally reduce to those for a conventional Bödewadt layer, as described in §3.8. From this we have $H_b(\infty) = 1.35$. It follows that the core rotation rate is

$$\Omega_c = 0.516 \ \Omega_f \left[\frac{\Omega_f w^2}{\nu}\right]^{1/3}. \tag{6.71}$$

Compare this with our prediction for one-dimensional laminar flow (6.60),

$$\frac{u_\theta}{r} = \frac{\Omega_f}{16}\left[\frac{\Omega_f R^2}{\nu}\right]\left[1 - \left(\frac{r}{R}\right)^2\right].$$

The dependence of u_θ on B is entirely different in the two cases, reflecting the different force balances. In the swirl-only flow, $\mathbf{J} \times \mathbf{B}$ is balanced by shear. In this second problem the primary balance in the core of the flow is between $\mathbf{J} \times \mathbf{B}$ and the Coriolis force. (This may be confirmed by tracing the origin of the terms in the azimuthal equation of motion.)

So this simple swirling flow is more complex than you might think! There are subtle and unexpected effects introduced by Ekman pumping. We shall return to this issue in Chapter 11, where we show that the balance between $\mathbf{J} \times \mathbf{B}$ and Coriolis forces is, in fact, typical of most flows encountered in practice.

6.4.3 Flows Resulting from Current Injection

There is a second way of driving motion in a conducting fluid. So far we have considered only currents which are induced in the fluid by rotation of the magnetic field. However, we can also inject current directly into a fluid, and the resulting Lorentz force will, in general, produce motion. The simplest example of this is the electromagnetic pump which was described in Chapter 1. Such a device consists of a duct in which mutually perpendicular magnetic and electric fields are arranged normal to the axis of the duct. The resulting Lorentz force, $\mathbf{J} \times \mathbf{B}$, is directed along the axis of the duct and this can be used to pump a conducting fluid. For example, sodium coolant is pumped around fast breeder nuclear reactors by this method. It turns out, however, that an understanding of this flow comes down to a careful

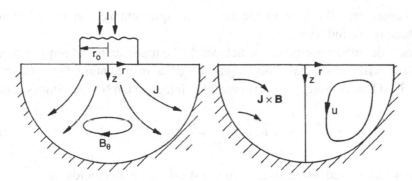

Figure 6.10 Geometry of the model problem.

consideration of the boundary layers, and so we shall postpone any discussion of this problem until the next section. Here we consider a configuration related to electric welding. The discussion is brief, but we shall return to this problem in Chapter 12, where we shall see that it has wider applications.

A useful model problem is the following. Suppose we have a liquid-metal pool which is hemispherical in shape, of radius R. The boundaries are assumed to be conducting and a current, I, is introduced into the pool by an electrode of radius r_0, which touches the surface. The entire geometry is axisymmetric and we use cylindrical polar coordinates (r, θ, z) with the origin at the pool's surface, as shown in Figure 6.10.

The poloidal current gives rise to an azimuthal field, B_θ, and the two are related by Ampère's law, according to which,

$$2\pi r B_\theta = \mu \int_0^r (2\pi r J_z)\,dr. \tag{6.72}$$

The interaction of $\mathbf{J}$ with B_θ gives rise to a Lorentz force and it is readily confirmed that

$$\mathbf{F} = \mathbf{J} \times \mathbf{B}/\rho = -\nabla\left(B_\theta^2/2\rho\mu\right) - [B_\theta^2/(\rho\mu r)]\hat{\mathbf{e}}_r. \tag{6.73}$$

Of course, the magnetic pressure merely augments the fluid pressure and does not influence the motion in the pool. We therefore write

$$\mathbf{F} = -\frac{B_\theta^2}{\rho\mu r}\,\hat{\mathbf{e}}_r, \tag{6.74}$$

on the understanding that p is augmented by $B_\theta^2/(2\mu)$. Clearly, this Lorentz force will drive a recirculating flow which converges at the surface (where B_θ is largest) and diverges near the base of the pool. The question is, can we estimate the magnitude of the induced flow?

We now describe a useful trick which we shall employ in the subsequent chapters. When we wish to estimate the magnitude of a recirculating flow induced by a prescribed Lorentz force, it is often useful to integrate the Navier–Stokes equation,

$$\frac{\partial \mathbf{u}}{\partial t} = \mathbf{u} \times \boldsymbol{\omega} - \nabla \left[\frac{p}{\rho} + \frac{u^2}{2} \right] + \nu \nabla^2 \mathbf{u} + \mathbf{F}, \qquad (6.75)$$

once around a closed streamline, C. In the steady state this yields

$$\oint_C \mathbf{F} \cdot d\mathbf{l} = -\nu \oint_C \nabla^2 \mathbf{u} \cdot d\mathbf{l}, \qquad (6.76)$$

since $(\mathbf{u} \times \boldsymbol{\omega}) \cdot d\mathbf{l} = 0$ and the gradient of Bernoulli's function integrates to zero. Evidently, there must be a global balance between the Lorentz force and the shear stresses. Physically, this arises because the work done by $\mathbf{F}$ on a fluid particle as it passes once around the streamline must be balanced by the (dissipative) work performed by the shear stresses acting on the same particle. If the two did not match, then the kinetic energy of the fluid particle would not be the same at the beginning and end of the integration, which is clearly not the case in a steady flow. We may use this integral equation to estimate $|\mathbf{u}|$.

Let us see where this leads in our model problem. We take C to be the bounding streamline, comprising the upper surface, the axis, and the curved boundary. Starting with the left-hand integral, and integrating in an anti-clockwise direction, we have

$$\oint \mathbf{F} \cdot d\mathbf{l} = \left[\int_0^R (B_\theta^2/\rho\mu r)\, dr \right]_{z=0} - \left[\int_0^R (B_\theta^2/\rho\mu r)\, dr \right]_{r^2+z^2=R^2}. \qquad (6.77)$$

The first of these integrals is readily evaluated since, from Ampère's law, the surface magnetic field is $2\pi r B_\theta = \mu I (r/r_0)^2$ for $r < r_0$ and $2\pi r B_\theta = \mu I$ for $r > r_0$. This yields

$$\left[\int_0^R (B_\theta^2/\rho\mu r)\, dr \right]_{z=0} = \frac{\mu I^2}{4\pi^2 \rho r_0^2} [1 - (r_0^2/2R^2)].$$

The second integral is more problematic. However, if $r_0 \ll R$, then the field in the vicinity of the boundary is approximately that due to a point source of current, and the corresponding magnetic field is readily shown to be

$$2\pi r B_\theta = \mu I [1 - z/(r^2 + z^2)^{1/2}], (r_0 \ll R).$$

This yields

$$\left[\int_0^R (B_\theta^2/\rho\mu r) \, dr \right]_{r^2+z^2=R^2} = \frac{\mu I^2}{4\pi^2 \rho R^2} [\ln 2 - 1/2].$$

Combining these expressions gives us

$$\oint \mathbf{F} \cdot d\mathbf{l} = \frac{\mu I^2}{4\pi^2 \rho} \left[\frac{1}{r_0^2} - \frac{\ln 2}{R^2} \right] = -\nu \oint \nabla^2 \mathbf{u} \cdot d\mathbf{l}. \tag{6.78}$$

For cases where r_0 does not satisfy $r_0 \ll R$ the factor of $\ln 2$ above will need modification. However, the details do not matter. The main point is that

$$\frac{\mu I^2}{4\pi^2 \rho r_0^2} \sim -\nu \oint \nabla^2 \mathbf{u} \cdot d\mathbf{l}. \tag{6.79}$$

Although we have performed the integration only for the bounding streamline, a similar relationship must hold for all streamlines which pass close to the electrode. For streamlines remote from the electrode we would expect

$$\frac{\mu I^2}{4\pi^2 \rho R^2} \sim -\nu \oint \nabla^2 \mathbf{u} \cdot d\mathbf{l}, \tag{6.80}$$

since r_0 ceases to be a relevant dimension in such cases. We can use these equations to estimate $|\mathbf{u}|$.

Suppose that the Reynolds number is not too high, say somewhat less than 20. Then there are no significant boundary layers on the outer wall. (Such layers usually start to form when Re > ~20.) The only region where high velocity gradients will form is near the electrode where the characteristic gradient in $\mathbf{F}$ is $|\mathbf{F}|/r_0$, and so we would expect local gradients in $\mathbf{u}$ to be of the order of $|\mathbf{u}|/r_0$. Elsewhere we would expect $\nabla^2 \mathbf{u}$ to scale as $|\mathbf{u}|/R^2$. If these statements are true, then our integral equations suggest

$$u \sim \frac{\mu I^2}{4\pi^2 \rho \nu r_0}, \quad \text{(near electrode)}$$

$$u \sim \frac{\mu I^2}{4\pi^2 \rho \nu R}. \quad \text{(elsewhere)}$$

We end this section by considering the idealised case where the outer boundary is removed and $r_0 \to 0$, so that we have a point electrode located on the surface of a semi-infinite fluid. Of course, this is of little practical significance, but it has been the subject of considerable attention in the literature because it turns out that there is an exact, self-similar solution for this flow. This solution is of the form

$$u \sim \frac{\mu I^2}{4\pi^2 \rho \nu (r^2 + z^2)^{1/2}} g(\varphi, \text{ Re}) \tag{6.81}$$

(see Moreau, 1990), where φ is the angle between the z-axis and the position vector, $\mathbf{x}$, and g is a function of φ and of $\text{Re} = u|\mathbf{x}|/\nu$. The similarity to our estimates above is reassuring. However, it would be wrong to place too much emphasis on this exact solution since, in many respects, it is atypical. It turns out that the absence of an outer boundary at large $|\mathbf{x}|$ means that the streamlines in this self-similar flow do not close on themselves, but merely converge towards the axis. The flow is therefore free from integral constraints of the form

$$\oint \mathbf{F} \cdot d\mathbf{l} = -\nu \oint \nabla^2 \mathbf{u} \cdot d\mathbf{l}.$$

We might anticipate, therefore, that there is a fundamental difference between this self-similar flow and that in which R is finite, and we shall confirm that this is so in Chapter 12.

6.5 Boundary Layers and Associated Duct Flows

6.5.1 Hartmann Boundary Layers

So far we have considered the influence of $\mathbf{J} \times \mathbf{B}$ on the interior of a flow only. We have not considered its effect on boundary layers. We close this chapter with a discussion of a phenomenon which received much attention in the early literature on liquid-metal MHD: the Hartmann layer. This is often discussed in the context of duct flows, but is really just a boundary-layer effect. The main point is that a steady magnetic field oriented at right angles to a boundary can completely transform the nature of the boundary layer, changing its characteristic thickness, for example.

Suppose we have rectilinear shear flow $u(y)\hat{\mathbf{e}}_x$ adjacent to a plane, stationary, surface. Far from the wall, the flow is uniform and equal to u_∞, but close to the wall the no-slip condition ensures some kind of boundary layer (see Figure 6.11). There is a uniform, imposed magnetic field $\mathbf{B} = B\hat{\mathbf{e}}_y$. Now $\mathbf{B} \cdot \boldsymbol{\omega} = 0$ and so (6.9) tells us $\nabla^2 V = 0$, implying that the electrostatic field is zero. We shall also assume that there is no externally imposed electric field, so $V = 0$. It then follows from (6.4) and (6.5) that $\mathbf{J} \times \mathbf{B} = -\sigma B^2 u \hat{\mathbf{e}}_x$, and so we have the usual magnetic damping force. The resulting Navier–Stokes equation,

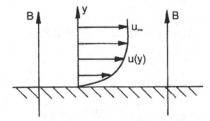

Figure 6.11 Hartmann flow.

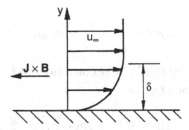

Figure 6.12 The Hartmann boundary layer.

$$\rho\nu\frac{\partial^2 u}{\partial y^2} - \sigma B^2 u = \frac{\partial p}{\partial x}, \tag{6.82}$$

may be transformed to

$$\frac{\partial^2}{\partial y^2}(u_\infty - u) - \frac{u_\infty - u}{\delta^2} = 0, \quad \delta = (\rho\nu/\sigma B^2)^{1/2}, \tag{6.83}$$

where u_∞ is the velocity remote from the boundary. The solution of (6.83) is evidently

$$u = u_\infty[1 - e^{-y/\delta}],$$

where δ is the characteristic boundary-layer thickness. We see that the velocity increases rapidly over a short distance from the wall. This boundary layer, which has a thickness of $\sim\delta$, is called a Hartmann layer (Figure 6.12). Note that the thickness of a Hartmann boundary layer is quite different from that of a conventional boundary layer.

We now consider the same flow, but between two stationary parallel plates located at $y = \pm w$. We also allow for the possibility of an imposed, uniform electric field, E_0, in the z-direction. Our equation of motion is now

$$\rho v \frac{\partial^2 u}{\partial y^2} - \sigma B^2 u = \frac{\partial p}{\partial x} + \sigma B E_0, \tag{6.84}$$

which has the solution

$$u = u_0 \left[1 - \frac{\cosh(y/\delta)}{\cosh(w/\delta)} \right], \tag{6.85}$$

where

$$\sigma B^2 u_0 = -\frac{dp}{dx} - \sigma E_0 B \tag{6.86}$$

defines u_0 and δ is given by (6.83). It is conventional to introduce the Hartmann number at this point, defined by

$$\mathrm{Ha} = w/\delta = Bw\sqrt{\sigma/\rho v}.$$

As noted in §4.1, Ha^2 represents the ratio of the Lorentz forces to the viscous forces. Our solution is, then,

$$u = u_0 \left[1 - \frac{\cosh(\mathrm{Ha}\, y/w)}{\cosh(\mathrm{Ha})} \right]. \tag{6.87}$$

It is instructive to look at the two limits, $\mathrm{Ha} \to 0$ and $\mathrm{Ha} \to \infty$. When Ha is very small we recover the parabolic velocity profile of conventional Poiseuille flow, $u = u_{\max} (1 - (y/w)^2)$. When Ha is very large, on the other hand, we find that exponential Hartmann layers form on both walls, separated by a core of uniform flow. All of the vorticity is then pushed to the boundaries. The two limits are shown in Figure 6.13.

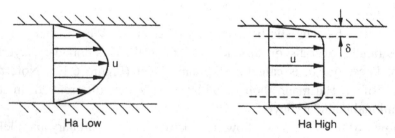

Figure 6.13 Duct flow at small and large Hartmann numbers.

6.5.2 Pumps, Propulsion and Projectiles

When Ha is large, Hartmann duct flow is characterised by a nearly uniform core velocity, $u \approx u_0$, and the two corresponding equations

$$J_z = \sigma(E_0 + u_0 B), \tag{6.88}$$

$$u_0 B = -E_0 - \frac{1}{\sigma B}\frac{dp}{dx}. \tag{6.89}$$

Note that we are free to choose the value of E_0, the external electric field. Depending on how we specify E_0, we obtain quite separate technological devices, as discussed at length in Shercliff, 1965, Moreau, 1990, and Müller & Bühler, 2001.

Suppose we choose $J_z = 0$, so that $E_0 = -u_0 B$. In this case there is no pressure gradient associated with B, the Lorentz force being zero. Such a device is called an MHD flowmeter, since E_0 may be measured to reveal u_0. Alternatively, if E_0 is zero, or small and positive, we have

$$J_z \approx \sigma u_0 B, \qquad \left|\frac{dp}{dx}\right| \approx \sigma B^2 u_0. \tag{6.90}$$

In this case we induce a current, but at the cost of a pressure drop. We are converting mechanical energy into electrical energy plus heat, and such a device is called a generator (Figure 6.14). This is the basis of MHD power generation, where hot ionised gas is propelled down a duct. The technological failure of MHD power generation, which was much publicised in scientific circles, is often attributed to the inability to develop refractory materials capable of withstanding the high temperatures involved (~3000K), rather than to any flaw in the MHD principle.

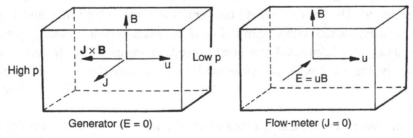

Figure 6.14 Principles of MHD generators and flowmeters.

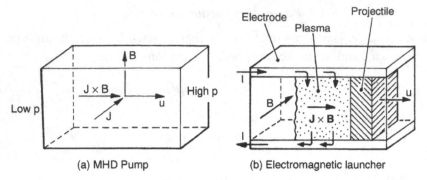

Figure 6.15 Principles of MHD pumps and guns.

If E_0 is negative and has a magnitude in excess of u_0B, the direction of J_z (and hence $\mathbf{J} \times \mathbf{B}$) is reversed. In this case, dp/dx is positive and we have a pump (Figure 6.15). Electrical energy is supplied to the device and this is converted into mechanical energy plus heat. MHD pumps are in common use, both in the metallurgical and the nuclear industries. Their obvious attraction is that they contain no moving parts and so, in principle, they are mechanically reliable. One can even combine an MHD generator with a pump to produce a so-called MHD flow-coupler. Here two ducts sit side by side, permeated by a common magnetic field, one producing electrical power for the other, which acts as a pump. One application is to transfer mechanical energy from one sodium loop to another in fast breeder reactors.

A variant of the MHD pump is the electromagnetic gun, sometimes called a rail gun or electromagnetic launcher. Here there is no externally applied magnetic field and so this is not a Hartmann flow. Rather, one relies on the magnetic field associated with the flow of current along the electrodes and through the fluid (plasma). It is readily confirmed that the interaction of $\mathbf{J}$ with its self field $\mathbf{B}$ induces a Lorentz force parallel to the electrodes, as shown in Figure 6.15. This is used to propel a plasma, ahead of which sits a non-conducting projectile. The advantage of such a device is that, as long as current is supplied, the projectile will accelerate. This contrasts with conventional guns, where movement of the projectile is associated with an expansion of the gas and hence a loss in pressure. Small masses have been accelerated up to speeds of around 7km/s in such devices.

Typically the electrodes are connected to a capacitor bank which delivers a current pulse of around $10^5 \rightarrow 10^6$ Amps in a period of a few milliseconds. In the first instance this vaporises a metal foil placed between the electrodes (rails) and so initiates a plasma. Current then enters the top rail, is syphoned off through the plasma and returns via the bottom rail (Figure 6.15). The resulting force accelerates the plasma along the duct, pushing the projectile ahead of it.

This simple idea is attractive to the extent that it can produce velocities much higher than those achievable by conventional means. It has been suggested that it might be used in fusion research, to create high impact velocities, as a laboratory tool to study high velocity projectiles and, of course, it has military applications. However, in practice it has three major drawbacks. First, the electrical power involved is substantial and this has to be delivered in a very short pulse. Considerable attention must be paid, therefore, to the storage and delivery of the electrical power. Second, the magnetic repulsion forces between the rails is very large and great care is required in the mechanical design of the gun, otherwise it is prone to self-destruction! Third, the plasma temperatures are very high, $\sim 2.5 \times 10^4$ K, and so there is a severe ablation of material from the inside surface of the duct. As a consequence, the plasma grows rapidly in size and weight, increasing the inertia of the propelled mass and reducing the projectile acceleration.

A variant of the electromagnetic gun, in which the projectile is removed, is the electromagnetic jet thruster. Here the device operates continuously: heating, ionising and propelling a plasma. Typically this has an annular geometry with a central cathode surrounded by a cylindrical anode. The gas is accelerated down the annular gap between the two, producing thrust. This has been proposed as a means of propelling space vehicles, its perceived advantage being its low fuel consumption.

There are many other variants of electromagnetic pumps and thrusters, including the much-publicised, but ill-fated, sea-water thruster for submarines (see §1.4). Some, such as the liquid-metal pump, are in common use. Others, such as the electromagnetic launcher, have yet to find any significant commercial application. In general it seems that the simplest, almost mundane, applications have faired best, while the more exotic suggestions have been largely confined to the realms of popular fiction.

Exercises

6.1 Consider the three-dimensional MHD jet shown in Figure 6.2(b). The imposed magnetic field is weak, in the sense that axial gradients in u are much smaller than the transverse gradients. Sketch the induced current distribution at any one cross-section of the jet and estimate, qualitatively, the distribution of $\mathbf{J} \times \mathbf{B}$. Hence explain why the jet elongates in the direction of $\mathbf{B}$ and also explain why a reverse flow is induced either side of the jet.

6.2 Consider the vortex shown in Figure 6.3. Sketch the induced current distribution (which is poloidal) and estimate, qualitatively, the distribution of $\mathbf{J} \times \mathbf{B}$. Show that this force induces a counter rotation in an annulus which surrounds the primary vortex.

6.3 The integration of (6.61) using the mixing length model of turbulence yields

$$\bar{u}_\theta/r = \Omega_f[(2\sqrt{2}\kappa)^{-1}\ln(R - r) + \text{ constant}].$$

The constant of integration is determined by the fact that, near the wall, the velocity profile must blend smoothly into the universal law of the wall

$$\bar{u}/V_* = \kappa^{-1}\ln(V_* y/\nu) + 5.5 , \quad y = R - r,$$

where $V_* = (\tau_w/\rho)^{1/2}$ is the shear velocity. This yields (6.62). When the surface is rough, however, the universal law of the wall changes to

$$\bar{u}/V_* = \kappa^{-1}\ln(y/k^*),$$

where k^* is a measure of the roughness height. Under these circumstances, show that (6.62) must be modified to

$$(\bar{u}_\theta/r)_0 = \Omega_f(2\sqrt{2}\kappa)^{-1}\ln(R/k^*).$$

6.4 When liquid metal is stirred in a hemispherical container by an azimuthal Lorentz force, Ekman layers are set up on the boundaries. Sketch the secondary flow induced by the Ekman pumping and estimate the azimuthal force balance in the core of the flow.

7

Dynamics at High Magnetic Reynolds Numbers

... and to those philosophers who pursue the inquiry (*of induction*) zealously yet continuously, combining experiment with analogy, suspicious of their preconceived notions, paying more respect to the fact than a theory, not too hasty to generalise, and above all things, willing at every step to cross-examine their own opinions, both by reasoning and experiment, no branch of knowledge can afford so fine and ready a field for discovery as this.

Faraday, *1837*

When R_m is large there is a strong influence of **u** on **B**, and so we obtain a two-way coupling between the velocity and magnetic fields. The tendency for **B** to be advected by **u**, which follows directly from Faraday's law of induction, coupled to Faraday's notion of tension in the field lines, results in a completely new phenomenon: the Alfvén wave. This two-way coupling also underpins our understanding of the origin of the earth's magnetic field and that of the sun.

High-R_m MHD is a substantial subject in its own right which has inspired many excellent monographs over the years, and so we cannot hope to provide a comprehensive coverage in an introductory textbook. Rather, our aim is to provide the beginner with a glimpse at some of the issues involved, offering a stepping stone to more serious study. The subject falls naturally into four main categories. First, there is the ability of magnetic fields to support internal waves, both Alfvén waves and so-called magnetostrophic waves. (The latter involves the Coriolis force, the former does not.) This topic is reasonably straightforward. The second category concerns the stability of a plasma that is threaded by a magnetic field, a topic which is motivated partly by the need for plasma confinement in fusion reactors, and partly by astrophysical applications, such as the stability of the flow through accretion discs. Here the interest lies in the stability of both static and non-static equilibria, and both areas are now reasonably well understood, at least as far as linear (small amplitude) stability is concerned. The third topic is geodynamo theory, which attempts to explain the maintenance of the earth's magnetic field (and also the magnetic fields of other planets) in terms of a self-excited fluid dynamo. Geodynamo theory is complex and incomplete, a situation which is not helped by the lack of observational data. So, despite the great advances that have been made, there remain many unresolved issues. Finally, there is astrophysical MHD, particularly topics such as star formation, accretion discs, and magnetic phenomena in and around the sun, such as field oscillations, sunspots, solar flares and the solar wind. Like geodynamo theory, the picture here is far from complete and many questions remain unanswered.

We shall cover the first two of these topics (waves and stability theory) in this chapter, as both are relatively clear cut, at least at an introductory level. We leave the more open-ended and expansive subjects for subsequent chapters, covering the geodynamo in Chapter 14 and astrophysical MHD in Chapter 15.

The layout of this chapter is as follows. We start with the simplest subject, that of wave theory. This divides naturally into two parts: Alfvén waves and magnetostophic waves, both of which play key roles in many astrophysical phenomena. Next we turn to the stability of MHD equilibria, both static and dynamic equilibria. Although the motivation here is plasma MHD, particularly the stability of fusion plasmas and accretion discs, we (artificially) restrict ourselves to incompressible fluids. The reason for this is simple: the algebra involved in developing stability theorems for even incompressible fluids is lengthy and somewhat tedious, and so it seems inappropriate in such a brief discussion to embrace all the additional complexities of compressibility.

Finally, perhaps it is worthwhile to comment on the notation employed in this chapter. Throughout we employ cylindrical polar coordinates, which we denote (r, θ, z). We make no use of spherical polar coordinates. When dealing with axisymmetric fields in cylindrical polar coordinates it is natural to divide a vector field, say a magnetic field **B**, into *azimuthal*, $(0, B_\theta, 0)$, and *poloidal*, $(B_r, 0, B_z)$, components. For example, the dipole-like external field of the earth is (more or less) poloidal, while the field induced by the current in a long straight wire is azimuthal. Axial symmetry demands that the curl of a poloidal field is azimuthal, while that of an azimuthal field is poloidal. Thus the current associated with a poloidal field, via Ampère's law, is azimuthal, while the current that supports an azimuthal magnetic field is poloidal. This is illustrated in Figure 7.1 for the particular case of the axisymmetric components of the current distribution and magnetic field in the core of the earth.

In the geophysical and astrophysical literature it is common to use a different terminology. The field is often divided into *toroidal* and *poloidal* components. When the field is axisymmetric, toroidal is equivalent to azimuthal. Occasionally the term meridional is substituted for poloidal. We shall not employ the terms toroidal or meridional.

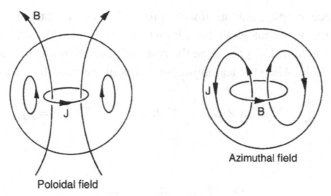

Figure 7.1 Poloidal and azimuthal and magnetic fields and the currents which support them.

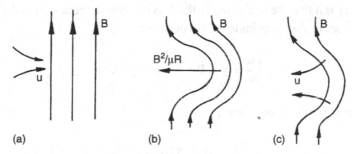

Figure 7.2 The formation of an Alfvén wave.

7.1 Alfvén Waves and Elsasser Variables

One of the remarkable properties of magnetic fields in MHD is that they can transmit transverse internal waves through an incompressible fluid, just like a plucked harp string. We have already discussed the physical origin of this phenomenon. It relies on the fact that the **B**-field and fluid are virtually frozen together when the electrical conductivity is high. To give an illustration, suppose that a portion of an initially uniform magnetic field is swept sideways by the lateral movement of the fluid (see Figure 7.2). The resulting curvature of the field line gives rise to a restoring force, $B^2/\mu R$, as discussed in §4.3. (Here R is the radius of curvature of the field line.) As the curvature increases, the restoring force rises and eventually the inertia of the fluid is overcome and the lateral movement is stopped. However, the Lorentz force is still present, and so the flow now reverses, carrying the field lines back with it. Eventually, the field lines return to their equilibrium position, only now the inertia of the fluid carries the field lines past the neutral point and the whole process starts in reverse. Oscillations then develop and this is called an Alfvén wave.

We now place our physical intuition on a formal mathematical basis. Suppose we have a uniform, steady magnetic field $\mathbf{B}_0$ which is perturbed by an infinitesimally small velocity field $\mathbf{u}$. Let $\mathbf{j}$ and $\mathbf{b}$ be the resulting perturbations in current density and magnetic field. Then the leading-order terms in the induction equation are

$$\frac{\partial \mathbf{b}}{\partial t} = \nabla \times (\mathbf{u} \times \mathbf{B}_0) + \lambda \nabla^2 \mathbf{b} , \qquad \nabla \times \mathbf{b} = \mu \mathbf{j}, \tag{7.1}$$

which yields

$$\frac{\partial \mathbf{j}}{\partial t} = \frac{1}{\mu}(\mathbf{B}_0 \cdot \nabla)\boldsymbol{\omega} + \lambda \nabla^2 \mathbf{j}. \tag{7.2}$$

We now consider the momentum of the fluid. Since $(\mathbf{u} \cdot \nabla)\mathbf{u}$ is quadratic in the small quantity $\mathbf{u}$ it may be neglected in the Navier–Stokes equation and so we have, to leading order in the amplitude of the perturbation,

$$\rho \frac{\partial \mathbf{u}}{\partial t} = \mathbf{j} \times \mathbf{B}_0 - \nabla p + \rho \nu \nabla^2 \mathbf{u}. \tag{7.3}$$

The equivalent vorticity equation is

$$\frac{\partial \boldsymbol{\omega}}{\partial t} = \frac{1}{\rho}(\mathbf{B}_0 \cdot \nabla)\mathbf{j} + \nu \nabla^2 \boldsymbol{\omega}. \tag{7.4}$$

We now eliminate $\mathbf{j}$ from (7.4) by taking the time derivative and then substitute for $\partial \mathbf{j}/\partial t$ using (7.2). After a little algebra we find

$$\left[\frac{\partial}{\partial t} - \lambda \nabla^2\right]\left[\frac{\partial}{\partial t} - \nu \nabla^2\right]\boldsymbol{\omega} = \frac{1}{\rho \mu}(\mathbf{B}_0 \cdot \nabla)^2 \boldsymbol{\omega}, \tag{7.5}$$

which supports plane-wave solutions of the form $\boldsymbol{\omega} = \hat{\boldsymbol{\omega}}\exp[j(\mathbf{k} \cdot \mathbf{x} - \varpi t)]$. A common situation in astrophysics and planetary interiors is $\lambda \gg \nu$, in which case we may drop the viscous term from (7.5). The dispersion relationship for plane waves is then

$$\varpi^2 + j\lambda k^2 \varpi = v_a^2 k_{//}^2, \tag{7.6}$$

where $v_a = \mathbf{B}_0/\sqrt{\rho\mu}$ is the Alfvén velocity and $k_{//}$ is the component of $\mathbf{k}$ parallel to $\mathbf{B}_0$.

There are three special cases of particular interest. When $\lambda = \nu = 0$ (a perfect fluid) we obtain $\varpi = \pm v_a k_{//}$, which represents the propagation of non-dispersive transverse waves with a phase velocity of v_a. When $\nu = 0$ and λ is small but finite, which is a good approximation for many astrophysical flows, as well as liquid-metals at high R_m, we find,

$$\varpi = \pm v_a k_{//} - (\lambda k^2/2)j. \tag{7.7}$$

This represents a plane wave propagating with phase velocity v_a and weakly damped by Ohmic dissipation (the high-R_m case in Figure 7.3).

Finally we consider the case of $\lambda \to \infty$, in the sense that the Lundquist number is small, $v_a/\lambda k \ll 1$. This characterises much of liquid-metal MHD at low R_m. Here we find the two roots,

$$\varpi = -j\lambda k^2, \qquad \varpi = -j\tau^{-1}\left(k_{//}/k\right)^2, \tag{7.8}$$

where τ is the magnetic damping time $\left(\sigma B^2/\rho\right)^{-1}$. The first of these solutions represents a disturbance which is rapidly eradicated by Ohmic dissipation. This is of little interest. However, the second solution represents a non-oscillatory disturbance which decays rather more slowly, on a time scale of τ (the low-R_m case in Figure 7.3). This is the origin of the pseudo-diffusive, low-R_m phenomenon discussed in Chapter 6, as illustrated in Figure 7.4 for the case of a low-R_m jet

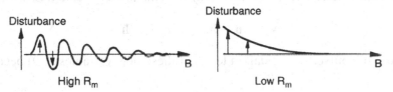

Figure 7.3 Damped Alfvén waves at high and low R_m. Note that the low R_m wave is non-oscillatory and corresponds to $v_a/\lambda k \ll 1$.

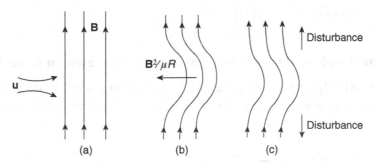

Figure 7.4 A submerged low-R_m jet acts on an imposed magnetic field and then spreads diffusively along the field lines. This can be understood in terms of Faraday tensions in the field lines.

that spreads diffusively along the magnetic field lines. (The diffusion arises through a gradual relaxation in the Faraday tension in the field lines.)

Let us now return to ideal fluids ($\lambda = \nu = 0$) and look for more general solutions of the linearised equations. We start by rewriting (7.1) and (7.3) in terms of the scaled field $\mathbf{h} = \mathbf{B}/\sqrt{\rho\mu}$,

$$\frac{\partial \mathbf{u}}{\partial t} = (\mathbf{h}_0 \cdot \nabla)\mathbf{h}, \quad \frac{\partial \mathbf{h}}{\partial t} = (\mathbf{h}_0 \cdot \nabla)\mathbf{u}, \tag{7.9}$$

where $\mathbf{h}_0$ is the imposed magnetic field and the pressure term is zero in the perturbed Euler equation because $\mathbf{u}$ and $\mathbf{h}$ are both solenoidal. Since Alfvén waves are non-dispersive, we might anticipate that they support D'Alembert-like solutions of a rather general form, and indeed it is readily confirmed that a generic solution of (7.9) is

$$\mathbf{u} = \mathbf{V}(\mathbf{x} \pm \mathbf{h}_0 t), \quad \mathbf{h} = \mathbf{h}_0 \pm \mathbf{V}(\mathbf{x} \pm \mathbf{h}_0 t), \tag{7.10}$$

where $\mathbf{V}$ is an arbitrary solenoidal vector function of it argument. Evidently, these represent disturbances of arbitrary shape which travel without change of shape along the imposed field at the Alfvén velocity, $\boldsymbol{v}_a = \mathbf{h}_0$, with the upper (lower) signs in (7.10) corresponding to backward (forward) traveling waves.

It is common to rewrite (7.9) and (7.10) in terms of the so-called Elsasser fields,

$$\mathbf{v}^+ = \mathbf{u} + \mathbf{h}, \quad \mathbf{v}^- = \mathbf{u} - \mathbf{h}, \tag{7.11}$$

which are, of course, solenoidal. In terms of these new variables (7.9) becomes

$$\frac{\partial \mathbf{v}^\pm}{\partial t} = \pm (\mathbf{h}_0 \cdot \nabla)\mathbf{v}^\pm, \tag{7.12}$$

while (7.10) confirms that $\mathbf{v}^-$ represents a wave travelling in the direction of $\mathbf{B}_0$, while $\mathbf{v}^+$ is a backward travelling wave.

7.2 Finite-Amplitude Alfvén Waves and the Conservation of Cross Helicity

Rather remarkably, the induction and Euler equations support *finite-amplitude* Alfvén waves of arbitrary shape in an ideal fluid. The governing equations for such fully non-linear waves, rewritten in terms of Maxwell stresses, are

$$\frac{\partial \mathbf{u}}{\partial t} + (\mathbf{u} \cdot \nabla)\mathbf{u} = -\nabla[(p + p_M)/\rho] + (\mathbf{h} \cdot \nabla)\mathbf{h}, \tag{7.13}$$

and

$$\frac{\partial \mathbf{h}}{\partial t} + (\mathbf{u} \cdot \nabla)\mathbf{h} = (\mathbf{h} \cdot \nabla)\mathbf{u}, \tag{7.14}$$

where $p_M = \mathbf{B}^2/2\mu$ is the magnetic pressure and $\mathbf{h}$ is the scaled field $\mathbf{h} = \mathbf{B}/\sqrt{\rho\mu}$. The symmetric nature of the non-linear inertial and Lorentz forces now plays a crucial role and leads to a great deal of cancellation when we perturb $\mathbf{B}$ about the base state $\mathbf{u} = 0$, $\mathbf{B} = \mathbf{B}_0$. In particular, it is readily confirmed (see Moffatt, 1978) that (7.13) and (7.14) support finite-amplitude waves of the form

$$\mathbf{u} = \mathbf{V}(\mathbf{x} \pm \mathbf{h}_0 t), \quad \mathbf{h} = \mathbf{h}_0 \pm \mathbf{V}(\mathbf{x} \pm \mathbf{h}_0 t), \tag{7.15}$$

which is identical to (7.10), except that $\mathbf{V}$ is now not only of arbitrary shape, but also of arbitrary amplitude. These represent finite-amplitude disturbances which travel without change of shape along an imposed field at the Alfvén velocity, with the upper sign in (7.15) representing backward travelling waves, and the lower sign forward travelling waves.

As with linear waves, a more symmetric form of the governing equations may be obtained by introducing Elsasser variables, $\mathbf{v}^\pm = \mathbf{u} \pm \mathbf{h}$. When rewritten in terms $\mathbf{v}^+$ and $\mathbf{v}^-$, (7.13) and (7.14) become

$$\frac{\partial \mathbf{v}^\pm}{\partial t} + (\mathbf{v}^\mp \cdot \nabla)\mathbf{v}^\pm = -\nabla[(p + p_M)/\rho], \tag{7.16}$$

which reduces to (7.12) in the linear regime, $|\mathbf{b}| \ll |\mathbf{B}_0|$. As before, $\mathbf{v}^-$ represents a wave travelling in the direction of $\mathbf{B}_0$, while $\mathbf{v}^+$ is a backward-travelling wave.

Now consider the situation where the finite-amplitude disturbance, and hence both $\mathbf{u}$ and $\mathbf{h} - \mathbf{h}_0$, are localised in space. It is convenient here to introduce modified Elsasser variables in which the mean field $\mathbf{h}_0$ is subtracted out: $\hat{\mathbf{v}}^\pm = \mathbf{u} \pm (\mathbf{h} - \mathbf{h}_0)$. Then (7.16) becomes

$$\frac{\partial \hat{\mathbf{v}}^\pm}{\partial t} + (\hat{\mathbf{v}}^\mp \cdot \nabla)\hat{\mathbf{v}}^\pm = \pm(\mathbf{h}_0 \cdot \nabla)\hat{\mathbf{v}}^\pm - \nabla[(p + p_M)/\rho]. \tag{7.17}$$

We now multiply (7.17) by $\hat{\mathbf{v}}^\pm$, integrate over all space, and invoke Gauss' theorem to convert all terms (with the exception of the first term on the left) into surface integrals. The localised nature of $\hat{\mathbf{v}}^\pm$ then demands that the surface integrals all vanish and we conclude that the volume integrals of $(\hat{\mathbf{v}}^+)^2$ and $(\hat{\mathbf{v}}^-)^2$ are individually conserved. Moreover, it is clear that $(\hat{\mathbf{v}}^+)^2 + (\hat{\mathbf{v}}^-)^2 = 2\left(\mathbf{u}^2 + (\mathbf{h} - \mathbf{h}_0)^2\right)$ and $(\hat{\mathbf{v}}^+)^2 - (\hat{\mathbf{v}}^-)^2 = 4\mathbf{u} \cdot (\mathbf{h} - \mathbf{h}_0)$. Evidently the global conservation of $(\hat{\mathbf{v}}^+)^2 + (\hat{\mathbf{v}}^-)^2$ is simply a manifestation of the conservation of energy of the disturbance. However, the global conservation of $\left(\hat{\mathbf{v}}^{+}\right)^{2} - \left(\hat{\mathbf{v}}^{-}\right)^{2}$ provides us with a second invariant,

$$h_{c} = \int_{V_{\infty}} \mathbf{u} \cdot (\mathbf{B} - \mathbf{B}_{0}) dV = \int_{V_{\infty}} \mathbf{u} \cdot \mathbf{b} dV = \text{constant}, \qquad (7.18)$$

which is the *cross helicity* of the disturbance. The conservation of cross helicity in an ideal fluid was introduced in §5.4.3 for cases where the mean magnetic field is zero, and the result above is a natural generalisation of this. The approximate (but not exact) conservation of helicity in real fluids is thought to play a crucial role in certain types of high-R_m turbulence, as discussed in Chapter 9.

With the notable exception of planetary dynamos, Alfvén waves are of little importance in liquid-metal MHD, since R_m is usually rather modest in such cases. However, they are of considerable importance in astrophysical MHD, where they provide an effective mechanism for propagating energy and momentum. For example, it has been suggested that they are responsible for transporting angular momentum away from the core of an interstellar cloud which is collapsing to form a star under the influence of self-gravitation. Finite-amplitude Alfvén waves also play a central role in our understanding of MHD turbulence at high R_m, with implications for the solar wind.

7.3 Colliding Alfvén Wave Packets and a Glimpse at Alfvénic Turbulence

Let us now consider the collision of oppositely travelling Alfvén wave packets of finite amplitude (Figure 7.5). We start by considering an ideal fluid ($\lambda = v = 0$) and by rewriting the finite-amplitude solutions (7.15) as

$$\mathbf{u} = \mathbf{V}^{F}(\mathbf{x} - \mathbf{h}_{0}t), \quad \mathbf{h} = \mathbf{h}_{0} - \mathbf{V}^{F}(\mathbf{x} - \mathbf{h}_{0}t), \qquad (7.19)$$

and

$$\mathbf{u} = \mathbf{V}^{B}(\mathbf{x} + \mathbf{h}_{0}t), \quad \mathbf{h} = \mathbf{h}_{0} + \mathbf{V}^{B}(\mathbf{x} + \mathbf{h}_{0}t), \qquad (7.20)$$

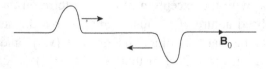

Figure 7.5 Alfvén waves travel without change of shape until they collide with oppositely travelling waves. These non-linear waves may have amplitudes of arbitrary magnitude.

where $\mathbf{V}^F$ and $\mathbf{V}^B$ are arbitrary solenoidal functions of their arguments, and the superscripts F and B stand for forward and backward travelling waves, respectively. Note that the cross helicity of the wavepackets, $\int \mathbf{u} \cdot \mathbf{b} dV$, is negative for forward-travelling waves and positive for backward-travelling waves. In terms of the Elsasser fields, $\hat{\mathbf{v}}^{\pm} = \mathbf{u} \pm (\mathbf{h} - \mathbf{h}_0)$, which evolve according to (7.17), we have $\hat{\mathbf{v}}^+ = 2\mathbf{V}^B$ and $\hat{\mathbf{v}}^- = 2\mathbf{V}^F$.

The most important property of (7.17) is that the non-linear term, $(\hat{\mathbf{v}}^{\mp} \cdot \nabla)\hat{\mathbf{v}}^{\pm}$, involves waves travelling in opposite directions. Thus two or more finite-amplitude Alfvén waves travelling in the same direction do not interact as we would expect, since we could simply reclassify a combination of such wave packets as a single non-linear wave travelling without change of shape. Evidently non-linear interactions are restricted to oppositely travelling waves.

Let us now consider what happens when two such wave packets collide. Suppose that $\hat{\mathbf{v}}^+$ and $\hat{\mathbf{v}}^-$ are localised disturbances. Then (7.17) tells us that $\int_{V_\infty} (\hat{\mathbf{v}}^+)^2 dV$ and $\int_{V_\infty} (\hat{\mathbf{v}}^-)^2 dV$ are both conserved throughout the collision, which is a manifestation of energy and cross helicity conservation. Consider now the case where our blob-like disturbances are initially distinct, so that $\hat{\mathbf{v}}^- = 2\mathbf{V}_1^F$ and $\hat{\mathbf{v}}^+ = 2\mathbf{V}_1^B$, but travel towards each other along $\mathbf{B}_0$. Initially there is no interaction between $\mathbf{V}_1^F$ and $\mathbf{V}_1^B$, but as they begin to collide the non-linear term $(\hat{\mathbf{v}}^{\mp} \cdot \nabla)\hat{\mathbf{v}}^{\pm}$ comes into play and so the wave packets are progressively distorted throughout the collision. In general, however, the wave packets will eventually pass through one another and emerge with new shapes, $\mathbf{V}_2^F$ and $\mathbf{V}_2^B$. At this point the non-linear interactions cease and the waves again propagate without change of shape. Before and after the collisions, when the two wave packets are non-overlapping, we have

$$\frac{1}{4}\int (\hat{\mathbf{v}}^-)^2 dV = \int (\mathbf{V}^F)^2 dV = \int \left[\frac{1}{2}\mathbf{u}^2 + \mathbf{b}^2/2\rho\mu\right]^F dV = -\frac{1}{\sqrt{\rho\mu}}\int [\mathbf{u} \cdot \mathbf{b}]^F dV,$$

$$(7.21)$$

$$\frac{1}{4}\int (\hat{\mathbf{v}}^+)^2 dV = \int (\mathbf{V}^B)^2 dV = \int \left[\frac{1}{2}\mathbf{u}^2 + \mathbf{b}^2/2\rho\mu\right]^B dV = \frac{1}{\sqrt{\rho\mu}}\int [\mathbf{u} \cdot \mathbf{b}]^B dV,$$

$$(7.22)$$

where the superscripts F and B indicate that we integrate over the forward (or backward) wave only. However, throughout the interaction $\int_{V_\infty} (\hat{\mathbf{v}}^+)^2 dV$ and $\int_{V_\infty} (\hat{\mathbf{v}}^-)^2 dV$ are individually conserved and so the waves emerge from the collision with their initial values of energy and cross helicity intact. The shapes of the two wave packets will change, however, as both disturbances are distorted during the collision by the non-linear interactions between the waves.

Now suppose that there exists a random sea of blob-like disturbances (wave-packets), some travelling forward and some travelling backward along the mean field, $\mathbf{B}_0$. They experience repeated collisions and it is reasonable to suppose that the energy in the blobs becomes increasingly spread over a wider range of scales during these collisions (as is typical of non-linear interactions), with some of the energy passed to smaller scales and some to larger scales. If we now allow for a small but finite magnetic diffusivity, the kinetic and magnetic energy which is passed to small scales will eventually be dissipated by Ohmic heating within thin current sheets. We now have a zero-order model of high-R_m turbulence evolving in the presence of a mean magnetic field, so-called *Alfvénic turbulence*.

For simplicity, suppose that $|\mathbf{b}| \ll |\mathbf{B}_0|$, so that the non-linear interactions are weak and many collisions are required for the energy in a given disturbance to reach the scale at which Ohmic dissipation occurs. During a single collision of two similarly sized disturbances, (7.17) suggests that the forward- and backward-travelling wave-packets distort each other through the non-linear term $(\hat{\mathbf{v}}^{\mp} \cdot \nabla)\hat{\mathbf{v}}^{\pm}$ by an amount

$$|\delta\hat{\mathbf{v}}^{\pm}| \sim |(\hat{\mathbf{v}}^{\mp} \cdot \nabla)\hat{\mathbf{v}}^{\pm}| \times (\text{wavepacket interaction time}).$$

We can rewrite this as

$$|\delta\hat{\mathbf{v}}^{\pm}| \sim |(\hat{\mathbf{v}}^{\mp} \cdot \nabla)\hat{\mathbf{v}}^{\pm}| \frac{\ell}{v_a},$$

where $v_a = |\mathbf{v}_a| = |\mathbf{h}_0|$ is the Alfvén wave speed and ℓ is the characteristic length scale of the colliding blobs. (We assume here that both disturbances can be characterised by a single length scale.) Thus we have the estimate

$$|\delta\hat{\mathbf{v}}^{\pm}| \sim \frac{|\hat{\mathbf{v}}^{\mp}||\hat{\mathbf{v}}^{\pm}|}{\ell} \frac{\ell}{v_a} = \frac{|\hat{\mathbf{v}}^{\mp}||\hat{\mathbf{v}}^{\pm}|}{v_a},$$

which suggests that the percentage change in $|\hat{\mathbf{v}}^{\pm}|$ per collision, $|\delta\hat{\mathbf{v}}^{\pm}|/|\hat{\mathbf{v}}^{\pm}|$, is proportional to $|\hat{\mathbf{v}}^{\mp}|/v_a$. This, in turn, suggests that the minority field (the weaker of $\hat{\mathbf{v}}^{+}$ and $\hat{\mathbf{v}}^{-}$) is more severely buffeted by each collision, and hence more rapidly destroyed. So, for example, if there is a local excess of energy in the forward travelling waves at some particular location, we might expect those waves to progressively eradicate the backward-travelling disturbances until all the wave-packets travel in the same direction. This hypothesised process, which has some observational support, is called *dynamic alignment*.

In short, any local bias towards forward- or backward-travelling waves will be progressively reinforced, until all the waves in a given region travel in the same direction, the non-linear interactions shut down, and we reach a local state in which

$$\mathbf{u} = (\mathbf{h} - \mathbf{h}_0) \quad \text{or else} \quad \mathbf{u} = -(\mathbf{h} - \mathbf{h}_0).$$

This is known as an *Alfvénic state* and it is characterised by the fact that the cross helicity is uniformly of one sign and all waves travel in the same direction.

7.4 Magnetostrophic Waves

There is second class of wave motion which is important in astrophysics and in planetary interiors, called *magnetostrophic waves*. It arises when there is both bulk rotation and a locally uniform magnetic field.

Consider a fluid which rotates at the rate $\mathbf{\Omega} = \Omega \hat{\mathbf{e}}_z$. We adopt a system of coordinates which rotates with the fluid and suppose that, in this rotating frame of reference, there is a steady, uniform magnetic field, $\mathbf{B}_0$. If we consider small perturbations in the rotating frame, the linearised equations of motion are similar to (7.1) and (7.4), except that we must incorporate the Coriolis force. For inviscid fluids these are

$$\frac{\partial \mathbf{b}}{\partial t} = (\mathbf{B}_0 \cdot \nabla)\mathbf{u} + \lambda \nabla^2 \mathbf{b}, \tag{7.23}$$

$$\frac{\partial \boldsymbol{\omega}}{\partial t} = \frac{1}{\rho}(\mathbf{B}_0 \cdot \nabla)\mathbf{j} + 2(\mathbf{\Omega} \cdot \nabla)\mathbf{u}, \tag{7.24}$$

and for ideal fluids ($\lambda = \nu = 0$) these combine to give

$$\left[\frac{\partial^2}{\partial t^2} - \frac{1}{\rho\mu}(\mathbf{B}_0 \cdot \nabla)^2\right]^2 \nabla^2 \mathbf{u} + (2\mathbf{\Omega} \cdot \nabla)^2 \frac{\partial^2 \mathbf{u}}{\partial t^2} = 0. \tag{7.25}$$

When $\mathbf{B}_0 = 0$ we recover inertial waves, with the frequency $\pm 2\mathbf{\Omega} \cdot \mathbf{k}/|\mathbf{k}|$, as discussed in §3.9.2, and when $\mathbf{\Omega} = 0$ we obtain Alfvén waves, whose frequency is $\pm \mathbf{B} \cdot \mathbf{k}/\sqrt{\rho\mu}$. We shall now show that (7.25) also embodies a third class of wave, known as magnetostrophic waves.

If we search for plane-wave solutions of (7.23) and (7.24) of the form $\mathbf{u} = \hat{\mathbf{u}}\exp[j(\mathbf{k} \cdot \mathbf{x} - \varpi t)]$, then we find

$$\hat{\mathbf{u}} = \mp\frac{\hat{\boldsymbol{\omega}}}{k}, \quad \hat{\mathbf{b}} = \mp\frac{\mu\hat{\mathbf{j}}}{k}, \quad \hat{\mathbf{u}} = -\frac{\varpi_B}{\varpi \mp \varpi_\Omega}\frac{\hat{\mathbf{b}}}{\sqrt{\rho\mu}}, \tag{7.26}$$

along with the dispersion relationship

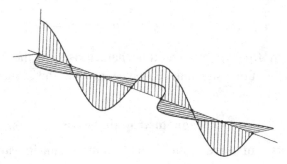

Figure 7.6 All plane-wave, monochromatic solutions of (7.25), including magnetostrophic waves, are circularly polarised waves in which the perturbation in
velocity rotates as the wave propagates. Like inertial waves, they have maximal
helicity.

$$\varpi(\varpi + j\lambda k^2) - \varpi_B^2 = \pm\varpi_\Omega(\varpi + j\lambda k^2), \tag{7.27}$$

where $\varpi_B = \mathbf{B}_0 \cdot \mathbf{k}/\sqrt{\rho\mu}$ and $\varpi_\Omega = 2\mathbf{\Omega} \cdot \mathbf{k}/k$. A great deal of useful structural
information follows from (7.26). For example, $\mathbf{u}$ and $\boldsymbol{\omega}$ are in phase and parallel,
as are $\mathbf{b}$ and $\mathbf{j}$, so that all plane-wave solutions of (7.25) have maximal kinetic and
magnetic helicity ($\mathbf{u} \cdot \boldsymbol{\omega}$ and $\mathbf{a} \cdot \mathbf{b}$), with the upper sign in (7.26) and (7.27)
corresponding to negative helicity. Moreover, the fact that $k\hat{\mathbf{u}} = \mp\hat{\boldsymbol{\omega}}$, irrespective
of the presence of a mean magnetic field, tells us that all plane-wave, monochro-
matic solutions of (7.25) have the same spatial structure for $\mathbf{u}$ (provided that $\mathbf{\Omega}$ is
non-zero). Thus the structure of these waves is identical to that of pure inertial
waves, i.e. a circularly polarised wave of maximal kinetic helicity (Figure 7.6).

Let us now ignore dissipation and set λ to zero. The dispersion relationship then
simplifies to

$$\varpi^2 \mp \varpi_\Omega\varpi = \varpi_B^2, \tag{7.28}$$

and the corresponding group velocity, $\mathbf{c}_g$, can be written as

$$[1 + (\varpi_B/\varpi)^2]\mathbf{c}_g = \pm[\mathbf{c}_g]_\Omega + 2(\varpi_B/\varpi)[\mathbf{c}_g]_B, \tag{7.29}$$

where $[\mathbf{c}_g]_B = \mathbf{B}_0/\sqrt{\rho\mu}$, $[\mathbf{c}_g]_\Omega = 2\mathbf{k} \times (\mathbf{\Omega} \times \mathbf{k})/k^3$ and $\pm[\mathbf{c}_g]_\Omega$ is the group velocity
of inertial waves. (Recall that the group velocity of a wave is the velocity at which
wave energy disperses from a localised disturbance in the form of wave radiation.)
Note that, if $\mathbf{B}_0$ is perpendicular to $\mathbf{\Omega}$, then

$$[1 + (\varpi_B/\varpi)^2]\mathbf{c}_g \cdot \mathbf{\Omega} = \pm[\mathbf{c}_g]_\Omega \cdot \mathbf{\Omega} = \pm 2k^{-3}[k^2\Omega^2 - (\mathbf{k} \cdot \mathbf{\Omega})^2],$$

and it follows that, like pure inertial waves, waves propagating in the direction of Ω carry negative helicity, while those propagating antiparallel to Ω carry positive helicity (see §3.9.2).

Now consider the special case of $|\varpi_\Omega| \gg |\varpi_B|$, so that, in some sense, rotation dominates over magnetic forces. In this case the dispersion relationship $\varpi^2 \mp \varpi_\Omega \varpi = \varpi_B^2$ yields two sets of solutions: $\varpi = \pm \varpi_\Omega$ and $\varpi = \mp \varpi_B^2/\varpi_\Omega$. The first of these is simply an inertial wave, which is to be expected. The second, however, is unexpected. It is a helical wave whose frequency is much less than either $|\varpi_\Omega|$ or $|\varpi_B|$. These slow, long-lived oscillations are called magnetostrophic waves, and they play an important role in planetary cores. The group velocity corresponding to $\varpi = \mp \varpi_B^2/\varpi_\Omega$ is

$$[1 + (\varpi_\Omega/\varpi_B)^2]\mathbf{c}_g = \pm[\mathbf{c}_g]_\Omega \mp 2(\varpi_\Omega/\varpi_B)[\mathbf{c}_g]_B, \qquad (7.30a)$$

or equivalently

$$\mathbf{c}_g = \varpi\left[\frac{2\mathbf{B}_0}{\mathbf{B}_0 \cdot \mathbf{k}} + \frac{\mathbf{k} \times (\mathbf{k} \times \Omega)}{k^2(\Omega \cdot \mathbf{k})}\right], \qquad (7.30b)$$

so that energy can propagate in the directions $\pm\mathbf{B}_0$ and $\pm\mathbf{k} \times (\mathbf{k} \times \Omega)$. Note that the component of the magnetostrophic group velocity which is parallel to Ω is much smaller than that of the equivalent inertial waves, reduced by a factor of $1 + (\varpi_B/\varpi)^2 = 1 + (\varpi_\Omega/\varpi_B)^2$. So these are slow, low-frequency waves whose timescales in planetary cores are typically measured in hundreds or thousands of years. By way of contrast, inertial waves can traverse the liquid core of the earth on a timescale of months.

7.5 The Energy Principle for Magnetostatic Equilibria in Ideal Fluids

One of the successes of high-R_m MHD lies in the area of stability theory. This has its roots, not in liquid-metal MHD, but rather in plasma physics, although we shall restrict our discussion to incompressible fluids. A question which is often asked in fluid mechanics is: 'Is a given equilibrium stable to small disturbances?' That is to say, if a static or non-static equilibrium is disturbed, will it evolve into a radically different form or will it remain close (in some sense) to its initial distribution. The method used most often to answer this question is so-called normal mode analysis. This proceeds by looking for small-amplitude disturbances which are of the separable form $\delta\mathbf{u}(\mathbf{x}, t) = \hat{\mathbf{u}}(\mathbf{x})e^{pt}$. When quadratic terms in the small disturbance are neglected the governing equations of motion become linear in $\delta\mathbf{u}$ and this defines an eigenvalue problem for the amplitude of the disturbance, $\hat{\mathbf{u}}(\mathbf{x})$.

The eigenvalues of this equation determine p, and the equilibrium is deemed to be unstable if any p can be found which has a real positive part. This works well when the geometry of the base configuration is particularly simple, possessing a high degree of symmetry, e.g. an equilibrium which is homogeneous in two directions. However, if there is any significant complexity to the base state (it is inhomogeneous in two or three directions) this procedure rapidly becomes very messy, requiring numerical methods to determine the eigenvalues.

In MHD an alternative method has been developed, which relies on the conservation of energy. This has the advantage that it may be readily applied to equilibria of arbitrary complexity, but it has two major shortcomings. First, it applies only to non-dissipative systems ($\lambda = \nu = 0$), which we might call *ideal MHD*. Second, when dealing with non-static equilibria (i.e. steady flows) it usually provides sufficient, but not necessary, conditions for stability. Thus, often a flow may be proved stable, but it cannot be shown to be unstable. (In the case of static equilibria, however, both necessary and sufficient conditions can usually be deduced.) We shall describe this energy method, along with some of its consequences, in the remainder of this chapter, starting with the simpler case of static equilibria. First, however, we shall discuss the motivation for developing special stability methods in MHD.

7.5.1 The Need for Stability in Plasma Confinement

In the 1950s the quest for controlled thermonuclear fusion began in earnest. This requires that the hot plasma be confined away from material surfaces and since these plasmas are good conductors, magnetic pressure seemed the obvious confinement mechanism. Such static equilibria are governed by the magnetostatic equation

$$\mathbf{J} \times \mathbf{B} = \nabla p, \tag{7.31}$$

which is analogous to the governing equation for steady Euler flows,

$$\boldsymbol{\omega} \times \mathbf{u} = -\nabla h, \qquad h = p/\rho + u^2/2. \tag{7.32}$$

(Note, however, the minus sign on the left of (7.32), a distinction which turns out to be important when it comes to stability considerations.) Evidently $\mathbf{J} \cdot \nabla p = \mathbf{B} \cdot \nabla p = 0$ and so the $\mathbf{J}$ and $\mathbf{B}$ lines lie on constant pressure surfaces. In a typical confinement scheme these surfaces take the form shown in Figure 7.7, being closed, nested tori.

When the magnetostatic equilibrium is axisymmetric, (7.31) takes a particularly simple form which allows the shape of the constant pressure surfaces, and the

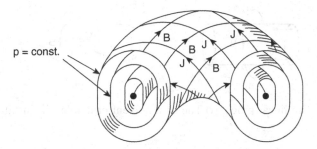

Figure 7.7 In magnetostatics the **B** and **J** lines sit on constant pressure surfaces.

distribution of **J** and **B** on those surfaces, to be determined. We start by dividing both **J** and **B** into azimuthal and poloidal components, i.e. $\mathbf{B} = \mathbf{B}_p + \mathbf{B}_\theta = (B_r, 0, B_z) + (0, B_\theta, 0)$ in cylindrical polar coordinates. Ampère's law then yields

$$\nabla \times \mathbf{B}_p = \mu \mathbf{J}_\theta , \qquad \nabla \times \mathbf{B}_\theta = \mu \mathbf{J}_p. \tag{7.33}$$

Moreover, axial symmetry ensures that $\nabla \cdot \mathbf{B}_p = 0$, and so we can introduce a *flux function* for $\mathbf{B}_p$, defined by

$$\mathbf{B}_p = \nabla \times [(\varphi/r)\hat{\mathbf{e}}_\theta] = \left(-\frac{1}{r}\frac{\partial \varphi}{\partial z}, \ 0, \ \frac{1}{r}\frac{\partial \varphi}{\partial r} \right), \tag{7.34}$$

where

$$\mu \mathbf{J}_\theta = \nabla \times \mathbf{B}_p = -r^{-1}\left(\nabla_*^2 \varphi \right)\hat{\mathbf{e}}_\theta, \tag{7.35}$$

and ∇_*^2 is the usual Laplacian-like operator,

$$\nabla_*^2 \equiv \frac{\partial^2}{\partial z^2} + r\frac{\partial}{\partial r}\left(\frac{1}{r}\frac{\partial}{\partial r} \right). \tag{7.36}$$

The magnetostatic equation (7.31) now yields: (i) $\mathbf{B}_p \cdot \nabla p = 0$, from which $p = p(\varphi)$; (ii) the azimuthal force balance $\mu r \mathbf{J}_p \times \mathbf{B}_p = \mathbf{B}_p \cdot \nabla(r B_\theta)\hat{\mathbf{e}}_\theta = 0$, which demands that $r B_\theta = \Gamma(\varphi)$ for some function Γ; and (iii) the poloidal force balance $\mathbf{J}_\theta \times \mathbf{B}_p + \mathbf{J}_p \times \mathbf{B}_\theta = \nabla p$. It is readily confirmed that, given $p = p(\varphi)$ and $r B_\theta = \Gamma(\varphi)$, the poloidal force balance reduces to the non-linear elliptic equation

$$\nabla_*^2 \varphi = -\Gamma \Gamma'(\varphi) - r^2 \mu p'(\varphi). \tag{7.37}$$

Evidently, the equivalent result for the steady Euler equation (7.32) is

Figure 7.8 The linear pinch. (i) The confinement principle, (ii) instability of the pinch.

$$\nabla_*^2 \psi = -\Gamma\Gamma'(\psi) + r^2 h'(\psi), \qquad \Gamma = r u_\theta, \tag{7.38}$$

where ψ is the Stokes streamfunction for the poloidal velocity. Equation (7.37) is referred to as the Grad–Shafranov equation in the plasma literature, while (7.38) is known as the Squire–Long equation in the fluids community, although perhaps it is worth noting that these equations were derived independently by several authors throughout the 1950s. In any event, (7.37) can be solved for the flux function if the functional forms of $p(\varphi)$ and $\Gamma(\varphi)$ are known. In principle, this allows one to determine the location of the constant pressure surfaces and the distributions of **J** and **B** on those surfaces. However, it does not tell us if the corresponding equilibrium is stable.

In any plasma confinement scheme one must arrange for the Lorentz force to point inward, towards the magnetic axis. Perhaps the simplest confinement system is that shown in Figure 7.8. An axial current is induced in the surface of the plasma, which is in the form of a cylinder, and the resulting azimuthal field creates a radial Lorentz force which is directed inward, since like currents attract. (Of course, to form a realisable confinement system, the cylinder should be deformed into a torus.) This configuration is known as the linear pinch, or *z*-pinch. Regrettably, it is unstable. Let I be the total current passing along the column. Then the surface field just outside the plasma is $B_\theta = \mu I/2\pi R$, where R is the radius of the column. If R locally decreases for some reason, say from R to $R - \delta R$, then B_θ rises by an amount $\delta B_\theta = B_\theta \delta R/R$. A 'sausage-mode' instability then develops because there is a rise in magnetic pressure, $\delta p_m = B_\theta \delta B_\theta/\mu$, at precisely those points where the radius reduces.

This sausage-mode instability may be stabilised by trapping a longitudinal magnetic field, B_z, within the plasma, which pushes radially outward. The idea is the following. If λ is very small then this longitudinal field is frozen into the plasma, so if R reduces locally to $R - \delta R$, the longitudinal magnetic field within the plasma will increase by an amount $\delta B_z = 2B_z\delta R/R$, the total magnetic flux, $\Phi = B_z\pi R^2$, remaining constant. The magnetic pressure due to B_z, which points outward, then increases by $\delta p_m = B_z\delta B_z/\mu = 2\,B_z^2\delta R/\mu R$, and this tends to counterbalance the

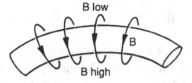

Figure 7.9 The kink instability.

rise in 'pinch pressure' $\delta p_m = B_\theta^2 \delta R / \mu R$. The column is then stable to axisymmetric disturbances provided that $B_z^2 > B_\theta^2 / 2$.

Unfortunately, this is not the end of the story. The column is unstable to non-axisymmetric disturbances even in the presence of a longitudinal field. This is known as the *kink instability*. Suppose that the column is bent slightly, as shown in Figure 7.9. The azimuthal field lines are pressed together on the concave side, and spaced out on the other side. Thus the external magnetic field, and hence the external magnetic pressure, is increased on the concave side of the column and reduced on the convex side. This produces a net sideways force which accentuates the initial disturbance.

In fact, confining plasmas using magnetic fields turns out to be altogether rather tricky, as discussed in Chapter 16. It is not just the *z*-pinch which is unstable. In the late 1950s, plasma physicists were faced with the problem of deciding which confinement schemes were stable and which were unstable. Conventional, normal-mode techniques seemed cumbersome and so a new stability theory was developed (primarily at Princeton), first for magnetostatic equilibria, such as that discussed above, and shortly afterwards for any steady solution of the equations of ideal MHD, static or non-static. This new method is based on the conservation of energy and in fact it is more in line with our intuitive notions of stability than conventional normal-mode analysis. For example, just as the stability of a ball sitting on an undulating surface is determined by whether its potential energy is a minimum or a maximum at equilibrium, so the stability of our magnetostatic equilibrium is determined by whether or not its magnetic energy, $E_B = \int (\mathbf{B}^2 / 2\mu) dV$, is a minimum at equilibrium. Unfortunately, though, the proof of these new stability theorems requires a great deal of vector calculus. Consequently, the proofs which follow are not for the impatient or the faint-hearted. The end result, though, is rewarding.

7.5.2 The Stability of Static Equilibria: A Variational Approach

To get an idea of how conservation of energy may be used in a stability analysis, we first consider the simplest problem of the magnetostatic equilibrium of an ideal, incompressible fluid. The fluid and magnetic field are both assumed to be contained in a volume V, bounded by the solid surface S, and the equilibrium is governed by

$$\mathbf{J}_0 \times \mathbf{B}_0 = \nabla p_0 , \qquad \mathbf{B}_0 \cdot d\mathbf{S} = 0. \qquad (7.39)$$

Here the subscript 0 indicates a steady, base configuration whose stability is in question, and $d\mathbf{S}$ is an element of the boundary, S. The use of the boundary condition $\mathbf{B}_0 \cdot d\mathbf{S} = 0$, which keeps the magnetic field confined to V, is convenient as it ensures that no magnetic energy escapes from the domain V.

Now suppose that our equilibrium configuration is slightly disturbed at $t = 0$, and that during this disturbance the magnetic field is frozen into the fluid. It is convenient to introduce the Lagrangian displacement, $\boldsymbol{\zeta}(\mathbf{x}, t)$, of a particle, p, from its equilibrium position, $\mathbf{x}_p(0)$:

$$\boldsymbol{\zeta}(\mathbf{x}, t) = \mathbf{x}_p(t) - \mathbf{x}_p(0), \qquad \mathbf{x} = \mathbf{x}_p(0). \qquad (7.40)$$

Following the initial disturbance some motion will ensue, perhaps in the form of Alfvén waves circulating around the field lines, or perhaps something more drastic. In any event, $\mathbf{B}$ will be frozen onto the fluid and the velocity field of the disturbance is related to the instantaneous particle displacement, $\boldsymbol{\zeta}$, by

$$\frac{\partial \boldsymbol{\zeta}}{\partial t} = \mathbf{u}(\mathbf{x} + \boldsymbol{\zeta}, \ t) = \mathbf{u}(\mathbf{x}, \ t) + (\boldsymbol{\zeta} \cdot \nabla)\mathbf{u}(\mathbf{x}, t) + \cdots. \qquad (7.41)$$

Let us now calculate the change in magnetic energy, E_B, which results from the particle displacement. The changes in E_B caused by the displacement of the field lines can be written as

$$E_B(\boldsymbol{\zeta}) = \int (\mathbf{B}^2/2\mu) dV = E_{B0} + \Delta E_B(\boldsymbol{\zeta}) = E_{B0} + \delta^1 E_B(\boldsymbol{\zeta}) + \delta^2 E_B(\boldsymbol{\zeta}) + \cdots.$$
$$(7.42)$$

Here $\delta^1 E_B(\boldsymbol{\zeta})$ and $\delta^2 E_B(\boldsymbol{\zeta})$ are the first- and second-order changes in E_B, $|\boldsymbol{\zeta}|$ being assumed small at all times. Note that, since $\mathbf{B}$ is frozen into the fluid, E_B depends only on the *instantaneous distribution* of the displacement field $\boldsymbol{\zeta}$, and not on the previous history of the motion. Thus $E_B(\boldsymbol{\zeta})$ is a function of time only to the extent that $\boldsymbol{\zeta}$ is time dependent. Now conservation of energy tells us that the sum of $E_B(\boldsymbol{\zeta})$ and the kinetic energy of the disturbance is conserved. Moreover, we shall see that the magnetic energy is stationary at equilibrium, $\delta^1 E_B(\boldsymbol{\zeta}) = 0$. It follows that, since $\partial \boldsymbol{\zeta}/\partial t = \mathbf{u}(\mathbf{x}, t)$ to leading order in $|\boldsymbol{\zeta}|$, we have, to order $|\boldsymbol{\zeta}|^2$,

$$\frac{1}{2} \int \rho \, \dot{\boldsymbol{\zeta}}^2 dV + \delta^2 E_B(\boldsymbol{\zeta}) = \Delta E_0, \qquad (7.43)$$

where ΔE_0 is the initial energy of the disturbance.

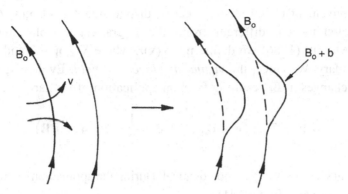

Figure 7.10 Perturbation of the field lines by the virtual velocity field **v**.

It is natural to take as our definition of stability the condition that the kinetic energy of the disturbance does not exceed the initial disturbance energy, ΔE_0. It follows that an equilibrium is stable if $\delta^2 E_B(\zeta)$ is positive for all possible displacement fields ζ, and so stability is ensured provided E_B is a minimum at equilibrium. This simple energy criterion is reassuringly familiar from rigid-body mechanics.

We now need to confirm that $\delta^1 E_B(\zeta)$ is indeed zero. We also need to find an explicit expression for $\delta^2 E_B(\zeta)$ in terms of ζ and $\mathbf{B}_0(\mathbf{x})$, so that we can test the stability of a given configuration. To that end we employ a trick, introduced by Moffatt (1986). We have seen that $E_B(\zeta)$ depends only on the instantaneous particle displacements. There are infinite ways in which a given particle can get from $\mathbf{x}_p(0)$ to $\mathbf{x}_p(t) = \mathbf{x}_p(0) + \zeta(\mathbf{x}_p, t)$, but since $E_B(\zeta)$ does not care about the route taken, we consider the simplest. Suppose that, for a short time τ, we apply an imaginary, steady velocity field $\mathbf{v}(\mathbf{x})$ to the fluid which shifts the particles from their equilibrium position $\mathbf{x}$ to $\mathbf{x} + \zeta$, carrying the field lines along with the particles (Figure 7.10). Since the fluid is incompressible our imaginary velocity field is solenoidal, $\nabla \cdot \mathbf{v} = 0$. Following Moffatt (1986), we shall refer to $\mathbf{v}(\mathbf{x})$ as a *virtual velocity field*.

Now the magnetic field lines are frozen into the fluid during the application of this virtual velocity field, and so we have

$$\frac{\partial \mathbf{B}}{\partial t_V} = \nabla \times (\mathbf{v} \times \mathbf{B}), \quad 0 < t_V < \tau, \tag{7.44}$$

where t_V is virtual time. This integrates to give

$$\mathbf{B}(\mathbf{x}, \tau) - \mathbf{B}_0(\mathbf{x}) = \nabla \times (\mathbf{v} \times \mathbf{B}_0)\tau + \frac{1}{2}\nabla \times (\mathbf{v} \times \delta^1 \mathbf{B})\tau + O(\tau^3). \tag{7.45}$$

It is now convenient to introduce a second displacement field, $\boldsymbol{\eta}(\mathbf{x}, t)$, which is closely related to, but different from, the Lagrangian displacement $\boldsymbol{\zeta}(\mathbf{x}, t)$. Following Moffatt (1986) we define $\boldsymbol{\eta} = \mathbf{v}(\mathbf{x})\tau$, where $\nabla \cdot \boldsymbol{\eta} = 0$ and $\boldsymbol{\eta} \cdot d\mathbf{S} = 0$ on the boundary. We call $\boldsymbol{\eta}$ the *virtual displacement field*. Evidently, in terms of $\boldsymbol{\eta}(\mathbf{x}, t)$, the changes in $\mathbf{B}_0$ resulting from an application of $\mathbf{v}(\mathbf{x})$ are

$$\delta^1 \mathbf{B} = \nabla \times (\boldsymbol{\eta} \times \mathbf{B}_0), \quad \delta^2 \mathbf{B} = \frac{1}{2} \nabla \times (\boldsymbol{\eta} \times \delta^1 \mathbf{B}). \tag{7.46}$$

Note that $\boldsymbol{\eta}(\mathbf{x}, t)$ and $\boldsymbol{\zeta}(\mathbf{x}, t)$ are not identical. During the application of our imaginary velocity field we have, from (7.41),

$$\frac{\partial \boldsymbol{\zeta}}{\partial t} = \mathbf{v}(\mathbf{x} + \boldsymbol{\zeta}) = \mathbf{v}(\mathbf{x}) + (\boldsymbol{\zeta} \cdot \nabla)\mathbf{v} + \cdots. \tag{7.47}$$

It follows that

$$\boldsymbol{\zeta} = \boldsymbol{\eta} + \frac{1}{2}(\boldsymbol{\eta} \cdot \nabla)\boldsymbol{\eta} + \cdots, \tag{7.48}$$

$$\boldsymbol{\eta} = \boldsymbol{\zeta} - \frac{1}{2}(\boldsymbol{\zeta} \cdot \nabla)\boldsymbol{\zeta} + \cdots. \tag{7.49}$$

Thus, the instantaneous Lagrangian particle displacement, $\boldsymbol{\zeta}(\mathbf{x})$, and the virtual displacement field, $\boldsymbol{\eta}(\mathbf{x})$, are equal only to first order. In any event, it turns out to be $\boldsymbol{\eta}(\mathbf{x})$, rather that $\boldsymbol{\zeta}(\mathbf{x})$, which provides the more convenient description of the disturbance, in part because $\boldsymbol{\eta}(\mathbf{x})$ is solenoidal, whereas $\boldsymbol{\zeta}(\mathbf{x})$ is solenoidal only at first order.

Let us now calculate $\delta^1 E_B$ and $\delta^2 E_B$ in terms of $\boldsymbol{\eta}(\mathbf{x})$. The first-order change in energy is

$$\delta^1 E_B = \frac{1}{\mu} \int (\mathbf{B}_0 \cdot \delta^1 \mathbf{B}) dV = \frac{1}{\mu} \int \left(\mathbf{B}_0 \cdot \nabla \times (\boldsymbol{\eta} \times \mathbf{B}_0) \right) dV, \tag{7.50}$$

which we have claimed is zero. To show that this is indeed the case we note that the integrand on the right may be rewritten as

$$\mathbf{B}_0 \cdot \nabla \times (\boldsymbol{\eta} \times \mathbf{B}_0) = (\boldsymbol{\eta} \times \mathbf{B}_0) \cdot \nabla \times \mathbf{B}_0 + \nabla \cdot [(\boldsymbol{\eta} \times \mathbf{B}_0) \times \mathbf{B}_0].$$

Rearranging the scalar triple product and expanding the vector triple product yields

$$\mathbf{B}_0 \cdot \nabla \times (\boldsymbol{\eta} \times \mathbf{B}_0) = -\mu(\mathbf{J}_0 \times \mathbf{B}_0) \cdot \boldsymbol{\eta} + \nabla \cdot \left[(\mathbf{B}_0 \cdot \boldsymbol{\eta})\mathbf{B}_0 - \mathbf{B}_0^2 \boldsymbol{\eta} \right].$$

The divergence integrates to zero by virtue of our boundary conditions, while (7.39) demands $\mathbf{J}_0 \times \mathbf{B}_0 = \nabla p_0$, and so

$$\delta^1 E_B = - \int (\boldsymbol{\eta} \cdot \nabla p_0) dV = - \oint p_0 \boldsymbol{\eta} \cdot d\mathbf{S} = 0. \tag{7.51}$$

The first-order change in magnetic energy is evidently zero, as anticipated above. The second-order change is

$$\delta^2 E_B(\boldsymbol{\eta}) = \frac{1}{2\mu} \int \left[(\delta^1 \mathbf{B})^2 + 2\mathbf{B}_0 \cdot \delta^2 \mathbf{B} \right] dV,$$

from which,

$$\delta^2 E_B(\boldsymbol{\eta}) = \frac{1}{2\mu} \int \left[\mathbf{b}^2 + \mathbf{B}_0 \cdot \nabla \times (\boldsymbol{\eta} \times \mathbf{b}) \right] dV, \tag{7.52}$$

where

$$\mathbf{b} = \delta^1 \mathbf{B} = \nabla \times (\boldsymbol{\eta} \times \mathbf{B}_0). \tag{7.53}$$

When combined with (7.43) this yields, to second order in the displacement field:

$$\frac{1}{2} \int \rho \dot{\boldsymbol{\eta}}^2 dV + \frac{1}{2\mu} \int \left[\mathbf{b}^2 + \mathbf{B}_0 \cdot \nabla \times (\boldsymbol{\eta} \times \mathbf{b}) \right] dV = \Delta E_0. \tag{7.54}$$

This is called the *energy principle* of magnetostatics. If, for a given $\mathbf{B}_0(\mathbf{x})$, it can be shown that $\delta^2 E_B(\boldsymbol{\eta}) > 0$ for all possible $\boldsymbol{\eta}$, then the magnetostatic equilibrium is linearly stable because the kinetic energy of the disturbance is bounded from above by the initial disturbance energy ΔE_0 (Bernstein et al., 1958). Note that, as formulated above, this provides only a sufficient condition for stability. However, it turns out that $\delta^2 E_B(\boldsymbol{\eta}) > 0$ provides both necessary and sufficient conditions for linear stability (see, for example, Biskamp, 1993). That is to say, a magnetostatic equilibrium is unstable if *any* $\boldsymbol{\eta}$ exists such that $\delta^2 E_B(\boldsymbol{\eta}) < 0$.

All of this is in accord with our intuitive notions of stability. We may think of $E_B(\boldsymbol{\eta})$ as a potential energy, in the sense that it is the conserved energy of an external force (the Lorentz force) acting on the fluid. Like a ball sitting on a hillside, the fluid (or ball) is in equilibrium if the potential energy is stationary, $\delta^1 E_B(\boldsymbol{\eta}) = 0$, and is stable if the potential energy is a minimum, $\delta^2 E_B(\boldsymbol{\eta}) > 0$ (Figure 7.11).

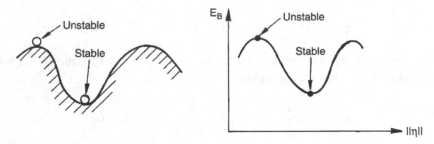

Figure 7.11 The analogy between magnetostatic and mechanical equilibria. Both are stable if the potential energy is a minimum at equilibrium.

Example 7.1 The Stability of Azimuthal Fields

As an example of the energy principle, let us return to Figure 7.8 and consider its application to the azimuthal field $\mathbf{B}_0 = (0, B_\theta(r), 0)$, in (r, θ, z) coordinates. Show that such a field is stable to axisymmetric disturbances if $(B_\theta/r)^2$ decreases with increasing radius, and stable to non-axisymmetric disturbances provided $(rB_\theta)^2$ decreases with increasing radius. Hence confirm that the equilibrium shown in Figure 7.8 is always unstable.   ∎

7.5.3 The Stability of Static Equilibria: A Direct Attack

Now there is a different, though ultimately equivalent, route to establishing this stability criterion. This alternative method proves more useful when working with non-static equilibria, and so we shall describe it in some detail. The idea is to develop a dynamic equation for the disturbance. This time we work, not with the virtual displacement field, $\boldsymbol{\eta}$, but rather with the Lagrangian displacement, $\boldsymbol{\zeta}$. In part, this is because it is more customary to use $\boldsymbol{\zeta}$, and in part because it seems more natural when using a dynamic, rather than variational, approach. Of course, to leading order in the amplitude of the disturbance, $\boldsymbol{\zeta}$ and $\boldsymbol{\eta}$ are equal, so the choice makes little difference.

In the dynamic approach we work only with first-order quantities, such as $\mathbf{b} = \delta^1 \mathbf{B}$, and discard all higher-order terms. The induction and momentum equations then give us the disturbance equations

$$\frac{\partial \mathbf{b}}{\partial t} = \nabla \times (\mathbf{u} \times \mathbf{B}_0), \tag{7.55}$$

$$\rho \frac{\partial \mathbf{u}}{\partial t} = \mathbf{j} \times \mathbf{B}_0 + \mathbf{J}_0 \times \mathbf{b} - \nabla p. \tag{7.56}$$

Here lowercase bold letters represent perturbed quantities, e.g. $\mathbf{J} = \mathbf{J}_0 + \mathbf{j}$, and quadratic terms in the disturbance, such as $\mathbf{u} \cdot \nabla \mathbf{u}$ or $\mathbf{j} \times \mathbf{b}$, are neglected. We also have, to leading order in $|\zeta|$,

$$\dot{\zeta}(\mathbf{x}, t) = \mathbf{u}(\mathbf{x}, t) , \qquad \nabla \cdot \zeta = 0 , \qquad \zeta \cdot d\mathbf{S} = 0. \tag{7.57}$$

The perturbation equations then give us

$$\mathbf{b} = \nabla \times (\zeta \times \mathbf{B}_0), \tag{7.58}$$

$$(\rho\mu)\ddot{\zeta} = (\nabla \times \mathbf{b}) \times \mathbf{B}_0 + (\nabla \times \mathbf{B}_0) \times \mathbf{b} - \mu\nabla p. \tag{7.59}$$

The first of these is a restatement of (7.53), since $\zeta = \eta$ to leading order. The second equation may be rewritten in a form familiar from rigid-body mechanics,

$$(\rho\mu)\ddot{\zeta} = \mathbf{F}(\zeta) - \mu\nabla p, \tag{7.60}$$

where the force

$$\mathbf{F}(\zeta) = (\nabla \times \mathbf{b}) \times \mathbf{B}_0 + (\nabla \times \mathbf{B}_0) \times \mathbf{b} , \qquad \mathbf{b} = \nabla \times (\zeta \times \mathbf{B}_0), \tag{7.61}$$

is linear in ζ and acts like a magnetic spring. The key question now is whether or not $\mathbf{F}(\zeta)$ does net work on the disturbance, thus fuelling an instability, or whether, like a mechanical spring, it acts to restore equilibrium. So let us evaluate the rate of working of this force. It is straightforward, but tedious, to show that the linear force operator $\mathbf{F}(\zeta)$ is self-adjoint, in the sense that

$$\int \zeta_1 \cdot \mathbf{F}(\zeta_2)dV = \int \zeta_2 \cdot \mathbf{F}(\zeta_1)dV. \tag{7.62}$$

We now multiply (7.60) by $\dot{\zeta}$ and invoke (7.62) in the form, $\zeta_1 = \zeta, \ \zeta_2 = \dot{\zeta}$. The result is an energy-like equation,

$$\frac{d}{dt}\frac{1}{2}\int \rho\dot{\zeta}^2 dV = \frac{d}{dt}\left[\frac{1}{2\mu}\int \mathbf{F}(\zeta) \cdot \zeta dV\right], \tag{7.63}$$

or equivalently,

$$\frac{1}{2}\int \rho\dot{\zeta}^2 dV - \frac{1}{2\mu}\int \mathbf{F}(\zeta) \cdot \zeta dV = \text{constant}. \tag{7.64}$$

The next step is to evaluate the second integral on the left, which is the net work done by the Lorentz force on the disturbed fluid. In fact, as we show below,

$$\frac{1}{2\mu}\int \mathbf{F}(\zeta) \cdot \zeta dV = -\frac{1}{2\mu}\int \left[\mathbf{b}^2 + \mathbf{B}_0 \cdot \nabla \times (\zeta \times \mathbf{b})\right] dV, \tag{7.65}$$

which, when combined with (7.64), takes us back to the energy principle (7.54).

The proof of (7.65) proceeds as follows. First we need a vector identity which follows directly from the equilibrium equation $\mathbf{J}_0 \times \mathbf{B}_0 = \nabla p_0$:

$$\mathbf{J}_0 \times [\nabla \times (\mathbf{q} \times \mathbf{B}_0)] + [\nabla \times (\mathbf{q} \times \mathbf{J}_0)] \times \mathbf{B}_0 = -\nabla(\mathbf{q} \cdot \nabla p_0), \tag{7.66}$$

where $\mathbf{q}$ is any solenoidal field. Next we take $\mathbf{q} = \zeta$ and rewrite (7.61) as

$$\mathbf{F}(\zeta) = \nabla \times [\nabla \times (\zeta \times \mathbf{B}_0) - \zeta \times (\nabla \times \mathbf{B}_0)] \times \mathbf{B}_0 + \nabla(\cdot), \tag{7.67}$$

from which

$$\mathbf{F}(\zeta) \cdot \zeta = -(\zeta \times \mathbf{B}_0) \cdot \nabla \times [\nabla \times (\zeta \times \mathbf{B}_0) - \zeta \times (\nabla \times \mathbf{B}_0)] + \zeta \cdot \nabla(\cdot). \tag{7.68}$$

After a little algebra we find,

$$\int \mathbf{F}(\zeta) \cdot \zeta dV = -\int \left[\mathbf{b}^2 + \mathbf{B}_0 \cdot \nabla \times [\zeta \times \mathbf{b}]\right] dV, \tag{7.69}$$

as required. When combined with (7.64) we recover the energy principle:

$$\frac{1}{2}\int \rho \, \dot{\zeta}^2 dV + \frac{1}{2\mu}\int \left[\mathbf{b}^2 + \mathbf{B}_0 \cdot \nabla \times (\zeta \times \mathbf{b})\right] dV = \text{constant} = \Delta E_0. \tag{7.70}$$

This more direct approach to the energy principle is in fact more commonly used, and was the method originally adopted by Bernstein et al. (1958). It is more cumbersome than the variational approach of §7.5.2, but generalises more readily to non-static equilibria. In this approach the second integral on the left of (7.70) is interpreted as (minus) the work done by the Lorentz force on the fluid, $W(\zeta) = -\frac{1}{2} \int \mathbf{F}(\zeta) \cdot \zeta dV$. Stability is then ensured if the Lorentz force does negative work on the disturbance for all possible perturbations.

The fact that the energy principle provides both necessary and sufficient conditions for stability can be interpreted as follows. If there exists a perturbation, $\zeta = \hat{\zeta}$, for which $\frac{1}{2} \int \mathbf{F}(\hat{\zeta}) \cdot \hat{\zeta} dV$ is positive, then there is nothing to prevent such a perturbations growing. That is to say, all kinematically permissible Lagrangian displacement fields (i.e. those satisfying $\nabla \cdot \zeta = 0$ and $\zeta \cdot d\mathbf{S} = 0$) are also dynamically accessible displacement fields. As we shall see, the same is not always true of non-static equilibria, where there often exist kinematically permissible displacement fields which are potentially destabilising, in the sense that the Lorentz and inertial forces do positive net work on the disturbance, yet the system remains linearly stable as those displacement fields are not dynamically accessible to a real perturbation.

7.6 An Energy-Based Stability Theorem for Non-Static Equilibria

Let us turn to non-static equilibria. Here the energy of the disturbance can come from either the stored magnetic energy or else from the kinetic energy of the base flow, and so the problem is altogether more complicated. Indeed, in general the energy method provides only sufficient conditions for stability in such cases. As before, we shall restrict ourselves to ideal, incompressible fluids, $\lambda = v = 0$, $\rho = $ constant, and to linear stability. To avoid carrying the constants ρ and μ throughout the analysis, we shall put $\rho = \mu = 1$. (In effect, we rescale $\mathbf{B}$ as $\mathbf{B}/\sqrt{\rho\mu}$.) Also, in the interests of simplicity, we shall take $\mathbf{B}$ to be confined to the fluid domain, V, so that no magnetic energy escapes through the bounding surface of V.

Using the rescaled magnetic field, the governing equations for ideal MHD become

$$\frac{\partial \mathbf{B}}{\partial t} = \nabla \times (\mathbf{u} \times \mathbf{B}), \qquad \mathbf{B} \cdot d\mathbf{S} = 0, \tag{7.71}$$

$$\frac{\partial \mathbf{u}}{\partial t} = \mathbf{u} \times \boldsymbol{\omega} + \mathbf{J} \times \mathbf{B} - \nabla h, \qquad \mathbf{u} \cdot d\mathbf{S} = 0, \tag{7.72}$$

where h is Bernoulli's function, $h = p + \frac{1}{2}u^2$. The equilibrium solutions are governed by

$$\mathbf{u}_0 \times \mathbf{B}_0 = \nabla D, \tag{7.73}$$

$$\mathbf{u}_0 \times \boldsymbol{\omega}_0 + \mathbf{J}_0 \times \mathbf{B}_0 = \nabla h_0, \tag{7.74}$$

for some function D. From (7.66) we see that the first of these yields the identity

$$\mathbf{u}_0 \times [\nabla \times (\mathbf{q} \times \mathbf{B}_0)] + [\nabla \times (\mathbf{q} \times \mathbf{u}_0)] \times \mathbf{B}_0 = -\nabla(\mathbf{q} \cdot \nabla D), \tag{7.75}$$

which is valid for any solenoidal vector $\mathbf{q}$. Similarly, it may be shown that (7.74) yields

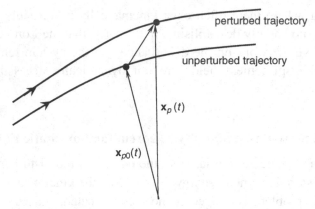

Figure 7.12 Unperturbed, $\mathbf{x}_{p0}(t)$, and perturbed, $\mathbf{x}_p(t)$, trajectories of a fluid particle.

$$\mathbf{u}_0 \times \nabla \times (\mathbf{q} \times \boldsymbol{\omega}_0) + \nabla \times (\mathbf{q} \times \mathbf{u}_0) \times \boldsymbol{\omega}_0 + \mathbf{J}_0 \times \nabla \times (\mathbf{q} \times \mathbf{B}_0)$$
$$+ \nabla \times (\mathbf{q} \times \mathbf{J}_0) \times \mathbf{B}_0 = -\nabla(\mathbf{q} \cdot \nabla h_0).$$
$$(7.76)$$

We shall not pause to prove these cumbersome relationships, but shall make use of them in the analysis which follows.

Let us now turn to the question of stability. Suppose that $\mathbf{u}_0$ and $\mathbf{B}_0$ are perturbed so that $\mathbf{u} = \mathbf{u}_0 + \delta\mathbf{u}$, $\boldsymbol{\omega} = \boldsymbol{\omega}_0 + \delta\boldsymbol{\omega}$, $\mathbf{B} = \mathbf{B}_0 + \mathbf{b}$ and $\mathbf{J} = \mathbf{J}_0 + \mathbf{j}$. As in the magnetostatic stability analysis, we take $\boldsymbol{\zeta}(\mathbf{x}, t)$ to be the Lagrangian displacement of a particle p from its position in the unperturbed flow (Figure 7.12). If $\mathbf{x}_{p0}(t)$ is the position vector of the particle in the equilibrium flow, and $\mathbf{x}_p(t)$ its trajectory in the perturbed flow, then this displacement is defined by

$$\boldsymbol{\zeta} = \mathbf{x}_p(t) - \mathbf{x}_{p0}(t). \qquad (7.77)$$

Now the rate of change of $\boldsymbol{\zeta}$, as seen by an observer at $\mathbf{x}_{p0}(t)$, is evidently

$$\frac{D\boldsymbol{\zeta}}{Dt} = \frac{\partial\boldsymbol{\zeta}}{\partial t} + (\mathbf{u}_0 \cdot \nabla)\boldsymbol{\zeta} = \mathbf{u}(\mathbf{x}_{p0} + \boldsymbol{\zeta}) - \mathbf{u}_0(\mathbf{x}_{p0}), \qquad (7.78)$$

which, given that $\mathbf{u} = \mathbf{u}_0 + \delta\mathbf{u}$, can be rewritten as

$$\frac{\partial\boldsymbol{\zeta}}{\partial t} + (\mathbf{u}_0 \cdot \nabla)\boldsymbol{\zeta} = \delta\mathbf{u}(\mathbf{x} + \boldsymbol{\zeta}) + [\mathbf{u}_0(\mathbf{x} + \boldsymbol{\zeta}) - \mathbf{u}_0(\mathbf{x})]. \qquad (7.79)$$

In the linear (small-amplitude) approximation, we have $\delta\mathbf{u}(\mathbf{x}+\boldsymbol{\zeta}) \approx \delta\mathbf{u}(\mathbf{x})$ and $\mathbf{u}_0(\mathbf{x}+\boldsymbol{\zeta}) - \mathbf{u}_0(\mathbf{x}) \approx (\boldsymbol{\zeta}\cdot\nabla)\mathbf{u}_0(\mathbf{x})$, and so it follows that $\delta\mathbf{u}$ and $\boldsymbol{\zeta}(\mathbf{x}, t)$ are related by

$$\frac{\partial\boldsymbol{\zeta}}{\partial t} + (\mathbf{u}_0\cdot\nabla)\boldsymbol{\zeta} = \delta\mathbf{u}(\mathbf{x}, t) + (\boldsymbol{\zeta}\cdot\nabla)\mathbf{u}_0(\mathbf{x}), \tag{7.80}$$

at least to leading order in $|\boldsymbol{\zeta}|$.

Henceforth we shall find it more convenient to work with the virtual displacement field $\boldsymbol{\eta}(\mathbf{x}, t)$, and since $\boldsymbol{\eta}(\mathbf{x}, t)$ and $\boldsymbol{\zeta}(\mathbf{x}, t)$ are equal to leading order, (7.80) becomes

$$\delta^1\mathbf{u}(\mathbf{x}, t) = \frac{\partial\boldsymbol{\eta}}{\partial t} + \nabla\times(\boldsymbol{\eta}\times\mathbf{u}_0), \tag{7.81}$$

to leading order in $|\boldsymbol{\eta}|$. A more careful analysis taken to second order in the virtual displacement field yields

$$\delta^2\mathbf{u}(\mathbf{x}, t) = \frac{1}{2}\nabla\times[\boldsymbol{\eta}\times\delta^1\mathbf{u}], \tag{7.82}$$

which rests on the fact that $\nabla\cdot\boldsymbol{\eta} = 0$. We note in passing that the equivalent expression in terms of $\boldsymbol{\zeta}(\mathbf{x}, t)$ is considerably more complex, since $\boldsymbol{\zeta}(\mathbf{x})$ is solenoidal to first order only.

We now seek an evolution equation for $\boldsymbol{\eta}$ using the momentum and induction equations. The linearised versions of these equations, which describe small-amplitude disturbances, are

$$\frac{\partial\mathbf{b}}{\partial t} = \nabla\times(\delta\mathbf{u}\times\mathbf{B}_0 + \mathbf{u}_0\times\mathbf{b}), \tag{7.83}$$

and

$$\frac{\partial}{\partial t}(\delta\mathbf{u}) = \mathbf{u}_0\times\delta\boldsymbol{\omega} + \delta\mathbf{u}\times\boldsymbol{\omega}_0 + \mathbf{j}\times\mathbf{B}_0 + \mathbf{J}_0\times\mathbf{b} - \nabla(\delta h). \tag{7.84}$$

We concentrate first on the induction equation. Introducing $\Delta\mathbf{B} = \mathbf{b} - \nabla\times(\boldsymbol{\eta}\times\mathbf{B}_0)$ and invoking (7.75) and (7.81), this can be rewritten as

$$\frac{\partial}{\partial t}(\Delta\mathbf{B}) = \nabla\times(\mathbf{u}_0\times\Delta\mathbf{B}).$$

Evidently, if we set $\Delta\mathbf{B} = 0$ at some initial instant, then $\Delta\mathbf{B} = 0$ for all subsequent time. If we restrict ourselves to perturbations of this type we then have

$$\mathbf{b} = \nabla \times (\boldsymbol{\eta} \times \mathbf{B}_0), \tag{7.85}$$

which is identical to (7.53). Setting $\Delta\mathbf{B} = 0$ in the initial condition is therefore equivalent to assuming that $\mathbf{B}$ is frozen into the fluid during the initial disturbance. As in our analysis of magnetostatic equilibria, we shall restrict ourselves to such initial conditions, so that (7.85) gives the first-order perturbation in $\mathbf{B}$.

Let us now turn to the momentum equation (7.84). On substituting for $\delta\mathbf{u}$ using (7.81) it becomes an evolution equation for $\boldsymbol{\eta}$,

$$\ddot{\boldsymbol{\eta}} + 2(\mathbf{u}_0 \cdot \nabla)\dot{\boldsymbol{\eta}} = \mathbf{F}(\boldsymbol{\eta}) + \nabla(\mathbf{u}_0 \cdot \dot{\boldsymbol{\eta}} - \delta h), \tag{7.86}$$

where the force

$$\mathbf{F}(\boldsymbol{\eta}) = (\nabla \times \mathbf{B}_0) \times \mathbf{b} + (\nabla \times \mathbf{b}) \times \mathbf{B}_0 - (\nabla \times \mathbf{u}_0) \times \hat{\mathbf{u}} - (\nabla \times \hat{\mathbf{u}}) \times \mathbf{u}_0 \tag{7.87}$$

is linear in $\boldsymbol{\eta}$, and $\hat{\mathbf{u}}$ is defined as

$$\hat{\mathbf{u}} = \nabla \times (\boldsymbol{\eta} \times \mathbf{u}_0). \tag{7.88}$$

Note that, while $\mathbf{b} = \nabla \times (\boldsymbol{\eta} \times \mathbf{B}_0)$ is the first-order change in $\mathbf{B}_0$, $\hat{\mathbf{u}}$ is not the first-order change in $\mathbf{u}_0$. On the contrary, (7.81) and (7.82) demand,

$$\delta^1\mathbf{u}(\mathbf{x}, t) = \dot{\boldsymbol{\eta}} + \hat{\mathbf{u}} \qquad \delta^2\mathbf{u}(\mathbf{x}, t) = \frac{1}{2}\nabla \times [\boldsymbol{\eta} \times (\dot{\boldsymbol{\eta}} + \hat{\mathbf{u}})], \tag{7.89}$$

so that $\delta^1\mathbf{u}(\mathbf{x}, t) = \hat{\mathbf{u}}$ only in the case of pseudo-static disturbances.

The anti-symmetric roles played by the vector pairs $(\mathbf{B}_0, \mathbf{b})$ and $(\mathbf{u}_0, \hat{\mathbf{u}})$ in $\mathbf{F}$ is intriguing, and we shall explain its significance shortly. In the meantime, we convert (7.86) into an energy equation by taking the dot product with $\dot{\boldsymbol{\eta}}$ and integrating over the domain V. Since $\mathbf{u}_0 \cdot d\mathbf{S} = \boldsymbol{\eta} \cdot d\mathbf{S} = 0$ on S, we obtain

$$\frac{d}{dt}\int \frac{1}{2}\dot{\boldsymbol{\eta}}^2 dV = \int \mathbf{F}(\boldsymbol{\eta}) \cdot \dot{\boldsymbol{\eta}} dV. \tag{7.90}$$

It turns out that, as in the magnetostatic case, the linear force operator $\mathbf{F}(\boldsymbol{\eta})$ is self-adjoint (Frieman and Rotenberg, 1960). Thus

$$\int \mathbf{F}(\boldsymbol{\eta}_1) \cdot \boldsymbol{\eta}_2 dV = \int \mathbf{F}(\boldsymbol{\eta}_2) \cdot \boldsymbol{\eta}_1 dV,$$

and so our energy equation integrates to give

$$e = \frac{1}{2} \int \dot{\boldsymbol{\eta}}^2 dV - \frac{1}{2} \int \mathbf{F}(\boldsymbol{\eta}) \cdot \boldsymbol{\eta} dV = \text{constant}, \tag{7.91}$$

which is a direct generalisation of the magnetostatic energy principle (7.64). Thus we may think of e as the initial energy of the disturbance, ΔE_0, as in (7.70). We shall see below that e may also be interpreted as the conserved Hamiltonian of the disturbance.

We now evaluate the second integral in (7.91) using (7.75) and (7.76). The calculation is lengthy but straight forward (see Davidson, 2000), and we find

$$\frac{1}{2} \int \mathbf{F}(\boldsymbol{\eta}) \cdot \boldsymbol{\eta} dV = \frac{1}{2} \int [\hat{\mathbf{u}}^2 + \mathbf{u}_0 \cdot \nabla \times (\boldsymbol{\eta} \times \hat{\mathbf{u}})] dV - \frac{1}{2} \int [\mathbf{b}^2 + \mathbf{B}_0 \cdot \nabla \times (\boldsymbol{\eta} \times \mathbf{b})] dV. \tag{7.92}$$

We recognise the second integral on the right as $\delta^2 E_B(\boldsymbol{\eta})$, the second-order perturbation in magnetic energy.

To summarise, our disturbance is constrained by the conservation equation

$$e = \frac{1}{2} \int \dot{\boldsymbol{\eta}}^2 dV + \frac{1}{2} \int [\mathbf{b}^2 + \mathbf{B}_0 \cdot \nabla \times (\boldsymbol{\eta} \times \mathbf{b})] \, dV - \frac{1}{2} \int [\hat{\mathbf{u}}^2 + \mathbf{u}_0 \cdot \nabla \times (\boldsymbol{\eta} \times \hat{\mathbf{u}})] \, dV, \tag{7.93}$$

where the constant e may be interpreted as the initial energy, or the conserved Hamiltonian, of the disturbance. This is often rewritten as

$$e = \int \frac{1}{2} \dot{\boldsymbol{\eta}}^2 dV + W(\boldsymbol{\eta}) = \text{constant}, \tag{7.94}$$

where

$$W(\boldsymbol{\eta}) = \frac{1}{2} \int [\mathbf{b}^2 + \mathbf{B}_0 \cdot \nabla \times (\boldsymbol{\eta} \times \mathbf{b})] \, dV - \frac{1}{2} \int [\hat{\mathbf{u}}^2 + \mathbf{u}_0 \cdot \nabla \times (\boldsymbol{\eta} \times \hat{\mathbf{u}})] \, dV. \tag{7.95}$$

Clearly the functional $W(\boldsymbol{\eta})$ is (minus) the work done on the disturbance by the Lorentz and inertial forces; $W(\boldsymbol{\eta}) = -\frac{1}{2} \int \mathbf{F}(\boldsymbol{\eta}) \cdot \boldsymbol{\eta} dV$.

Let us take $\int \frac{1}{2} \dot{\boldsymbol{\eta}}^2 dV$ as a measure of our disturbance. If $W(\boldsymbol{\eta})$ can be shown to be positive for all possible choices of $\boldsymbol{\eta}$, then $\int \frac{1}{2} \dot{\boldsymbol{\eta}}^2 dV$ is bounded from above by e and the equilibrium is then stable, at least in the sense that $\int \frac{1}{2} \dot{\boldsymbol{\eta}}^2 dV$ is bounded for all time (Frieman and Rotenberg, 1960). Thus we have a sufficient, though not necessary, condition for stability. Physically, this corresponds to a situation where the net rate of working of the inertial and Lorentz forces on the disturbance is negative, $\int \mathbf{F}(\boldsymbol{\eta}) \cdot \boldsymbol{\eta} dV < 0$, irrespective of the shape of the perturbation. In such a situation the disturbance cannot acquire the energy to grow. Notice, however, that (7.94) does not furnish a necessary condition for stability, and indeed it is easy to construct examples in which $W(\boldsymbol{\eta})$ is indefinite in sign and yet the flow is linearly stable. This corresponds to situations where the net rate of working of $\mathbf{F}(\boldsymbol{\eta})$ on the disturbance is positive for certain *kinematically admissible* disturbances, i.e. disturbances that satisfy $\nabla \cdot \boldsymbol{\eta} = 0$ and $\boldsymbol{\eta} \cdot d\mathbf{S} = 0$ on the boundary, but where such disturbances turn out to be *dynamically inaccessible*, and so cannot trigger an instability.

It turns out that $W(\boldsymbol{\eta})$ is always indefinite in sign to short-wavelength, three-dimensional disturbances unless $\mathbf{B}_0$ and $\mathbf{u}_0$ are co-linear (see Exercise 7.1). This greatly limits the use of (7.93) in practice. Nevertheless, (7.93) is a remarkably general result which has been rediscovered many times by alternative means. It covers magnetostatics, ideal MHD, and inviscid flows in the absence of a magnetic field. While there are relatively few three-dimensional flows for which $W(\boldsymbol{\eta})$ can be shown to be positively definite, there are many two-dimensional and axisymmetric flows whose (two-dimensional or axisymmetric) stability can be established through this criterion.

One of the intriguing features of (7.93) is the form of the third integral. Comparing it with the second integral, we see that it represents the second-order change in kinetic energy which would occur if the *streamlines* were frozen into the fluid. Yet it appears in (7.93) with a minus sign in front of it, and in any event the **u**-lines, unlike the **B**-lines, are not frozen into the fluid. The physical significance of this integral is discussed in Davidson (2000, 2013), and perhaps it is worth digressing for a moment to summarise this discussion. The key is to shift to a Hamiltonian framework and use Lagrange's equations.

Let us change notation and use T for kinetic energy and V for the magnetic energy, which we now think of as the potential energy of an external force applied to the fluid. In addition, we introduce $L = T - V$ for the Lagrangian and the symbols d^1 and d^2 to represent the first- and second-order perturbations to any vector field which arise from the 'frozen-in' displacement of its field-lines. (In the case of the magnetic field, a d-variation is the same as a δ- perturbation.) It turns out that, under a *d-perturbation*, in which both the **B**-lines and the **u**-lines are frozen into the fluid, $L = T - V$ is function of $\boldsymbol{\eta}$, $\mathbf{B}_0$ and $\mathbf{u}_0$ only. Moreover, it is easily shown that $L(\boldsymbol{\eta})$ is stationary at equilibrium, $d^1 L(\boldsymbol{\eta}) = 0$, and when $L(\boldsymbol{\eta})$ is a maximum ($d^2 L(\boldsymbol{\eta}) < 0$ for all admissible $\boldsymbol{\eta}$) that equilibrium is linearly stable, as is evident from (7.93). (See Davidson, 2000, for a discussion.)

The significance of the freezing of the **u**-lines into the fluid during such a variation is also discussed in Davidson (2000). The point is that such

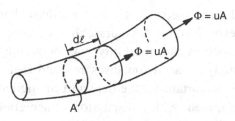

Figure 7.13 A stream-tube perturbed by a *d*-variation conserves the volume, $(\delta\ell)A$, of a short element $\delta\ell$ of the tube, as well as the flux $\Phi = uA$. Thus the time, $\delta t = \delta\ell/u$, for fluid to pass through the element, $\delta\ell$, is also conserved.

a variation creates a new set of particle trajectories which have the property that their time of flight is the same as for the original trajectories. The simplest way to see this is to think of a stream-tube which is frozen into the fluid during *d*-variation. In particular, consider a short material element of the tube with length $\delta\ell$ and cross-sectional area A (Figure 7.13). During a *d*-variation both the volume of that element, $(\delta\ell)A$, and the flux of **u** along the tube, $\Phi = uA$, are conserved. (This flux conservation is equivalent to Alfvén's theorem applied to a **B**-field in ideal MHD, or Kelvin's theorem applied to the vorticity field in an Euler flow.) It follows that $\delta\ell/u$ is conserved for each short element of the stream-tube during a *d*-variation, which in turn is the time δt taken for fluid to pass through that element.

Summing all such incremental transit times from one end of a stream-tube to the other, we conclude that the time of flight of all fluid trajectories is conserved during a *d*-variation. Of course, this is exactly the kind of perturbation normally associated with Hamilton's principle, and indeed the fact that L is stationary in a *d*-variation at equilibrium, $d^1 L(\eta) = 0$, can be shown to follow directly from Hamilton's principle, or equivalently, Lagrange's equations. Moreover, the constant e in (7.93) can be shown to be the conserved Hamiltonian of the perturbed motion. But the underlying reason why, within a Hamiltonian framework, stability is ensured when L is a maximum, $d^2 L(\eta) < 0$, remains a mystery, although in a strictly mathematical sense it clearly follows from (7.93).

7.7 The Chandrasekhar–Velikhov Instability in Rotating MHD

In order to illustrate the utility of the generalised energy principle (7.93), we now focus on one particular instability which, although well known for over 50 years, has recently attracted considerable attention. This is the observation that a rotating fluid, which is centrifugally stable by Rayleigh's criterion, can be destabilised by an axial magnetic field. This instability, which is sometimes called the magneto-rotational instability (or MRI for short) was discovered by

Velikhov in 1959 and then independently by Chandrasekhar in 1960. Although an entire chapter is devoted to it in Chandrasekhar's classic 1960s textbook, the astrophysical significance of the instability was only fully appreciated in the 1990s, when those studying accretion discs realised that it could trigger the turbulence which is so important for the transport of angular momentum across such discs. Let us start, then, with a discussion of accretion discs and the key role of instabilities within these discs.

7.7.1 The Magnetic Destabilisation of Rotating Flow

Magnetic fields play a central role in the dynamics of those accretion discs that are hot enough to be efficient electrical conductors. A classical problem in the modelling of accretion discs is the observation that, as mass spirals inward towards the central star, it must shed its angular momentum by turbulent diffusion (Figure 7.14). However, such discs are in a Keplerian orbit around the central star, and so the variation with radius of the angular velocity, $\Omega(r)$, and angular momentum density, $\Gamma(r)$, are

$$\Omega(r) = \sqrt{GM/r^3}, \qquad \Gamma(r) = r^2\Omega(r) = \sqrt{GMr}, \qquad (7.96)$$

where M is the mass of the central star and G the universal gravitational constant. According to Rayleigh's centrifugal criteria, such profiles are stable to axisymmetric disturbances. That is, in the purely hydrodynamic problem, a necessary and

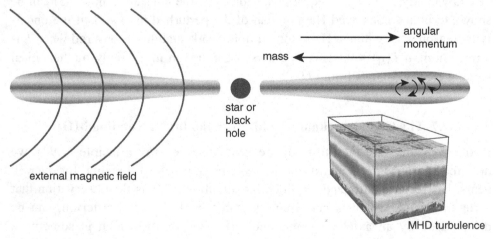

Figure 7.14 Schematic representation of an accretion disc. (Courtesy of Phil Armitage.)

sufficient condition for axisymmetric disturbances to be centrifugally stable is that Rayleigh's discriminant,

$$\Phi(r) = \frac{1}{r^3}\frac{d\Gamma^2}{dr},\tag{7.97}$$

is positive for all r. The question then arises as to what generates the turbulence required to diffuse angular momentum across the disc, thus allowing the plasma to shed its angular momentum as it spirals radially inward.

A partial answer to this is to found in Chandrasekhar (1961), who considered the model problem $\mathbf{u}_0 = (0,\ r\Omega(r),\ 0)$ and $\mathbf{B}_0 = (0, 0, B_0)$ in (r, θ, z) coordinates. The axial field, B_0, is uniform and the fluid is taken as ideal ($\lambda = \nu = 0$). A conventional normal mode analysis for axisymmetric disturbances then reveals that the presence of the axial field modifies Rayleigh's criterion, imposing the more stringent test

$$r\frac{d\Omega^2}{dr} > 0,\tag{7.98}$$

as a sufficient (but not necessary) condition for stability. While the angular velocity profile in a disc passes Rayleigh's criterion, $\Phi(r) > 0$, it fails (7.98), and so the suggestion is that the axial magnetic fields that thread through such discs might provide the much-needed destabilisation mechanism. Moreover, Chandrasekhar's analysis reveals that criterion (7.98) continues to hold as the magnetic field becomes vanishingly weak, a rather surprising result that provoked Chandrasekhar to observe, 'In some ways it is remarkable that we do not recover Rayleigh's criterion, $\Phi(r) > 0$, in the limit of zero magnetic field.'

Perhaps it is worth taking a moment to identify the destabilisation mechanism. In the linear approximation, the azimuthal force balance, which is $D\Gamma/Dt = 0$ in the axisymmetric hydrodynamic case, becomes

$$\rho\left(\frac{\partial(r\delta u_\theta)}{\partial t} + \delta u_r\frac{d\Gamma}{dr}\right) = -rj_rB_0 = \frac{r}{\mu}\frac{\partial b_\theta}{\partial z}B_0,\tag{7.99}$$

where, as usual, we write $\mathbf{u} = \mathbf{u}_0 + \delta\mathbf{u}$ and use lowercase symbols for perturbed magnetic quantities. The governing equation for an azimuthal field in an axisymmetric flow, on the other hand, is discussed in §5.5.3, and is

$$\frac{D}{Dt}\left(\frac{B_\theta}{r}\right) = (\mathbf{B}_p \cdot \nabla)\left(\frac{u_\theta}{r}\right),\tag{7.100}$$

where $\mathbf{B}_p$ is the poloidal magnetic field. We recognise the source term on the right as resulting from differential rotation, whereby gradients in angular velocity spiral up the poloidal field lines (Figure 5.8). In the present context this equation reduces to

$$\frac{\partial b_\theta}{\partial t} = r\frac{d\Omega}{dr}b_r + B_0\frac{\partial \delta u_\theta}{\partial z}. \tag{7.101}$$

Finally, the perturbation in radial magnetic field is governed by

$$\frac{\partial b_r}{\partial t} = B_0\frac{\partial \delta u_r}{\partial z}. \tag{7.102}$$

It turns out that in the weak-field, marginal-stability limit we may simplify these equations by dropping the time derivative in (7.99), because disturbances are non-oscillatory and grow slowly, and also the axial gradient in (7.101). (See, for example, Davidson, 2013.) The resulting equations for the disturbance are then

$$\delta u_r\frac{1}{r}\frac{d\Gamma}{dr} = \frac{B_0}{\rho\mu}\frac{\partial b_\theta}{\partial z}, \tag{7.103}$$

$$\frac{\partial b_r}{\partial t} = B_0\frac{\partial \delta u_r}{\partial z}, \tag{7.104}$$

$$\frac{\partial b_\theta}{\partial t} = r\frac{d\Omega}{dr}b_r, \tag{7.105}$$

which combine to give

$$\left[\Phi(r)\frac{\partial^2}{\partial t^2} - r\frac{d\Omega^2}{dr}\frac{B_0^2}{\rho\mu}\frac{\partial^2}{\partial z^2}\right]\delta u_r = 0. \tag{7.106}$$

Evidently (7.106) predicts an instability whenever $\Phi(r) > 0$ and $\Omega'(r) < 0$, which is the essence of Chandrasekhar's findings.

More importantly, (7.103) through (7.105) expose the instability mechanism. Suppose we have an axisymmetric radial jet which perturbs the axial magnetic field, as shown in Figure 7.15. The outward radial movement, δu_r, sweeps out a radial field b_r from B_0, as described by (7.104). The differential rotation $\Omega'(r)$ then spirals out b_r to create an azimuthal field b_θ in accordance with (7.105). However, the magnetic field lines tend to resist this shearing and set up an opposing

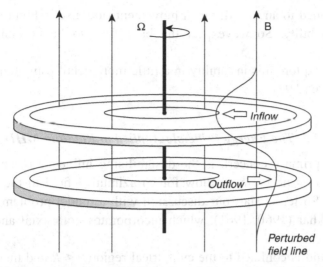

Figure 7.15 The mechanism of MRI. (Adapted from a figure by Cary Forest.)

force, $-j_r B_0 \hat{\mathbf{e}}_\theta = \mu^{-1} B_0 (\partial b_\theta / \partial z) \hat{\mathbf{e}}_\theta$, and hence an opposing torque, $-r j_r B_0 \hat{\mathbf{e}}_z = \mu^{-1} r B_0 (\partial b_\theta / \partial z) \hat{\mathbf{e}}_z$, which appears on the right of (7.103), a torque which is balanced by the Coriolis force on the left. However, the direction of this torque, and hence the direction of the balancing Coriolis force, depends crucially on the sign of $d\Omega^2 / dr$.

Consider first the case where $d\Omega^2 / dr > 0$. Then fluid which has been displaced radially outward (bottom disc in Figure 7.15) is sheared in the positive θ direction and so the field lines, which act like elastic bands in tension, exert an opposing force on the fluid in the *negative* θ direction. From (7.103), we see that the Coriolis force which balances this negative force requires a negative radial velocity. (We assume here that $\Phi > 0$ and so $d\Gamma^2 / dr > 0$.) The fluid then tends to return to its initial radial position, resulting in stable oscillations. On the other hand, if the initial radial movement is inward (top disc in Figure 7.15) the azimuthal shearing of the bent magnetic field lines is such that the Lorentz force is now in the *positive* θ direction, and the matching Coriolis force requires a positive radial velocity. Once again we find that the fluid tends to return to its initial radial position, implying stable oscillations. We conclude, therefore, that the flow is stable if $d\Omega^2 / dr > 0$.

However, if $d\Omega^2 / dr < 0$ the situation is quite different. Outward-moving fluid now results in the bent field lines being sheared in the *negative* θ direction, creating a *positive* Lorentz torque. This, in turn, is balanced by a Coriolis force which now requires a positive radial velocity in accordance with (7.103). Thus the initial outward radial movement is reinforced and we have an instability. Similar arguments applied to an inward radial movement (top disc in Figure 7.15) again results in instability. So we expect the base flow to be unstable whenever $d\Omega^2/dr < 0$.

Let us now explore this instability in a little more detail using the generalised energy principle (7.93).

7.7.2 The Energy Principle Applied to Rotating MHD

Although our primary concern is the destabilising influence of an axial magnetic field, it does no harm to allow for an azimuthal field in the equilibrium configuration. So let us start our discussion with a model problem, introduced by Chandrasekhar (1960, 1961), which incorporates both axial and azimuthal fields.

A perfect fluid is confined to the cylindrical region $r < R$ and threaded by the magnetic field $\mathbf{B}_0 = (0, B_\theta(r), B_0)$, in (r, θ, z) coordinates. The axial field, B_0, is taken to be uniform and the azimuthal field, $B_\theta(r)$, is supported by an axial current. The fluid is rotating, with velocity distribution $\mathbf{u}_0 = (0, r\Omega(r), 0)$. We now perturb the flow and use the energy principle (7.93) to provide a sufficient condition for stability. As in §7.6, we simplify the algebra by taking $\rho = \mu = 1$. We also restrict ourselves to small-amplitude, axisymmetric disturbances.

We start by noting that (7.85), (7.89), (7.94) and (7.95) give the perturbed magnetic field, velocity perturbation and initial disturbance energy, ΔE_0, for any axisymmetric disturbance in terms of the virtual displacement field, $\boldsymbol{\eta}$:

$$\mathbf{b} = \nabla \times (\boldsymbol{\eta} \times \mathbf{B}_0) = B_0 \frac{\partial \boldsymbol{\eta}}{\partial z} - r\eta_r \frac{d}{dr}\left(\frac{B_\theta}{r}\right)\hat{\mathbf{e}}_\theta, \qquad (7.107)$$

$$\delta \mathbf{u} = \dot{\boldsymbol{\eta}} + \hat{\mathbf{u}} = \dot{\boldsymbol{\eta}} - r\eta_r \Omega'(r)\hat{\mathbf{e}}_\theta, \qquad (7.108)$$

$$\Delta E_0 = \frac{1}{2}\int \dot{\boldsymbol{\eta}}^2 dV + W(\boldsymbol{\eta}) = \text{constant}, \qquad (7.109)$$

and

$$W(\boldsymbol{\eta}) = \frac{1}{2}\int \left[\mathbf{b}^2 + \mathbf{B}_0 \cdot \nabla \times (\boldsymbol{\eta} \times \mathbf{b})\right] dV - \frac{1}{2}\int \left[\hat{\mathbf{u}}^2 + \mathbf{u}_0 \cdot \nabla \times (\boldsymbol{\eta} \times \hat{\mathbf{u}})\right] dV. \qquad (7.110)$$

The flow is stable when $W(\boldsymbol{\eta})$ is positive for all possible $\boldsymbol{\eta}$, and so our task is to evaluate $W(\boldsymbol{\eta})$. Since $\hat{\mathbf{u}} = -r\eta_r\Omega'(r)\hat{\mathbf{e}}_\theta$, and $\mathbf{b}$ is given by (7.107), the kinetic and magnetic contributions to (7.110) can be written as

$$\frac{1}{2}\int[(\hat{\mathbf{u}})^2 + \mathbf{u}_0 \cdot \nabla \times (\boldsymbol{\eta} \times \hat{\mathbf{u}})]\,dV = -\frac{1}{2}\int r\eta_r^2 \frac{d\Omega^2}{dr}\,dV, \qquad (7.111)$$

and

$$\delta^2 E_B = \frac{1}{2}B_0^2\int(\partial\boldsymbol{\eta}/\partial z)^2 dV - \frac{1}{2}\int r\eta_r^2\frac{d}{dr}(B_\theta/r)^2 dV + 2B_0\int\frac{\eta_r B_\theta}{r}\frac{\partial\eta_\theta}{\partial z}\,dV, \quad (7.112)$$

from which we obtain $W(\boldsymbol{\eta})$. We see immediately the importance for stability of the term $d\Omega^2/dr$ in (7.111). There are different ways of combining these integrals, but two of the more useful forms are

$$W(\boldsymbol{\eta}) = \frac{1}{2}\int \eta_r^2\left[r\frac{d\Omega^2}{dr} - \frac{1}{r^3}\frac{d}{dr}(rB_\theta)^2\right]dV + \frac{1}{2}B_0^2\int\left(\frac{\partial\boldsymbol{\eta}_p}{\partial z}\right)^2 dV$$
$$+ \frac{1}{2}\int\left[\frac{2\eta_r B_\theta}{r} + B_0\frac{\partial\eta_\theta}{\partial z}\right]^2 dV, \qquad (7.113)$$

and

$$W(\boldsymbol{\eta}) = \frac{1}{2}\int r\eta_r^2\frac{d}{dr}\left[\Omega^2 - \left(\frac{B_\theta}{r}\right)^2\right]dV + \frac{1}{2}B_0^2\int\left(\frac{\partial\boldsymbol{\eta}}{\partial z}\right)^2 dV + 2B_0\int\frac{\partial\eta_\theta}{\partial z}\frac{\eta_r B_\theta}{r}\,dV. \qquad (7.114)$$

Here $\boldsymbol{\eta}_p$ is the usual poloidal component of $\boldsymbol{\eta}$, $\boldsymbol{\eta}_p = (\eta_r, 0, \eta_z)$.

Of course, these expressions simplify considerably when we consider specific cases, as we do below. Before doing this, however, it will prove useful for our subsequent discussion to write down the azimuthal component of the perturbed momentum and induction equations. It is readily confirmed that, in the presence of a mean azimuthal field, (7.99) generalises to

$$\frac{\partial}{\partial t}\delta u_\theta = -\frac{1}{r}\frac{d}{dr}(r^2\Omega)\delta u_r + B_0\frac{\partial b_\theta}{\partial z} + \frac{b_r}{r}\frac{d}{dr}(rB_\theta), \qquad (7.115)$$

while (7.100) gives

$$\frac{\partial b_\theta}{\partial t} + r\frac{d}{dr}\left(\frac{B_\theta}{r}\right)\delta u_r = r\frac{d\Omega}{dr}b_r + B_0\frac{\partial\delta u_\theta}{\partial z}. \qquad (7.116)$$

7.7.3 The Destabilising Influence of an Azimuthal Field

Many axisymmetric stability criteria follow from (7.113) and (7.114). We start with the situation in which there is no axial magnetic field, $B_0 = 0$, so that $\mathbf{B}_0 = (0, B_\theta(r), 0)$. Then Equation (7.107) demands $b_r = 0$ and hence, with the help of (7.108), Equation (7.115) reduces to

$$\delta u_\theta = -\frac{1}{r}\frac{d}{dr}\left(r^2\Omega\right)\eta_r, \tag{7.117}$$

while (7.108) gives $\dot{\eta}_\theta = -2\Omega\eta_r$.

The most convenient form of $W(\boldsymbol{\eta})$ in this situation is (7.114). Combining this with $\dot{\eta}_\theta = -2\Omega\eta_r$ yields

$$\Delta E_0 = \frac{1}{2}\int \dot{\boldsymbol{\eta}}_p^2 dV + \frac{1}{2}\int \eta_r^2\left[\Phi(r) - r\frac{d}{dr}(B_\theta/r)^2\right]dV, \tag{7.118}$$

where Φ is Rayleigh's discriminant. Evidently $\int \dot{\boldsymbol{\eta}}_p^2 dV$ is bounded, and the flow therefor stable, when the second integral on the right of (7.118) is positive. Note that magnetostatic equilibria are stable to axisymmetric disturbances whenever $(B_\theta/r)^2$ decreases with increasing radius, which is consistent with the magneto-static criterion of Bernstein et al. (See Example 7.1 at the end of §7.5.2). More generally, the equilibrium is stable whenever, for all r,

$$\Phi(r) - r\frac{d}{dr}(B_\theta/r)^2 > 0. \tag{7.119}$$

(Actually, (7.119) provides both necessary and sufficient conditions for axisymmetric stability, as shown in Chandrasekhar, 1961.) Evidently, Rayleigh's criterion is a special case of (7.119).

Stability criterion (7.119) may be generalised to non-axisymmetric disturbances of the equilibrium configuration $\mathbf{u}_0 = (0, r\Omega(r), 0)$, $\mathbf{B}_0 = (0, B_\theta(r), 0)$. In such cases we find,

$$2W(\boldsymbol{\eta}) = \int \frac{\eta_r^2}{r^3}\frac{d}{dr}[r^4\Omega^2 - r^2 B_\theta^2]dV + \int \left[\left(\frac{B_\theta}{r}\right)^2 - \Omega^2\right]\left[\left(\frac{\partial\eta_r}{\partial\theta}\right)^2 + \left(\frac{\partial\eta_z}{\partial\theta}\right)^2\right]$$
$$+ \left(\frac{\partial\eta_\theta}{\partial\theta} + 2\eta_r\right)^2\right] dV,$$

(Davidson, 2013). Clearly $W(\boldsymbol{\eta})$ is positive for all possible $\boldsymbol{\eta}$, and the flow stable, provided that $\Omega^2 r^2 < B_\theta^2$ (so called sub-Alfvénic flow) and

$$\Phi(r) - \frac{1}{r^3}\frac{d}{dr}(rB_\theta)^2 > 0. \tag{7.120}$$

For magnetostatic equilibria, three-dimensional stability is ensured whenever $(rB_\theta)^2$ decreases with increasing radius.

7.7.4 The Destabilising Influence of an Axial Field

We now turn to the case where $\mathbf{B}_0 = (0, 0, B_0)$. Once again we focus on axisymmetric disturbances. When $B_\theta = 0$, Equation (7.107) yields $\mathbf{b} = B_0\,\partial\boldsymbol{\eta}/\partial z$, so (7.114) simplifies to

$$\Delta E_0 = \frac{1}{2}\int(\dot{\boldsymbol{\eta}}^2 + \mathbf{b}^2)dV + \frac{1}{2}\int r\eta_r^2\frac{d\Omega^2}{dr}dV. \tag{7.121}$$

The equilibrium is now stable provided that

$$r\frac{d\Omega^2}{dr} > 0. \tag{7.122}$$

The most surprising feature of (7.122) is that, for $B_0 \to 0$, we do not recover Rayleigh's criterion. Rather, a very weak magnetic field can destabilise a flow which is otherwise stable to axisymmetric disturbances. This apparent paradox, and its resolution, is discussed in, for example, Balbus and Hawley (1998) or Davidson (2013). In brief, the paradox is resolved by the fact that the growth rate of disturbances of *fixed* wavenumber tends to zero along with B_0. Of course, in any real system there is a limited range of wavelengths that can be realised.

Finally, we consider the more general case of $\mathbf{B}_0 = (0, B_\theta(r), B_0)$ and $\mathbf{u}_0 = (0, r\Omega(r), 0)$. As above, we restrict ourselves to axisymmetric disturbances. Here (7.113) tells us that $W(\boldsymbol{\eta})$ is positive for all $\boldsymbol{\eta}$ whenever

$$r\frac{d\Omega^2}{dr} - \frac{1}{r^3}\frac{d}{dr}(rB_\theta)^2 > 0, \tag{7.123}$$

which provides a sufficient, but not necessary, condition for stability.

Note that we cannot use the energy principle to establish stability criteria for non-axisymmetric disturbances in this case. This restriction also applies to the situation $\mathbf{B}_0 = (0, 0, B_0)$. The reason is that a finite B_z means that $\mathbf{u}_0$ and $\mathbf{B}_0$ are not co-linear, and in such cases $W(\boldsymbol{\eta})$ is always indefinite in sign to three-dimensional disturbances. (See Exercise 7.1 at the end of this chapter).

Table 7.1 *Sufficient conditions for the linear stability of* $\mathbf{u} = (0, r\Omega, 0)$, $\mathbf{B} = (0, B_\theta, B_0)$.

	$B_0 \neq 0, B_\theta = 0$	$B_\theta \neq 0, B_0 = 0$	$B_0 \neq 0, B_\theta \neq 0$
Stable to axisymmetric modes	$r\dfrac{d\Omega^2}{dr} > 0$	$\Phi(r) - r\dfrac{d}{dr}(B_\theta/r)^2 > 0$ (necessary and sufficient)	$r\dfrac{d\Omega^2}{dr} - \dfrac{1}{r^3}\dfrac{d}{dr}(rB_\theta)^2 > 0$
Stable to three-dimensional modes	No criteria. $\mathbf{B}_0$ and $\mathbf{u}_0$ nonparallel.	$\Phi(r) - \dfrac{1}{r^3}\dfrac{d}{dr}(rB_\theta)^2 > 0$ (and $\Omega^2 r^2 < B_\theta^2$)	No criteria. $\mathbf{B}_0$ and $\mathbf{u}_0$ nonparallel.

The various criteria we have discussed are summarised in Table 7.1. We have provided only a brief overview of the Chandrasekhar–Velikhov instability, focussed on axisymmetric disturbances and designed to bring out the utility of (7.93) for establishing stability criteria. Those interested in more details could do worse that consult Chandrasekhar (1961) for a classical account, or else Balbus and Hawley (1998) for an astrophysical interpretation.

7.8 From MHD to Euler Flows: The Kelvin–Arnold Theorem

The stability analysis of §7.6 applies equally to Euler flows in the absence of a magnetic field, and so it is natural to find points of contact with the corresponding variational theorems which have been nurtured in the strictly hydrodynamic literature. Perhaps the most important of these is the Kelvin–Arnold theorem.

In 1887 Kelvin noted that the kinetic energy of a steady Euler flow is stationary under perturbations in which the vortex lines are frozen into the fluid. He also noted that such flows are stable provided that their kinetic energy is an extremum (a maximum or a minimum) under such a perturbation, but that no conclusions could be drawn about stability when the equilibrium represented a saddle point in energy. Kelvin stated it thus:

The condition for steady motion of an incompressible inviscid fluid filling a finite fixed portion of space ... is that, with given vorticity, the energy is a thorough maximum, or a thorough minimum, or a minimax. The further condition of stability is secured, by consideration of energy alone, for any case of steady motion for which the energy is a thorough maximum or a thorough minimum; because when the boundary is held fixed the energy is of necessity constant. But the mere consideration of energy does not decide the question of stability for any case of steady motion in which the energy is a minimax.

Kelvin's insightful observations languished in relative obscurity until 1966, when Arnold's extension of Kelvin's variational principle (Arnold, 1966) popularised the original theorem and provided a systematic means of calculating the second variation in kinetic energy.

Now we have already noted that the magnetostatic stability criterion presented by Bernstein et al. is a special case of (7.93), and it is natural to ask if the Kelvin–Arnold theorem is also a special case. It turns out that it is. When $\mathbf{B}_0 = 0$, the conserved Hamiltonian in (7.93) can be rewritten, with the aid of (7.88) and (7.89), as

$$e = \frac{1}{2} \int \left[(\delta^1 \mathbf{u})^2 + 2\mathbf{u}_0 \cdot \delta^2 \mathbf{u} \right] dV = constant, \qquad (7.124)$$

where $\delta^1 \mathbf{u}$ and $\delta^2 \mathbf{u}$ are obtained by uncurling

$$\delta^1 \boldsymbol{\omega} = \nabla \times (\boldsymbol{\eta} \times \boldsymbol{\omega}_0), \quad \delta^2 \boldsymbol{\omega} = \tfrac{1}{2} \nabla \times \left(\boldsymbol{\eta} \times \delta^1 \boldsymbol{\omega} \right) \qquad (7.125)$$

(Davidson, 2000). If we compare these expressions with (7.46) and (7.52), it is evident that e in (7.124) is the second-order change in kinetic energy under a perturbation in which the ω-lines are frozen into the fluid (a so-called *isovortical perturbation*). Of course, this is a physically realistic perturbation since the ω-lines are indeed frozen into the fluid in the absence of $\mathbf{B}_0$. Moreover, it is readily confirmed that, as Kelvin anticipated, the first-order change in kinetic energy is zero under such a perturbation. Thus, according to Kelvin, (7.124) holds the key to stability, with stability ensured when e is positive or negative definite.

We can indeed use (7.124) to provide a sufficient condition for the stability of steady Euler flows. The idea is the following. Provided that $e(\boldsymbol{\eta})$ is positive or negative definite, it can be used like a Lyaponov functional to bound the growth of any disturbance. For example, suppose that $\|\delta \mathbf{u}\|$ is some integral measure of the disturbance, then the flow will be unstable if $\|\delta \mathbf{u}\|$ grows despite the conservation of e, and so a prerequisite for instability is that $|e|/\|\delta \mathbf{u}\|^2 \to 0$. Consequently, if there exists a bound of the form $|e| \geq \lambda \|\delta \mathbf{u}\|^2$ for all $\delta \mathbf{u}$, λ being a constant, then the flow cannot become unstable. When e is positive or negative definite then such a bound can usually be found and hence, as Kelvin foresaw, the flow is stable.

A modern rephrasing of Kelvin's 1887 statement might be: *The kinetic energy of an Euler flow is stationary at equilibrium under a perturbation in which the vortex lines are frozen into the fluid, and a sufficient (though not necessary) condition for the stability of a steady Euler flow is that its kinetic energy is a minimum or a maximum under such a perturbation.* This is known as the Kelvin–Arnold variational principle, and stability criteria are readily derived using this theorem. Indeed, examples of two-dimensional and axisymmetric stability criteria are discussed below in Exercises 7.2 and 7.3.

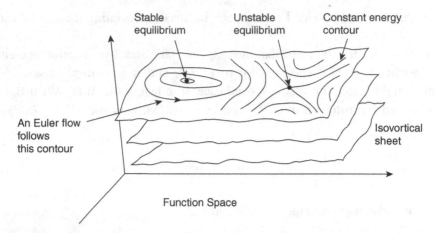

Figure 7.16 Euler flows evolve on an isovortical sheet in function space. They follow constant energy contours on such a sheet, and so steady flows are stationary points on the sheet while extrema in energy are necessarily stable.

The Kelvin–Arnold theorem is often illustrated in cartoon fashion as shown in Figure 7.16. This represents (in a somewhat symbolic way) the function space of all solenoidal velocity fields which satisfy $\mathbf{u} \cdot d\mathbf{S} = 0$ on the boundary. The idea is that an Euler flow evolves within such a function space while preserving its vortex-line topology, as well as its energy. So it is natural to divide such a function space into lower dimensional sub-domains in which the vorticity fields are all linked by a smooth, volume-preserving displacement field, $\mathbf{\eta}$. Such sub-domains are often referred to as *isovortical sheets* and an Euler flow is constrained to follow a constant energy contour along such a 'sheet'. Thus steady Euler flows are stationary points with respect to energy on isovortical sheets.

Notions of stability follow directly from such a cartoon. For example, suppose the flow is perturbed isovortically onto an adjacent energy contour. Then it will move off along that contour, confined to it for all time. It follows that energy extrema represent stable Euler flows, since the perturbed flow always stays close (in some sense) to the stationary point in function space. By contrast, saddle points represent potentially unstable flows, since the conservation of energy places no restriction on the migration of the flow from the stationary point in function space.

Exercises

7.1 *Short-wavelength disturbances applied to non-static equilibria in ideal MHD.*
It is well known that (7.93) cannot provide sufficient conditions for stability

to three-dimensional disturbances unless $\mathbf{B}_0$ and $\mathbf{u}_0$ are everywhere parallel. Consider a perturbation whose wavelength is very small by comparison with the characteristic length-scale for $\mathbf{B}_0$ and $\mathbf{u}_0$. Show that $W(\boldsymbol{\eta})$ reduces, in this case, to

$$W(\boldsymbol{\eta}) = \frac{1}{2} \int \left[(\mathbf{B}_0 \cdot \nabla \boldsymbol{\eta})^2 - (\mathbf{u}_0 \cdot \nabla \boldsymbol{\eta})^2 \right] \, dV.$$

Consider the effect of varying the orientation of the wavenumber, $\mathbf{k}$, of the disturbance and show that $W(\boldsymbol{\eta})$ is always indefinite in sign if $\mathbf{B}_0$ and $\mathbf{u}_0$ are non-parallel.

7.2 *The Kelvin–Arnold principle applied to two-dimensional flow.*
Consider a steady, inviscid, two-dimensional flow, $\mathbf{u}_0 \cdot \nabla \omega_0 = 0$, in which the spatial gradients in ω_0 are everywhere non-zero. Show that $\omega_0 = \omega_0(\Psi_0)$, where Ψ_0 is the two-dimensional streamfuction. Let $\varphi = \delta^1 \Psi$ be the first-order change in Ψ_0 under an isovortical perturbation of the vortex lines and show that (7.124) reduces to

$$e = \frac{1}{2} \int \left[(\nabla \varphi)^2 - (\nabla^2 \varphi)^2 \Big/ \omega'_0(\psi_0) \right] \, dV.$$

Such flows are stable by the Kelvin–Arnold principle provided that $\omega'_0(\psi_0) < 0$, or else $\omega'_0(\psi_0) > 0$ and the second term in the energy integral is dominant for all possible φ. Now show that the latter case requires the minimum eigenvalue, λ, of

$$\nabla^2 \varphi + \lambda \omega'_0(\psi_0)\varphi = 0,$$

to be greater than unity.

7.3 *Rayleigh's centrifugal stability criterion and the Kelvin–Arnold principle.*
Use (7.124) to show that the flow $\mathbf{u} = (0, r\Omega(r), 0)$, in (r, θ, z) coordinates, is stable to axisymmetric disturbances provided that

$$\Phi(r) = \frac{1}{r^3} \frac{d}{dr} \left(r^2 \Omega \right)^2 > 0.$$

8

An Introduction to Turbulence

You asked, 'What is this transient pattern?'
If we tell the truth of it, it will be a long story;
It is a pattern that came up out of an ocean
And in a moment returned to that ocean's depth.

(Omar Khayyam)

Turbulence is not an easy subject. Our understanding of it is limited, and those bits we do understand are arrived at through detailed and difficult calculation. G.K. Batchelor gave some hint of the difficulties when, in the introduction to his 1953 monograph, he wrote:

It seems that the surge of progress which began immediately after the war has now largely spent itself, and there are signs of a temporary dearth of new ideas. . . we have got down to the bedrock difficulty of solving non-linear partial differential equations.

(G.K. Batchelor, 1953)

Little has changed since 1953. Nevertheless, it is hard to avoid the subject of turbulence in MHD, since the Reynolds number, even in metallurgical MHD, is invariably large. (Liquid metals have a kinematic viscosity somewhat less than that of air.) So at some point we simply have to bite the bullet and do what we can. This chapter and the next are intended as an introduction to the subject, providing a springboard for those who wish to take it up seriously. In order not to demotivate the novice, we have tried to keep the mathematical difficulties to a minimum. Consequently, only schematic outlines are given of certain standard derivations. For example, deriving the standard form for the velocity correlation tensors in isotropic turbulence can seem like hard work. Such derivations are well documented elsewhere[1] and so there seems little point in giving a blow-by-blow description here. Rather, we have concentrated on getting across the main physical ideas.

---

[1] Detailed accounts of hydrodynamic turbulence may be found in, say, Tennekes and Lumley (1972) and Davidson (2015), while MHD turbulence is discussed at length in Biskamp (2003) and Davidson (2013).

Now the sceptic might say: if the theory of turbulence is so hard, why bother with it at all? After all, we now have powerful computers available to us, which can compute both the mean flow and the motion of every turbulent eddy. The experimentalist Stan Corrsin had one answer to this. Having estimated the computing resources required to simulate even the most modest of turbulent flows, and shown them to be well beyond the capacity at that time, he made the following whimsical comment:

The foregoing estimate (*of computing power*) is enough to suggest the use of analogue instead of digital computation; in particular, how about an analogue consisting of a tank of water? *(Corrsin, 1961)*

Corrsin said this in 1961, but actually it is still pertinent today. Despite the great advances which have occurred in computational fluid dynamics, half a century later our capacity to simulate accurately turbulent flows by computation is still rather poor, restricted to simple geometries and to modest Reynolds numbers. The problem, as you shall discover shortly, is that turbulent flows contain, at any instant, eddies (vortical structures) which have a wide range of sizes and time scales, from the large and slow to the tiny and fast. It is difficult to capture this wide spectrum of eddies in a numerical simulation, especially in the complex geometries of interest to engineers, or at the large Reynolds numbers encountered in geophysical and astrophysical flows.

As a prelude to discussing MHD turbulence, it seems prudent to summarise first the simpler features of conventional turbulence. So this chapter provides an introduction to conventional hydrodynamic turbulence, and we defer our discussion of MHD turbulence until Chapter 9. We start with a short historical introduction.

8.1 An Historical Interlude

At times water twists to the northern side, eating away the base of the bank; at times it overthrows the bank opposite on the south; at times it leaps up swirling and bubbling to the sky; at times revolving in a circle it confounds its course... Thus without any rest it is ever removing and consuming whatever borders upon it. Going thus with fury it is *turbulent* and destructive.

Leonardo da Vinci

So began man's study of turbulent fluid motion.

Our historical survey begins with Newton and the ideas of viscosity and eddy viscosity. The relationship between shear stresses and gradients in mean velocity has been a recurring theme in turbulence theory. In the laminar context this was established in 1687 by Newton who, in *Principia*, hypothesised that the resistance to relative movement in parts of a fluid are 'proportional to the velocity with which the parts of fluid are separated from one another', i.e. the relative rate of sliding of layers within the fluid. The constant of proportionality is, of course, the coefficient of viscosity. Newton's idea of internal friction was somewhat overlooked by the eighteenth-century mathematicians and it languished until 1823 when Navier, and a little later Stokes, in 1845, introduced viscous forces into the equations of hydrodynamics.

Shortly after the introduction of Newton's law of viscosity questions were raised as to the uniformity of the viscosity. For example, in 1851 Saint-Venant speculates that

If Newton's assumption,..., which consists in taking interior friction proportional to the speed of the fluid elements sliding against one another, can be applied approximately to the set of points of a given fluid section, all the known facts lead us to infer that the coefficient of this proportionality should increase with the size of transverse sections; this may be explained up to a point by noticing that the fluid elements are not progressing parallel to each other with regularly graded velocities, and that ruptures, eddies and other complex and oblique motions, which must strongly effect the magnitude of frictions, are formed.

There is clearly some embryonic notion of turbulence and of *eddy viscosity* here, albeit confused with molecular action. This was pursued by both Reynolds and Boussinesq, the latter being Saint-Venant's student. Boussinesq came first, noting that turbulence must greatly increase the (eddy) viscosity because

...the (*turbulent*) friction experienced, being caused by finite sliding between adjacent layers, will be much larger than would be the case should velocities vary in a continuous way.   *(Boussinesq, 1870)*

Shortly after, Reynolds' classic paper on pipe flow appeared (1883). This clearly differentiates between laminar and turbulent flow, and identifies the key role played by $Re = u\ell/v$ in determining which state prevails. Later, Reynolds reaffirmed the idea of an eddy viscosity while introducing the notion that the fluid velocity might be decomposed into a mean and fluctuating component, the latter giving rise to the fictitious, time-averaged shear stresses which now bear Reynolds' name. Reynolds used the term *sinuous* to describe the appearance of turbulence.

By 1925, Prandtl clearly recognised the analogy between the turbulent transport of momentum (through turbulent eddies) and the laminar shear stress caused by molecular motion, as predicted by the kinetic theory of gases. He introduced the mixing length model of turbulence described in Chapter 3, which had some notable successes at the time (e.g. the log-law of the wall) but is now regarded as flawed. (The problem is that there is no real separation of length scales between the turbulent fluctuations and gradients in mean velocity, as required by a mixing length theory. In fact, most result deduced by mixing length can also be deduced by purely dimensional arguments.)

A breakthrough in turbulence theory came with the pioneering work of G.I. Taylor in the early 1930s, who for the first time fully embraced the need for a statistical approach to the subject. He introduced the idea of the velocity correlation tensor $Q_{ij}(\mathbf{r}, t) = \langle u'_i(\mathbf{x})u'_j(\mathbf{x} + \mathbf{r})\rangle$, a generalisation of the Reynolds stress, which is now the common currency of turbulence theory. (The angled brackets here represent an ensemble average, as distinct from the overbar used in Chapter 3, which denotes a time average[2].) The quantity $Q_{ij}(\mathbf{r})$ tells us about the degree to which the fluctuating component of motion, $\mathbf{u}'$, is statistically correlated at two points separated by a distance $|\mathbf{r}|$. A strong correlation implies that there are vortices which span the gap $|\mathbf{r}|$. Conversely, if $Q_{ij}(\mathbf{r})$ is very small then $\mathbf{x}$ and $\mathbf{x} + \mathbf{r}$ are more or less statistically independent. Thus $Q_{ij}(\mathbf{r})$ contains information about the structure of the turbulence.

Taylor also promoted the useful idealisation of statistically homogeneous and isotropic turbulence. This initiative was pursued by the Hungarian engineer von Kármán, who showed that, with the help of the symmetry implied by isotropy, Q_{ij} could be expressed in terms of a single scalar function, $f(|\mathbf{r}|)$, and that the Navier–Stokes equation could be manipulated into the form $\partial f/\partial t = (\ldots)$. At last there was the possibility of making rigorous, quantitative predictions about turbulence. Unfortunately, the right-hand side of this equation includes new terms such as triple velocity correlations of the form $\langle u'_i(\mathbf{x})u'_j(\mathbf{x})u'_k(\mathbf{x} + \mathbf{r})\rangle$. Consequently, it is rarely possible to make rigorous predictions for the evolution of f. Nevertheless, in certain circumstances the triple correlations can be finessed away, and so Kármán's equation, now called the Kármán–Howarth equation, can provide useful information.

The statistical theory of turbulence was greatly developed in the (then-) USSR in the 1940s, particularly by Kolmogorov and his student Obukhov. These researchers followed the lead of Richardson, who had pointed out that a vast range of scales (eddy sizes) exist in a typical turbulent flow and that viscosity influences only the smallest of these eddies. Kolmogorov quantified the idea of the energy cascade, in which eddies continually pass their kinetic energy onto smaller and smaller vortices, until viscosity destroys that energy. This allowed him to predict how the energy of a turbulent flow is distributed amongst the various eddy sizes. Great strides were made and by 1950 a physical and mathematical picture of homogeneous turbulence had emerged which is little different today. However, this picture is not entirely deductive, but relies rather on certain (plausible) physical

---

[2] In statistically steady turbulence an ensemble average is equivalent to a time average, while in statistically homogeneous turbulence an ensemble average is the same as a volume average.

assumptions based on empirical evidence. Sometimes these assumptions are satisfied, and sometimes they are not.

Turbulence plays a key part in MHD. Virtually all laboratory and industrial flows are turbulent. Moreover, random, non-linear interactions (which we might loosely label as turbulence) are an essential ingredient of planetary dynamo theory, while turbulence is needed in astrophysical MHD to explain, for example, how the plasma in an accretion disc can shed its angular momentum as it spirals radially inward towards the central star, or how the flux tube reconnections which trigger solar flares can occur, reconnections which are so hard to account for in terms of the vanishingly small molecular diffusivity. Comparing the development of turbulence with the laws of electromagnetism we see that turbulence was rather a late developer, reflecting the formidable difficulties inherent in tackling a non-linear, random process (Table 8.1). Even today there is no universal 'theory of turbulence'. We have a few theoretical results relating to various idealised configurations (e.g. Kolmogorov's theory of the small scales, or Prandtl's log-law for the mean velocity near a wall), and a great deal of experimental data. Sometimes, but not always, the two coincide. Of course, it is when theory and experiment differ, and we try to reconcile those differences, that we learn the most. As with most of fluid mechanics, our understanding of turbulence has developed through a careful

Table 8.1

Theory of turbulence		Electricity and magnetism	
		11th century	Compass
		1269	Peregrinus: magnetic poles
1500s	Leonardo's first observations	1600	Gilbert: geomagnetism
		1750s	Coulomb: action at a distance
		1820s	Ampère: forces on currents
		1831	Faraday: induction; concept of fields
1860s	Boussinesq: eddy viscosity	1860s	Maxwell's equations
1880s	Reynolds: two types of flow, turbulent stresses	1889	Hertz: electromagnetic waves
1904	Prandtl: boundary layers		
1920s	Prandtl: mixing length theory		
1930s	Taylor, von Kármán: statistical theory of turbulence		
1941	Kolmogorov: theory of small scales		
	1942: Beginning of MHD – Alfvén waves discovered		

assessment of the experimental evidence – which brings us back to Leonardo da Vinci's observations.

One cannot help but be struck by the similarities between Reynolds' idea of two motions, a mean forward motion and a turbulent vortical motion, and his observation of the sinuous nature of turbulence in a pipe, and Leonardo da Vinci's note in 1513:

Observe the motion of the surface of water, which resembles the behaviour of hair, which has two motions, of which one depends on the weight of the strands, the other on the line of its revolving; thus water makes revolving eddies, one part of which depends upon the impetus of the principle current, and the other depends on the incident and reflected motions.

Did Leonardo foresee what we now call the Reynolds decomposition?

8.2 The Structure of Turbulent Flows: Richardson's Cascade

Let us start with a traditional question in turbulence theory. Suppose we have a statistically steady turbulent flow; say, flow in a pipe. Then the turbulent eddies are continually subject to viscous dissipation yet the energy of the turbulence does not, on average, change. Where does the turbulence energy come from? Of course, in some sense it comes from the mean flow. The traditional way of quantifying this relies on the idea of dividing the flow into two distinct parts, a mean component and a turbulent motion, and then examining the exchange of energy between the two: which brings us back to the idea of a Reynolds stress.

You are already familiar with the concept of the Reynolds stress, which we shall now label as τ_{ij}^R. In Chapter 3 we showed that, when we time-average the Navier–Stokes equation in a turbulent flow, the presence of the turbulence gives rise to additional stresses, $\tau_{ij}^R = -\rho \overline{u'_i u'_j}$, which act on the mean flow. Here the prime on $\mathbf{u'}$ indicates that this is a fluctuating component of velocity, $\mathbf{u'} = \mathbf{u} - \bar{\mathbf{u}}$, and the overbar signifies a time average. Now these Reynolds stresses give rise to a net force acting on the mean flow, $f_i = \partial \tau_{ij}^R / \partial x_j$, and if the rate of working of this force, $f_i \bar{u}_i$, is negative, then the mean flow must lose mechanical energy to the agent which supplies the force, i.e. to the turbulence. We say that mechanical energy, usually kinetic energy, is transferred from the mean flow to the turbulence. This is why the turbulence in a pipe, say, does not die away; the viscous dissipation of turbulent eddies is matched by the rate of working of f_i.

Of course, this is all a little artificial, in the sense that we have just one fluid and one flow. All we are saying is that when we decompose $\mathbf{u}$ into $\bar{\mathbf{u}}$ and $\mathbf{u'}$ then the total kinetic energy, which is conserved in the absence of viscosity, is likewise divided between $\frac{1}{2}\bar{\mathbf{u}}^2$ and $\frac{1}{2}\overline{\mathbf{u'}^2}$. When $f_i \bar{u}_i$ is negative, energy is transferred from $\bar{\mathbf{u}}$ to $\mathbf{u'}$. Physically this corresponds to the stretching of turbulent eddies (blobs of vorticity) by the mean flow, a stretching process that increases the kinetic energy of the turbulent eddies. Now we can write $f_i \bar{u}_i$ as

$$f_i \bar{u}_i = \frac{\partial}{\partial x_j} \left[\bar{u}_i \tau_{ij}^R \right] - \tau_{ij}^R \bar{S}_{ij}; \qquad \bar{S}_{ij} = \frac{1}{2} \left[\frac{\partial \bar{u}_i}{\partial x_j} + \frac{\partial \bar{u}_j}{\partial x_i} \right]. \qquad (8.1)$$

Here $\bar{S}_{ij}$ is the strain-rate tensor introduced in §3.2. The first term on the right of (8.1) is just the divergence of $\bar{u}_i \tau_{ij}^R$. In a finite, closed domain, in which $\bar{u}_j$ is zero on the boundary, or else in a statistically homogeneous turbulent flow, this term integrates to zero. Thus the net rate of transfer of mechanical energy to the turbulence is usually just the volume integral of $\tau_{ij}^R \bar{S}_{ij}$, which is sometimes called the *deformation work*. Usually $\tau_{ij}^R \bar{S}_{ij}$ is a positive quantity, reflecting the tendency for the mean flow to stretch the turbulent vorticity field, thus increasing its energy. So a finite strain-rate in the mean flow tends to keep the turbulence alive. Note that there are no viscous effects involved in this transfer of energy (if Re is large): it is a non-dissipative process. The next question, therefore, is where does this turbulent energy go?

If we have a steady-on-average flow in a pipe, say, then there is a continual energy transfer from the mean flow, via $\tau_{ij}^R \bar{S}_{ij}$, to the turbulence. However, the turbulence in such a situation will be statistically steady and so this energy must be dissipated somehow. Ultimately, of course, it is viscosity which destroys the mechanical energy of the eddies. However, when Re is large, the viscous stresses acting on the large eddies are negligible, so there must be some rather subtle process at work. This leads to the idea of the energy cascade, a concept first proposed by the British meteorologist L.F. Richardson in the 1920s.

It is an empirical observation that any turbulent flow comprises 'eddies' (blobs of vorticity) which have a wide range of sizes. That is to say, there is always a wide spectrum of length scales, velocity gradients and so forth, as illustrated by Leonardo's sketch of turbulence in a pool of water (Figure 8.1). Richardson's idea is that the largest eddies, which are typically created by instabilities in the mean flow, are rather transient and rapidly pass their energy onto yet smaller vortices. These smaller eddies then, in turn, pass energy onto even smaller vortices and so on. There is a continual *cascade* of energy from the large scale down to the small (Figure 8.2). Richardson's key observation is that this flux of energy from large to small scales is usually a multistep process, involving a hierarchy of eddy sizes. Thus we refer to Richardson's cascade as a *multistage* flux of energy through *scale-space*. This flux should not, of course, be confused with the flux of energy from place to place in physical space. Rather, it is a transfer of energy from large to small eddies (large to small blobs of vorticity) which coexist in the same region of space.

It should be emphasised that viscosity plays no part in this cascade. That is, when Re is large (based on $\mathbf{u}'$ and a typical eddy size), then the viscous stresses acting on

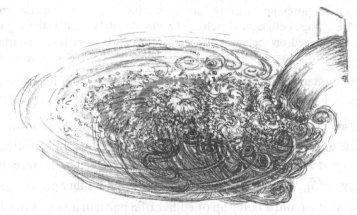

Figure 8.1 Copy of Leonardo's sketch of water flowing into a pool. Note the different scales of motion. (Courtesy of F.C. Davidson.)

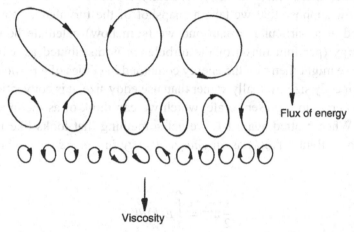

Figure 8.2 A schematic representation of Richardson's energy cascade showing a flux of energy from large to small scale.

the larger eddies are negligible. The whole process is simply driven by inertia. The cascade is halted, however, when the eddies become so small that the Reynolds number based on the small-scale eddy size is of the order of unity. That is, the very smallest eddies are dissipated by viscous forces, and for the viscous forces to be significant we need a Reynolds number of order unity. We may think of viscosity as providing a dustbin for energy at the very end of the cascade. In this sense the viscous forces are passive in nature, mopping up whatever energy is fed downward from above. This concept of a progressive, multistep energy cascade from large to small eddies was nicely summed up by Richardson in his parody of Swift's Fleas Sonnet:

One gets a similar impression when making a drawing of a rising cumulus from a fixed point; the details change before the sketch can be completed. We realise that big whirls have little whirls, which feed on their velocity, and little whirls have lesser whirls and so to viscosity.   *(Richardson, 1922)*

Let us try to quantify this process. Let ℓ and u' be typical length and velocity scales for the larger eddies. We might, for example, define u' through $(u')^2 = \overline{(u'_x)^2}$ or $\overline{(u'_y)^2}$. Also, let $\varepsilon = 2\nu \overline{S'_{ij} S'_{ij}}$ be the rate of dissipation of mechanical energy per unit mass due to viscosity acting on the small-scale eddies. In statistically steady turbulence ε must also equal the rate at which energy is fed to the turbulence from the mean flow, $\tau_{ij}^R \overline{S}_{ij}$. If it did not, the turbulence would either gain or lose energy. In fact, if we are to avoid a build-up of eddies of a particular size, ε must equal the rate at which energy is passed down the cascade at any point within that cascade. Let Π be the rate at which energy (per unit mass) is passed down the cascade, i.e. the *scale-space energy flux*. Then, symbolically, we have $\Pi = \varepsilon$.

Let us now suppose that we take a snapshot of the turbulence at a particular location and at a particular instant and we (somehow) calculate how the total kinetic energy (per unit mass) of the turbulence is distributed as a function of eddy size. We might then plot the energy contained in eddies of a particular size as a function of eddy size. Actually, rather than use eddy size, it is conventional to use a wavenumber, k, to represent scale, which we can think of as the inverse of the eddy size. When plotted against k, we get something that looks like Figure 8.3, where we may think of $E(k)$ as an *energy density in scale space*, which has the property

$$\frac{1}{2}\overline{\mathbf{u}'^2} = \int_0^\infty E(k)dk. \tag{8.2}$$

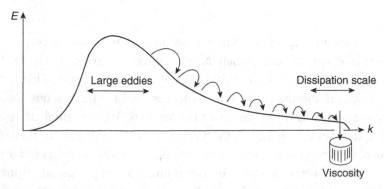

Figure 8.3 The energy cascade represented in a plot of energy density versus wavenumber. (From Davidson, 2015, by permission of Oxford University Press.)

(We shall give formal definitions of E and k in §8.3.2, where we discuss how Fourier analysis may be used to distinguish between the energy held at different scales.)

It is important to remember that there is dissipation only at the smallest scales, and so the energy flux, Π, has to be the same at all points between the large and the small eddies. That is, suppose that $\Pi_\ell = \Pi(k = \ell^{-1})$ and $\Pi(k)$ are the fluxes of energy to small scales measured at the large and intermediate scales, respectively. Then statistical equilibrium demands $\Pi_\ell = \Pi(k)$. Also, the flux of energy near the top of the cascade must equal the rate at which energy (per unit mass) enters the turbulence, and so we also have $\Pi_\ell = \tau_{ij}^R \overline{S}_{ij}/\rho$. Combining these expressions yields

$$\tau_{ij}^R \overline{S}_{ij}/\rho = \Pi_\ell = \Pi(k) = \varepsilon, \tag{8.3}$$

for statistically steady turbulence. Now it is an empirical observation that the rate of extraction of energy (per unit mass) from the large eddies to the energy cascade is of the order of

$$\Pi_\ell \sim (u')^3/\ell. \tag{8.4}$$

This is not a trivial result. As we shall see, it turns out to be pivotal in many theories of turbulence. Physically, it states that the largest eddies pass on a significant fraction of their energy to smaller scales on a time scale of ℓ/u', their turnover time.

Let us now try to determine the size of the smallest eddies. Let v and η be the characteristic velocity and length scales of the smallest structures in the flow. There are two things we can say about these eddies. First, we have $v\eta/\nu \sim 1$. That is, rather like a boundary layer, the size of the small eddies automatically adjusts to make the viscous forces an order-one quantity. Second, the energy dissipation rate per unit mass, which in a laminar flow is $\varepsilon = 2\nu S_{ij} S_{ij}$, must be of order $\varepsilon \sim \nu(v/\eta)^2$.

Let us now summarise everything we know about the energy cascade:

(i) The process is inviscid except at the smallest scales and so, in statistically steady turbulence,

$$\Pi_\ell = \Pi(k) = \varepsilon;$$

(ii) Empirically it is observed that energy is extracted from the large scales on a timescale of the large-eddy turnover time, so that

$$\Pi_\ell \sim (u')^3/\ell. \tag{8.5}$$

(iii) The smallest scales must satisfy

$$v\eta/\nu \sim 1, \quad \varepsilon \sim \nu(v/\eta)^2. \tag{8.6}$$

We may eliminate either η or v from (8.6) and then use the fact of $\varepsilon \sim \Pi_\ell \sim (u')^3/\ell$ to express η or v in terms of the large-scale parameters. Following this procedure we find

$$\eta = \left(\nu^3/\varepsilon\right)^{1/4} \sim \left(u'\ell/\nu\right)^{-3/4}\ell \sim \mathrm{Re}^{-3/4}\ell, \tag{8.7}$$

$$v = (\nu\varepsilon)^{1/4} \sim \left(u'\ell/\nu\right)^{-1/4}u' \sim \mathrm{Re}^{-1/4}u'. \tag{8.8}$$

Here Re is based on the large-scale velocity and length scales and the expressions $\eta = \left(\nu^3/\varepsilon\right)^{1/4}$ and $v = (\nu\varepsilon)^{1/4}$ are usually used to *define* η and v. By contrast, the estimates $\eta/\ell \sim \mathrm{Re}^{-3/4}$ and $v/u' \sim \mathrm{Re}^{-1/4}$ are used to *estimate the magnitude* of η and v. Suppose, for example, that $\mathrm{Re} \sim 10^4$ and $\ell \sim 1\mathrm{cm}$, which is not untypical in a wind-tunnel experiment. Then $\eta \sim 0.01$ mm, which is very small! There is, therefore, a large spectrum of eddy sizes in a typical turbulent flow, and it is this which makes turbulence so difficult to simulate numerically.

The quantities v and η are known as the *Kolmogorov microscales* of velocity and length, whereas ℓ, the size of the large eddies, is known as the *integral scale* of the turbulence. (It is possible to give a more precise definition of ℓ, which we do later.)

There is something else of interest to be extracted from these simple estimates. Eliminating ν from (8.6), and then equating ε to $\Pi(k \sim \eta^{-1})$, we find that the rate at which energy cascades downward at the tail end of the energy cascade is

$$\Pi(k \sim \eta^{-1}) \sim v^3/\eta. \tag{8.9}$$

Compare this with (8.5). The implication is that the smallest eddies, just like the largest ones, pass on their energy on a timescale of their turnover time, $\tau = \eta/v$. As we shall see, this is characteristic of all eddy sizes, from ℓ down to η. Moreover, (8.7) and (8.8) give

$$\tau \sim \mathrm{Re}^{-1/2}\ell/u' \ll \ell/u'. \tag{8.10}$$

So the characteristic timescale for the evolution of the small eddies is very much faster than the turnover time of the large eddies. Things happen very rapidly at the small scales. The relationships between the smallest and largest scales are listed in Table 8.2.

Table 8.2

Dimension	Relationship of Kolmogorov scales to large scales
Length	$\eta = \left(\nu^3/\varepsilon\right)^{1/4} \sim \left(u'\ell/\nu\right)^{-3/4}\ell \sim \mathrm{Re}^{-3/4}\ell$
Velocity	$\upsilon = \left(\nu\varepsilon\right)^{1/4} \sim \left(u'\ell/\nu\right)^{-1/4}u' \sim \mathrm{Re}^{-1/4}u'$
Time	$\tau = \left(\nu/\varepsilon\right)^{1/2} \sim \left(u'\ell/\nu\right)^{-1/2}\ell/u' = \mathrm{Re}^{-1/2}\ell/u'$

8.3 Kinematic Preliminaries (for Homogeneous Turbulence)

We now introduce some of the statistical diagnostics which are commonly used to characterise the structure of a turbulent velocity field. In the interests of simplicity, we restrict ourselves to statistically homogeneous turbulence in which there is no mean velocity. The absence of a mean shear means that there is no transfer of energy to the turbulence from the mean flow, and so the turbulence simply decays with time. A good approximation to this is the turbulence created by a grid in a wind tunnel. In such a situation we can no longer say that the flux of energy down through the cascade is independent of eddy size, as the energy of eddies of various sizes changes with time. So (8.3) must be relaxed to read

$$\Pi_\ell \sim u'^3/\ell \sim \Pi(k) \sim \varepsilon, \tag{8.11}$$

where $\varepsilon = 2\nu\overline{S'_{ij}S'_{ij}}$. Note that $2\nu S_{ij}S_{ij}$ differs from $\nu(\nabla \times \mathbf{u})^2 = \nu\omega^2$ by a divergence, $2\nu S_{ij}S_{ij} = \nu\omega^2 + \nabla \cdot (\sim)$, and so $\nu\omega^2$ is often used as a proxy for the dissipation in homogeneous turbulence, as discussed below.

Since we assume no mean flow, we shall drop the prime on $\mathbf{u}'$ to indicate a fluctuating velocity component, and instead reserve the prime for quite another purpose. Also, since the turbulence is not statistically steady, we must abandon time averages and adopt instead ensemble averages (or, equivalently, volume averages), which we denoted by angled brackets. So, for example, the kinetic energy of the turbulence is $\frac{1}{2}\langle\mathbf{u}^2\rangle$, and $\langle\mathbf{u}\rangle = 0$. Note that the operations of taking an ensemble average and differentiation commute, so in homogeneous turbulence $\langle\nabla \cdot (\sim)\rangle = \nabla \cdot \langle\sim\rangle = 0$, and hence $\varepsilon = \nu\langle\omega^2\rangle$.

There are two distinct approaches to the kinematics of homogeneous turbulence, one based in Fourier space and one in real space. We begin with the latter.

8.3.1 Correlation Functions and Structure Functions

Correlation functions and *structure functions* are real-space diagnostics which are closely related but are used for rather different purposes. Correlation functions tell us how fluctuations at one position, $\mathbf{x}$, are correlated to fluctuations at some distant location, $\mathbf{x}' = \mathbf{x} + \mathbf{r}$, whereas structure functions are usually used to estimate the scale-by-scale distribution of energy across a range of eddy sizes. We begin with correlation functions.

The velocity field induced by a blob of vorticity (i.e. an eddy) is determined by the Biot–Savart law, and so the velocity distribution at points adjacent to a vortex is largely determined by that vortex. Thus eddies (blobs of vorticity) of characteristic size s will introduce a significant statistical correlation into the velocity field on the scale of s. In order to detect this, we introduce the two-point, second-order velocity correlation tensor

$$Q_{ij}(\mathbf{r}, t) = \langle u_i(\mathbf{x})u_j(\mathbf{x} + \mathbf{r})\rangle = \langle u_i(\mathbf{x})u_j(\mathbf{x}')\rangle, \tag{8.12}$$

where $\langle \sim \rangle$ indicates an ensemble average. Note that, because of statistical homogeneity, Q_{ij} is independent of $\mathbf{x}$ and depends only on time and on the displacement vector $\mathbf{r}$. Frequently the more compact notation $Q_{ij}(\mathbf{r}) = \langle u_i u'_j \rangle$ is employed, where the prime on $\mathbf{u}$ is now used to indicate a velocity measured at $\mathbf{x}'$ and the time dependence is understood. Evidently, Q_{ij} provides a measure of how well $\mathbf{u}$ is correlated to $\mathbf{u}'$. For $|\mathbf{r}| < \ell$ the two points will be strongly correlated by the large eddies, whereas Q_{ij} will be small for $|\mathbf{r}| \gg \ell$.

Two special cases of Q_{ij} are the longitudinal and transverse correlation functions, $f(r)$ and $g(r)$, defined by

$$Q_{xx}(r\hat{\mathbf{e}}_x) = \langle u_x^2\rangle f(r), Q_{xx}(r\hat{\mathbf{e}}_y) = \langle u_x^2\rangle g(r). \tag{8.13}$$

The longitudinal function, $f(r)$, though not the transverse function, $g(r)$, is normally observed to fall monotonically from unity down to zero as r increases, and so typically $f(r)$ remains positive. Consequently, the integral scale in a particular direction is conventionally defined via the integral of the appropriate longitudinal correlation function,

$$\ell = \int_0^\infty f(r)dr. \tag{8.14}$$

The correlation tensor, $Q_{ij}(\mathbf{r})$, has a number of useful kinematic properties which follow from simple geometrical considerations and from incompressibility. For example,

$$Q_{ij}(\mathbf{r}) = Q_{ji}(-\mathbf{r}), \ \partial Q_{ij}/\partial r_i = \partial Q_{ij}/\partial r_j = 0. \tag{8.15}$$

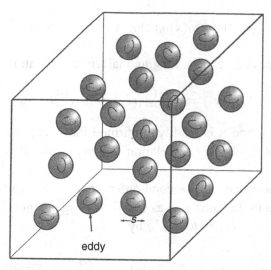

Figure 8.4 An artificial field of isotropic turbulence composed of a random sea of Gaussian eddies.

The first property in (8.15) is a geometrical consequence of the definition of $Q_{ij}(\mathbf{r})$, while the second follows from the fact that $\partial(\sim)/\partial r_i$ operating on an average becomes $-\partial(\sim)/\partial x_i$, the operations $\langle\sim\rangle$ and $\partial(\sim)/\partial x_i$ commute, and $\nabla\cdot\mathbf{u} = 0$.

In order to picture how Q_{ij} is related to the distribution of eddies present in a field of turbulence, it is helpful to consider the following model problem. Consider a 'generic' eddy that has a velocity field which takes the simple form $\mathbf{u} = \boldsymbol{\Omega} \times \mathbf{x} \exp[-2\mathbf{x}^2/s^2]$, where $\boldsymbol{\Omega}$ is a characteristic angular velocity and s the scale of the eddy. Now picture a sea of such Gaussian eddies of fixed size s which are randomly but uniformly distributed in space and with random orientation (Figure 8.4). The resulting velocity field is statistically isotropic and homogeneous. It is readily confirmed (see Davidson, 2015) that the resulting longitudinal correlation function and associated two-point vector correlation are given by

$$\langle u_x^2\rangle f(r) = \tfrac{1}{3}\langle \mathbf{u}^2\rangle \exp[-r^2/s^2], \tag{8.16}$$

$$\langle \mathbf{u} \cdot \mathbf{u}'\rangle = \langle \mathbf{u}^2\rangle\left[1 - \frac{2}{3}\frac{r^2}{s^2}\right]\exp[-r^2/s^2]. \tag{8.17}$$

Equation (8.14) now gives $\ell = s\sqrt{\pi}/2$ and so, as expected, the integral scale is of order s, and the two-point correlations fall off on the same length scale.

It is convenient to introduce one additional velocity correlation: the third-order, two-point velocity correlation tensor, S_{ijk}. This is defined as

$$S_{ijk}(\mathbf{r}, t) = \langle u_i(\mathbf{x})u_j(\mathbf{x})u_k(\mathbf{x}+\mathbf{r})\rangle = \langle u_i u_j u'_k\rangle,$$

and a common special case is the longitudinal triple correlation

$$S_{xxx}(r\hat{\mathbf{e}}_x) = \langle u_x^2 u'_x\rangle(r\hat{\mathbf{e}}_x) = \langle u_x^2\rangle^{3/2} K(r). \tag{8.18}$$

Evidently $S_{ijk} = S_{jik}$ and it is readily confirmed that incompressibility demands $\partial S_{ijk}/\partial r_k = 0$. This third-order correlation will prove useful when we turn from kinematics to dynamics.

Finally we introduce the second-order, two-point vorticity correlation, $\langle \omega_i \omega'_j\rangle(\mathbf{r})$, defined in the same way as $\langle u_i u'_j\rangle$. Noting that $\boldsymbol{\omega} = \nabla \times \mathbf{u}$, it is simple to show that $\langle \omega_i \omega'_j\rangle$ is related to $\langle u_i u'_j\rangle$ by

$$\langle \omega_i \omega'_j\rangle(\mathbf{r}) = -\delta_{ij}\nabla^2 Q_{kk}(\mathbf{r}) + \frac{\partial^2 Q_{kk}(\mathbf{r})}{\partial r_i \partial r_j} + \nabla^2 Q_{ji}(\mathbf{r}), \tag{8.19}$$

(see, for example, Batchelor, 1953) of which an important special case is

$$\langle \boldsymbol{\omega} \cdot \boldsymbol{\omega}'\rangle(\mathbf{r}) = -\nabla^2\langle \mathbf{u} \cdot \mathbf{u}'\rangle. \tag{8.20}$$

We now turn from correlation tensors to structure functions. These are usually defined in terms of the *longitudinal velocity increment* $\delta v = u_x(\mathbf{x}+r\hat{\mathbf{e}}_x) - u_x(\mathbf{x})$. For example, the second-order longitudinal structure function is defined as

$$\langle(\delta v)^2\rangle(r) = \langle[u_x(\mathbf{x}+r\hat{\mathbf{e}}_x) - u_x(\mathbf{x})]^2\rangle. \tag{8.21}$$

Clearly this is related to $f(r)$ by

$$\langle(\delta v)^2\rangle(r) = 2\langle u_x^2\rangle[1 - f(r)], \tag{8.22}$$

and satisfies,

$$\langle(\delta v)^2\rangle(r \to 0) = \langle(\partial u_x/\partial x)^2\rangle r^2, \tag{8.23}$$

$$\langle(\delta v)^2\rangle(r \to \infty) = 2\langle u_x^2\rangle. \tag{8.24}$$

This structure function is usually interpreted as a measure of the cumulative kinetic energy held by eddies of characteristic size r or less. The logic is as follows. Those eddies which are much larger than r exhibit negligible differences in velocity on the scale of r, and so contribute more or less equally to u_x and u'_x as they are swept past the two points $\mathbf{x}$ and $\mathbf{x}'$. Conversely, eddies much smaller than r may contribute to u_x or to u'_x as they are swept past $\mathbf{x}$ or $\mathbf{x}'$, but not to both points simultaneously. In short, we have a sort of crude low-pass filter which suppresses contributions from eddies larger than r. This suggests that, in some zero-order sense, we might interpret $\frac{1}{4}\langle(\delta v)^2\rangle(r)$ as the cumulative contribution to $\frac{1}{2}\langle u_x^2\rangle$ from all eddies smaller than scale r, with $\frac{1}{4}\langle(\delta v)^2\rangle \to \frac{1}{2}\langle u_x^2\rangle$ as $r \to \infty$, in accordance with (8.24). For the particular case of isotropic turbulence we have $\frac{3}{4}\langle(\delta v)^2\rangle(r \to \infty) = \frac{1}{2}\langle \mathbf{u}^2\rangle$. The same logic then leads us to interpret $\frac{3}{4}\langle(\delta v)^2\rangle(r)$ as the cumulative kinetic energy held below scale r, with $\partial[\frac{3}{4}\langle(\delta v)^2\rangle(r)]/\partial r$ acting as a sort of energy density in scale space.

Although this interpretation of $\langle(\delta v)^2\rangle$ is commonplace in the turbulence literature, it is easy to see that it is too simplistic. A more careful analysis proceeds as follows. From (8.24) we see that eddies whose size, s, is much smaller than r make a contribution to $\langle(\delta v)^2\rangle(r)$ of $2\langle u_x^2\rangle_s$. On the other hand (8.23) tells us that eddies much larger than r make a contribution to $\langle(\delta v)^2\rangle(r)$ of $\langle(\partial u_x/\partial x)^2\rangle_s r^2$. It follows that

$$\langle(\delta v)^2\rangle(r) \approx \sum_{s<r} 2\langle u_x^2\rangle_s + r^2 \sum_{s>r} \langle(\partial u_x/\partial x)^2\rangle_s, \qquad (8.25)$$

where the subscript s indicates that we are considering contributions to $\langle u_x^2\rangle$ or $\langle(\partial u_x/\partial x)^2\rangle$ from eddies of scale s. In the case of statistically isotropic turbulence it can be shown that $\langle(\partial u_x/\partial x)^2\rangle = \langle\omega^2\rangle/15$, where $\langle\omega^2\rangle/2$ is the *enstrophy* introduced in §3.4. So (8.25) becomes, for isotropic turbulence,

$$\frac{3}{4}\langle(\delta v)^2\rangle(r) \approx \sum_{s<r} \frac{1}{2}\langle \mathbf{u}^2\rangle_s + \frac{r^2}{10} \sum_{s>r} \frac{1}{2}\langle\omega^2\rangle_s, \qquad (8.26)$$

or,

$$\frac{3}{4}\langle(\delta v)^2\rangle(r) \approx [\text{energy held in eddies of size } s<r]$$

$$+ \frac{r^2}{10}[\text{enstrophy held in eddies of size } s>r]. \qquad (8.27)$$

Clearly, $\langle(\delta v)^2\rangle(r)$ is an odd diagnostic, mixing information about energy and enstrophy, as well as information from all scales. Sometimes the major contribution to $\frac{3}{4}\langle(\delta v)^2\rangle(r)$ comes from the first term on the right of (8.27), and the naive interpretation of $\partial\left[\frac{3}{4}\langle(\delta v)^2\rangle(r)\right]/\partial r$ acting as an energy density in scale space is then a reasonable rule of thumb. On other occasions, however, the second term on the right of (8.27) is important, and this simplistic interpretation of $\langle(\delta v)^2\rangle$ breaks down. In any event, $\langle(\delta v)^2\rangle(r)$ was the diagnostic of choice for Kolmogorov and, while overshadowed by spectral methods for many decades, it is now popular once again.

Note that higher-order structure functions may also be introduced. The structure function of order p is defined as $\langle(\delta v)^p\rangle(r)$. A particularly important case is the third-order structure function, $\langle(\delta v)^3\rangle(r)$, which plays an important role in dynamics, as we shall see.

8.3.2 Spectral Analysis

We now show that the Fourier transform provides an alternative means of distinguishing between the energy held at different scales in a turbulent flow. As above, we restrict the discussion to homogeneous turbulence.

The Fourier transform is useful in turbulence because it acts like a filter which differentiates, at least in a crude sense, between the different scales which are present in a random signal. This filtering property of the Fourier transform can be most easily understood in the context of a one-dimensional signal. Consider the Fourier transform pair for a real, one-dimensional function $f(x)$,

$$F(k) = \frac{1}{2\pi} \int_{-\infty}^{\infty} f(x)e^{-jkx}dx, \tag{8.28}$$

$$f(x) = \int_{-\infty}^{\infty} F(k)e^{jkx}dk, \tag{8.29}$$

for which we have the autocorrelation theorem

$$2\pi|F(k)|^2 = \text{Transform}\left[\int_{-\infty}^{\infty} f(x)f(x+r)dx\right]. \tag{8.30}$$

Now suppose that $f(x)$ is a random signal of finite length, which is composed of fluctuations of different characteristic scales. This signal may be smoothed over some intermediate length, L, as follows. For each value of x, we replace $f(x)$ by its local average obtained by integrating f from $x - L$ to $x + L$ and then dividing by $2L$. In practice this may be achieved using the so-called box filter, defined as $B^L(r) = (2L)^{-1}$

for $|r| < L$ and $B^L(r) = 0$ for $|r| > L$. The smoothed function, $f^L(x)$, is then defined as the convolution of $f(x)$ with $B^L(r)$:

$$f^L(x) = \int_{-\infty}^{\infty} B^L(r)f(x+r)dr. \tag{8.31}$$

Clearly $f^L(x)$ is a smoothed version of f, with fluctuations on scales smaller that L suppressed. Now the convolution theorem gives the transform of $f^L(x)$ as

$$F^L(k) = \frac{\sin(kL)}{kL}F(k). \tag{8.32}$$

Ignoring the weak oscillations in $\sin(kL)/kL$ for $k > \pi/L$, we may regard $F^L(k)$ as a truncated version of $F(k)$, with a cut-off at around $k = \pi/L$. We conclude that the operation of smoothing $f(x)$ over the scale L in real space is roughly equivalent to discarding the high-k (i.e. $k > \pi/L$) part of $F(k)$. In this sense the Fourier transform may be thought of as a kind of filter, differentiating between fluctuations of different scale. However, it is an imperfect filter, since $\sin(kL)/kL$ does not provide a very sharp cutoff in k-space. In any event, loosely speaking, we may think of fluctuations of scale L contributing to $F(k)$ predominantly at wavenumber $k = \pi/L$.

In turbulence, the filtering property of the Fourier transform is usually combined with the idea of an autocorrelation. Consider a signal of zero mean, $f(x)$, which is statistically homogeneous in x. In such cases the autocorrelation is usually defined in terms of the convergent integral

$$T(r) = \frac{1}{2X} \int_{-X}^{X} f(x)f(x+r)dx, \tag{8.33}$$

where X is taken to be much larger than any length scale associated with $f(x)$. Consider, for example, the simple case

$$f(x) = \sum_i A_i \sin(k_i x + \varphi_i). \tag{8.34}$$

Then (8.33) yields, for $kX \rightarrow \infty$,

$$T(r) \sim \sum_i A_i^2 \cos(k_i r). \tag{8.35}$$

If we take the Fourier transform of $T(r)$ we get a set of delta functions at the discrete wavenumbers k_i, with amplitudes proportional to A_i^2. In this simple example, then, the process of forming an autocorrelation and then taking its Fourier transform is a way of extracting the amplitudes of the different Fourier modes contained in the original signal. This property of the autocorrelation is captured by the autocorrelation theorem (8.30).

The quantity $|F(k)|^2$ in (8.30) is referred to as the power spectrum of $f(x)$, and it is widely used in turbulence. It is important to note, however, that (8.30) and (8.35) tell us that all the information about the phases of the various Fourier modes in $f(x)$ is absent from both the autocorrelation function and the corresponding power spectrum. Consequently we cannot reconstruct the original signal from a knowledge of its power spectrum alone, and in fact very different signals can yield the same power spectra.

Let us now apply the Fourier transform to turbulence. The first step is to introduce the three-dimensional transform pair

$$F(\mathbf{k}) = \frac{1}{(2\pi)^3} \int f(\mathbf{x}) e^{-j\mathbf{k} \cdot \mathbf{x}} d\mathbf{x}, \tag{8.36}$$

$$f(\mathbf{x}) = \int F(\mathbf{k}) e^{j\mathbf{k} \cdot \mathbf{x}} d\mathbf{k}. \tag{8.37}$$

Note that, if the function $f(\mathbf{x})$ happens to have spherical symmetry, i.e. $f = f(r)$ where $r = |\mathbf{r}|$, then $F(\mathbf{k})$ also has spherical symmetry: $F = F(k)$, $k = |\mathbf{k}|$. In such cases we may integrate over the polar angles in (8.36) and (8.37) to give

$$F(k) = \frac{1}{2\pi^2} \int_0^\infty r^2 f(r) \frac{\sin(kr)}{kr} dr, \tag{8.38}$$

$$f(r) = 4\pi \int_0^\infty k^2 F(k) \frac{\sin(kr)}{kr} dk. \tag{8.39}$$

The all-important *spectral tensor*, $\Phi_{ij}(\mathbf{k})$, is defined as the Fourier transform of the two-point velocity correlation, $Q_{ij}(\mathbf{r})$:

$$\Phi_{ij}(\mathbf{k}) = \frac{1}{(2\pi)^3} \int Q_{ij}(\mathbf{r}) e^{-j\mathbf{k} \cdot \mathbf{r}} d\mathbf{r}, \tag{8.40}$$

$$Q_{ij}(\mathbf{r}) = \int \Phi_{ij}(\mathbf{k})e^{j\mathbf{k}\cdot\mathbf{r}}d\mathbf{k}. \tag{8.41}$$

Since $Q_{ij}(\mathbf{r}) = Q_{ij}(-\mathbf{r})$ we have Hermitian symmetry, $\Phi_{ij}(\mathbf{k}) = \Phi_{ji}^*(\mathbf{k})$, where $*$ represents a complex conjugate, and so the diagonal components of $\Phi_{ij}(\mathbf{k})$ are all real. In fact it is readily confirmed, using the three-dimensional version of the autocorrelation theorem, that they are real and positive. Hence (8.41) tells us that

$$\langle \mathbf{u}^2 \rangle = \int \Phi_{ii}(\mathbf{k})d\mathbf{k}, \qquad \Phi_{ii}(\mathbf{k}) > 0. \tag{8.42}$$

In order to avoid the three-dimensional integral in (8.42), it is natural to introduce the so-called *energy spectrum, E(k)*, defined as the integral of $\frac{1}{2}\Phi_{ii}(\mathbf{k})$ over a spherical surface of radius $k = |\mathbf{k}|$ in **k**-space. It follows from this definition that

$$\frac{1}{2}\langle \mathbf{u}^2 \rangle = \int_0^\infty E(k)dk, \quad E(k) > 0. \tag{8.43}$$

If we recall the filtering property of the Fourier transform, we may think of $E(k)$ as a spectral energy density, in the sense that $E(k)$ is the contribution to $\frac{1}{2}\langle \mathbf{u}^2 \rangle$ that comes from eddies of scale $s \sim \pi/k$. As we shall see in §8.4, Kolmogorov's theory of the small scales tells us that kinetic energy is distributed in scale space such that E takes the form $E(k) \sim \varepsilon^{2/3}k^{-5/3}$ in the range $\pi/\ell \ll k \ll \pi/\eta$.

It is instructive at this point to return to Figure 8.4, which represents an isotropic, random sea of Gaussian eddies of fixed size s. This has a velocity correlation given by (8.17), and using (8.38) and (8.40) it is simple to calculate the corresponding energy spectrum. (See, for example, Davidson, 2015). After a little algebra we find

$$E(k) = \frac{\frac{1}{2}\langle \mathbf{u}^2 \rangle s}{12\sqrt{\pi}}(ks)^4 \exp[-k^2 s^2/4]. \tag{8.44}$$

This exhibits a sharp peak at $k = \sqrt{8}/s \sim \pi/s$, which is reassuring. Note, however, that these eddies of fixed size s have made a broadband contribution to $E(k)$, spread over all k.

The caveat above has important repercussions. Suppose we construct an artificial field of turbulence composed a random sea of Gaussian eddies whose sizes, s, vary from the Kolmogorov microscale to the integral scale, $\eta < s < \ell$. Then we will find that each size-range makes a contribution to $E(k)$ centred around $k \sim \pi/s$. However, each scale also contributes to $E(k)$ across the full range of k, as illustrated by (8.44). Thus $E(k)$ has a finite contribution of the form $E(k) \sim k^4$ in the range $0 < k \ll \pi/\ell$, which is the sum of all the low-k tails of the different spectra corresponding to the various eddy sizes. Similarly, $E(k)$ is finite, if exponentially small, in the range $\pi/\eta < k < \infty$. But there are no eddies in either of these two ranges, and so we cannot interpret $E(k)$ as the energy of eddies of scale π/k when $k < \pi/\ell$ or else $k > \pi/\eta$. Clearly the statement

$$E(k) \sim \text{kinetic energy of eddies of size } \pi/k,$$

is an imperfect interpretation of $E(k)$.

So far we have focussed on velocity fields and their kinetic energy, and so we now turn to vorticity and to enstrophy. It is clear from

$$\langle \boldsymbol{\omega} \cdot \boldsymbol{\omega}' \rangle (\mathbf{r}) = -\nabla^2 \langle \mathbf{u} \cdot \mathbf{u}' \rangle \tag{8.45}$$

that the enstrophy density corresponding to $E(k)$ is simply $k^2 E(k)$, and so

$$\frac{1}{2} \langle \boldsymbol{\omega}^2 \rangle = \int_0^\infty k^2 E(k) dk. \tag{8.46}$$

Since $E(k) \sim \varepsilon^{2/3} k^{-5/3}$ for $\pi/\ell \ll k \ll \pi/\eta$, the enstrophy spectrum in the same range scales as $k^2 E(k) \sim \varepsilon^{2/3} k^{1/3}$. These enstrophy and energy spectra are shown schematically in Figure 8.5. Evidently most of the energy is held at large scales, while the bulk of the enstrophy is found near the Kolmogorov microscale, η. As Re $\to \infty$ the energy-dominated and enstrophy-dominated eddies become increasing separated in scale, both with respect to length and time scales. As we shall see, this leads to a partial statistical decoupling of the dynamics of these two groups of eddies.

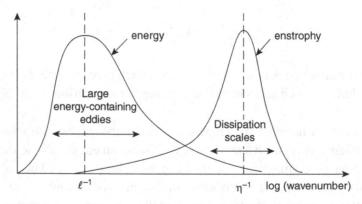

Figure 8.5 The spectral energy and enstrophy densities at large Re.

8.3.3 The Special Case of Statistically Isotropic Turbulence

We now follow the lead of G.I. Taylor and consider the simplifications which arise when the turbulence is statistically isotropic. That is, we consider turbulence which is statistically invariant under rotations of the frame of reference and under changes from a right-handed to a left-handed coordinate system (i.e. reflectional symmetry). We shall not pause to prove the various kinematic results given below, but rather refer the reader to the standard texts.

Actually, the large scales in a turbulent flow are almost invariably strongly anisotropic, either because of the way the turbulence is generated, or else because of the environment in which it develops. However, it turns out that the anisotropy present at the large scales is often only weakly imposed on the small scales. This is a consequence of the partial statistical decoupling of the large and small eddies at large Re, a decoupling which becomes more pronounced as Re increases. Thus the smaller scales in high-Re turbulence are, at least approximately, statistically isotropic. This small-scale isotropy (despite the large scale anisotropy) is referred to as *local isotropy*, and it seems natural to take advantage of this symmetry.

Consider the extreme case where all scales, large and small, are statistically isotropic. The integral scales u and ℓ are defined in the usual way as

$$\ell = \int_0^\infty f(r)dr, \quad u^2 = \frac{1}{3}\langle \mathbf{u}^2 \rangle, \tag{8.47}$$

where $f(r)$ is the longitudinal correlation function, (8.13). Our primary interest lies in the consequences of isotropy for two-point statistics. Here the combination of statistical isotropy and conservation of mass, $\nabla \cdot \mathbf{u} = 0$, leads to a number of important simplifications, as detailed in, for example, Batchelor (1953). Let us start with real, as distinct from spectral, space.

It turns out that, in isotropic turbulence, the longitudinal and transverse velocity correlations, $f(r)$ and $g(r)$, defined by (8.13), are related though $\nabla \cdot \mathbf{u} = 0$: $g = f + \frac{1}{2}rf'(r)$. More importantly, using isotropy and (8.15), the two-point velocity correlation tensor, $Q_{ij}(\mathbf{r}) = \langle u_i u_j' \rangle$, can be expressed entirely in terms of the longitudinal correlation function,

$$Q_{ij}(\mathbf{r}) = \frac{u^2}{2r}\left[(r^2 f)' \delta_{ij} - f'(r)r_i r_j \right], \tag{8.48}$$

and in particular,

$$Q_{ii}(r) = \langle \mathbf{u} \cdot \mathbf{u}' \rangle(r) = \frac{u^2}{r^2}\frac{\partial}{\partial r}(r^3 f). \tag{8.49}$$

The situation is similar for the third-order correlation, $S_{ijk}(\mathbf{r}) = \langle u_i u_j u'_k \rangle$. Here isotropy, symmetry in the indices i and j, and continuity in the form $\partial S_{ijk}/\partial r_k = 0$, allows us to express $S_{ijk}(\mathbf{r})$ solely in terms of the third-order longitudinal correlation function,

$$S_{xxx}(r\hat{\mathbf{e}}_x) = \langle u_x^2(\mathbf{x}) u_x(\mathbf{x} + r\hat{\mathbf{e}}_x) \rangle = u^3 K(r).$$

In particular we find

$$S_{ijk}(\mathbf{r}) = u^3 \left[\frac{K - rK'}{2r^3} r_i r_j r_k + \frac{2K + rK'}{4r}(r_i\delta_{jk} + r_j\delta_{ik}) - \frac{K}{2r}r_k\delta_{ij} \right]. \qquad (8.50)$$

This ability to express the tensors $Q_{ij}(\mathbf{r})$ and $S_{ijk}(\mathbf{r})$ in terms of the two scalar functions $f(r)$ and $K(r)$ represents a considerable simplification. Note that (8.50) demands that $S_{iik}(\mathbf{r}) = \langle \mathbf{u}^2 u'_k \rangle = 0$, which is a special case of the more general result that $\langle Tu'_k \rangle = 0$ in isotropic turbulence, where T is any scalar property of the fluid. Thus, for example, the two-point pressure-velocity correlation is zero,

$$\langle pu'_k \rangle = 0, \qquad (8.51)$$

which will turn out to be useful when we discuss dynamics.

The second- and third-order longitudinal structure functions can also be written in terms of $f(r)$ and $K(r)$. Noting that $\langle u_i u'_j u'_k \rangle(\mathbf{r}) = -\langle u_j u_k u'_i \rangle(\mathbf{r})$ in isotropic turbulence, we have

$$\langle (\delta v)^2 \rangle(r) = \langle [u_x(\mathbf{x} + r\hat{\mathbf{e}}_x) - u_x(\mathbf{x})]^2 \rangle = 2u^2[1 - f(r)], \qquad (8.52)$$

$$\langle (\delta v)^3 \rangle(r) = \langle [u_x(\mathbf{x} + r\hat{\mathbf{e}}_x) - u_x(\mathbf{x})]^3 \rangle = 6u^3 K(r), \qquad (8.53)$$

and for small r these yield

$$u^2 f(r) = u^2 - \frac{1}{2}\langle (\partial u_x/\partial x)^2 \rangle r^2 + O(r^4), \qquad (8.54)$$

$$6u^3 K(r) = \langle (\partial u_x/\partial x)^3 \rangle r^3 + O(r^5). \qquad (8.55)$$

Moreover, in isotropic turbulence it turns out that we have the two kinematic relationships

$$\langle (\partial u_x/\partial x)^2 \rangle = \frac{1}{15}\langle \omega^2 \rangle, \qquad (8.56)$$

$$\langle (\partial u_x / \partial x)^3 \rangle = -\frac{2}{35} \langle \omega_i \omega_j S_{ij} \rangle. \tag{8.57}$$

It follows that the *skewness* of $\partial u_x / \partial x$, defined as $S_0 = \langle (\partial u_x / \partial x)^3 \rangle / \langle (\partial u_x / \partial x)^2 \rangle^{3/2}$, can be written in the form

$$S_0 = -\frac{6\sqrt{15} \langle \omega_i \omega_j S_{ij} \rangle}{7 \langle \omega^2 \rangle^{3/2}}. \tag{8.58}$$

This allows us to relate the mean rate of generation of enstrophy by vortex-line stretching, $\langle \omega_i \omega_j S_{ij} \rangle$, to the skewness, S_0, a quantity which is readily measured in experiments or in numerical simulations. (See Equation (3.30) for the physical significance of $\omega_i \omega_j S_{ij}$.)

Returning to the longitudinal correlation, $f(r)$, (8.54) and (8.56) combine to give

$$u^2 f(r) = u^2 - \frac{1}{30} \langle \omega^2 \rangle r^2 + O(r^4), \tag{8.59}$$

which is usually rewritten as

$$f(r) = 1 - \frac{r^2}{2\lambda^2} + O(r^4). \tag{8.60}$$

The length λ is called the *Taylor microscale* and is defined as

$$\lambda^2 = \frac{5 \langle \mathbf{u}^2 \rangle}{\langle \omega^2 \rangle} = \frac{15 \nu u^2}{\varepsilon}. \tag{8.61}$$

Since $\varepsilon \sim u^3 / \ell$ we have $\lambda^2 \sim 15 \nu \ell / u$, which in turn yields $\lambda / \ell \sim \sqrt{15} \mathrm{Re}^{-1/2}$ and $\lambda / \eta \sim \sqrt{15} \mathrm{Re}^{1/4}$, where $\mathrm{Re} = u\ell/\nu$. Hence, when Re is large, λ is an intermediate length scale that satisfies $\eta \ll \lambda \ll \ell$.

Let us now turn to Fourier space and to the spectral tensor $\Phi_{ij}(\mathbf{k})$. Here isotropy allows us to write $\Phi_{ii} = \Phi_{ii}(k)$, while a combination of isotropy and incompressibility yields

$$\Phi_{ij}(\mathbf{k}) = \frac{1}{2} \Phi_{ii}(k) \left[\delta_{ij} - \frac{k_i k_j}{k^2} \right]. \tag{8.62}$$

Since Φ_{ii} is a function of $k = |\mathbf{k}|$ only, the definition of $E(k)$ now reduces to $E(k) = \frac{1}{2} \Phi_{ii}(k) [4\pi k^2]$, and it follows that we can rewrite $\Phi_{ij}(\mathbf{k})$ in terms of $E(k)$ and $\mathbf{k}$ alone:

$$\Phi_{ij}(\mathbf{k}) = \frac{E(k)}{4\pi k^2} \left[\delta_{ij} - \frac{k_i k_j}{k^2} \right]. \tag{8.63}$$

As both Φ_{ii} and Q_{ii} are spherically symmetric functions, we can use (8.38) and (8.39) to relate $\Phi_{ii}(k)$, and hence $E(k)$, to $Q_{ii}(r) = \langle \mathbf{u} \cdot \mathbf{u}' \rangle$. This yields the transform pair

$$E(k) = \frac{1}{\pi} \int_0^\infty \langle \mathbf{u} \cdot \mathbf{u}' \rangle kr \sin(kr) dr, \tag{8.64}$$

$$\langle \mathbf{u} \cdot \mathbf{u}' \rangle (r) = 2 \int_0^\infty E(k) \frac{\sin(kr)}{kr} dk, \tag{8.65}$$

which is probably the most important link between real-space and Fourier-space diagnostics in isotropic turbulence.

Since we now know the precise relationship between $\langle \mathbf{u} \cdot \mathbf{u}' \rangle$ and $E(k)$, at least in isotropic turbulence, it is of interest to re-examine the link between $\langle (\delta v)^2 \rangle$ and $E(k)$, as both diagnostics are often used to give a scale-by-scale measure of kinetic energy. The first point to note is that (8.27), rewritten in terms of $E(k)$, becomes

$$\frac{3}{4} \langle (\delta v)^2 \rangle (r) \approx \int_{\pi/r}^\infty E(k) dk + \frac{r^2}{10} \int_0^{\pi/r} [k^2 E(k)] \, dk. \tag{8.66}$$

This estimate is based on (8.23) and (8.24), and on an extremely simple geometrical picture, so let us see if we can do any better using definition (8.65). To this end, we note that (8.49) can be rewritten as

$$\frac{1}{r^2} \frac{\partial}{\partial r} [r^3 \langle (\delta v)^2 \rangle] = 2 \langle \mathbf{u}^2 \rangle - 2 \langle \mathbf{u} \cdot \mathbf{u}' \rangle,$$

which may be combined with (8.65) to yield

$$\frac{3}{4} \langle (\delta v)^2 \rangle = \int_0^\infty E(k) H(kr) dk, \tag{8.67}$$

$$H(x) = 1 + 3x^{-2} \cos x - 3x^{-3} \sin x. \tag{8.68}$$

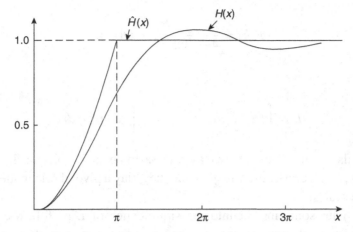

Figure 8.6 Shape of $H(x)$ and its approximation $\hat{H}(x)$.

The function $H(x)$ has the shape shown in Figure 8.6. It takes the form $H(x) \to x^2/10$ for $x \to 0$ and $H \to 1$ for $x \to \infty$.

A plausible approximation to $H(x)$ is

$$\hat{H}(x) = (x/\pi)^2, \quad x < \pi, \qquad\qquad \hat{H}(x) = 1, \quad x > \pi,$$

as shown in Figure 8.6. If we replace $H(x)$ by $\hat{H}(x)$ in (8.67) we obtain

$$\frac{3}{4}\langle(\delta v)^2\rangle(r) \approx \underbrace{\int_{\pi/r}^{\infty} E(k)dk}_{\substack{\text{energy in eddies}\\\text{of size } r \text{ or less}}} + \underbrace{\frac{r^2}{\pi^2}\int_0^{\pi/r}[k^2 E(k)]\,dk}_{\substack{\text{enstrophy in eddies}\\\text{of size } r \text{ or greater}}}, \qquad (8.69)$$

which is reassuringly similar to the geometrically motivated statement (8.66).

We conclude this section with a brief discussion of the large scales (i.e. the low-k end of the spectrum), a story we shall pick up again in §8.6. Suppose that the long-range vorticity correlations, $\langle\omega_i\omega'_j\rangle_\infty$, decay rapidly with r, say exponentially fast. Then, because of $\langle\omega\cdot\omega'\rangle(\mathbf{r}) = -\nabla^2\langle\mathbf{u}\cdot\mathbf{u}'\rangle$, the vector correlation $\langle\mathbf{u}\cdot\mathbf{u}'\rangle_\infty$ must also decay rapidly with r. (Here the subscript ∞ indicates $r \to \infty$.) We may then expand (8.64) to give

$$E(k) = \frac{k^2}{4\pi^2}\int\langle\mathbf{u}\cdot\mathbf{u}'\rangle d\mathbf{r} - \frac{k^4}{24\pi^2}\int\langle\mathbf{u}\cdot\mathbf{u}'\rangle r^2 d\mathbf{r} + O(k^6),$$

or equivalently

$$E(k) = \frac{Lk^2}{4\pi^2} + \frac{Ik^4}{24\pi^2} + O(k^6), \tag{8.70}$$

where

$$L = \int \langle \mathbf{u} \cdot \mathbf{u}' \rangle d\mathbf{r}, \qquad I = -\int \langle \mathbf{u} \cdot \mathbf{u}' \rangle r^2 d\mathbf{r}. \tag{8.71}$$

The integrals L and I are known as the Saffman and Loitsyansky integrals, respectively, and we shall show in §8.6 that they play a pivotal role in the dynamics of the large scales.

We may gain some insight into the significance of L as follows. Ensemble averages are equivalent to volume averages, and so we can rewrite L as

$$L = \lim_{V \to \infty} \left\langle \left[\int_V \mathbf{u} dV \right]^2 \right\rangle \Big/ V. \tag{8.72}$$

Evidently L is a measure of the square of the linear momentum held in some large control volume, V. It is important to realise at this point that our frame of reference is one in which $\langle \mathbf{u} \rangle = 0$, or equivalently

$$\lim_{V \to \infty} \left[\int_V \mathbf{u} dV \right] \Big/ V = 0.$$

However, this constraint is not enough to ensure that L is zero, and indeed the central limit theorem suggests that, in general, L will be finite. So, in the absence of some dynamical constraint that forces L to be zero, we might expect $E(k \to 0) \sim Lk^2$, which is known as a *Saffman spectrum*. In §8.6 we shall show that, as a direct consequence of linear momentum conservation, L is an invariant in freely decaying turbulence. Note that combining (8.49) with (8.71) yields

$$L = 4\pi \int_0^\infty \langle \mathbf{u} \cdot \mathbf{u}' \rangle r^2 dr = 4\pi [r^3 u^2 f(r)]_\infty, \tag{8.73}$$

so that a Saffman spectrum requires $f_\infty \sim r^{-3}$, despite the more rapid decay of $\langle \mathbf{u} \cdot \mathbf{u}' \rangle_\infty$. This idiosyncratic feature of Saffman turbulence turns out to be a consequence of the Biot–Savart law, as we shall show.

Now suppose that, because of the way the turbulence is initiated, the eddies have relatively little linear momentum in our preferred frame of reference (i.e. the reference frame in which $\langle \mathbf{u} \rangle = 0$). In particular, suppose that, immediately following the initiation of the turbulence, $\int_V \mathbf{u}dV$ grows more slowly with V than $V^{1/2}$. Then L will be zero at $t = 0$, and since L is an invariant, it stays zero. In such a situation (8.70) suggests that $E(k \to 0) \sim Ik^4$, which is sometimes called a *Batchelor spectrum*. We shall see in §8.6 that, just as $L \sim \langle[\int_V \mathbf{u}dV]^2\rangle/V$ in a Saffman spectrum, so

$$I \sim \left\langle \left[\int_V \mathbf{x} \times \mathbf{u}dV\right]^2 \right\rangle \Big/ V$$

in a Batchelor spectrum (with the important caveat that $\langle u_i u'_j \rangle_\infty$ decays rapidly with r). Thus I is related to the angular momentum of the eddies. Both types of turbulence, $E(k \to 0) \sim Lk^2$ and $E(k \to 0) \sim Ik^4$, are observed in computer simulations and both have been measured in grid turbulence. (See Davidson, 2013, for a review.) It is the initial conditions which determine which is realised in any given situation.

8.4 Kolmogorov's Theory of the Small Scales

Kolmogorov published his now famous theory in 1941 (Kolmogorov, 1941a). He subsequently refined the theory twenty years later (Kolmogorov, 1962) and in recent years the precise nature of that refinement has possibly received more attention than the original theory itself. However, in many ways it is the original theory that constitutes the more remarkable result, as it yields one of the few near-universal and robust sets of predictions in turbulence, predictions that are well supported by experiment, at least for low-order statistics. Indeed, Kolmogorov's 1941 theory is perhaps rivalled only by the log-law of the wall in terms of universality, robustness and empirical support. This is all the more remarkable as it is based on a number of rather tentative assumptions whose validity is not so obvious. So let us start with those inspired assumptions.

Kolmogorov's starting point was Richardson's idea of an energy cascade, and in particular the notion that, at high Re, this multistep cascade would lead to a loss of information and to a partial decoupling of the large and small scales. This decoupling, which is hinted at in Figure 8.5, rests in part on the notion that information is lost as energy is passed down to small scales through a hierarchy of interactions, and in part because the large and small eddies have very different length and time scales, and so their dynamics are not well correlated, i.e. the large and small scales do not communicate directly[3]. This partial decoupling led Kolmogorov to the

---

[3] This is true only in three-dimensional turbulence. It is not true in two-dimensional turbulence where the large and small scales communicate directly, with the bulk of the straining of the small scales coming directly from the largest eddies. (See, for example, Davidson, 2013.)

hypothesis that, for very large Re, the small scales should be (i) isotropic, despite the anisotropy of the large scales (so-called *local isotropy*) and (ii) of *universal form*, exhibiting the same statistical behaviour in boundary layers, jets, wakes and so forth. He further suggested that the small scales are in *statistical equilibrium* with respect to the instantaneous state of the large eddies. This final assumption is based on the estimates shown in Table 8.2, where we note that the small scales have a characteristic time scale which, at large Re, is very fast by comparison with that of the large eddies. Thus it seems plausible that the small eddies, which rely on the large vortices for their energy, adjust very rapidly to any changes in the large scales and, given an instantaneous large-scale energy flux Π, may be considered to be in approximate statistical equilibrium with respect to that flux.

These three assumptions, of local isotropy, universality, and statistical equilibrium, led to Kolmogorov's theory being labelled as the *universal equilibrium theory of the small scales*, and to the scales which satisfy $r \ll \ell$ being called the *universal equilibrium range*. Given these various assumptions, Kolmogorov took the bold step of assuming that the small eddies know about the largest scales only to the extent that they establish an instantaneous energy flux, $\Pi(t)$, a flux which looks quasi-steady as far as the small scales are concerned. The key assumption here is that all the other information about the large scales is lost in the cascade. This inspired guess is the hallmark of the 1941 theory.

Let us now think about the consequences of these assumptions for the statistics of the longitudinal velocity increment, $\delta v = u_x(\mathbf{x} + r\hat{\mathbf{e}}_x) - u_x(\mathbf{x})$, in the range $r \ll \ell$. If we set aside the inconvenient fact that δv contains information about the large-scale vorticity (see Equation (8.69)) and adopt the crude rule of thumb that δv filters out the influence of scales much larger than r, we might imagine that the statistics of δv depend on only Π, r and ν for $r \ll \ell$. This is the basis of Kolmogorov's so-called first similarity hypothesis, which states:

When Re is large and $r \ll \ell$, the statistical properties of δv have a universal form that depends on only $\varepsilon = 2\nu\langle S_{ij}S_{ij}\rangle$, r and ν.

It is important to note that the dissipation, $\varepsilon = 2\nu\langle S_{ij}S_{ij}\rangle$, is being used here as a proxy for the energy flux, Π, which we are allowed to do because statistical equilibrium means that $\Pi = \varepsilon$ in the equilibrium range. The use of ε rather than Π is convenient since the dissipation, but not the energy flux, is readily measured in an experiment.

Given that the Kolmogorov microscales are $v = (\nu\varepsilon)^{1/4}$ and $\eta = (\nu^3/\varepsilon)^{1/4}$, we can rephrase the first similarity hypothesis to say that the statistics of δv at small scales depends on v, η and r only. Dimensional analysis then demands

$$\langle (\delta v)^2 \rangle = v^2 H(r/\eta), \qquad r \ll \ell, \tag{8.74}$$

whose spectral space equivalent is

$$E(k) = v^2 \eta \, F(k\eta), \qquad k \gg \ell^{-1}. \tag{8.75}$$

Here H and F are, according to the theory, universal functions, i.e. the same for any type of flow.

Let us now restrict attention to the range of scales $\eta \ll r \ll \ell$, which is called the *inertial subrange*. Since these eddies are well removed from the dissipation scales we would not expect v to be a relevant parameter, which leads to Kolmogorov's second similarity hypothesis:

When Re is large and $\eta \ll r \ll \ell$, the statistical properties of δv have a universal form which is uniquely determined by r and $\varepsilon = 2v\langle S_{ij}S_{ij}\rangle$ alone.

The need to eliminate v from (8.74) fixes the functional form of H and we obtain

$$\langle (\delta v)^2 \rangle = \beta \varepsilon^{2/3} r^{2/3}, \qquad \eta \ll r \ll \ell, \tag{8.76}$$

where β is a universal constant. This is Kolmogorov's *two-thirds law*, and it is well supported by experiments, which suggest $\beta \approx 2.0$. The spectral equivalent of (8.76), which Kolmogorov did not discuss (although his student Obukhov did), is evidently

$$E(k) = \alpha \varepsilon^{2/3} k^{-5/3}, \qquad \eta \ll k^{-1} \ll \ell, \tag{8.77}$$

which is also well supported by experiment and is known as the *five-thirds law*. The Fourier transform relating $E(k)$ to $\langle (\delta v)^2 \rangle$ yields, in the limit of Re $\to \infty$, $\beta = 1.315\alpha$, from which $\alpha \approx 1.5$.

Since Kolmogorov's two similarity hypotheses are framed in terms of the statistics of δv, we may also consider higher-order moments, and the second similarity hypothesis yields, for $p \geq 3$,

$$\langle (\delta v)^p \rangle = \beta_p \varepsilon^{p/3} r^{p/3}, \qquad \eta \ll r \ll \ell, \tag{8.78}$$

where the β_p are again universal constants. Unlike (8.76), this has no spectral equivalent. One notable special case is the third-order structure function, whose form in the inertial subrange can be deduced directly from the Navier–Stokes equation and is

$$\langle (\delta v)^3 \rangle = -\frac{4}{5}\varepsilon r, \qquad \eta \ll r \ll \ell. \tag{8.79}$$

This exact result, which we shall derive from first principles in the next section, is called the *four-fifths law* and it lends independent support to Kolmogorov's 1941 theory.

Perhaps it is worth making a few observations at this point. First, while the experimental evidence in support of (8.76) and (8.77) is compelling, and (8.79) is exact, the experimental data increasingly deviates from the power law $\langle (\delta v)^p \rangle \sim r^{p/3}$ as p gets larger. This discrepancy in the higher-order moments has led to much speculation, with Kolmogorov's own interpretation (Kolmogorov, 1962) setting the tone of the debate. (Kolmogorov, and nearly all subsequent researchers, attribute this discrepancy in the exponent $p/3$ to spatial intermittency of the energy flux in the inertial subrange.) Second, an alternative derivation of the two-thirds law, usually attributed to Kolmogorov's student Obukhov, is to assume that at each scale in the inertial subrange, r, the eddies pass on a significant proportion of their kinetic energy, v_r^2, on the time scale of their turnover time, r/v_r. This yields $\Pi \sim v_r^3/r \sim \varepsilon$, from which we obtain $v_r^2 \sim \varepsilon^{2/3} r^{2/3}$. If we are willing to make the estimate $\langle (\delta v)^2 \rangle (r) \sim v_r^2$, then we recover (8.76). Third, Kolmogorov himself had an alternative derivation of (8.76), in which he starts from the four-fifths law and them proposes that the skewness of the velocity increment, $S(r) \sim \langle (\delta v)^3 \rangle / \langle (\delta v)^2 \rangle^{3/2}$, is independent of scale (Kolmogorov, 1941b). The justification for assuming that $S(r)$ is independent of scale is unclear in Kolmogorov's paper, but perhaps it can be justified on the grounds that the dynamics are in some sense statistically self-similar across the inertial subrange. Fourth, there was already strong experimental evidence in the German literature for the scaling $\langle (\delta v)^2 \rangle \sim r^{2/3}$ (Gödecke, 1935), so there may be an element of retrospective rationalisation as well as genuine prediction in the 1941 theory.

Finally we note that, shortly after Kolmogorov published his theory, Landau pointed out that it cannot be strictly correct, an objection that appears in the 1944 Russian edition of Landau and Lifshitz's *Fluid Dynamics,* and subsequently turns up as a footnote in the English translation (Landau and Lifshitz, 1959). In effect, Landau's objection is that the spatially averaged dissipation per unit mass, ε, is not a meaningful parameter on which to base a cascade-like theory because random spatial variations in the large scales will lead to random spatial variations in the energy flux, Π. Thus there is not one cascade and one flux, but rather many cascades at many points, all proceeding at different rates and with different fluxes. Landau observes that we must average over all of these cascades to get a statistical law of the form (8.78), and notes that the result of that averaging cannot be universal, because the large-scale structure of the turbulence is itself non-universal, being different for boundary layers, wakes etc. Simple toy problems can be constructed which confirm Landau's concerns (see, for example, Davidson, 2013), and these show that the β_p's in (8.78) are indeed non-universal, being a function of the degree of large-scale spatial intermittency of the energy flux, Π.

In effect, Landau was attacking the universality of the prefactors β_p in (8.78), although he was not questioning the power-law exponent $p/3$. However, it turns out that, at least for the low-order moments, the variations in β_p due to large-scale spatial intermittency are very small, probably so small that they are undetectable in most laboratory experiments. Nevertheless, Landau's comments about the consequences of spatial intermittency greatly influenced Kolmogorov. In his 1962 refined theory of the small scales, Kolmogorov considered the consequences spatial intermittency of the energy flux in the *inertial subrange* (rather than the large scales) and showed that such intermittency can modify the power-law exponent $p/3$ in (8.78). The predicted change to the exponents is very small for the low-order moments (again, at the noise level in laboratory experiments) but is significant at higher order.

8.5 The Kármán–Howarth Equation

Kolmogorov's theory is phenomenological, and stands or falls based on the legitimacy of its underlying assumptions. We now take a more formal approach and ask what a statistically averaged version of the Navier–Stokes equation might yield. This leads us to the all-important Kármán–Howarth equation.

8.5.1 The Kármán–Howarth Equation and the Closure Problem

We continue to restrict ourselves to homogeneous turbulence in which there is no mean flow and, as before, we define the integral scales through (8.47) and write $\mathbf{u}' = \mathbf{u}(\mathbf{x}')$ where $\mathbf{x}' = \mathbf{x} + \mathbf{r}$. In order to convert the Navier–Stokes equation into statistical form we write

$$\frac{\partial u_i}{\partial t} = -\frac{\partial}{\partial x_k}(u_i u_k) - \frac{\partial}{\partial x_i}\left(\frac{p}{\rho}\right) + \nu \nabla_x^2 u_i,$$

$$\frac{\partial u'_j}{\partial t} = -\frac{\partial}{\partial x'_k}\left(u'_j u'_k\right) - \frac{\partial}{\partial x'_j}\left(\frac{p'}{\rho}\right) + \nu \nabla_{x'}^2 u'_j,$$

and multiply the first by u'_j and the second by u_i. We now note that u'_j is independent of $\mathbf{x}$ while u_i is independent of $\mathbf{x}'$. Adding the equations and then averaging, while remembering that the processes of taking averages and differentiation commute, we obtain

$$\frac{\partial}{\partial t}\langle u_i u'_j\rangle = -\frac{\partial}{\partial x_k}\langle u_i u_k u'_j\rangle - \frac{\partial}{\partial x'_k}\langle u_i u'_j u'_k\rangle - \frac{\partial}{\partial x_i}\left\langle\frac{p u'_j}{\rho}\right\rangle - \frac{\partial}{\partial x'_j}\left\langle\frac{p' u_i}{\rho}\right\rangle$$
$$+ \nu\langle\nabla_x^2\left(u_i u'_j\right) + \nabla_{x'}^2\left(u_i u'_j\right)\rangle.$$

Note that there are no body forces (e.g. buoyancy or Lorentz forces) in these equations, so we are considering freely decaying turbulence. We now recall that $\partial/\partial x_i$ acting on an average can be replaced by $-\partial/\partial r_i$ and $\partial/\partial x'_j$ by $\partial/\partial r_j$, and so we obtain

$$\frac{\partial}{\partial t}\langle u_i u'_j\rangle = \frac{\partial}{\partial r_k}\left[\langle u_i u_k u'_j\rangle - \langle u_i u'_j u'_k\rangle\right] + \frac{\partial}{\partial r_i}\left\langle\frac{pu'_j}{\rho}\right\rangle - \frac{\partial}{\partial r_j}\left\langle\frac{p'u_i}{\rho}\right\rangle + 2\nu\nabla^2\langle u_i u'_j\rangle.$$

(8.80)

Finally, we restrict ourselves to isotropic turbulence, so that (8.51) gives $\langle p\mathbf{u'}\rangle = 0$, while reflectional symmetry ensures $\langle u_i u'_j u'_k\rangle(\mathbf{r}) = -\langle u_j u_k u'_i\rangle(\mathbf{r})$. Our dynamic equation now reduces to a simple evolution equation for the two-point velocity correlation $Q_{ij}(r,\,t)$:

$$\frac{\partial Q_{ij}}{\partial t} = \frac{\partial}{\partial r_k}\left[S_{ikj} + S_{jki}\right] + 2\nu\nabla^2 Q_{ij}.$$

(8.81)

It is more convenient to work with $Q_{ii}(r) = \langle\mathbf{u}\cdot\mathbf{u'}\rangle$ than $Q_{ij}(r,\,t)$, since the evolution equation for $\langle\mathbf{u}\cdot\mathbf{u'}\rangle$ is simpler and in any event the two are related through (8.48) and (8.49) in isotropic turbulence. So we set $i=j$ in (8.81) and use (8.50) to evaluate S_{iki}. This yields the famous Kármán–Howarth equation,

$$\frac{\partial}{\partial t}\langle\mathbf{u}\cdot\mathbf{u'}\rangle = \frac{1}{r^2}\frac{\partial}{\partial r}\frac{1}{r}\frac{\partial}{\partial r}\left[r^4 u^3 K(r)\right] + 2\nu\nabla^2\langle\mathbf{u}\cdot\mathbf{u'}\rangle.$$

(8.82)

This is typically rewritten in variety of related forms. For example, using (8.49) to rewrite $\langle\mathbf{u}\cdot\mathbf{u'}\rangle$ in terms of the longitudinal correlation function, $f(r)$, and then integrating with respect to r, yields the longitudinal form of the equation:

$$\frac{\partial}{\partial t}\left[u^2 f\right] = \frac{1}{r^4}\frac{\partial}{\partial r}\left[r^4 u^3 K(r)\right] + 2\nu\frac{1}{r^4}\frac{\partial}{\partial r}\left[r^4 u^2 f'(r)\right].$$

(8.83)

Alternatively, if $\langle(\delta\mathbf{u})^2\delta\mathbf{u}\rangle$ is the third-order structure function associated with the vector velocity increment $\delta\mathbf{u} = \mathbf{u}(\mathbf{x}+\mathbf{r}) - \mathbf{u}(\mathbf{x})$, then (8.82) may be rewritten as

$$\frac{\partial}{\partial t}\langle\mathbf{u}\cdot\mathbf{u'}\rangle = \frac{1}{2}\nabla\cdot\left[\langle(\delta\mathbf{u})^2\delta\mathbf{u}\rangle\right] + 2\nu\nabla^2\langle\mathbf{u}\cdot\mathbf{u'}\rangle,$$

(8.84)

which has the advantage that it is also valid for anisotropic turbulence (see, for example, Davidson, 2013). In any event, whichever form of the Kármán–Howarth equation we adopt, we hit the so-called *closure problem of turbulence* in that, to predict the evolution of $\langle\mathbf{u}\cdot\mathbf{u'}\rangle(r,t)$, we need to know about the behaviour of the triple correlation $K(r, t)$, or equivalently the third-order structure function $\langle (\delta \mathbf{u})^2 \delta \mathbf{u} \rangle$.

So let us try to get an equation for the third-order correlations. Adopting a procedure similar to that which led to (8.81), we find that, at least symbolically,

$$\frac{\partial S_{ijk}}{\partial t} = -\frac{1}{\rho} \frac{\partial}{\partial r_k} \langle u_i u_j p' \rangle - \frac{1}{\rho} \left\langle u'_k \left(u_i \frac{\partial p}{\partial x_j} + u_j \frac{\partial p}{\partial x_i} \right) \right\rangle + \frac{\partial}{\partial r} \langle uuuu \rangle + \nu(\sim), \quad (8.85)$$

where $\nu(\sim)$ represent a viscous term and $\partial \langle uuuu \rangle / \partial r$ gradients in fourth-order, two-point velocity correlations. Evidently, in order to predict the triple correlations we need to know about the fourth-order velocity correlations. However, two-point correlations of the form $\langle uuuu \rangle$ are governed by an evolution equation which involves the fifth-order correlations, and so it goes on. In short, as a direct consequence of choosing to work with statistical quantities we have ended up with an unclosed hierarchy of equations which, if truncated, involves more unknowns than equations. This is the closure problem of turbulence and it tells us that, except in exceptional circumstances (some of which are discussed below), we cannot make rigorous predictions about statistical quantities. Rather, in order to make some form of prediction we usually need to truncate this hierarchy and introduce some additional information that closes the system, an approach which constitutes a so-called *turbulence closure model*. Such closure schemes, which are largely *ad-hoc* in nature, are discussed at length in many texts, and we will not pause to describe them here.

Despite the closure problem, (8.82) yields a great deal of useful information, as we now discuss. For example, we can determine the temporal evolution of the Saffman and Loitsyansky integrals, $L = \int \langle \mathbf{u} \cdot \mathbf{u}' \rangle d\mathbf{r}$ and $I = -\int \langle \mathbf{u} \cdot \mathbf{u}' \rangle r^2 d\mathbf{r}$, introduced in (8.71). Integrating (8.82) over all $\mathbf{r}$ gives, after integration by parts,

$$\frac{dL}{dt} = 4\pi \left[\frac{1}{r} \frac{\partial}{\partial r} [r^4 u^3 K] \right]_\infty + 8\pi \nu \left[r^2 \frac{\partial}{\partial r} \langle \mathbf{u} \cdot \mathbf{u}' \rangle \right]_\infty, \quad (8.86)$$

$$\frac{dI}{dt} = 4\pi \left[2[r^4 u^3 K] - r \frac{\partial}{\partial r} [r^4 u^3 K] \right]_\infty + 8\pi \nu \left[2r^3 \langle \mathbf{u} \cdot \mathbf{u}' \rangle - r^4 \frac{\partial}{\partial r} \langle \mathbf{u} \cdot \mathbf{u}' \rangle \right]_\infty - 12\nu L,$$
$$(8.87)$$

where, as usual, the subscript ∞ indicates $r \to \infty$. As we shall see in §8.6, in isotropic turbulence $\langle \mathbf{u} \cdot \mathbf{u}' \rangle_\infty$ and K_∞ fall off as $\langle \mathbf{u} \cdot \mathbf{u}' \rangle_\infty \sim r^{-4}$ and $K_\infty \sim r^{-4} + O(r^{-5})$, or faster, and so for isotropic turbulence we obtain

$$L = \int \langle \mathbf{u} \cdot \mathbf{u}' \rangle d\mathbf{r} = \text{constant}, \qquad (8.88)$$

$$\frac{dI}{dt} = 8\pi \left[r^4 u^3 K(r) \right]_\infty - 12\nu L. \qquad (8.89)$$

Evidently L is an invariant, while the temporal properties of I depends on the behaviour of K at large separation. We shall return to these expressions in §8.6 where we shall see that the behaviour of L and I dominates the dynamics of the large scales.

8.5.2 The Four-Fifths Law

Another important consequence of the Kármán–Howarth equation is that it provides us with the four-fifths law (Kolmogorov, 1941b), as we now show. We start by rewriting the longitudinal version of the Kármán–Howarth equation,

$$\frac{\partial}{\partial t} [u^2 f] = \frac{1}{r^4} \frac{\partial}{\partial r} \left[r^4 u^3 K(r) \right] + 2\nu \frac{1}{r^4} \frac{\partial}{\partial r} \left[r^4 u^2 f'(r) \right],$$

in terms of structure functions. Noting that $\langle (\delta v)^2 \rangle = 2u^2(1 - f)$ and $\langle (\delta v)^3 \rangle = 6u^3 K$ in isotropic turbulence, we obtain

$$-\frac{1}{2}\frac{\partial}{\partial t}\langle(\delta v)^2\rangle - \frac{2\varepsilon}{3} = \frac{1}{6r^4}\frac{\partial}{\partial r}\left[r^4\langle(\delta v)^3\rangle\right] - \nu\frac{1}{r^4}\frac{\partial}{\partial r}\left[r^4\frac{\partial}{\partial r}\langle(\delta v)^2\rangle\right]. \qquad (8.90)$$

Consider now the universal equilibrium range, $r \ll \ell$, for which $\langle (\delta v)^2 \rangle = \beta \varepsilon^{2/3} r^{2/3}$ in the inertial subrange and $\langle (\delta v)^2 \rangle \sim (\varepsilon/\nu)\eta^2 \sim \varepsilon^{2/3}\eta^{2/3}$ at the Kolmogorov scales. It follows from Kolmogorov's two-thirds law and the estimate $\langle (\delta v)^2 \rangle (r = \eta) \sim \varepsilon^{2/3}\eta^{2/3}$ that

$$\left| \frac{\partial}{\partial t} \langle (\delta v)^2 \rangle \right| \sim \left(\frac{r}{\ell} \right)^{2/3} \varepsilon \ll \varepsilon$$

for $r \ll \ell$, and so we may neglect the time derivative on the left of (8.90). The resulting equation may be integrated with respect to r to give

$$\langle (\delta v)^3 \rangle = -\frac{4}{5}\varepsilon r + 6\nu \frac{\partial}{\partial r}\langle(\delta v)^2\rangle, \; r \ll \ell, \qquad (8.91)$$

which is known as *Kolmogorov's equation of the universal equilibrium range*. The absence of a time derivative in (8.91) is a reflection of the fact that the small scales are in statistical equilibrium with respect to the instantaneous large-scale energy flux, despite the slow decay of the large scales. In the inertial subrange, viscous effects may be neglected and Kolmogorov's equation reduces to

$$\langle (\delta v)^3 \rangle = -\frac{4}{5}\varepsilon r, \ \eta \ll r \ll \ell, \tag{8.92}$$

which is the famous four-fifths law. (This law may also be derived directly from the principle of linear momentum conservation applied to a small control volume within the turbulence, as discussed in Davidson, 2015.)

We note in passing that (8.91) may be rewritten as

$$6\nu \frac{\partial}{\partial r}\langle (\delta v)^2 \rangle - S(r)\langle (\delta v)^2 \rangle^{3/2} = \frac{4}{5}\varepsilon r, \ r \ll \ell, \tag{8.93}$$

where $S(r)$ is the skewness of $\delta v(r)$, $S = \langle (\delta v)^3 \rangle / \langle (\delta v)^2 \rangle^{3/2}$. In the inertial subrange this skewness is independent of r and fixed by a combination of the two-thirds and four-fifths laws: $S_{IR} = -\frac{4}{5}\beta^{-3/2} \approx -0.28$. In fact the skewness is observed to remain fairly uniform across the equilibrium range, varying from $S_{IR} \approx -0.28$ in the inertial subrange to a maximum of $S \approx -0.4$ at $r = 0$.

One of the simplest, oldest and most reliable turbulence closure models for the equilibrium range, which is usually attributed to Obukhov, is to assume that $S = S_{IR}$ throughout the universal equilibrium range. If we introduce the dimensionless variables

$$V(x) = \frac{\langle (\delta v)^2 \rangle}{\beta (15\beta)^{1/2} v^2}, \ x = \frac{r}{(15\beta)^{3/4}\eta},$$

and adopt this closure assumption, then (8.93) simplifies to

$$\frac{1}{2}\frac{dV}{dx} + V^{3/2} = x. \tag{8.94}$$

This is readily integrated, subject to the boundary conditions $V(x \to 0) = x^2$ and $V(x \to \infty) = x^{2/3}$, to yield $\langle (\delta v)^2 \rangle$ and $\langle (\delta v)^3 \rangle$ across the entire equilibrium range. The resulting predictions are found to compare surprisingly well with recent direct numerical simulations, as discussed in Davidson, 2015.

8.5.3 Spectral Dynamics

Often it is convenient to consider cascade dynamics in spectral space, which we now do, restricting the discussion to isotropic turbulence. Recalling the definition of $E(k, t)$ in isotropic turbulence, the Kármán–Howarth equation (8.82) is readily Fourier transformed to give

$$\frac{\partial E}{\partial t} = T(k, t) - 2\nu k^2 E, \tag{8.95}$$

where

$$T(k, t) = \frac{k}{\pi} \int_0^\infty \frac{1}{r} \frac{\partial}{\partial r} \left[\frac{1}{r} \frac{\partial}{\partial r} [r^4 u^3 K] \right] \sin(kr) dr. \tag{8.96}$$

The function $T(k, t)$ is called the spectral energy transfer, and it represents the redistribution of energy in scale-space by inertial forces, with Fourier space acting as a proxy for scale-space. Two alternative forms of $T(k, t)$ are

$$T(k, t) = \frac{k^4}{\pi} \int_0^\infty \frac{\sin(kr) - kr \cos(kr)}{(kr)^3} \frac{\partial}{\partial r} [r^4 u^3 K] dr, \tag{8.97}$$

$$T(k, t) = \frac{\partial}{\partial k} \left\{ \frac{1}{\pi} \int_0^\infty \left[\left(\frac{1}{r} \frac{\partial}{\partial r} \right)^3 [r^4 u^3 K] \right] \sin(kr) dr \right\}, \tag{8.98}$$

both of which follow directly from (8.96) through integration by parts and noting the limits $K(r \to 0) = O(r^3)$ and $K(r \to \infty) \leq O(r^{-4})$.

From (8.97) we see that

$$T(k \to 0) = \frac{k^4}{3\pi} [r^4 u^3 K]_\infty + O(k^6), \tag{8.99}$$

which may be combined with (8.95) and the low-k spectral expansion of $E(k, t)$, Equation (8.70), to yield evolution equations for the Saffman and Loitsyansky integrals, $L = \int \langle \mathbf{u} \cdot \mathbf{u}' \rangle d\mathbf{r}$ and $I = - \int \langle \mathbf{u} \cdot \mathbf{u}' \rangle r^2 d\mathbf{r}$. Of course, the end result is identical to (8.88) and (8.89), with the Saffman integral an invariant and Loitsyansky integral potentially time dependent, depending on whether or not $[r^4 u^3 K]\infty$ is finite. On the other hand (8.98) allows us to introduce the so-called *spectral energy flux*, $\Pi_E(k, t)$, defined as

$$\Pi_E(k,t) = -\frac{1}{\pi}\int_0^\infty \left[\left(\frac{1}{r}\frac{\partial}{\partial r} \right)^3 [r^4 u^3 K] \right] \sin(kr)dr. \tag{8.100}$$

This has the properties

$$T(k,t) = -\frac{\partial \Pi_E}{\partial k}, \tag{8.101}$$

$$\Pi_E(k=0) = \Pi_E(k \to \infty) = 0, \tag{8.102}$$

which yields

$$\int_0^\infty T dk = 0. \tag{8.103}$$

This confirms that inertial forces cannot create or destroy kinetic energy, but rather redistribute energy in scale-space. Our spectral energy equation, (8.95), now becomes,

$$\frac{\partial E}{\partial t} = -\frac{\partial \Pi_E}{\partial k} - 2\nu k^2 E. \tag{8.104}$$

Integrating (8.104) from $k=0$ to some intermediate k tells us that $\Pi_E(k,t)$ represents the flux of kinetic energy across wavenumber k from small to large k, i.e. from large to small scales.

In the equilibrium range (8.104) may be simplified by noting that $\partial E/\partial t$ is negligible, for precisely the same reason that $\partial \langle (\delta v)^2 \rangle/\partial t$ may be neglected in (8.91). It follows that

$$0 = -\frac{\partial \Pi_E}{\partial k} - 2\nu k^2 E, \quad k \gg \ell^{-1}, \tag{8.105}$$

which is the spectral analogue of Kolmogorov's equation (8.91) for the equilibrium range. Integrating (8.105) from some intermediate wavenumber within the inertial subrange to $k \to \infty$, and remembering that

$$\varepsilon = \nu \langle \omega^2 \rangle = 2\nu \int_0^\infty k^2 E(k)dk, \tag{8.106}$$

yields

$$\Pi_E(k) = \varepsilon, \ \eta \ll k^{-1} \ll \ell. \tag{8.107}$$

So, in the inertial subrange our spectral energy flux is identical to our real-space energy flux, $\Pi_E(k) = \Pi(r) = \varepsilon$, as it must be. In fact we can deduce that $\Pi_E = \varepsilon$ in the inertial subrange directly from definition (8.100). The procedure is as follows. For $\text{Re} \to \infty$ the inertial range dominates the shape of $\langle (\delta v)^3 \rangle$, and hence that of $u^3 K$, across nearly all scales. So in this limit we may substitute for $u^3 K$ in (8.100) using the four-fifths law, and definition (8.100) then yields $\Pi_E = \varepsilon$, in line with (8.107).

Historically, closure models of isotropic turbulence have been mostly developed in spectral space, and so there is a vast literature devoted to turbulence viewed through the particular lens of the Fourier transform. The advantages of such an approach are that certain mathematical operations become simpler, and that spectral space provides a convenient (though sometimes misleading) proxy for scale-space. The primary disadvantage, which is admittedly artificial but nevertheless ubiquitous in the homogeneous turbulence literature, is that the Fourier mode progressively displaces the vortex as the basic building block of turbulence. In any event, spectral closure models dominate large parts of the turbulence literature, and at one time there was widespread hope that they might 'solve' the problem of homogeneous turbulence, in that reliable predictions could be made for a wide class of flows. Such hopes have long since been dashed, and although the spectral closures can mimic quite well the dynamics of the universal equilibrium range (they are 'tuned' to do so), they have proven unreliable when it comes to the dynamics of the large scales, as we now discuss.

8.6 Freely Decaying Turbulence

We now turn to the dynamics of the large scales, our understanding of which has had a somewhat turbulent history. We shall focus on the case of isotropic turbulence. Readers interested in the more subtle anisotropic case will find a detailed discussion in Davidson (2013).

8.6.1 Saffman versus Batchelor Turbulence: Two Canonical Energy Decays Laws

In §8.3.3 we saw that performing a Taylor expansion of definition (8.64) about $k = 0$, and assuming that $\langle \mathbf{u} \cdot \mathbf{u}' \rangle_\infty$ decays sufficiently rapidly as $r \to \infty$ yields

$$E(k) = \frac{Lk^2}{4\pi^2} + \frac{Ik^4}{24\pi^2} + O(k^6), \tag{8.108}$$

where

$$L = \int \langle \mathbf{u} \cdot \mathbf{u}' \rangle d\mathbf{r}, \qquad I = -\int \langle \mathbf{u} \cdot \mathbf{u}' \rangle r^2 d\mathbf{r} \qquad (8.109)$$

are the Saffman and Loitsyansky integrals, respectively. In §8.6.2 we shall comment on how fast $\langle \mathbf{u} \cdot \mathbf{u}' \rangle_\infty$ actually falls off with r, but in the meantime we note that expansion (8.108) is justified provided that $\langle \mathbf{u} \cdot \mathbf{u}' \rangle_\infty \leq O(r^{-6})$. However, if $\langle \mathbf{u} \cdot \mathbf{u}' \rangle_\infty = O(r^{-4})$, say, we need to use a truncated Taylor expansion with remainder, and in such a case only the first term in the expansion survives. We also noted in §8.3.3 that

$$L = 4\pi \int_0^\infty \langle \mathbf{u} \cdot \mathbf{u}' \rangle r^2 dr = 4\pi [r^3 u^2 f(r)]_\infty, \qquad (8.110)$$

and

$$L = \lim_{V \to \infty} \left\langle \left[\int_V \mathbf{u} dV \right]^2 \right\rangle \bigg/ V. \qquad (8.111)$$

At first sight (8.110) seems inconsistent with $\langle \mathbf{u} \cdot \mathbf{u}' \rangle_\infty \leq O(r^{-4})$, though in fact (8.49) tells us that it is not. Indeed a kinematically admissible energy spectrum is

$$E(k) = \frac{2\langle \mathbf{u}^2 \rangle s}{\sqrt{\pi}} k^2 s^2 \exp[-k^2 s^2], \qquad (8.112)$$

$$\langle \mathbf{u} \cdot \mathbf{u}' \rangle = \langle \mathbf{u}^2 \rangle \exp\left[-\frac{1}{4} r^2 / s^2 \right], \quad \ell = (3/2)\sqrt{\pi} s, \qquad (8.113)$$

which corresponds to a random sea of vortex rings of scale s. In this simple example L is clearly non-zero, yet $\langle \mathbf{u} \cdot \mathbf{u}' \rangle_\infty$ is exponentially small. More importantly, we conclude from (8.111) that L is a measure of the square of the linear momentum held in some large control volume, V. The central limit theorem then suggests that, in general, L will be finite, despite the constraint imposed by insisting that we adopt a frame of reference in which $\langle \mathbf{u} \rangle = 0$, i.e.

$$\lim_{V \to \infty} \left[\int_V \mathbf{u} dV \right] \bigg/ V = 0.$$

So, in the absence of some dynamical constraint that forces L to be zero, we would expect to find $E(k \to 0) \sim Lk^2$, which is known as Saffman turbulence (after Saffman, 1967).

Moreover, in §8.5.1 we used the Kármán–Howarth equation, along with the assumption that the triple correlations fall off as $K_\infty \sim r^{-4} + O(r^{-5})$ (which we will justify in the next section), to show that

$$L = \int \langle \mathbf{u} \cdot \mathbf{u}' \rangle d\mathbf{r} = \text{constant} , \tag{8.114}$$

$$\frac{dI}{dt} = 8\pi [r^4 u^3 K(r)]_\infty . \tag{8.115}$$

The invariance of L in (8.114) turns out to be a direct consequence of the law of linear momentum conservation, in the sense that $\mathbf{P} = \int_V \mathbf{u} dV$ in (8.111) is conserved except to the extent that pressure forces act on the surface of the large control volume V, or else there is a net momentum flux out through that surface. However, both these surface effects turn out to have negligible influence on L in the limit of $V \to \infty$. (See Saffman, 1967, for a brief discussion of these surface effects, or else Davidson, 2010, for a more detailed analysis of both the isotropic and anisotropic cases.)

Since L is an invariant, turbulence which starts out with a Saffman spectrum retains that spectrum. On the other hand, if the initial conditions are such that $L = 0$ at $t = 0$, then L must remain zero and we would expect $E(k \to 0) \sim Ik^4$, which is known as Batchelor turbulence after Batchelor and Proudman (1956). Prior to Saffman (1967), it was generally assumed that all homogeneous turbulence was of the form $E(k \to 0) \sim Ik^4$, and it was Saffman's seminal paper that provided convincing arguments that $E(k \to 0) \sim Lk^2$ is not only a distinct possibility, but indeed very likely in many circumstances.

The existence and behaviour of the integrals L and I have played a crucial role in our understanding of the dynamics of the large scales. Prior to Proudman and Reid (1954) it was not only widely believed that all spectra were of the from $E(k \to 0) \sim Ik^4$, but also that $K_\infty < O(r^{-4})$, so that I is an invariant. This supposed invariance of I allowed Kolmogorov (1941c) to predict the rate of decay of energy in fully developed turbulence. The point is that, once fully developed, the large scales in isotropic turbulence are observed to evolve in a self-similar way, so that the conservation of I yields

$$I = - \int \langle \mathbf{u} \cdot \mathbf{u}' \rangle r^2 d\mathbf{r} \sim u^2 \ell^5 = \text{constant}. \tag{8.116}$$

Moreover, in fully developed isotropic turbulence we have the empirical, yet well-founded, law

$$\frac{du^2}{dt} = -A\frac{u^3}{\ell}, \qquad A \approx \text{constant}, \tag{8.117}$$

where $A \approx 0.4$ when the integral scales u and ℓ are defined according to (8.47) and Re is large. Combining (8.117) with $u^2\ell^5 = \text{constant}$ yields Kolmogorov's decay laws for fully developed Batchelor turbulence,

$$\frac{u^2}{u_0^2} = \left[1 + \frac{7A}{10}\left(\frac{u_0 t}{\ell_0}\right)\right]^{-10/7}, (E \sim Ik^4), \tag{8.118}$$

$$\frac{\ell}{\ell_0} = \left[1 + \frac{7A}{10}\left(\frac{u_0 t}{\ell_0}\right)\right]^{2/7}, (E \sim Ik^4), \tag{8.119}$$

where u_0 and ℓ_0 are the initial values of u and ℓ. A quarter of a century later, Saffman (1967) did the same thing for $E(k \to 0) \sim Lk^2$ turbulence. Noting that L is an invariant, and combining (8.117) with $u^2\ell^3 = \text{constant}$, gives Saffman's decay laws for fully developed $E \sim Lk^2$ turbulence,

$$\frac{u^2}{u_0^2} = \left[1 + \frac{5A}{6}\left(\frac{u_0 t}{\ell_0}\right)\right]^{-6/5}, (E \sim Lk^2), \tag{8.120}$$

$$\frac{\ell}{\ell_0} = \left[1 + \frac{5A}{6}\left(\frac{u_0 t}{\ell_0}\right)\right]^{2/5}, (E \sim Lk^2). \tag{8.121}$$

Thus we see that the temporal evolution of the large scales depends critically on the form of the spectrum at small k, and this is determined by the initial conditions. Decay laws (8.120) and (8.121) are observed in, for example, the grid turbulence experiments of Krogstad and Davidson (2010), and in the $E(k \to 0) \sim Lk^2$ numerical simulations of Davidson, Okamoto and Kaneda (2012). Indeed, in the case of the numerical simulations, Saffman's decay laws are clearly observed even when the turbulence is decidedly anisotropic. On the other hand, the conservation of I and the associated decay laws (8.118) and (8.119) are realised (after an initial transient) in the $E(k \to 0) \sim Ik^4$ numerical simulations of Ishida, Davidson and Kaneda, 2006, as shown in Figures 8.7 and 8.8.

This observed conservation of I, when contrasted with (8.115), implies that the long-range correlation $K_\infty \sim r^{-4}$ dies out as the turbulence matures. This might be contrasted with the popular eddy-damped quasi-normal Markovian (EDQNM) spectral closure model, which *builds in* a finite $K_\infty \sim r^{-4}$ correlation at all times.

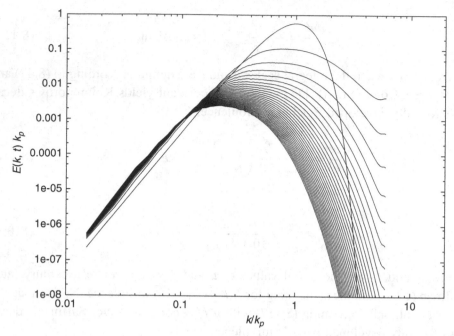

Figure 8.7 The numerical simulations of $E \sim Ik^4$ turbulence in a large periodic domain. As t increases, I tends to a constant. (From Ishida, Davidson and Kaneda, 2006.)

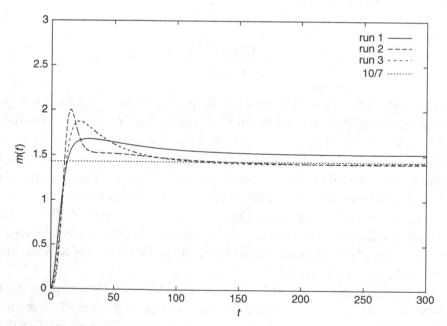

Figure 8.8 The decay exponent m in the power law $u^2 \sim t^{-m}$. In runs 2 and 3 $m \to 10/7$ once the turbulence is fully developed. However, in run 1 Re is very low (Re = 62.5) and so $m > 10/7$ due to the direct action of viscous stresses on the large eddies. (From Ishida, Davidson and Kaneda, 2006.)

Indeed the magnitude of $[r^4 u^3 K(r)]_\infty$ in EDQNM is set by an arbitrary model parameter. In short, the EDQNM closure enforces a predetermined time dependence of I whose magnitude is in some sense arbitrary, and this in turn enforces a fixed deviation from Kolmogorov's decay exponents.

In the light of predictions (8.110) and (8.115),

$$L = 4\pi [r^3 u^2 f(r)]_\infty, \quad \frac{dI}{dt} = 8\pi [r^4 u^3 K(r)]_\infty,$$

it seems crucial to understand how the long-range velocity correlations, $\langle \mathbf{u} \cdot \mathbf{u}' \rangle_\infty$, f_∞ and K_∞, fall off with r in homogeneous turbulence. So let us now turn to these long-range correlations.

8.6.2 Long-Range Interactions in Turbulence

The long-range statistical correlations in homogeneous turbulence are discussed at length in Batchelor and Proudman (1956), Saffman (1967) and Davidson (2013). Here we merely summarise the key points. It turns out that the defining feature of Saffman turbulence is that individual eddies possess, in some mean statistical sense, a significant linear impulse, defined as $\mathbf{L} = \frac{1}{2} \int_V \mathbf{x} \times \boldsymbol{\omega} dV$, where the volume V encloses the vorticity of the eddy. In such cases the Biot–Savart law tells us that the far-field velocity associated with a given eddy is

$$\mathbf{u}_\infty(\mathbf{x}) = \frac{1}{4\pi} (\mathbf{L} \cdot \nabla) \nabla(1/r) + O(r^{-4}), \tag{8.122}$$

and so the eddies cast long shadows, with $|\mathbf{u}_\infty| = O(r^{-3})$. In anisotropic turbulence, this leads directly to long-range velocity correlations of the form $\langle u_i u'_j \rangle_\infty = O(r^{-3})$ and $\langle u_i u_j u'_k \rangle_\infty = O(r^{-3})$, and hence to a finite value of L through (8.110), as discussed in Saffman (1967). However, when the symmetries of isotropy are imposed, they kill the leading-order terms in $\langle \mathbf{u} \cdot \mathbf{u}' \rangle_\infty$ and $\langle u_i u_j u'_k \rangle_\infty$, though not in general in $\langle u_i u'_j \rangle_\infty$, leaving $\langle \mathbf{u} \cdot \mathbf{u}' \rangle_\infty \leq O(r^{-4})$ and $\langle u_i u_j u'_k \rangle_\infty \leq O(r^{-4})$. Indeed (8.113) provides an example where $\langle \mathbf{u} \cdot \mathbf{u}' \rangle_\infty$ is exponentially small, despite the correlation $\langle u_i u'_j \rangle_\infty = O(r^{-3})$. Note that $\langle u_i u_j u'_k \rangle_\infty \leq O(r^{-4})$ is consistent with the usual assumption of $K_\infty \sim r^{-4} + O(r^{-5})$ in isotropic turbulence. Note also that $\langle \mathbf{u} \cdot \mathbf{u}' \rangle_\infty \leq O(r^{-4})$ leaves the first term in expansion (8.108) intact.

In the case of Batchelor turbulence, a typical eddy has no significant linear impulse (in some mean statistical sense), and so we do not expect $\langle u_i u_j u'_k \rangle_\infty = O(r^{-3})$. Rather, what happens is that a typical eddy throws out a long-range pressure field that falls off as $p_\infty \sim r^{-3}$. This, in turn, offers the possibility of pressure-velocity correlations of the form $\langle u_i u_j p' \rangle_\infty = O(r^{-3})$, as shown in Figure 8.9 and as discussed in Batchelor and Proudman (1956). These then appear

Table 8.3 *Long-range velocity correlations in Saffman and Batchelor turbulence.*

	Saffman turbulence (anisotropic)	Saffman turbulence (isotropic)	Batchelor turbulence (anisotropic)	Batchelor turbulence (isotropic)
$\langle u_i u'_j \rangle_\infty$	$O(r^{-3})$	$O(r^{-3})$	$\leq O(r^{-5})$	$\leq O(r^{-6})$
$\langle \mathbf{u} \cdot \mathbf{u}' \rangle_\infty$	$\leq O(r^{-3})$	$\leq O(r^{-4})$	$\leq O(r^{-5})$	$\leq O(r^{-6})$
$\langle u_i u_j u'_k \rangle_\infty$	$O(r^{-3})$	$\leq O(r^{-4})$	$\leq O(r^{-4})$	$\leq O(r^{-4})$

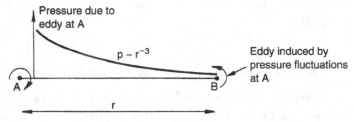

Figure 8.9 A schematic representation of Batchelor's long-range pressure effects.

as source terms on the right of (8.85), with the potential to produce long-range triple correlations of the form

$$\langle u_i u_j u'_k \rangle_\infty = u^3 a_{ijk}(r/\ell)^{-4} + O(r^{-5}). \tag{8.123}$$

(Here the a_{ijk} are dimensionless pre-factors.) This happens in both isotropic and anisotropic turbulence and is consistent with the usual assumption that $K_\infty \sim r^{-4} + O(r^{-5})$ in isotropic turbulence.

It is important to note, however, that (8.123) is in effect a kinematic result, in the sense that there is no rigorous dynamical theory that can predict the magnitude of the coefficients a_{ijk}, which could even be zero in certain situations. Indeed, the numerical simulations of Ishida, Davidson and Kaneda (2006) suggest that a_{xxx} is negligibly small in fully developed turbulence, so that the long-range triple correlation $K_\infty \sim r^{-4}$ can be ignored. In any event, the Kármán–Howarth equation, combined with $\langle u_i u_j u'_k \rangle_\infty \leq O(r^{-4})$, demands $\langle u_i u'_j \rangle_\infty \leq O(r^{-5})$ in anisotropic turbulence, though when isotropy is imposed the leading-order term is killed and we obtain the weaker correlation $\langle u_i u'_j \rangle_\infty \leq O(r^{-6})$. The long-range correlation $\langle \mathbf{u} \cdot \mathbf{u}' \rangle_\infty \leq O(r^{-6})$ is sufficiently weak as to leave the first two terms in expansion (8.108) intact.

These results are summarised in Table 8.3. From the point of view of §8.6.1, the key point is that the Biot–Savart law enforces $\langle u_i u'_j \rangle_\infty = O(r^{-3})$ in Saffman turbulence, which in turn yields a finite value of L. Moreover, there is a slow fall-off of the triple correlations in isotropic turbulence, $\langle u_i u_j u'_k \rangle_\infty = u^3 a_{ijk}(r/\ell)^{-4}$.

In principle this makes Loitsyansky's integral time-dependent in Batchelor turbulence, though in practice the simulations of Ishida, Davidson and Kaneda (2006) suggest that, in fully developed turbulence, the long-range triple correlation $K_\infty \sim r^{-4}$ can be ignored, ensuring that I is conserved.

8.6.3 Landau's Theory: the Role of Angular Momentum Conservation

Since linear momentum conservation underpins conservation of Saffman's invariant in $E \sim k^2$ turbulence, it is natural to ask what lies behind conservation of Loitsyansky's integral in $E \sim k^4$ turbulence, at least in those cases where $\langle u_i u_j u'_k \rangle_\infty < O(r^{-4})$. Landau suggested that angular momentum conservation might provide the key (Landau and Lifshitz, 1959). However, it turns out that Landau's analysis is flawed. In this section we summarise Landau's arguments, while in §8.6.4 we point out the difficulties and provide some hint as to how these problems may be overcome.

Landau considered the case of *inhomogeneous* turbulence evolving in a large, *closed* domain, V, as shown in Figure 8.10. The fact that the domain is closed turns out to be of crucial importance to Landau's theory, and creates considerable problems when we try to recast that theory in terms of strictly homogeneous turbulence.

We start by integrating the identity

$$(\mathbf{x} \times \mathbf{u}) \cdot (\mathbf{x}' \times \mathbf{u}') = 2\mathbf{x} \cdot \mathbf{x}'(\mathbf{u} \cdot \mathbf{u}') - u'_i x'_j \nabla \cdot [x_i x_j \mathbf{u}]$$

over a closed volume, V. This yields an expression for the square of the angular momentum, $\mathbf{H} = \int_V (\mathbf{x} \times \mathbf{u}) dV$, held in V,

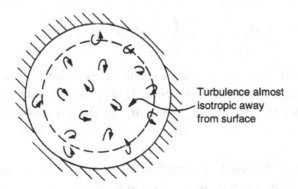

Figure 8.10 Landau's thought experiment.

$$\mathbf{H}^2 = \left[\int_V \mathbf{x} \times \mathbf{u} dV \right]^2 = \iint_{VV} 2\mathbf{x} \cdot \mathbf{x}'(\mathbf{u} \cdot \mathbf{u}') d\mathbf{x}' d\mathbf{x}. \tag{8.124}$$

(Here we have used the fact that $\mathbf{u} \cdot d\mathbf{S} = 0$ on the surface of V to deduce (8.124).) Since $u_i = \nabla \cdot (x_i \mathbf{u})$, we also have $\int_V \mathbf{u} dV = 0$ for any closed domain, and so (8.124) may be rewritten in the form

$$\mathbf{H}^2 = -\iint_{VV} (\mathbf{x}' - \mathbf{x})^2 (\mathbf{u} \cdot \mathbf{u}') d\mathbf{x}' d\mathbf{x}. \tag{8.125}$$

The final step is to ensemble average (8.125):

$$\langle \mathbf{H}^2 \rangle = \int_V \left[-\int_{V*} r^2 \langle \mathbf{u} \cdot \mathbf{u}' \rangle d\mathbf{r} \right] d\mathbf{x}. \tag{8.126}$$

Now the shape of the domain of integration of the inner integral, V^*, depends on the location of the outer variable, $\mathbf{x}$, within V. However, this difficulty may be overcome if we assume that $V \gg \ell^3$ and take $\langle \mathbf{u} \cdot \mathbf{u}' \rangle_\infty$ to fall off rapidly with separation, r, say, exponentially fast. Then for all points $\mathbf{x}$ which are remote from the boundary the inner integral in (8.126) can be replaced by an integral over *all* $\mathbf{r}$. This is a good approximation for all points $\mathbf{x}$ within V, except those which lie within a distance O (ℓ) from the surface. It follows that, in the limit of $V/\ell^3 \to \infty$, we obtain

$$\langle \mathbf{H}^2 \rangle / V = -\int r^2 \langle \mathbf{u} \cdot \mathbf{u}' \rangle d\mathbf{r} \quad . \tag{8.127}$$

This might be compared with (8.111) for $E \sim k^2$ turbulence, i.e.

$$\langle \mathbf{P}^2 \rangle / V = L = \int \langle \mathbf{u} \cdot \mathbf{u}' \rangle d\mathbf{r}, \quad \mathbf{P} = \int_V \mathbf{u} dV. \tag{8.128}$$

It seems, therefore, that there is indeed a link between Loitsyansky's integral and the angular momentum of the turbulence, $\mathbf{H}$. However Landau's analysis makes explicit use of the fact that $\mathbf{P} = \int_V \mathbf{u} dV = 0$, and this is possible only because the domain is closed. This will cause us fundamental problems when we try to adapt the analysis to strictly isotropic turbulence.

So far we have not used conservation of angular momentum to explain the invariance of I in those cases where the long-range triple correlations are weak, $(r^4 K)_\infty \approx 0$. To do this we need to consider the particular situation in which the closed domain is spherical, of radius R. In such a case $\mathbf{H}$ is conserved for each realisation of the flow, in the sense that the viscous stresses on the surface of V have a negligible influence on $\mathbf{H}$ in the limit of $(R/\ell) \to \infty$. Expression (8.127) then tells us that the invariance of I is indeed a consequence of angular momentum conservation, at least for turbulence evolving in a large closed domain.

8.6.4 Problems with Landau's Theory and Its Partial Resolution

Unfortunately, the arguments above rapidly start to unravel when we try to recast these ideas in terms of a large, open domain embedded within a field of isotropic turbulence. The central problem is that if V is a large, spherical control volume of radius R, say V_R, and $\mathbf{P} = \int_{V_R} \mathbf{u}\, dV$ is the instantaneous linear momentum held within V_R, then it is readily shown that, for isotropic $E \sim k^4$ turbulence,

$$\langle \mathbf{P}^2 \rangle = 4\pi^2 R^2 u^2 \int_0^\infty r^3 f(r)\, dr. \tag{8.129}$$

(The derivation of (8.129) is spelt out in Davidson, 2013.) Thus there is always some residual linear momentum in a large, open control volume, and this is very different from the closed domain considered by Landau. In Saffman turbulence, (8.111) tells us that this linear momentum is of order $\langle \mathbf{P}^2 \rangle \sim LR^3 \sim (u^2 \ell^3) R^3$, while in Batchelor turbulence, (8.129) demands $\langle \mathbf{P}^2 \rangle \sim (u^2 \ell^4) R^2$. It turns out that this residual linear momentum dominates the angular momentum within V_R, leading to the scaling $\langle \mathbf{H}^2 \rangle = \langle \mathbf{P}^2 \rangle R^2 \sim (u^2 \ell^4) R^4$ in Batchelor ($E \sim Ik^4$) turbulence, rather than the $\langle \mathbf{H}^2 \rangle \sim IR^3$ scaling suggested by Landau's prediction (8.127). (Again, the details are given in Davidson, 2013.) So it would seem that open and closed domains have fundamentally different properties, and this prevents us from simply transferring Landau's analysis across to homogeneous turbulence. In particular, we lose the direct connection between $\langle \mathbf{H}^2 \rangle$ and Loitsyansky's integral, $I = - \int r^2 \langle \mathbf{u} \cdot \mathbf{u}' \rangle d\mathbf{r}$, suggested by (8.127).

It turns out, however, that it *is* possible to adapt Landau's analysis to strictly homogeneous turbulence, although the approach is not straightforward. The problem turns out to be associated with the surface of the spherical control volume, V_R, which is embedded in the field of homogeneous turbulence. This surface dissects eddies (blobs of vorticity) and as a consequence leads to an unexpectedly high contribution to $\mathbf{P} = \int_{V_R} \mathbf{u}\, dV$ from the surface, a contribution which in turn leads to the scaling $\langle \mathbf{H}^2 \rangle = \langle \mathbf{P}^2 \rangle R^2 \sim (u^2 \ell^4) R^4$ (See the discussion in Davidson, 2013). To avoid this phenomenon we must follow the suggestion of Kolmogorov and consider weighted integrals of the angular momentum density

$\mathbf{x} \times \mathbf{u}$ over all space, integrals that are centred on the volume V_R, but do not involve a spherical control surface.

The argument proceeds as follows. We start by introducing the vector potential for $\mathbf{u}$, defined in the usual way by $\nabla \times \mathbf{A} = \mathbf{u}$ and $\nabla \cdot \mathbf{A} = 0$. Then $\langle \mathbf{A} \cdot \mathbf{A}' \rangle$ is related to $\langle \mathbf{u} \cdot \mathbf{u}' \rangle$ by $\langle \mathbf{u} \cdot \mathbf{u}' \rangle = -\nabla^2 \langle \mathbf{A} \cdot \mathbf{A}' \rangle$ and to Loitsyansky's integral by

$$I = -\int r^2 \langle \mathbf{u} \cdot \mathbf{u}' \rangle d\mathbf{r} = \int r^2 \nabla^2 \langle \mathbf{A} \cdot \mathbf{A}' \rangle d\mathbf{r}. \tag{8.130}$$

Since $\langle \mathbf{u} \cdot \mathbf{u}' \rangle_\infty \leq O(r^{-6})$ in isotropic Batchelor turbulence, we have $\langle \mathbf{A} \cdot \mathbf{A}' \rangle_\infty \leq r^{-4}$, and so (8.130) integrates by parts to yield $I = 6 \int \langle \mathbf{A} \cdot \mathbf{A}' \rangle d\mathbf{r}$, from which

$$I = \underset{V \to \infty}{Lim} \frac{6}{V} \left\langle \left[\int_V \mathbf{A} dV \right]^2 \right\rangle. \tag{8.131}$$

This is the key result. Crucially, if V is taken to be a large, spherical control volume of radius R, say V_R, then $\int_{V_R} \mathbf{A} dV$ can be written as a weighted integral over all space of the angular momentum density, $\mathbf{x} \times \mathbf{u}$, with the weighting centred on V_R:

$$\int_{V_R} \mathbf{A} dV = \frac{1}{\sqrt{6}} \int_{V_\infty} (\mathbf{x} \times \mathbf{u}) W(|\mathbf{x}|/R) d\mathbf{x} = \frac{1}{\sqrt{6}} \hat{\mathbf{H}}_R \tag{8.132}$$

(Davidson, 2009). Here V_∞ is an integral over all space, (8.132) defines the weighted angular momentum integral, $\hat{\mathbf{H}}_R$, and W is the weighting function

$$W = \sqrt{2/3}, \quad for \ |\mathbf{x}| \leq R ; \qquad W = \sqrt{2/3}(|\mathbf{x}|/R)^{-3}, \quad for \ |\mathbf{x}| > R.$$

Thus we have

$$I = -\int r^2 \langle \mathbf{u} \cdot \mathbf{u}' \rangle d\mathbf{r} = \underset{R \to \infty}{Lim} \langle \hat{\mathbf{H}}_R^2 \rangle / V_R, \tag{8.133}$$

which is the homogeneous equivalent of Landau's inhomogeneous result (8.127). A more detailed analysis is then required to show that the invariance of I in the absence of the long-range triple correlation $K_\infty \sim r^{-4}$ follows from angular momentum conservation, though the details are spelt out in Davidson (2009).

9

MHD Turbulence at Low and High Magnetic Reynolds Numbers

> Science is nothing without generalisations. Detached and ill-assorted
> facts are only raw material, and in the absence of a theoretical solvent,
> have little nutritive value.
>
> Rayleigh, *1884*

Magnetohydrodynamic turbulence is very different from conventional hydrodynamic turbulence in two respects. First, an imposed magnetic field typically drives
a strong statistical anisotropy at the large scales, with the large-scale eddies
elongated in the direction of the imposed magnetic field. Second, the existence of
ideal helical invariants (magnetic helicity and cross helicity) leads to the approximate conservation of magnetic and cross helicity in those turbulent flows which
have low diffusivities, and the restrictions imposed by the need to conserve these
quadratic invariants tends to drive MHD turbulence towards particular end states.
Both the tendency to develop anisotropy and the preference for particular end states
are closely related to the ability of a conducting fluid threaded by a mean magnetic
field to sustain Alfvén waves.

We will discuss both of these phenomena (large-scale anisotropy and
a preference for particular end states) in this chapter, starting with the anisotropy
created by an imposed magnetic field.

9.1 The Growth of Anisotropy at Low and High R_m

Let us consider the influence of a uniform, imposed magnetic field on the evolution
of freely evolving turbulence. We start by returning to the model problem discussed
in §6.3, extending it, with the help of Landau's ideas (see §8.6.3), into a statistical
theory.

Suppose that a conducting fluid is held in an insulated sphere of radius R and
volume V_R (Figure 9.1). The sphere sits in a uniform, imposed field $\mathbf{B}_0$, so that the
total magnetic field is $\mathbf{B} = \mathbf{B}_0 + \mathbf{b}$, where $\mathbf{b}$ is associated with the currents induced

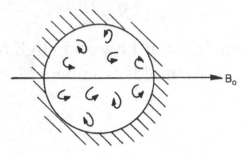

Figure 9.1 MHD turbulence evolving in a closed sphere. (Compare this with Landau's thought experiment shown in Figure 8.10.)

by **u** within the sphere. For simplicity, we take the fluid to be inviscid (we shall put viscosity back in later). However, we place no restriction on R_m, nor on the interaction parameter which we define here to be $N = \sigma B_0^2 \ell / \rho u$, ℓ and u being the integral scales of the turbulence. When R_m is small we have $|\mathbf{b}| \ll |\mathbf{B}_0|$, but in general $|\mathbf{b}|$ may be as large as, or possibly even larger than, $|\mathbf{B}_0|$. At $t = 0$ the fluid is vigorously stirred and then left to itself. The questions of interest are (i) 'Can we characterise the anisotropy introduced into the turbulence by $\mathbf{B}_0$?' and (ii) 'How does the kinetic energy of the turbulence decay?'

We shall attack the problem in precisely the same way as in §6.3. We start by noting that the global torque exerted on the fluid by the Lorentz force is

$$\mathbf{T} = \int_{V_R} \mathbf{x} \times (\mathbf{J} \times \mathbf{B}_0)dV + \int_{V_R} \mathbf{x} \times (\mathbf{J} \times \mathbf{b})dV. \tag{9.1}$$

However, a closed system of currents produce zero net torque when they interact with their self-field, **b**. (This follows from conservation of angular momentum.) It follows that the second integral on the right is zero. We now transform the first integral using the identity

$$2\mathbf{x} \times [\mathbf{G} \times \mathbf{B}_0] = [\mathbf{x} \times \mathbf{G}] \times \mathbf{B}_0 + \mathbf{G} \cdot \nabla[\mathbf{x} \times (\mathbf{x} \times \mathbf{B}_0)]. \tag{9.2}$$

Setting $\mathbf{G} = \mathbf{J}$ and invoking Gauss' theorem we find

$$\mathbf{T} = \left\{ \frac{1}{2} \int (\mathbf{x} \times \mathbf{J})dV \right\} \times \mathbf{B}_0 = \mathbf{m} \times \mathbf{B}_0, \tag{9.3}$$

where **m** is the net dipole moment of the current distribution. Evidently the global angular momentum evolves according to

$$\rho \frac{d\mathbf{H}}{dt} = \mathbf{T} = \mathbf{m} \times \mathbf{B}_0; \qquad \mathbf{H} = \int_{V_R} (\mathbf{x} \times \mathbf{u}) dV. \tag{9.4}$$

By implication, $\mathbf{H}_{//}$ is conserved. This, in turn, gives a lower bound on the total energy of the system,

$$E = \int_{V_\infty} (\mathbf{b}^2/2\mu) dV + \int_{V_R} (\rho \mathbf{u}^2/2) dV \geq \int_{V_R} (\rho \mathbf{u}_\perp^2/2) dV \geq \rho \mathbf{H}_{//}^2 \bigg/ 2 \int_{V_R} \mathbf{x}_\perp^2 dV, \tag{9.5}$$

which follows from the Schwarz inequality in the form $\mathbf{H}_{//}^2 \leq \int \mathbf{u}_\perp^2 dV \int \mathbf{x}_\perp^2 dV$, as discussed in §6.3. (As usual, we use the subscripts // and $\perp$ to indicates components parallel and perpendicular to $\mathbf{B}_0$.) However, the energy declines due to Joule dissipation according to

$$\frac{dE}{dt} = \frac{d}{dt} \int_{V_R} \left(\frac{1}{2}\rho \mathbf{u}^2\right) dV + \frac{d}{dt} \int_{V_\infty} (\mathbf{b}^2/2\mu) dV = -\frac{1}{\sigma} \int_{V_R} \mathbf{J}^2 dV. \tag{9.6}$$

We appear to have the makings of a paradox. One component of angular momentum is conserved, requiring that E is non-zero, yet energy is dissipated as long as $\mathbf{J}$ is finite. The only way out of this paradox is for the turbulence to evolve towards an end state in which $\mathbf{J} = 0$, yet the kinetic energy is non-zero, satisfying (9.5). However, if $\mathbf{J} = 0$ then Ohm's law reduces to $\mathbf{E} = -\mathbf{u} \times \mathbf{B}_0$, while Faraday's law requires that $\nabla \times \mathbf{E} = \mathbf{0}$. It follows that, at large times, $\nabla \times (\mathbf{u} \times \mathbf{B}_0) = (\mathbf{B}_0 \cdot \nabla)\mathbf{u} = \mathbf{0}$, and so $\mathbf{u}$ becomes independent of $\mathbf{x}_{//}$ as $t \to \infty$. The final state is therefore strictly two-dimensional, of the form $\mathbf{u}_\perp = \mathbf{u}_\perp(\mathbf{x}_\perp)$, $\mathbf{u}_{//} = \mathbf{0}$. In short, the turbulence ultimately reaches a state which consists of one or more columnar eddies, each aligned with $\mathbf{B}_0$. Note that all of the components of $\mathbf{H}$, other than $\mathbf{H}_{//}$, are destroyed during this evolution.

This kind of anisotropy is a consequence of Alfvén wave propagation and is ubiquitous in MHD turbulence, as illustrated in the simulations shown in Figure 9.2.

At low R_m, this transition will occur on the time scale of $\tau = \left(\sigma B_0^2/\rho\right)^{-1}$, the magnetic damping time. For our thought experiment (Figure 9.1) the proof is straightforward. At low R_m, the current density is governed by

$$\mathbf{J} = \sigma(-\nabla V + \mathbf{u} \times \mathbf{B}_0), \tag{9.7}$$

and so the global dipole moment, $\mathbf{m}$, is

(a)

(b)

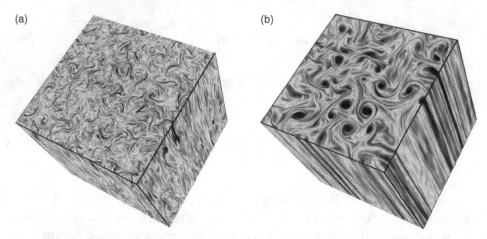

Figure 9.2 Axial vorticity contours in forced, high-R_m turbulence with an imposed magnetic field. The imposed field is modest on the left and strong on the right. (Courtesy of Alex Alexakis.)

$$\mathbf{m} = \frac{1}{2} \int \mathbf{x} \times \mathbf{J} dV = (\sigma/2) \int \mathbf{x} \times (\mathbf{u} \times \mathbf{B}_0) dV - (\sigma/2) \oint (V\mathbf{x}) \times d\mathbf{S}.$$

The surface integral vanishes while the volume integral transforms, with the aid of (9.2) in the form $\mathbf{G} = \mathbf{u}$, to give

$$\mathbf{m} = (\sigma/4)\mathbf{H} \times \mathbf{B}_0.$$

Substituting into (9.4) we obtain

$$\frac{d\mathbf{H}}{dt} = -\frac{\mathbf{H}_\perp}{4\tau}, \qquad \tau^{-1} = \sigma B_0^2/\rho, \tag{9.8}$$

and so $\mathbf{H}_{//}$ is conserved (as expected) while $\mathbf{H}_\perp$ decays exponentially on the time scale 4τ.

In summary then, whatever the initial condition, and for any R_m or N, the flow evolves towards the anisotropic state,

$$\mathbf{u}_\perp = \mathbf{u}_\perp(\mathbf{x}_\perp), \qquad \mathbf{H}_{//} = \mathbf{H}_{//}(t = 0), \qquad \mathbf{H}_\perp = \mathbf{u}_{//} = \mathbf{b} = \mathbf{J} = 0. \tag{9.9}$$

Of course, in these arguments we have ignored v and hence the process of energy removal via the energy cascade. In reality, for a finite v, the predicted growth of anisotropy will occur only if the turbulence lives for long enough and this, in turn, requires a Lorentz force of magnitude $\mathbf{J} \times \mathbf{B} \sim \rho(\mathbf{u} \cdot \nabla)\mathbf{u}$ or larger, i.e. $N \geq O(1)$. The conservation of $\mathbf{H}_{//}$, on the other hand, is a robust result that applies for any

N and any R_m. In any event, from the point of view of turbulence theory, the two most important points are (i) $\mathbf{B}_0$ can introduce severe anisotropy into the turbulence and (ii) $\mathbf{H}_{//}$ is conserved during the decay.

Following Landau's (imperfect) arguments of §8.6.3, the latter point implies that

$$\langle \mathbf{H}^2_{//} \rangle = -\int \int \mathbf{r}^2_\perp \langle \mathbf{u}_\perp \cdot \mathbf{u}'_\perp \rangle d^3\mathbf{r} \; d^3\mathbf{x} = \text{constant}, \qquad (9.10)$$

where $\mathbf{r} = \mathbf{x}' - \mathbf{x}$ (see Equation 8.126). If we are free to ignore the long-range statistical correlations discussed in §8.6.2 then, for as long as $R \gg \ell$, this suggests the existence of the invariant

$$I_{//} = \langle \mathbf{H}^2_{//} \rangle / V = -\int \mathbf{r}^2_\perp \langle \mathbf{u}_\perp \cdot \mathbf{u}'_\perp \rangle d^3\mathbf{r} = \text{constant}, \qquad (9.11)$$

which is Loitsyansky's integral for MHD turbulence. (When $\mathbf{B}_0 = \mathbf{0}$, and the turbulence is isotropic, we can replace (9.11) by (8.127).) Of course, when we try to rework the derivation of (9.11) in the context of strictly homogeneous turbulence, we run into all the problems outlined in §8.6.4. However, the remedy offered in §8.6.4 also applies in MHD turbulence and so the invariant (9.11) does indeed exist for turbulence in which the long-range triple correlations are weak (Davidson, 1997, 2009). This is typically the case in fully developed MHD turbulence of the form $E(k \to 0) \sim k^4$, as discussed below. However, when $E(k \to 0) \sim k^2$, we must replace (9.11) by a Saffman-like invariant.

9.2 Loitsyansky and Saffman-like Invariants for MHD Turbulence at Low R_m

Let us take a moment to consider the behaviour of integrals like $I_{//}$ in homogeneous, MHD turbulence under the assumption that the velocity correlations decay rapidly with separation, specifically as $\langle u_i u'_j \rangle_\infty < O(r^{-5})$ and $\langle u_i u_j u'_k \rangle_\infty < O(r^{-4})$. We restrict ourselves to low R_m and follow the arguments of Okamoto, Davidson and Kaneda (2010).

Rather than start with angular momentum conservation, in the spirit of Landau, we turn to the generalised Kármán–Howarth equation,

$$\frac{\partial}{\partial t} \nabla^2 \langle u_i u'_j \rangle = [\text{N.L.}] - \frac{2}{\tau} \frac{\partial^2}{\partial r^2_{//}} \langle u_i u'_j \rangle, \quad \tau^{-1} = \sigma B_0^2 / \rho, \qquad (9.12)$$

where the viscous terms have been ignored and the symbol [N.L.] indicates any term which involves the triple correlations, $\langle uuu' \rangle$, or else pressure-velocity correlations. The final term on the right of (9.12) reflects the low-R_m form of the solenoidal-rotational part of the Lorentz force:

$$\nabla^2(\mathbf{J} \times \mathbf{B}_0/\rho) = -\frac{1}{\rho}\nabla \times (\mathbf{B}_0 \cdot \nabla \mathbf{J}) = -\frac{1}{\rho}\left(\mathbf{B}_0 \cdot \nabla(\nabla \times \mathbf{J})\right)$$

$$= -\frac{\sigma}{\rho}(\mathbf{B}_0 \cdot \nabla)^2\mathbf{u} = -\frac{1}{\tau}\frac{\partial^2 \mathbf{u}}{\partial x_{//}^2}.$$

It should be said immediately that, in a given situation, we do not know in advance if $\langle u_i u'_j \rangle_\infty < O(r^{-5})$ and $\langle u_i u_j u'_k \rangle_\infty < O(r^{-4})$, and indeed Table 8.3 suggests that in general it is not the case. However, our experience with isotropic turbulence suggests that, after an initial transient, the pre-factors which multiply the leading-order far-field velocity correlations can be very small (Ishida, Davidson and Kaneda, 2006), so small that they may be ignored. It seems natural, therefore, to explore the consequences of ignoring the $\langle u_i u'_j \rangle_\infty \sim O(r^{-5})$ and $\langle u_i u_j u'_k \rangle_\infty \sim O(r^{-4})$ tails, and then compare the resulting predictions with the available evidence.

In Okamoto, Davidson and Kaneda (2010) it is shown that, provided $\langle u_i u'_j \rangle_\infty < O(r^{-5})$ and $\langle u_i u_j u'_k \rangle_\infty < O(r^{-4})$, (9.12) does indeed integrate to yield

$$I_{//} = -\int r_\perp^2 \langle \mathbf{u}_\perp \cdot \mathbf{u}'_\perp \rangle d\mathbf{r} = \text{constant}, \tag{9.13}$$

$$(E(k \to 0) \sim k^4 \text{ MHD turbulence}),$$

just as Landau's angular momentum argument suggests. The same conclusion can be reached by examining the equations in Fourier space. Expanding the exponential in definition (8.40) as a power series in $\mathbf{k} \cdot \mathbf{r}$, the spectral tensor, $\Phi_{ij}(\mathbf{k})$, takes the form

$$(2\pi)^3\Phi_{ij}(\mathbf{k}) = \int \langle u_i u'_j \rangle d\mathbf{r} - jk_m \int r_m \langle u_i u'_j \rangle d\mathbf{r} - \frac{1}{2}\int (\mathbf{k} \cdot \mathbf{r})^2 \langle u_i u'_j \rangle d\mathbf{r} + O(k^3).$$

$$\tag{9.14}$$

However, the first term on the right may be converted into a surface integral, which vanishes under the assumption that $\langle u_i u'_j \rangle_\infty$ decays faster than r^{-3}, while the second term also goes to zero if we restrict ourselves to $i = j$, since homogeneity tells us that the integrand is then odd in $\mathbf{r}$. Finally, expanding $(\mathbf{k} \cdot \mathbf{r})^2$ in the third integral we find

$$32\pi^3 \Phi_\perp(\mathbf{k} \to 0) = -k_\perp^2 \int \mathbf{r}_\perp^2 \langle \mathbf{u}_\perp \cdot \mathbf{u}'_\perp \rangle d\mathbf{r} - 2k_{//}^2 \int r_{//}^2 \langle \mathbf{u}_\perp \cdot \mathbf{u}'_\perp \rangle d\mathbf{r} + O(k^3).$$

$$(9.15)$$

In terms of $I_{//}$, we have

$$32\pi^3 \Phi_\perp(k_{//} = 0, k_\perp \to 0) = I_{//} k_\perp^2. \qquad (9.16)$$

Of course this all assumes that expansion (9.14) is justified, which turns out to require $\langle u_i u'_j \rangle_\infty < O(r^{-5})$, i.e. the same prerequisite for (9.13) to hold. Turning to dynamics, the transform of (9.12) is

$$\frac{\partial \Phi_{ij}}{\partial t} = [\text{N.L.}] - \frac{2}{\tau} \frac{k_{//}^2}{k^2} \Phi_{ij}. \qquad (9.17)$$

Now, in the absence of the long-range triple correlation $\langle u_i u_j u'_k \rangle_\infty \sim r^{-4}$, the non-linear terms in (9.17) are of order k^3 (Batchelor, 1953). So, if the inertial terms on the right of (9.17) can be ignored at $O(k^2)$, we have

$$\Phi_{ij}(\mathbf{k}, t) = \Phi_{ij}(t = 0)\exp\left[-\frac{k_{//}^2}{k^2} \frac{2t}{\tau}\right] + O(k^3), \qquad (9.18)$$

and in particular,

$$\Phi_\perp(k_{//} = 0, k_\perp \to 0) = \text{constant}. \qquad (9.19)$$

Combining this with (9.16) we recover (9.13), and in particular we recover (9.13) under the same restrictions of $\langle u_i u'_j \rangle_\infty < O(r^{-5})$ and $\langle u_i u_j u'_k \rangle_\infty < O(r^{-4})$.

One advantage of this spectral approach is that it brings out the relationship between $I_{//}$, and $\Phi_\perp (\mathbf{k} \to 0)$. To this end it is convenient to introduce the quantity $J_{//}$, defined as

$$J_{//} = \operatorname*{Lim}_{k_\perp \to 0} \frac{32\pi^3 \Phi_\perp(k_{//} = 0, \ k_\perp)}{k_\perp^2}. \qquad (9.20)$$

Then, on the assumption that $\langle u_i u'_j \rangle_\infty < O(r^{-5})$ and $\langle u_i u_j u'_k \rangle_\infty < O(r^{-4})$ in fully developed MHD turbulence, we have

$$I_{//} = J_{//} = \text{constant}. \qquad (9.21)$$

(fully-developed, $E(k \to 0) \sim k^4$ turbulence)

Equation (9.21) is important because $I_{//} \sim u^2 \ell_\perp^4 \ell_{//}$, so if the large scales evolve in a self-similar way, we have the powerful constraint that

$$u^2 \ell_\perp^4 \ell_{//} = \text{constant}, \tag{9.22}$$

in fully developed, $E(k \to 0) \sim k^4$ turbulence. This, in turn, allows us to predict the rate of decay of kinetic energy, as discussed in the next section.

We now turn to the numerical simulations of Okamoto, Davidson and Kaneda (2010), which were performed in a large periodic domain. (In the simulations $L_{BOX}/\ell = 100 \to 200$, where L_{BOX} is the domain size.) Here the initial energy spectrum is isotropic and has the form $E \sim k^4 \exp[-2(k/k_p)^2]$, where k_p is the wavenumber at which $E(k)$ is a maximum. These numerical experiments show that, after an initial transient, we do indeed find that $I_{//} = J_{//}$, in accordance with (9.21). Moreover, Figure 9.3 shows $J_{//}(t)/J_{//}(0)$ and $dJ_{//}/dt$ in various simulations. Here, time is normalised by the initial eddy turnover time, T, and the various curves are labelled with $\text{Re}(t = 0)$, N_0 and k_p, where $\text{Re} = u\ell/\nu$ and N_0 is the initial value of the interaction parameter, $N = \sigma B_0^2 \ell/\rho u$. Evidently, $J_{//}$ is indeed constant after a transient, consistent with (9.21).

In summary then, the numerical evidence strongly supports the Landau-like prediction that, as a consequence of angular momentum conservation,

$$I_{//} = - \int r_\perp^2 \langle \mathbf{u}_\perp \cdot \mathbf{u}'_\perp \rangle d\mathbf{r} \approx J_{//} \approx \text{constant}, \tag{9.23}$$

$$(\text{fully-developed, } E(k \to 0) \sim k^4 \text{ turbulence}),$$

at least in fully developed MHD turbulence. The implication is that, as in conventional hydrodynamic turbulence, the $\langle u_i u'_j \rangle_\infty \sim O(r^{-5})$ and $\langle u_i u_j u'_k \rangle_\infty \sim O(r^{-4})$ tails are very weak and can be ignored in the fully developed state.

In the discussion above we have restricted ourselves to $E(k \to 0) \sim k^4$ turbulence, where angular momentum conservation lies behind the existence of the integral invariant $I_{//}$. We now turn to $E(k \to 0) \sim k^2$ turbulence, where linear momentum conservation yields the integral invariants (see §8.6.1). Once again, our starting point is the low-R_m version of the Kármán–Howarth equation

$$\frac{\partial}{\partial t} \nabla^2 \langle u_i u'_j \rangle = [\text{N.L.}] - \frac{2}{\tau} \frac{\partial^2}{\partial r_{//}^2} \langle u_i u'_j \rangle, \tag{9.24}$$

and its transform

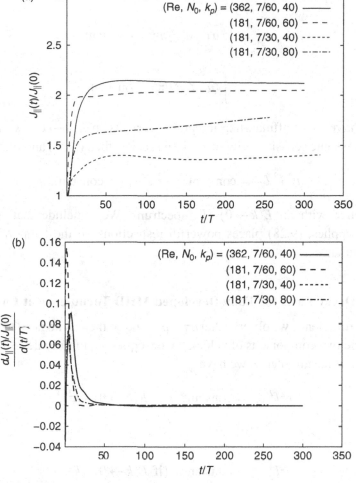

Figure 9.3 The time dependence of $J_{//}$. (a) $J_{//}(t)/J_{//}(0)$ and (b) $(dJ_{//}/dt)\left(T/J_{//}(0)\right)$. (From Okamoto, Davidson and Kaneda, 2010.)

$$\frac{\partial}{\partial t}\Phi_{ij} = [N.L.] - \frac{2}{\tau}\frac{k_{//}^2}{k^2}\Phi_{ij}. \qquad (9.25)$$

Now we know from Saffman (1967) that the non-linear terms in (9.25) are no greater than $O(k)$ for $E(k \to 0) \sim k^2$ turbulence, and so these do not influence the temporal evolution of Φ_{ij} $(k \to 0)$. It follows that

$$\Phi_{\perp}(k_{//} = 0, k_{\perp} \to 0) = \text{constant}, \quad \Phi_{//}(k_{//} = 0, k_{\perp} \to 0) = \text{constant},$$

which in real space translates to (Davidson, 2010):

$$L_\perp = \int \langle \mathbf{u}_\perp \cdot \mathbf{u'}_\perp \rangle d\mathbf{r} = \text{constant}, \tag{9.26}$$

$$L_{//} = \int \langle u_{//} u'_{//} \rangle d\mathbf{r} = \text{constant}. \tag{9.27}$$

Thus we have two Saffman-like integral invariants in $E(k \to 0) \sim k^2$ turbulence. If we now assume self-similarity of the large scales, then (9.26) and (9.27) demand

$$u_\perp^2 \ell_\perp^2 \ell_{//} = \text{constant}, \quad u_{//}^2 \ell_\perp^2 \ell_{//} = \text{constant}, \tag{9.28}$$

in turbulence with an $E(k \to 0) \sim k^2$ spectrum. We conclude that, when self-similarity applies, (9.28) places powerful restrictions on the temporal evolution of the turbulence.

9.3 Decay Laws for Fully Developed MHD Turbulence at Low R_m

In low-R_m turbulence we observe that $u_\perp^2 \sim u_{//}^2$, and so there is no need to distinguish between the two components of velocity. Consequently, in fully developed, homogeneous, low-R_m turbulence we have

$$u^2 \ell_\perp^2 \ell_{//} = \text{constant} \quad (\text{if } E(k \to 0) \sim k^2), \tag{9.29}$$

and

$$u^2 \ell_\perp^4 \ell_{//} = \text{constant} \quad (\text{if } E(k \to 0) \sim k^4), \tag{9.30}$$

assuming, of course, that the large scales evolve in a self-similar manner. We can use these integral constraints to estimate the temporal evolution of u^2, $\ell_\perp$ and $\ell_{//}$, provided that we have an expression relating the energy dissipation rate to the integral scales. This is discussed in Davidson (1997), who notes that the curl of the low-R_m form of Ohm's law yields $\nabla \times \mathbf{J} = \sigma \mathbf{B}_0 \cdot \nabla \mathbf{u}$, and hence $|\mathbf{J}| \sim (\ell_\perp / \ell_{//}) \sigma B_0 u$. Thus we can estimate the Joule dissipation per unit mass as

$$\frac{\langle \mathbf{J}^2 \rangle}{\rho \sigma} \sim \left(\frac{\ell_\perp}{\ell_{//}} \right)^2 \frac{u^2}{\tau}, \tag{9.31}$$

where $\tau = \left(\sigma B_0^2 / \rho \right)^{-1}$ is the Joule dissipation time and $u^2 = \frac{1}{3} \langle \mathbf{u}^2 \rangle$. Moreover, the usual high-Re estimate of the viscous dissipation is $\nu \langle \boldsymbol{\omega}^2 \rangle \sim u^3 / \ell$. Combining these scaling laws suggests that the energy equation,

$$\frac{d}{dt}\frac{1}{2}\langle \mathbf{u}^2 \rangle = -\nu\langle \boldsymbol{\omega}^2 \rangle - \langle \mathbf{J}^2 \rangle/\rho\sigma,$$

can be modelled as

$$\frac{du^2}{dt} = -\alpha\frac{u^3}{\ell_\perp} - \beta\left(\frac{\ell_\perp}{\ell_{//}}\right)^2\frac{u^2}{\tau}, \tag{9.32}$$

where α and β are constants of order unity.

At this point it is convenient to introduce the interaction parameter, $N = \sigma B_0^2 \ell_\perp/\rho u = \ell_\perp/u\tau$. It is evident that (9.32) is consistent with what we already know about the limits of large and small N. For example, for $N \to 0$ the magnetic field plays no role, and (9.32) reverts back to the conventional form of the dissipation law for isotropic turbulence. We then recover the classical decay laws of $u^2 \sim t^{-6/5}$ and $u^2 \sim t^{-10/7}$ for Saffman and Batchelor turbulence, respectively (see §8.6.1). On the other hand, inertia is negligible when $N \to \infty$, and the first term on the right of (9.32) may be ignored. The corresponding version of (9.32) may be combined with either (9.29) or (9.30), along with the observation that $\ell_\perp = $ constant in large N flows, to give the usual high-N decay laws,

$$u^2 = u_0^2[1 + 2\beta t/\tau]^{-1/2}, \quad \ell_{//} = \ell_0[1 + 2\beta t/\tau]^{1/2}, \tag{9.33}$$

where u_0 and ℓ_0 are the initial values of u and ℓ. (These laws are discussed in §6.2.3.)

Evidently the decay laws for large and small N are readily recovered from (9.32). For intermediate values of N, however, there is a problem. There are now two equations (9.32 plus 9.29 or 9.30) but three unknowns: u^2, $\ell_\perp$ and $\ell_{//}$. To close the system, Davidson (1997) suggested the relationship

$$\frac{d}{dt}(\ell_{//}/\ell_\perp)^2 = 2\beta/\tau, \tag{9.34}$$

which interpolates between the exact results of $N \to 0$ and $N \to \infty$. The model Equations (9.32) and (9.34) have been tested against direct numerical simulations by Okamoto, Davidson and Kaneda (2010) and found to be good approximations in fully developed MHD turbulence.

We now combine (9.32) and (9.34) with the invariants (9.29) and (9.30). Integrating from isotropic initial conditions, and introducing the scaled time

$$\hat{t} = 1 + 2\beta(t/\tau), \tag{9.35}$$

we find that the decay laws for Saffman $(E \sim k^2)$ turbulence are

$$\frac{u^2}{u_0^2} = \left[1 + \frac{5\alpha}{9\beta} \frac{\left(\hat{t}^{3/4} - 1\right)}{N_0} \right]^{-6/5} \hat{t}^{-1/2},\tag{9.36}$$

$$\frac{\ell_{/\!/}}{\ell_0} = \left[1 + \frac{5\alpha}{9\beta} \frac{\left(\hat{t}^{3/4} - 1\right)}{N_0} \right]^{2/5} \hat{t}^{1/2},\tag{9.37}$$

where N_0 is the initial value of the interaction parameter. The equivalent results for $E \sim k^4$ turbulence turn out to be

$$\frac{u^2}{u_0^2} = \left[1 + \frac{7\alpha}{15\beta} \frac{\left(\hat{t}^{3/4} - 1\right)}{N_0} \right]^{-10/7} \hat{t}^{-1/2},\tag{9.38}$$

$$\frac{\ell_{/\!/}}{\ell_0} = \left[1 + \frac{7\alpha}{15\beta} \frac{\left(\hat{t}^{3/4} - 1\right)}{N_0} \right]^{2/7} \hat{t}^{1/2}.\tag{9.39}$$

The corresponding evolution equations for $\ell_\perp/\ell_0$ for either Saffman $(E \sim k^2)$ or Batchelor $(E \sim k^4)$ turbulence may be deduced from (9.29) and (9.30).

Estimates (9.36) through (9.39) reduce to the Saffman $(u^2 \sim t^{-6/5})$ and Kolmogorov $(u^2 \sim t^{-10/7})$ decay laws for $N_0 \ll 1$, and to (9.33) for $N_0 \gg 1$, as they must. For $N_0 \sim 1$, on the other hand, they yield $u^2 \sim t^{-7/5}$ and $\ell_{/\!/} \sim t^{4/5}$ in $E \sim k^2$ turbulence, and $u^2 \sim t^{-11/7}$ and $\ell_{/\!/} \sim t^{5/7}$ in $E \sim k^4$ turbulence.

Predictions (9.38) and (9.39) were tested in the numerical simulations of Okamoto, Davidson and Kaneda (2010). For example, Figure 9.4 shows the temporal evolution of u^2/u_0^2, $\ell_{/\!/}/\ell_0$, and $\ell_\perp/\ell_0$, all normalised by the corresponding model predictions (9.38) and (9.39). (As in Figure 9.3, the curves are labelled with $\mathrm{Re}(t = 0)$, N_0 and k_p, where $\mathrm{Re} = u\ell/\nu$ and N_0 is the initial value of the interaction parameter.) In each case the normalised plots show clear plateaus at large times, lending support to the predictions.

9.4 The Spontaneous Growth of a Seed Field at High R_m: Batchelor's Criterion

We now turn to high-R_m turbulence, $R_m = u\ell/\lambda \gg 1$, and consider the case where the imposed field, $\mathbf{B}_0$, is zero. We are interested in whether or not a weak 'seed' field,

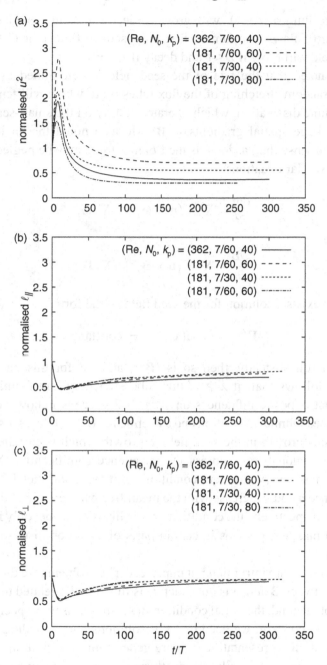

Figure 9.4 Computed values of u^2/u_0^2, $\ell_{//}/\ell_0$, and $\ell_\perp/\ell_0$, normalised by the model predictions (9.38) and (9.39). (From Okamoto, Davidson and Kaneda 2010.)

present in the fluid at $t = 0$, will grow or decay in homogeneous turbulence. An intriguing (if flawed) argument first proposed by Batchelor (1950) suggests that a seed field will grow if $\lambda < \nu$ and decay if $\lambda > \nu$.

Batchelor noted that the fate of the seed field is determined by the balance between the random stretching of the flux tubes by **u**, which will tend to increase $\langle \mathbf{B}^2 \rangle$, and Ohmic dissipation, which operates mainly on the small-scale flux tubes (which have large spatial gradients in **B**). He also noted the analogy between $\boldsymbol{\omega}$ and **B** in the sense that, as long as the Lorentz force may be neglected, they are governed by similar equations:

$$\frac{\partial \boldsymbol{\omega}}{\partial t} = \nabla \times (\mathbf{u} \times \boldsymbol{\omega}) + \nu \nabla^2 \boldsymbol{\omega}, \tag{9.40}$$

$$\frac{\partial \mathbf{B}}{\partial t} = \nabla \times (\mathbf{u} \times \mathbf{B}) + \lambda \nabla^2 \mathbf{B}. \tag{9.41}$$

If $\lambda = \nu$, there exists a solution for the seed field of the form

$$\mathbf{B}(\mathbf{x}) = \alpha \, \boldsymbol{\omega}(\mathbf{x}), \quad \alpha = \text{constant}. \tag{9.42}$$

So, if $\langle \boldsymbol{\omega}^2 \rangle$ is quasi-steady, then so is $\langle \mathbf{B}^2 \rangle$, at least for this particular initial condition. It follows that, if $\lambda = \nu$, flux tube stretching and Ohmic dissipation have equal but opposite influences on $\langle \mathbf{B}^2 \rangle$. If λ exceeds ν, however, we would expect enhanced Ohmic dissipation and a decline in $\langle \mathbf{B}^2 \rangle$, while $\lambda < \nu$ should lead to the spontaneous growth in the seed field, a growth which is curtailed only when $\mathbf{J} \times \mathbf{B}$ is large enough to suppress the turbulence significantly. (Note that the threshold $\lambda = \nu$ is a very stringent condition. In most liquid metals, for example, $\nu/\lambda \sim 10^{-6}$. Since σ and ν increase with the mean free path lengths of the charge and mass carriers, respectively, the condition $\lambda < \nu$ is likely to be met only in very sparse astrophysical plasmas, perhaps in certain parts of accretion discs or in the interstellar medium.)

These arguments are intriguing but imperfect. The problems are three-fold. First, the analogy between **B** and $\boldsymbol{\omega}$ is not exact: $\boldsymbol{\omega}$ is functionally related to **u** in a way in which **B** is not. Second, the initial condition $\mathbf{B}(\mathbf{x}) = \alpha\boldsymbol{\omega}(\mathbf{x})$ is very specific, and in the light of the first concern, we need to ask if solutions corresponding to this initial condition are at all representative of more general initial conditions. Third, if the turbulence is to be statistically steady then a forcing term must appear in the vorticity equation representing some kind of mechanical stirring (which is required to keep the turbulence alive). Since the corresponding term is absent in the induction equation the analogy between **B** and ω is again broken. One can circumvent this third objection by considering freely decaying turbulence. Unfortunately, this also leads to difficulties, since the turbulence will die out on a timescale of ℓ/u, and if $R_m = u_0\,\ell_0/\lambda$ is large, this implies we can get a transient growth in $\langle \mathbf{B}^2 \rangle$ only for times of order $\eta/v < t < \ell/u$, or equivalently

$$\eta/v < t \ll \ell^2/\lambda, \tag{9.43}$$

i.e. less than the Ohmic time scale, ℓ^2/λ. (Here η and v are the Kolmogorov scales.)

The precise conditions under which $\langle \mathbf{B}^2 \rangle$ will spontaneously grow are still debated. In §9.5 we shall consider magnetic field generation in randomly forced statistically steady turbulence. We shall see that the magnetic Prandtl number, $Pr_m = v/\lambda$, does indeed play a critical role, and that in general it is much easier for a seed magnetic field to grow when $Pr_m > 1$, just as Batchelor foresaw. However, for the moment, we shall restrict the discussion to freely decaying turbulence, and to the conditions under which a transient growth in $\langle \mathbf{B}^2 \rangle$ may be realised. In such turbulence we already know from Batchelor (1950) that the criterion for transient growth is $Pr_m > 1$, at least for the particular initial condition of $\mathbf{B}(\mathbf{x}) = \alpha\omega(\mathbf{x})$, and so the key question is whether or not this rather specific initial condition is representative of a broader class of initial **B**-fields.

The problem of computing the evolution of a weak magnetic field, in which we ignored any back reaction from the Lorentz force on the fluid, is called the passive-vector problem (by analogy with a passive scalar), and it has been pursued by a number of authors. For example, Ohkitani (2002) took the initial power spectra for **B** and ω to be the same, though the two fields were otherwise different at $t = 0$. His primary finding is that **B** and ω behave differently at $Pr_m = 1$, with magnetic field-line stretching generally stronger than vortex-line stretching. However, if the initial conditions are constrained to resemble $\mathbf{B}(\mathbf{x}) = \alpha\omega(\mathbf{x})$, then the rate of growth of $\langle \mathbf{B}^2 \rangle$ is inhibited, with $\mathbf{B} = \alpha\omega$ representing an approximate minimum in magnetic field stretching. In short, Batchelor's analysis is too simplistic, and there is not a sharp transition at $Pr_m = 1$ for the growth of a seed field. Rather, the form of the initial conditions must also be taken into account and when this is done a transient growth in magnetic energy is indeed possible for $Pr_m < 1$.

9.5 Magnetic Field Generation in Forced, Non-Helical Turbulence at High R_m

We now examine in more detail the turbulent generation of magnetic fields at high R_m. As in §9.4, we take the externally imposed field to be zero. Perhaps the first point to note is that, historically, there have been several classes of related problems which have attracted attention, with each class designed to answer rather different questions. This subdivision has arisen, at least in part, because the way that a magnetic field at high R_m responds to fluid motion depends critically on whether or not the flow, or the magnetic field, is strongly helical, as we now discuss.

9.5.1 Different Categories of Magnetic Field Generation

In Chapter 14, where we consider planetary dynamos, we shall see that the flow in planetary cores is highly helical, and that this kinetic helicity is key to maintaining a quasi-steady, large-scale magnetic field (typically a dipole field aligned with the rotation axis). It so happens that the flow is also irregular and random, which we might call turbulence; a randomness which is driven by the non-linear interactions between the buoyancy, velocity and magnetic fields (though not by $\mathbf{u} \cdot \nabla \mathbf{u}$, which is completely negligible in the core of the earth). One of the most striking features of planetary dynamos is that, despite the turbulent nature of the flow which drives the dynamo, the resulting magnetic field has most of its magnetic energy at the largest available scale (the size of the planetary core) and this large scale field is usually relatively simple and quasi-steady. Thus, in planetary dynamo theory, we seek not only to explain how fluid motion driven by buoyancy can generate magnetic energy, but also why that magnetic field has such a simple, quasi-steady, large-scale structure. Indeed, it is the second of these two questions which has proven to be more difficult to answer, though it is now widely believed that this self-organisation of the large-scale magnetic field occurs because the flow is highly helical, with a strong spatial segregation of that helicity (say left-handed spirals in the north and right-handed spirals in the south – see Chapter 14). In short, one of the main objectives of planetary (and solar) dynamo theory is to explain the ability of the saturated magnetic field to self-organise into a simple, large-scale structure, and how this self-organisation relates to the spatial segregation of the kinetic helicity in the flow.

By contrast, a second group of theoreticians, mostly from plasma physics, are interested in understanding the conditions under which magnetic energy can grow by field-line stretching in a conducting, turbulent fluid which is subject to continuous agitation. The interest here lies in the spontaneous growth, and subsequent saturation, of magnetic energy, and not so much in the ability of the large-scale field to self-organise into a simple, steady form. In this second class of studies, which we might refer to as MHD turbulence (as distinct from planetary or solar dynamo theory), there is often no spatial segregation of the mean helicity, or indeed any mean helicity at all, and so the resulting magnetic field is usually disorganised, unsteady, and of a scale no greater than the turbulent motion that generated it. Moreover, the turbulence in these studies is often taken to be Kolmogorov-like, with inertial cascades driven by $\mathbf{u} \cdot \nabla \mathbf{u}$, a situation which cannot exist in a planetary core where $\mathbf{u} \cdot \nabla \mathbf{u}$ is tiny by comparison with the buoyancy and Coriolis forces. It is this second class of problem, that of field-line stretching in Kolmogorov-like turbulence, which we address here. Our discussion of planetary and solar dynamo theory is postponed until Chapters 14 and 15.

To these two distinct classes of study, we might add a third – that of *freely decaying* MHD turbulence at large R_m. Here the interest lies mostly in the end state to which the turbulence evolves, and again helicity plays a crucial role, though this time it is the helicity trapped in the magnetic field which is important. In particular, MHD turbulence at large R_m tends to retain is magnetic helicity and cross helicity. It turns out that the limitations imposed by the need to conserve this helicity tend to drive the turbulence towards particular end states, those end states being very different depending on whether magnetic helicity or cross helicity is dominant. We shall turn to this third topic in §9.6.

A common feature of studies of MHD turbulence at high R_m is the need to identify the microscale of the magnetic field, η_λ, since the behaviour the turbulence can be very different depending on whether the magnetic microscale is greater or smaller than the Kolmogorov microscale, η. So let us take a moment to estimate this microscale, whose relative value clearly depends on Pr_m.

Consider first the case where $\mathrm{Pr}_m \ll 1$, which characterises liquid metals and dense plasmas. Here $\mathbf{B}$ diffuses more readily than the vorticity, $\boldsymbol{\omega}$, and so η_λ will be larger than the Kolmogorov scale, η. The stretching of the field lines is therefore achieved by eddies which are somewhat larger than the Kolmogorov scale, which we might take to lie in the inertial subrange. The defining feature of η_λ is then $v_\lambda \eta_\lambda / \lambda \sim 1$, so that diffusion can compete with advection, where v_λ is the characteristic fluctuating velocity at the scale η_λ. From Kolmogorov's two-thirds law we have $v_\lambda \sim \varepsilon^{1/3} \eta_\lambda^{1/3}$ and it follows that $\varepsilon^{1/3} \eta_\lambda^{4/3} / \lambda \sim 1$. Since $\eta \sim \left(\nu^3 / \varepsilon\right)^{1/4}$, we can rewrite this as

$$\eta_\lambda / \eta \sim (\lambda / \nu)^{3/4} = \mathrm{Pr}_m^{-3/4}, \quad \mathrm{Pr}_m \ll 1. \tag{9.44}$$

Eddies below this scale are ineffective at stretching the **B**-lines because magnetic diffusion dominates for scales below η_λ.

The case of $\mathrm{Pr}_m \gg 1$ is rather different. Here the diffusion of **B** is less rapid than that of the vorticity, and so η_λ will be smaller than the Kolmogorov scale, $\eta_\lambda < \eta$. The stretching of the magnetic field is therefore dominated by η-scale eddies whose velocity field looks relatively smooth on the scale of η_λ. (Recall that the strain-rate is largest at the Kolmogorov scale, so that eddies much larger than η are less effective at stretching **B**.) These Kolmogorov-sized eddies will organise the magnetic flux into tubes and ribbons, as discussed in §5.5.4. The characteristic thickness of these tubes and ribbons is $\eta_\lambda \sim \sqrt{\lambda / \alpha}$, where α is the characteristic strain-rate of the velocity doing the stretching. Taking α to be the strain-rate of the Kolmogorov eddies, $\alpha \sim \upsilon/\eta$, where υ is the Kolmogorov velocity, we have $\eta_\lambda \sim \sqrt{\lambda\eta/\upsilon}$. However, the Kolmogorov scales satisfy $\upsilon\eta/\nu\sim 1$, and so our estimate of η_λ may be rewritten as

$$\eta_\lambda/\eta \sim \sqrt{\lambda/\nu} \;\; = \mathrm{Pr}_m^{-1/2}, \quad \mathrm{Pr}_m \gg 1, \tag{9.45}$$

which is clearly different from (9.44).

Armed with these simple estimates of η_λ, we now turn to the issue of magnetic field stretching, and to a popular kinematic model of that stretching, called the Kazantsev model.

9.5.2 *A Kinematic Model for Field Generation in Forced, Non-Helical Turbulence*

Let us restrict ourselves to isotropic turbulence in which there is no externally imposed magnetic field and the random seed field is weak. We begin by deriving an evolution equation for the two-point magnetic field correlation, $\langle B_i B'_j\rangle(\mathbf{r}, t)$, where $\mathbf{r} = \mathbf{x}' - \mathbf{x}$. We start with the induction equation applied at $\mathbf{x}$ and $\mathbf{x}'$. Multiplying $\partial\mathbf{B}'/\partial t$ by $\mathbf{B}$ and $\partial\mathbf{B}/\partial t$ by $\mathbf{B}'$, adding and then averaging, we find

$$\frac{\partial}{\partial t}\langle \mathbf{B} \cdot \mathbf{B}'\rangle = \frac{\partial}{\partial r_k}\langle B'_k(\mathbf{u}' \cdot \mathbf{B}) - B_k(\mathbf{u} \cdot \mathbf{B}')\rangle - \frac{\partial}{\partial r_k}\langle (u'_k - u_k)\,\mathbf{B}\cdot\mathbf{B}'\rangle$$
$$+ 2\lambda\nabla^2\langle \mathbf{B}\cdot\mathbf{B}'\rangle. \tag{9.46}$$

In §8.3.3 we noted that, in isotropic turbulence, correlations like $\langle p\mathbf{B}'\rangle$ and $\langle p\mathbf{u}'\rangle$, where p is a scalar, are zero because $\mathbf{B}$ and $\mathbf{u}$ are solenoidal. So we can add the term $\langle(\mathbf{u}' \cdot \mathbf{B}')B_k - (\mathbf{u} \cdot \mathbf{B})B'_k\rangle$ to the right-hand side of our two-point evolution equation, which allows us to replace $\mathbf{u}$ by the velocity increment, $\delta\mathbf{u} = \mathbf{u}' - \mathbf{u}$:

$$\frac{\partial}{\partial t}\langle \mathbf{B}\cdot\mathbf{B}'\rangle = \nabla \cdot \langle(\delta\mathbf{u}\cdot\mathbf{B})\mathbf{B}' + (\delta\mathbf{u}\cdot\mathbf{B}')\mathbf{B} - \delta\mathbf{u}(\mathbf{B}\cdot\mathbf{B}')\rangle + 2\lambda\nabla^2\langle\mathbf{B}\cdot\mathbf{B}'\rangle. \tag{9.47}$$

Next we note that combining isotropy with $\nabla\cdot\mathbf{B}=0$ allows us to write $\langle\mathbf{B}\cdot\mathbf{B}'\rangle(\mathbf{r}, t)$ in terms of the longitudinal correlation function,

$$\langle B_x B'_x\rangle(r\hat{\mathbf{e}}_x) = \langle B_x^2\rangle f_B(r, t) = B^2 f_B(r, t). \tag{9.48}$$

That is to say, by analogy with (8.49), we have,

$$\langle\mathbf{B}\cdot\mathbf{B}'\rangle = \frac{B^2}{r^2}\frac{\partial}{\partial r}\left(r^3 f_B\right). \tag{9.49}$$

Moreover, in isotropic turbulence, the triple correlations on the right of (9.47) all point in the direction of $\mathbf{r}$ (see, for example, Davidson, 2013), and so (9.47) simplifies to

$$\frac{\partial}{\partial t} B^2 f_B = T(r,t) + 2\lambda B^2 \frac{1}{r^4} \frac{\partial}{\partial r}\left(r^4 \frac{\partial f_B}{\partial r}\right), \tag{9.50}$$

where

$$T(r,t) = \left[\langle(\delta\mathbf{u}\cdot\mathbf{B})\,\mathbf{B}'\rangle + \langle(\delta\mathbf{u}\cdot\mathbf{B}')\,\mathbf{B}\rangle - \langle\delta\mathbf{u}(\mathbf{B}\cdot\mathbf{B}')\rangle\right]\cdot\frac{\mathbf{r}}{r^2}. \tag{9.51}$$

This bears an obvious similarity to the longitudinal form of the Kármán–Howarth equation, (8.83).

To make progress, we need to know more about the mixed triple correlation $\langle\delta u_i B_j B'_k\rangle$, which in turn depends upon the joint statistics of $\mathbf{B}$ and $\delta\mathbf{u}$. A popular model of magnetic field generation in non-helical, isotropic turbulence is the so-called Kazantsev model. In this model the velocity field is taken to have Gaussian statistics and to be delta-correlated (white noise) in time. It is then possible to show that (see, for example, Tobias, Cattaneo, and Boldyrev, 2013),

$$T(r,t) = B^2 \left[\delta\kappa\frac{\partial^2 f_B}{\partial r^2} + \frac{1}{r^4}\frac{\partial}{\partial r}\left(r^4(\delta\kappa)\right)\frac{\partial f_B}{\partial r} + \frac{1}{r^4}\frac{\partial}{\partial r}\left(r^4\frac{\partial(\delta\kappa)}{\partial r}\right)f_B\right], \tag{9.52}$$

where $\delta\kappa$ is related to the longitudinal velocity structure function, $\langle(\delta v)^2\rangle$, being proportional to $\langle(\delta v)^2\rangle$ times a velocity decorrelation time, τ_c, i.e. $\delta\kappa \sim \langle(\delta v)^2\rangle\tau_c$.

Combining (9.50) with (9.52) allows us to examine the effect of the statistics of $\langle(\delta v)^2\rangle$ on those of $\langle\mathbf{B}\cdot\mathbf{B}'\rangle$. Let us consider the cases of low and high Pr_m separately. For $Pr_m \ll 1$ the microscale η_λ, and hence the associated magnetic energy, sits in the inertial range, so we might take the decorrelation time to be the eddy turnover time, $\tau_c \sim r/\langle(\delta v)^2\rangle^{1/2}$, where $\langle(\delta v)^2\rangle \sim \varepsilon^{2/3}r^{2/3}$. This then gives us the estimate $\delta\kappa \sim r\langle(\delta v)^2\rangle^{1/2} \sim \varepsilon^{1/3}r^{4/3}$. With this specification of $\delta\kappa$, the Kazantsev model predicts that magnetic field amplification is in fact possible, but only for relatively large values of R_m. The power spectrum for the magnetic field is then peaked at η_λ and the growth rate scales on the turnover time of the inertial range eddies of size η_λ.

Conversely, for $Pr_m \gg 1$, we know that the velocity field is smooth on the scale of η_λ, with $\langle(\delta v)^2\rangle$ proportional to r^2: $\langle(\delta v)^2\rangle \sim (v^2/\eta^2)r^2$. If we now take the decorrelation time to be the Kolmogorov timescale, $\tau_c = \eta/v$, we have the alternative estimate of $\delta\kappa \sim (\eta/v)\langle(\delta v)^2\rangle \sim (v/\eta)r^2$. In this case the model predicts an exponential growth in magnetic energy even for modest values of R_m. The magnetic energy grows across the range of scales $\eta_\lambda < r < \eta$, with a power spectrum peaked at η_λ and a growth rate proportional to the Kolmogorov time scale.

9.5.3 The Role of the Magnetic Reynolds and Magnetic Prandtl Numbers

The specific predictions of the Kazantsev model must be treated with considerable caution, as the assumed statistics of the velocity field are not at all physical, and are chosen mostly through a desire to obtain analytical results. Nevertheless, the predictions of the model are similar in spirit to the findings of Ohkitani (2002), which we discussed in §9.4. In both cases a picture emerges in which Batchelor (1950) was correct to the extent that spontaneous magnetic field generation is much easier for $\mathrm{Pr}_m > 1$ than for $\mathrm{Pr}_m < 1$. However, the specific initial condition considered by Batchelor, that of $\mathbf{B}(\mathbf{x}) = \alpha\boldsymbol{\omega}(\mathbf{x})$, is overly pessimistic and minimises magnetic field stretching. Thus magnetic field generation does not suddenly shut down as Pr_m drops below $\mathrm{Pr}_m = 1$, as suggested in Batchelor (1950). Rather, the Kazantsev model suggests that magnetic fields may continue to be maintained for $\mathrm{Pr}_m < 1$, though it is much harder to do so, in the sense that a relatively large value of R_m is required to maintain a growth.

These general observations are supported by many of the direct numerical simulations of MHD turbulence at high R_m, some of which are discussed in Schekochihin et al. (2007) and Tobias, Cattaneo, and Boldyrev (2013). The picture that emerges from these numerical experiments is the following. For both low and high Pr_m there is a critical value of R_m below which magnetic field generation cannot be sustained, and it is useful to think of a parameter space in which the critical value of R_m is plotted as a function of Pr_m. The critical value of R_m is modest (perhaps around $R_m \sim 80$) when $\mathrm{Pr}_m > 1$, but rises rapidly as we drop below $\mathrm{Pr}_m = 1$, approaching a value of around $R_m \sim 400$ by the time we reach $\mathrm{Pr}_m = 0.1$. (Of course, the exact values of R_m depend on precisely how the integral scale appearing in the definition of R_m is defined in a particular simulation.) It is still uncertain what happens for values of Pr_m much below 0.1, as this regime is not readily accessed by the direct numerical simulations.

A typical example of the results of the simulations, taken from Schekochihin et al. (2007), is shown in Figure 9.5. This shows the computed growth rate γ of $\langle \mathbf{B}^2 \rangle(t)$ as a function of Pr_m, with $\gamma > 0$ indicating a growth in magnetic energy. Several sets of data are shown, corresponding to different values of R_m, ranging from $R_m = 60$ to $R_m = 830$. For $\mathrm{Pr}_m = 1$ all values in excess of $R_m \sim 100$ give rise to a growing magnetic field, while a minimum value of $R_m \sim 400$ is required for growth at $\mathrm{Pr}_m = 0.1$. The filled symbols represent direct numerical simulations of MHD turbulence, while the open symbols correspond to so-called hyper-viscous

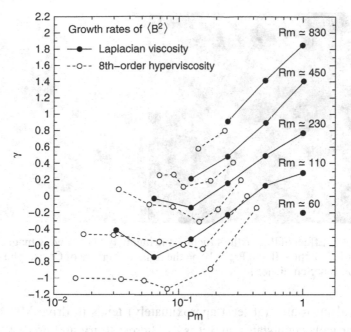

Figure 9.5 Computed growth rates for the magnetic energy in MHD turbulence as a function of magnetic Prandtl number and magnetic Reynolds number. (From Schekochihin et al. (2007). © IOP Publishing & Deutsche Physikalische Gesellschaft. CC BY-NC-SA.)

simulations, in which the small viscous scales, which are computationally expensive to resolve, are modelled.

The effect of Pr_m on the structure and intensity of a magnetic field is illustrated in Figure 9.6, which is taken from Federrath et al. (2014). This shows growing magnetic fields at $Pr_m = 0.1$ and $Pr_m = 10$ in forced turbulence at Re = 1600. In both cases the magnetic field is undergoing transient growth and the two panels correspond to the same elapsed time. At $Pr_m = 0.1$ the magnetic field is diffuse and weak, while at $Pr_m = 10$ it exhibits sharper gradients and is more intense, since the growth rate is faster at large Pr_m.

9.6 Unforced, Helical Turbulence at High Magnetic Reynolds Numbers

Let us return to freely decaying turbulence and consider the consequences of having a significant amount of helicity, either magnetic or cross helicity, in the initial conditions. We shall see that, whenever Re and R_m are both large, the existence of these ideal helical invariants leads to the approximate conservation of magnetic and cross helicity in MHD turbulence. Moreover, the need to conserve

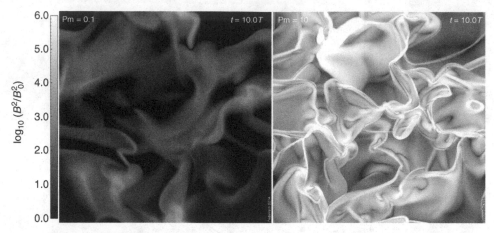

Figure 9.6 Effect of Pr_m on the structure and intensity of a growing magnetic field. $\mathrm{Pr}_m = 0.1$ is on the left and $\mathrm{Pr}_m = 10$ on the right. (Courtesy of Christoph Federrath and Dominik Schleicher.)

these helical invariants (at least approximately) tends to drive MHD turbulence towards particular end states, and it is of interest to try and predict those states.

9.6.1 Ideal Invariants and Selective Decay

In §5.4 we noted that, in addition to conserving energy, the equations of ideal MHD admit two helical invariants:

$$\text{magnetic helicity,} \quad h_m = \int \mathbf{A} \cdot \mathbf{B} dV , \tag{9.53}$$

$$\text{cross helicity,} \quad h_c = \int \mathbf{u} \cdot \mathbf{B} dV . \tag{9.54}$$

However, when we allow for finite diffusivities, these are no longer conserved. For example, for a localised disturbance evolving in an infinite domain it is readily shown that

$$\frac{d}{dt} \int_{V_\infty} \mathbf{A} \cdot \mathbf{B} dV = -\frac{2}{\sigma} \int_{V_\infty} \mathbf{J} \cdot \mathbf{B} dV, \tag{9.55}$$

$$\frac{d}{dt} \int_{V_\infty} \mathbf{u} \cdot \mathbf{B} dV = -\mu(\lambda + \nu) \int_{V_\infty} \mathbf{J} \cdot \boldsymbol{\omega} dV, \tag{9.56}$$

or equivalently, in homogeneous turbulence,

$$\frac{d}{dt}\langle \mathbf{A} \cdot \mathbf{B} \rangle = -\frac{2}{\sigma}\langle \mathbf{J} \cdot \mathbf{B} \rangle, \tag{9.57}$$

$$\frac{d}{dt}\langle \mathbf{u} \cdot \mathbf{B} \rangle = -\mu(\lambda + \nu)\langle \mathbf{J} \cdot \mathbf{\omega} \rangle. \tag{9.58}$$

To these we might add the energy equation

$$\frac{dE}{dt} = \frac{d}{dt}\langle \tfrac{1}{2}\rho \mathbf{u}^2 \rangle + \frac{d}{dt}\langle \mathbf{B}^2/2\mu \rangle = -\rho\nu\langle \mathbf{\omega}^2 \rangle - \frac{1}{\sigma}\langle \mathbf{J}^2 \rangle, \tag{9.59}$$

where $E = \langle \tfrac{1}{2}\rho \mathbf{u}^2 + \mathbf{B}^2/2\mu \rangle$ in the energy density.

Note that the diffusive terms on the right of (9.55) and (9.56) are not sign-definite, and so they can generate as well as destroy helicity as they facilitate reconnections of the vortex lines and magnetic field lines. Note also that (9.56) is readily generalised to the case where there is a mean field, $\mathbf{B}_0$. Thus if $\mathbf{B} = \mathbf{B}_0 + \mathbf{b}$, and $\mathbf{b}$ is a localised disturbance, then

$$\frac{d}{dt}\int_{V_\infty} \mathbf{u} \cdot \mathbf{b}dV = -\mu(\lambda + \nu)\int_{V_\infty} \mathbf{J} \cdot \mathbf{\omega}dV, \tag{9.60}$$

while the associated energy equation is

$$\frac{d}{dt}\int_{V_\infty} \left(\tfrac{1}{2}\rho \mathbf{u}^2\right)dV + \frac{d}{dt}\int_{V_\infty} \left(\mathbf{b}^2/2\mu\right)dV = -\rho\nu\int_{V_\infty} \mathbf{\omega}^2 dV - \frac{1}{\sigma}\int_{V_\infty} \mathbf{J}^2 dV. \tag{9.61}$$

In MHD turbulence an important question is, 'What happens to these three ideal quadratic invariants when the diffusivities are small but finite?' We know from conventional hydrodynamic turbulence that the rate of dissipation of energy is finite and independent of Re in the limit of $\nu \to 0$. Similarly, in three-dimensional MHD turbulence we expect that, in the limit of small but finite diffusivities, the rate of dissipation of energy will remain finite and independent of λ and ν. This is achieved through the formation of thin, intense current sheets and vortex tubes, which act as centres of dissipation and whose size automatically adjusts to ensure that the small-scale dissipation matches the flux of energy from the large scales. It is natural to ask what happens to the magnetic and cross helicities in the same limit. We shall see that, unlike energy, they are approximately conserved, and so we might characterise freely decaying MHD turbulence as a system that dissipates energy subject to the constraint of conserving helicity. This is known as *selective decay*.

There are two important special cases – one in which the initial condition is dominated by magnetic helicity, and one in which it is dominated by cross helicity – and it turns out that the turbulence in these two situations behaves quite differently. Let us start with the case where the initial fields have negligible cross helicity but a relatively large amount of magnetic helicity.

9.5.2 Taylor Relaxation

Let us suppose that we have a significant amount of magnetic helicity, h_m, in the initial condition, and consider (9.57) and (9.59) in the limit of large R_m. The Schwarz inequality tells us that $\langle \mathbf{J} \cdot \mathbf{B} \rangle^2 \leq \langle \mathbf{J}^2 \rangle \langle \mathbf{B}^2 \rangle$, and so from (9.57) we can place an upper bound on the rate of change of magnetic helicity,

$$\frac{1}{\mu} \left| \frac{d}{dt} \langle \mathbf{A} \cdot \mathbf{B} \rangle \right| = \frac{1}{\mu} \left| \frac{dh_m}{dt} \right| \leq 2\lambda \sqrt{\sigma |\dot{E}|} \sqrt{2\mu E} = \sqrt{8\lambda} \, |\dot{E}|^{1/2} E^{1/2}, \qquad (9.62)$$

where $\dot{E}$ is the rate of change of energy per unit volume. It is useful to define a characteristic initial integral scale, ℓ_0, and a reconnection time scale, τ_{rec}, via the expressions

$$\ell_0 = \frac{|h_m(t = 0)|}{\mu E(t = 0)}, \qquad \left| \frac{dh_m}{dt} \right| = \frac{|h_m(t = 0)|}{\tau_{rec}}. \qquad (9.63)$$

Inequality (9.62) can then be rewritten as

$$\tau_{rec} \frac{|\dot{E}|}{E} \geq \frac{\ell_0^2}{8\lambda \tau_{rec}}. \qquad (9.64)$$

Now consider the limit of $\lambda \rightarrow 0$. From the discussion of reconnection within current sheets given in §5.5.5, we expect that the reconnection time will scale with λ as $\tau_{rec} \sim \lambda^{-p}$, $0 \leq p \leq 1/2$, in the limit of $\lambda \rightarrow 0$ (so-called 'fast reconnection'). In such cases we have $|\dot{E}| \gg E/\tau_{rec}$ for $R_m \gg 1$, and we conclude that the rate of decay of energy is much faster than the rate of change of magnetic helicity.

Actually, a similar conclusion may be obtained directly from (9.62) without the need to invoke the notion of a reconnection time. Physically, the idea is that the rate of dissipation of energy remains finite in the limit of $\lambda \rightarrow 0$, through the formation of thin, intense current sheets (Figure 9.7), while (9.62) tells us that the rate of change of magnetic helicity tends to zero in the same limit. So, as $\lambda \rightarrow 0$, we have an order-one rate of destruction of energy, subject to the approximate conservation of magnetic helicity.

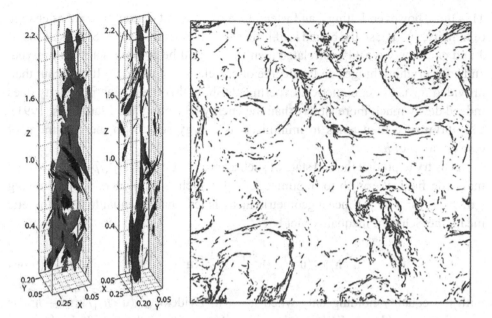

Figure 9.7 Current sheets in a numerical simulation of high-R_m turbulence with a magnetic field imposed in the z direction. From Zhdankin et al. (2014), courtesy of Stas Boldyrev. The image on the right is a horizontal section through the flow.

The ultimate fate of such turbulence is then determined by minimising energy subject to the conservation of h_m. In bounded domains this presents us with a well-defined variational problem, in which $\mathbf{u} = 0$ and the magnetic energy is minimised subject to the constraint of h_m = constant. In §5.4.2 we saw that such a minimisation leads to the force-free magnetic field $\nabla \times \mathbf{B} = \alpha\mathbf{B}$, where α^{-1} is of the order of the domain size. Thus freely decaying, helical turbulence in a closed domain evolves towards a large-scale, force-free, magnetostatic field. This is known as *Taylor relaxation* after the seminal paper by J.B. Taylor (1974).

9.6.3 Dynamic Alignment and Alfvénic States

We now turn to the opposite case, in which the initial magnetic helicity is small but the cross helicity is large. In the light of (9.62) it is natural to ask if the rate of change of h_c can be similarly bounded from above. Using the inequality $\langle \mathbf{J} \cdot \boldsymbol{\omega} \rangle^2 \leq \langle \mathbf{J}^2 \rangle \langle \boldsymbol{\omega}^2 \rangle$, (9.58) yields

$$\sqrt{\rho/\mu}\left|\frac{d}{dt}\langle \mathbf{u} \cdot \mathbf{B} \rangle\right| \leq \sqrt{\rho\mu}(\lambda + \nu)\sqrt{\sigma|\dot{E}|}\,\sqrt{|\dot{E}|/\rho\nu} = \frac{(\lambda + \nu)}{\sqrt{\lambda\nu}}\,|\dot{E}|. \qquad (9.65)$$

This time, however, letting λ and ν tend to zero does not restrict the rate of change of helicity, $\langle \mathbf{u} \cdot \mathbf{B} \rangle$, so the bound is of little value. Nevertheless, (9.58) tells us that if $\mathbf{J}$ and $\boldsymbol{\omega}$ are only weakly correlated then $\langle \mathbf{u} \cdot \mathbf{B} \rangle$ will be approximately conserved, and so everything hinges on the degree of correlation of $\mathbf{J}$ and $\boldsymbol{\omega}$. It turns out that, provided R_m is large, numerical experiments show that the magnetic helicity does indeed decay much more slowly than the energy (Stribling and Matthaeus, 1991). The turbulence then tends to minimises its energy subject to the constraint of conservation of h_c.

Let us try to predict the end state of such a process. Let $\mathbf{h} = \mathbf{B}/\sqrt{\rho\mu}$ be the scaled magnetic field. We wish to minimise $E/|h_c|$, which is equivalent to maximising $|\langle \mathbf{u} \cdot \mathbf{h} \rangle|/\frac{1}{2}\langle u^2 + h^2 \rangle$. Since a geometric mean is less than or equal to an arithmetic mean, the Schwarz inequality yields

$$|\langle \mathbf{u} \cdot \mathbf{h} \rangle| \le \langle u^2 \rangle^{1/2} \langle h^2 \rangle^{1/2} \le \frac{1}{2}\left[\langle u^2 \rangle + \langle h^2 \rangle\right], \qquad (9.66)$$

with the equality holding only when $\mathbf{u} = \pm\mathbf{h}$. Evidently, $|\langle \mathbf{u} \cdot \mathbf{h} \rangle|/\frac{1}{2}\langle u^2 + h^2 \rangle$ is a maximum, and hence $E/|h_c|$ a minimum, when $\mathbf{u} = \mathbf{h}$, or else $\mathbf{u} = -\mathbf{h}$. In short, we expect MHD turbulence which has a relatively large amount of initial cross helicity to tend towards a state in which $\mathbf{u}$ and $\mathbf{h}$ are aligned. Such an alignment of the two fields is called an *Alfvénic state*, since it corresponds to a sea of finite-amplitude Alfvén waves, as discussed in §7.3.

In summary, then, there are two distinct states to which freely decaying MHD turbulence might relax, and which path the turbulence takes depends on the initial conditions. A strong initial magnetic helicity and a weak cross helicity favour Taylor relaxation to a the force-free magnetostatic field, $\nabla \times \mathbf{B} = \alpha\mathbf{B}$, $\mathbf{u} = 0$. On the other hand, a large initial cross helicity tend to yield the Alfvénic state $\mathbf{u} = \mathbf{h}$ or $\mathbf{u} = -\mathbf{h}$. (When the magnetic helicity and cross helicity are both strong, a more complex situation arises, as discussed in Stribling and Matthaeus, 1991.)

The usual rationalisation for the emergence of an Alfvénic state in the presence of a strong cross helicity involves the collision of finite-amplitude Alfvén waves, as discussed in §7.3. This is called *dynamic alignment* and it is best explained for the case where there is a mean magnetic field, $\mathbf{B} = \mathbf{B}_0 + \mathbf{b}$. So let us consider the collision of oppositely travelling Alfvén wave packets of finite amplitude (Figure 9.8).

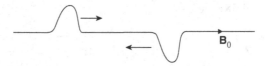

Figure 9.8 Finite-amplitude Alfvén waves travel without change of shape until they collide with oppositely travelling waves.

For an ideal fluid these wave packets take the form

$$\mathbf{u} = \mathbf{V}^F(\mathbf{x} - \mathbf{h}_0 t), \quad \mathbf{h} = \mathbf{h}_0 - \mathbf{V}^F(\mathbf{x} - \mathbf{h}_0 t), \qquad (9.67)$$

and

$$\mathbf{u} = \mathbf{V}^B(\mathbf{x} + \mathbf{h}_0 t), \quad \mathbf{h} = \mathbf{h}_0 + \mathbf{V}^B(\mathbf{x} + \mathbf{h}_0 t), \qquad (9.68)$$

where $\mathbf{V}^F$ and $\mathbf{V}^B$ are arbitrary functions of their arguments, and the superscripts F and B stand for forward- and backward-travelling waves, respectively. In terms of the Elsasser fields, $\hat{\mathbf{v}}^\pm = \mathbf{u} \pm (\mathbf{h} - \mathbf{h}_0)$, we have $\hat{\mathbf{v}}^+ = 2\mathbf{V}^B$ and $\hat{\mathbf{v}}^- = 2\mathbf{V}^F$.

Such wave packets evolve according to (7.17), whose most important feature is that the non-linear term, $(\hat{\mathbf{v}}^\mp \cdot \nabla)\hat{\mathbf{v}}^\pm$, involves only waves travelling in opposite directions. That is, non-linear interactions are restricted to oppositely travelling wave packets as illustrated in Figure 9.9. When two such wave packets collide,

Figure 9.9 Colliding Alfvén wave packets. In the top image the wave packets in the far field are propagating from right to left towards the near field, and there is another wave packet in the near field that propagates from left to right. Before the collision the wave packets are relatively smooth. The bottom image shows the collision. Broadband turbulence develops and high harmonics are now evident in the field lines. (Courtesy of Jean Perez and Stas Boldyrev.)

(7.17) tells us that, in an ideal fluid, $\int_{V_\infty}(\hat{\mathbf{v}}^+)^2 dV$ and $\int_{V_\infty}(\hat{\mathbf{v}}^-)^2 dV$ are both conserved throughout the collision. As a consequence, the wave packets emerge from the collision with their initial values of energy and cross helicity intact (see the discussion in §7.3), although both packets are distorted during the collision by the non-linear interactions.

Let us now consider the consequences of introducing a small but finite magnetic diffusivity, and of allowing for multiple collisions. Suppose that there exists a random sea of such wave packets travelling backward and forward along the mean field, $\mathbf{B}_0$, with $|\mathbf{b}| \ll |\mathbf{B}_0|$. The wave packets experience multiple collisions and as a consequence their energy becomes increasingly spread across a wider range of scales, with some of the energy passed to smaller scales where it is eradicated by Ohmic dissipation within thin-current sheets. If the non-linear interactions are weak, then multiple collisions are required for the energy to reach the scale at which Ohmic dissipation occurs. During a single collision, the forward- and backward-travelling wave packets distort each other through the non-linear term $(\hat{\mathbf{v}}^\mp \cdot \nabla)\hat{\mathbf{v}}^\pm$ by an amount

$$|\delta\hat{\mathbf{v}}^\pm| \sim |(\hat{\mathbf{v}}^\mp \cdot \nabla)\hat{\mathbf{v}}^\pm| \times \text{(packet interaction time)} \sim |(\hat{\mathbf{v}}^\mp \cdot \nabla)\hat{\mathbf{v}}^\pm| \frac{\ell}{v_a}, \quad (9.69)$$

where $v_a = |v_a|$ is the Alfvén wave speed and ℓ is the typical scale of the colliding wave packets. Thus we have the estimate

$$|\delta\hat{\mathbf{v}}^\pm| \sim \frac{|\hat{\mathbf{v}}^\mp||\hat{\mathbf{v}}^\pm|}{\ell} \frac{\ell}{v_a} = \frac{|\hat{\mathbf{v}}^\mp||\hat{\mathbf{v}}^\pm|}{v_a}. \quad (9.70)$$

The key point about (9.70) is that the fractional change in $|\hat{\mathbf{v}}^\pm|$ per collision, $|\delta\hat{\mathbf{v}}^\pm|/|\hat{\mathbf{v}}^\pm|$, is proportional to $|\hat{\mathbf{v}}^\mp|/v_a$, and so the weaker of the two wave packets is more severely distorted in each collision. So, if there is a local excess of energy in, say, the forward-travelling waves at some particular location, we might expect those waves to progressively eradicate the backward-travelling disturbances until all the wave packets travel in the same direction. This is the basis of the principle of *dynamic alignment*. It asserts that any local bias towards forward- or backward-travelling waves will be progressively reinforced until all the waves in a given region travel in the same direction. At this point the non-linear interactions shut down, and we reach a local state in which $\mathbf{u} = (\mathbf{h} - \mathbf{h}_0)$ or else $\mathbf{u} = -(\mathbf{h} - \mathbf{h}_0)$. This is the Alfvénic state predicted by minimising energy, subject to the conservation of cross helicity.

Part III

Applications in Engineering and Materials

Maxwell, like every other pioneer who does not live to explore the country he opened out, had not had time to investigate the most direct means of access to the country, or the most systematic way of exploring it. This has been reserved for Oliver Heaviside to do. Maxwell's treatise is cumbered with the debris of his brilliant lines of assault... Oliver Heaviside has cleared those away, has opened up a direct route, has made a broad road, and has explored a considerable tract of country.

George FitzGerald, *1893*

Oliver Heaviside assails the 'Cambridge mathematicians'... As to Heaviside and the mathematicians, the 'Heaviside Calculus' of a decade ago has become the 'Theory of Laplace Transforms', and the 'Heaviside Layer' has acquired the more scientific name of 'Ionosphere'.

Den Hartog, *1948*

Self-taught and stubbornly independent, the brilliant scientist Oliver Heaviside was nothing if not colourful. Famous for his battles with the Cambridge mathematician P.G. Tait over the introduction of vector analysis (Tait preferred quaternions, Heaviside vector calculus), he was also one of the early pioneers of electromagnetism. Indeed, it was Heaviside who first clearly identified, from the myriad of detail in Maxwell's treatise, the key role played by the four 'Maxwell equations'. He was also an intensely practical man, fascinated by the growing discipline of electrical engineering.

Over the years, a great variety of potential applications of MHD have captured the imagination of engineers and applied scientists, sometimes to great effect and sometimes not. Undoubtedly the greatest impact of MHD in engineering has been in the metallurgical industries. In part this is because the interaction of intense magnetic fields with molten metal is an intrinsic feature of many metallurgical processes, such as in the plasma-arc melting of specialist aerospace alloys, or in the electrolytic conversion of bauxite to aluminium, which is a massive worldwide consumer of electrical power. However, MHD is also important because magnetic fields offer the unique ability to reach into the interior of a molten metal and control its motion. For many years the steel industry has taken advantage of this in a wide variety of casting operations, sometimes to stir up the solidifying metal in order to refine its crystal structure, and sometimes to suppress turbulence in the liquid in order to avoid the entrainment of pollutants from the surface of the casting. The four chapters that follow provide a partial overview of these metallurgical applications. We hope that Heaviside would have approved.

10

The World of Metallurgical MHD

10.1 The History of Electrometallurgy

When Faraday first made public his remarkable discovery that a magnetic flux produces an EMF, he was asked, 'What use is it?' His answer was: 'What use is a new-born baby?' Yet think of the tremendous practical applications his discovery has led to... Modern electrical technology began with Faraday's discoveries. The useless baby developed into a prodigy and changed the face of the earth in ways its proud father could never have imagined.

R.P. Feynman, 1964

There were two revolutions in the application of electricity to industrial metallurgy. The first, which occurred towards the end of the nineteenth century, was a direct consequence of Faraday's discoveries. The second took place around 80 years later. We start with Faraday.

The discovery of electromagnetic induction revolutionised almost all of nineteenth-century industry, and none more so than the metallurgical industries. For example, until 1854, aluminium could be produced from alumina only in small batches by various chemical means. The arrival of the dynamo transformed everything, sweeping aside those inefficient chemical processes. At last it was possible to produce aluminium continuously by electrolysis. Robert Bunsen – he of the 'burner' fame (though actually it was Faraday who invented the burner) – was the first to experiment with this method in 1854. By the 1880s the technique had been refined into a commercial process by Charles Martin Hall and Paul Héroult, a process which is little changed today (see Figure 10.1).

In the steel industry, electric furnaces for melting and alloying iron began to appear around 1900. There were two types: arc furnaces and induction furnaces (Figure 10.2). Industrial-scale arc furnaces made an appearance as early as 1903. (The first small-scale furnaces were designed by von Siemens in 1878.) These used

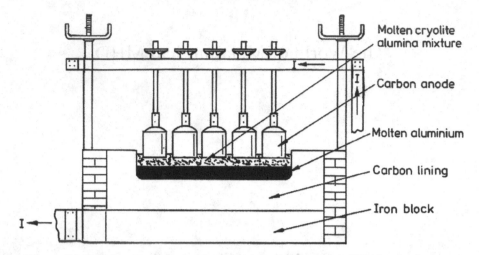

Molten cryolite alumina mixture

Carbon anode

Molten aluminium

Carbon lining

Iron block

Figure 10.1 Early twentieth-century aluminium reduction cell (side view).

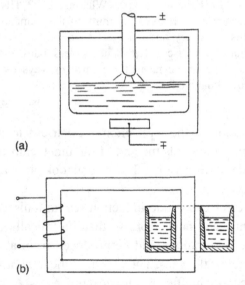

(a)

(b)

Figure 10.2 (a) An early arc-furnace. (b) An early induction furnace.

an electric arc, which was made to play on the molten metal surface as a means of heating the metal. Modern vacuum-arc remelting furnaces are a direct descendant of this technology. The first induction furnace, which used an AC magnetic field (rather than an arc) to heat the steel, was designed by Ferranti in 1887. Shortly thereafter, commercial induction furnaces became operational in the United States. Thus, by the turn of the century, electromagnetic fields were already an integral part of industrial metallurgy. However, their use was restricted essentially to heating and melting and to electrolysis. The next big step, which was the application of electromagnetic fields to casting, was not to come for another 80 years or so.

The great flurry of activity and innovation in electrometallurgy which began at the end of the nineteenth century progressively gave way to a process of consolidation throughout much of the twentieth. Things began to change, however, in the 1970s and 1980s. The steel industry was revolutionised by the commercial development of continuous casting, which displaced traditional batch casting methods. Around the same time the worldwide growth in steelmaking, particularly in Japan, increased international competition, while successive oil crises in the 1970s focused attention on the growing cost of energy. Once again, the time was ripe for innovative technologies.

There was another reason for change. The aerospace industry was making increasing demands on quality. A single, microscopic, non-metallic particle trapped in, say, a turbine blade can lead to a fatigue crack and perhaps ultimately to the catastrophic failure of an aircraft engine. New techniques were needed to control and monitor the level of non-metallic inclusions in castings.

Metallurgists set about rethinking and redesigning traditional casting and melting processes. However, increasing demands on cost, purity and control meant that traditional methods were no longer adequate. Just like their predecessors a century earlier, they found an unexpected ally in the electromagnetic field, and myriad electromagnetic technologies emerged. Metallurgical MHD, which had been sitting in the wings since the turn of the century, suddenly found itself centre-stage.

Magnetic fields provide a versatile, non-intrusive means of controlling the motion of liquid metals. They can repel liquid-metal surfaces, dampen unwanted motion and excite motion in otherwise still liquid. In the 1970s, metallurgists began to recognise the versatility of the Lorentz force and magnetic fields are now routinely used to heat, pump, stir, stabilise, repel and levitate liquid metals.

Metallurgical applications of MHD represent a union of two very different technologies – industrial metallurgy and electrical engineering – and it is intriguing to note that Faraday was a major contributor to both. It will come as no surprise to learn that, on Christmas Day 1821, Faraday built the first primitive electric motor, and of course his discovery of electromagnetic induction (in 1831) marked the beginning of modern electrical technology. However, Faraday's contributions to metallurgy are, perhaps, less well known. Not only did his researches into electrolysis help pave the way for modern aluminium production, but his work on alloy steels, which began in 1819, was well ahead of his time. In fact, as far back as 1820 he was making razors from a non-rusting platinum steel as gifts for his friends. As noted by Robert Hadfield,

Faraday was undoubtedly the pioneer of research on special alloy steel; and had he not been so much in advance of his time in regard to general metallurgical knowledge and industrial requirements his work would almost certainly have led immediately to practical development.

It is interesting to speculate how Faraday would have regarded the fusion of two of his favourite subjects – chemistry and electromagnetism – in a single endeavour.

In any event, that unlikely union of sciences has indeed occurred. The phrase *metallurgical MHD* was probably coined at an IUTAM conference held in Cambridge in 1982 (Moffatt and Proctor, 1984). In the years immediately preceding this conference, magnetic fields were beginning to make their mark in casting, but those applications which did exist seemed rather disparate. This conference forged a science from these diverse, embryonic technologies, and three decades later we have a reasonably complete understanding of these complex processes. Unfortunately, much of this research has failed to find its way into textbooks and monographs, but rather is scattered across various conference proceedings and journal papers. One of the purposes of Part III of this book, therefore, is to give some sense of the breadth of the industrial developments and of our attempts to understand and quantify these complex flows.

10.2 An Overview of the Role of Magnetic Fields in Materials Processing

In order to give a flavour of the kind of applications that MHD has found in materials processing, we now briefly consider five representative examples. These are

(i) *magnetic stirring* of a partially solidified ingot by a rotating magnetic field
(ii) the *magnetic damping* of jets and vortices in castings
(iii) motion arising from the *injection of current* into a liquid-metal pool
(iv) *interfacial instabilities* which arise when a current is passed between two conducting fluids, particularly in aluminium reduction cells
(v) the *magnetic levitation and heating* induced by high-frequency magnetic fields.

The first four of these topics will be considered in detail in Chapters 11–13, while the fifth topic is discussed at some length in Moffatt and Proctor (1984).

The hallmark of all these processes is that R_m is invariably very small. Consequently, Part III of this text rests heavily on the material of Chapter 6. Although these five processes may be unfamiliar in the metallurgical context, they all have simple mechanical analogues, each of which would have been familiar to Faraday.

- *Magnetic stirring* (the first topic) is nothing more than a form of induction motor, where the liquid metal takes the place of the rotor.
- *Magnetic damping* takes advantage of the fact that the relative motion between a conductor and a magnetic field tends to induce a current in the conductor whose Lorentz force opposes the relative motion. This is the second of our topics.
- *Current injected* into a metal block causes that block to pinch in on itself (parallel currents attract each other) and the same is true if current passes through a liquid-metal pool. Sometimes the pinch forces caused by the injection of current can be balanced by fluid pressure; at other times they induce motion in the pool.
- The *magnetic levitation* of small metallic objects is also quite familiar. It relies on the fact that an induction coil carrying a high-frequency current will tend to induce opposing currents in any adjacent conductor. Opposite currents repel each other and so the conductor is repelled by the induction coil. What is true of solids is also true of liquids. Thus a 'basket' composed of a high-frequency induction coil can be used to levitate liquid-metal droplets.

Let us now examine each of these processes in a little more detail, placing them in a metallurgical context. *Magnetic stirring* is the name given to the generation of a swirling flow by a rotating magnetic field (see Figure 10.3). This is routinely used in casting operations to homogenise the liquid zone of a partially solidified ingot. In effect, the liquid metal acts as the rotor of an induction motor, and the resulting motion has a profound influence on the metallurgical structure of the ingot, produc-ing a fine-grained product with little or no porosity.

From the perspective of a fluid dynamicist, this turns out to be a study in Ekman pumping. That is, Ekman layers form on the boundaries, and the resulting Ekman pumping (a secondary, poloidal motion which is superimposed on the primary swirling flow) is the primary mechanism by which heat, chemical species and

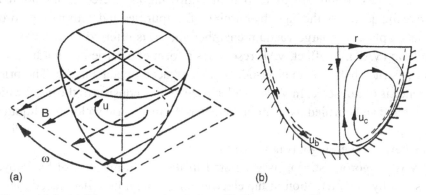

Figure 10.3 (a) Electromagnetic stirring. (b) Ekman pumping.

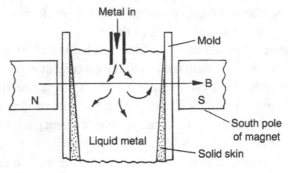

Figure 10.4 Magnetic damping.

angular momentum are redistributed within the pool. Magnetic stirring is discussed in Chapter 11.

In contrast, magnetic fields are used in other casting operations to suppress unwanted motion. Here we take advantage of the ability of a static magnetic field to convert kinetic energy into heat via Joule dissipation. This is commonly used, for example, to suppress turbulence within submerged jets which feed casting moulds. If unchecked, this turbulence can disrupt the free surface of the liquid, leading to the entrainment of oxides or other contaminants from the surface (see Figure 10.4). It turns out, however, that although the Lorentz force associated with a static magnetic field destroys kinetic energy, it cannot create or destroy linear or angular momentum. A study of *magnetic damping*, therefore, often comes down to the question, 'How does a flow manage to dissipate kinetic energy while preserving its linear and angular momentum?' The answer to this question furnishes a great deal of useful information, and we look at the damping of jets and vortices in Chapter 11.

In yet other metallurgical processes, an intense DC current is used to fuse metal. An obvious (small-scale) example of this is electric welding. At a much larger scale, intense currents are used to melt entire ingots. Here the intention is to improve the quality of the ingot by burning off impurities and eliminating porosity. This takes place in a large vacuum chamber and so is referred to as Vacuum-Arc Remelting (VAR). In effect, VAR resembles a form of electric-arc welding, where an arc is struck between an electrode and an adjacent metal surface. The primary difference is one of scale. In VAR the electrode, which consists of the ingot which is to be melted and purified, is ~1 m in diameter. As in electric welding, a liquid pool builds up beneath the electrode as it melts, and this pool eventually solidifies to form a new, cleaner, ingot (Figure 10.5).

In VAR, vigorous stirring is generated in the liquid-metal pool by buoyancy forces and by the interaction of the electric current with its self-induced magnetic

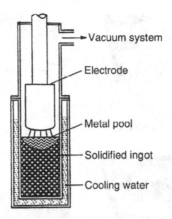

Figure 10.5 Vacuum-arc remelting (VAR).

field. This motion, which has a major influence on the liquid-metal temperature distribution, and hence on the metallurgical structure of the recast material, is still only partly understood. It appears that there is delicate balance between the Lorentz forces, which tend to drive a poloidal flow which converges at the surface, and the buoyancy forces associated with the relatively cool solidification front. (The buoyancy-driven motion is also poloidal, but opposite in direction to the Lorentz-driven flow.) Modest changes in current can transform the motion from a buoyancy-dominated flow, which diverges at the surface, to a Lorentz-dominated motion. This change in flow regime is accompanied by a dramatic change in temperature distribution and of ingot structure (Figure 10.6). This is discussed at length in Chapter 12.

We now give a brief account of an intriguing and unusual form of instability which has bedevilled the aluminium industry for many decades. As we shall see, the solution to this problem is finally in sight and the potential for energy savings is enormous. The instability arises in electrolysis cells which are used to reduce alumina to aluminium. These cells consist of broad but shallow layers of electrolyte and liquid aluminium, with the electrolyte (called cryolite) lying on top. Alumina is dissolved in the cryolite and a large current, perhaps 300 k Amps, passes vertically downward through the two liquid layers, continually reducing the oxide to metal (see Figure 10.7). Small droplets of aluminium form in the electrolyte layer, droplets which then sediment out to feed the liquid aluminium pool below.

The process is highly energy-intensive, largely because of the high electrical resistance of the electrolyte. For example, in the USA, around 5 per cent of all generated electricity is used for aluminium production. It has long been known that stray magnetic fields can destabilise the aluminium-electrolyte interface, in effect, by amplifying interfacial gravity waves. In order to avoid this instability the

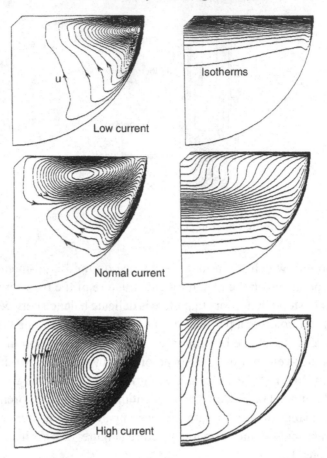

Figure 10.6 Changes in flow pattern (left) and temperature field (right) with current in VAR. Buoyancy-driven flow dominates for low currents and Lorentz forces for high currents. Most commercial units operate at the transition point from one flow regime to the other.

electrolyte layer must be maintained at a thickness above some critical threshold, and this carries with it a severe energy penalty.

This instability has been the subject of intense research for over forty years, and the underlying mechanisms have finally been identified. Of course, with hindsight, they turn out to be simple. The instability depends on the fact that the interface can support interfacial gravity waves. A tilting of the interface causes a perturbation in current, $\mathbf{j}$, as shown in Figure 10.8. Excess current is drawn from the anode at points where the highly resistive layer of electrolyte thins, and less current is drawn where the layer thickens. The resulting perturbation in current shorts through the highly conducting aluminium layer, leading to a large horizontal current in the aluminium. This, in turn, interacts with the

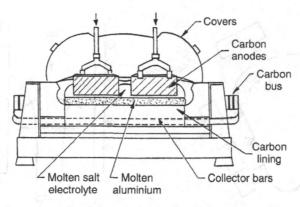

Figure 10.7 A modern aluminium reduction cell (end view).

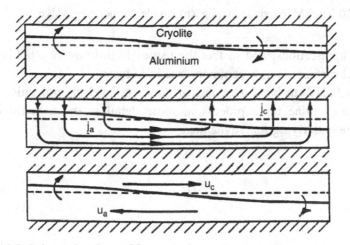

Figure 10.8 Schematic of unstable waves in a reduction cell.

vertical component of the background magnetic field to produce a Lorentz force which is directed into the page. It is readily confirmed that two such sloshing motions which are mutually perpendicular can feed on each other, the Lorentz force from one driving the motion of the other. The result is an instability. This instability is discussed at length in Chapter 13.

A quite different application of MHD in metallurgy is *magnetic levitation*. This relies on the fact that a high-frequency induction coil will repel any adjacent conducting material by inducing opposing currents in the adjacent conductor. (Opposite currents repel each other.) Thus a 'basket' formed from a high-frequency induction coil can be used to levitate and melt highly reactive metals (Figure 10.9a), or a high-frequency solenoid can be used to form a magnetic valve which modulates the flow of a liquid-metal jet (see Figure 10.9b). There are many

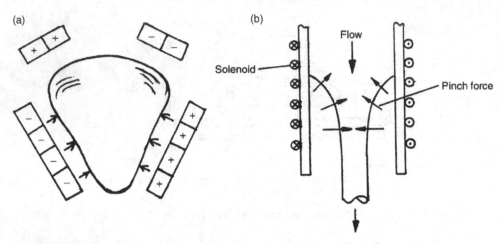

Figure 10.9 (a) Magnetic levitation. (b) An electromagnetic valve.

other such applications, and in fact the use of high-frequency fields to support liquid-metal surfaces is now commonplace in industry.

This concludes our brief overview of the metallurgical applications of MHD. We shall re-examine these process in some detail in the next three chapters, exposing the underlying fluid dynamics.

11

The Generation and Suppression of Motion in Castings

> It will emerge from dark and gloomy caverns, casting all human races
> into great anxiety, peril and death. It will take away the lives of many;
> with this men will torment each other with many artifices, traductions and
> treasons. O monstrous creature, how much better it would be if you were
> to return to hell.
>
> *(Leonardo da Vinci on the extraction and casting of metals.)*

11.1 Magnetic Stirring Using Rotating Fields

Liquid metals freeze in much the same way as water. First, snowflake-like crystals form, and as these multiply and grow a solid emerges. This solid, however, can be far from homogeneous. Just as a chef preparing ice-cream has to beat and stir the partially solidified cream to break up the crystals and release any trapped gas, so many steelmakers have to stir partially solidified ingots to ensure a fine-grained, homogeneous product. The preferred method of stirring is electromagnetic, and has been dubbed the 'electromagnetic teaspoon'. We shall describe this process shortly. First, however, it is necessary to say a little about commercial casting operations.

11.1.1 Casting, Stirring and Metallurgy

Man has been casting metals for quite some time. Iron blades, perhaps 5000 years old, have been found in Egyptian pyramids, and by 1000 B.C. we find Homer mentioning the working and hardening of steel blades. Until relatively recently, all metal was cast by a batch process involving pouring the melt into closed moulds. However, today the bulk of aluminium and steel is cast in a continuous fashion as indicated in Figure 11.1. In brief, a solid ingot is slowly withdrawn from a liquid-metal pool, the pool being continuously replenished from above. In the case of steel, which has a low thermal diffusivity, the pool is long and deep, resembling a long liquid-metal column. For aluminium, however, the pool is roughly

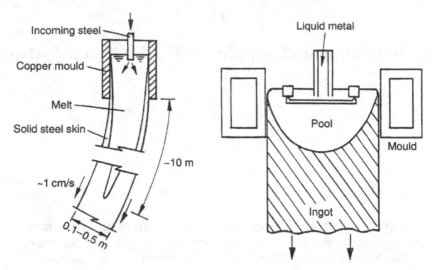

Figure 11.1 (a) Continuous casting of steel. (b) Continuous casting of aluminium.

hemispherical in shape, perhaps 0.5 m in diameter. Casting speeds are of the order of a few mm/s.

Unfortunately, ingots cast in this manner are far from homogeneous. For example, during solidification alloying elements tend to segregate out of the host material, giving rise to inhomogeneities in the final structure. This is referred to as macro-segregation. (Macro-segregation was a recognised problem in casting as far back as 1540, when Biringuccio noted problems arising from macro-segregation in the production of gun barrels.) Moreover, small cavities can form on the ingot surface or near the centre-line. Surface cavities are referred to as blow holes or pin holes and arise from the formation of gas bubbles (CO or N_2 in the case of steel). Centre-line porosity, on the other hand, is associated with shrinkage of the metal during freezing.

All of these defects can be alleviated, to some degree, by stirring the liquid pool (Birat and Chone, 1992, Takeuchi et al., 1992). This is most readily achieved using a rotating magnetic field, as shown in Figure 11.2. Stirring has the added benefit of promoting the nucleation and growth of *equi-axed crystals* (crystals like snow-flakes) at the expense of *dendritic crystals* (those like fir-trees) which are aniso-tropic and generally undesirable (Figure 11.3). In addition to these metallurgical benefits, it has been found that stirring has a number of incidental operational advantages, such as allowing higher casting temperatures and faster casting rates (Marr, 1984).

The perceived advantages of magnetic stirring led to a widespread implementation of this technology in the 1980s, particularly in the steel industry. In fact, by

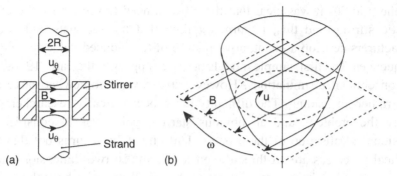

Figure 11.2 (a) The magnetic stirring of steel. (b) The magnetic stirring of aluminium.

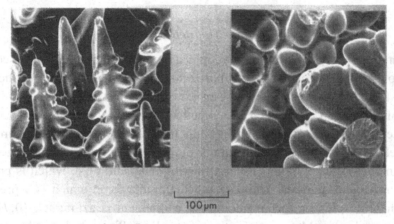

Figure 11.3 A comparison of steel crystals taken from unstirred (left) and stirred (right) castings. The unstirred casting has a classical dendritic structure.

1985, some 20 per cent of slab casters (casters producing steel ingots with a wide cross-section) and 50 per cent of bloom casters (casters producing medium-sized steel ingots) had incorporated magnetic stirring. However, this was not the end of the story. While some manufacturers reported significant benefits, others encountered problems. For example, in steel-making excessive stirring can lead to the entrainment of oxides from the free surface and to a thinning of the solid steel shell at the base of the mould. This latter phenomenon is particularly dangerous as it can lead to a rupturing of the solid skin. Different problems were encountered in the aluminium industry. Here it was found that, in certain alloys, macrosegregation was aggravated (rather than reduced) by stirring, possibly because centrifugal forces tend to separate out crystal fragments of different density, and hence different composition.

By the mid-80s it was clear that there was a need to rationalise the effects of magnetic stirring and this, in turn, required that metallurgists and equipment manufacturers develop a quantitative picture of the induced velocity field and its consequences. The first, simple models began to appear in the early 1980s, usually based on computer simulations. However, these were somewhat naïve and the results rather misleading. The difficulty arose because early researchers tried to simplify the problem, and an obvious starting point was to consider a two-dimensional idealisation of the process. Unfortunately, it turns out that the key dynamical processes are all three-dimensional, and so two-dimensional idealisations of magnetic stirring are inadequate. We shall describe both the early two-dimensional models and their more realistic three-dimensional counterparts in the subsequent sections.

11.1.2 The Magnetic Teaspoon

There are many ways of inducing motion in a liquid metal pool. The most common means of stirring is to use a rotating, horizontal magnetic field, an idea which dates back to 1932. The field acts rather like an induction motor, with the liquid taking the place of the rotor (Figure 11.4). In practice, a rotating magnetic field may be generated in a variety of ways, each producing a different spatial structure for **B**. (The field is never perfectly uniform nor purely horizontal.) However, the details do not matter. The key point is that a rotating magnetic field, which is predominantly horizontal, induces a time-averaged Lorentz force which is a prescribed function of position and whose dominant component is azimuthal; $(0, F_\theta, 0)$ in (r, θ, z) coordinates. Moreover, friction is usually sufficiently large that the swirl induced in the liquid is much slower than the rate of rotation of the magnetic field, and so the Lorentz force is independent of the velocity of the metal.

Perhaps the important questions are (i) 'How does the induced velocity scale with the Lorentz force or magnetic field strength?', (ii) 'Are there significant secondary flows $(u_r, 0, u_z)$?' and (iii) 'Does the induced swirl $(0, u_\theta, 0)$ have

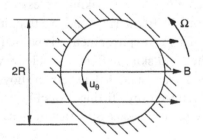

Figure 11.4 A one-dimensional model of stirring.

a spatial structure which mimics the spatial variations of the applied Lorentz force (i.e. strong swirl in regions where F_θ is intense and weak swirl where F_θ is low)?' To cut a long story short, the answers to these questions turn out to be

(i) $u_\theta \sim B$.
(ii) The secondary flows are intense and dominate the dynamics of the liquid metal.
(iii) u_θ does *not* mimic the spatial variations in the Lorentz force.

It is the subtle, yet critical, role played by the secondary flows which makes this problem more interesting from a fluids perspective than it might otherwise be.

The first step in predicting the intensity and distribution of the induced swirl is to determine the Lorentz force. Fortuitously, the magnitude and distribution of the time-averaged Lorentz force is readily calculated. There are two reasons for this. First, the magnetic field associated with the current induced in the liquid metal is almost always negligible by comparison with the imposed field, **B**. That is, in a frame of reference rotating with the applied field, we may apply the low R_m approximation. (See, for example, Davidson and Hunt, 1987.) Faraday's law then gives, in the laboratory frame,

$$\nabla \times \mathbf{E} = -\partial \mathbf{B}_0/\partial t, \tag{11.1}$$

where $\mathbf{B}_0$ is the known, imposed magnetic field. Second, the induced velocities are generally so low (by comparison with the rate of rotation of the **B**-field) that Ohm's law reduces to $\mathbf{J} = \sigma\mathbf{E}$. Consequently, **E** (and hence **J**) may be found directly by uncurling (11.1) and the Lorentz force follows. In fact, we have already seen an example of just such a calculation in §6.4.1. Here we evaluated the time-averaged Lorentz force generated by a perfectly uniform magnetic field rotating about an infinitely long, liquid-metal column. The time-averaged force is (see Equation (6.58))

$$\mathbf{F} = \frac{1}{2}\sigma B^2 \Omega r \hat{\mathbf{e}}_\theta, \tag{11.2}$$

where Ω is the field rotation rate, r is the radial coordinate and we use (r, θ, z) coordinates with an origin at the centre of the liquid-metal column. The restrictions on this expression are

$$u_\theta \ll \Omega R \leq \lambda/R , \qquad \lambda = (\mu\sigma)^{-1}, \tag{11.3}$$

where R is the radius of the column. However, these conditions are almost always satisfied in practice. The first inequality, $u_\theta \ll \Omega R$, is precisely the condition required to ignore $\mathbf{u} \times \mathbf{B}$ in Ohm's law, while the second, $\Omega R \leq \lambda/R$, is equivalent to saying that R_m (based on ΩR) is small, so that the induced magnetic field is negligible.

Of course, for more complicated spatial distributions of $\mathbf{B}$ we cannot use (11.2). Nevertheless, for any rotating field with a predominantly one-dimensional character (as in Figure 11.4), the Lorentz force is primarily azimuthal and on dimensional grounds it must be of order $\sigma B^2 \Omega R$ (provided that (11.3) is satisfied). Moreover, for fields which are symmetric about a vertical plane through the origin, the Lorentz force must vanish on the axis. It follows that rotating, symmetric magnetic fields which satisfy (11.3) will induce a time-averaged force of the form

$$\mathbf{F} = \left[\tfrac{1}{2} \sigma \Omega B_0^2 r \right] f(r/R, z/R) \, \hat{\mathbf{e}}_\theta. \tag{11.4}$$

Here B_0 is some characteristic field strength and f is a function of order unity whose spatial distribution depends on that of $\mathbf{B}$ and whose exact form can be determined by uncurling (11.1). When $\mathbf{B}$ is uniform, (11.2) tells us that $f = 1$. We now consider the dynamical consequences of this force.

11.1.3 Simple Models of Stirring

The earliest attempts to quantify magnetic stirring consisted of taking a transverse slice through the problem. That is, the axial variations of $\mathbf{F}$ in (11.4) were neglected, the sides of the pool were considered to be vertical, and boundary layers at the top or the bottom of the pool were ignored. In effect, this represents uniform stirring of a long, deep, circular column of radius R. Although this is a natural first step, it turns out that this idealisation is quite misleading, as we shall see.

These z-independent models are characterised by the fact that $\mathbf{F}$ drives a one-dimensional swirl flow $u_\theta(r)$. There are no mean inertial forces, so rings of fluid simply slide over each other like onion rings, driven by F_θ and resisted by laminar and turbulent shear stresses (Figure 11.5). The Reynolds-averaged Navier–Stokes equation reduces to a balance between the applied Lorentz force and azimuthal shear:

$$\tau_{r\theta} = \rho \nu r \frac{d}{dr} \left(\frac{\bar{u}_\theta}{r} \right) - \rho \overline{u'_r u'_\theta} = -r^{-2} \int_0^r F_\theta r^2 \, dr. \tag{11.5}$$

Here ν is the viscosity, $\mathbf{u}'$ represents the fluctuating component of velocity, and the overbar denotes a time average. In fact, we have already met this problem in §6.4.1, where we used a simple mixing length model to estimate the Reynolds stress $\overline{u'_r u'_\theta}$.

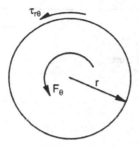

Figure 11.5 When the induced swirl is one-dimensional, of the form u_θ (r), the torque induced by the Lorentz force is balanced by shear stresses.

Integration of (11.5) is then straightforward. The results are best expressed in terms of a quantity Ω_f, defined by

$$\Omega_f^2 = \sigma\Omega B^2/\rho. \tag{11.6}$$

When **B** is uniform and $f = 1$, Equation (11.5) yields

$$\Omega_0 = (\overline{u}_\theta/r)_{r=0} = \Omega_f[\Omega_f R^2/16\nu], \tag{11.7}$$
$$\text{(laminar flow)}$$

and

$$\Omega_0 = (\overline{u}_\theta/r)_{r=0} = \Omega_f\left\{\frac{1}{2\sqrt{2}\kappa}\ln\left(\frac{\Omega_f R^2}{\nu}\right) + 1.0\right\},$$

$$\text{(turbulent flow, smooth boundary)} \tag{11.8}$$

where $\kappa = 0.4$ is Kármán's constant. These correspond to (6.60) and (6.62), respectively. When the surface at $r = R$ is rough and dendritic, rather than smooth, the mixing length estimate of $\overline{u'_r u'_\theta}$ must be modified slightly. The required modification is well known in hydraulics and it turns out that, if k^* is the typical roughness height, then (11.8) becomes

$$\Omega_0 = (\overline{u}_\theta/r)_{r=0} = \Omega_f\left\{\frac{1}{2\sqrt{2}\kappa}\ln\left(\frac{R}{k^*}\right)\right\}$$

$$\text{(turbulent flow, rough boundary)}. \tag{11.9}$$

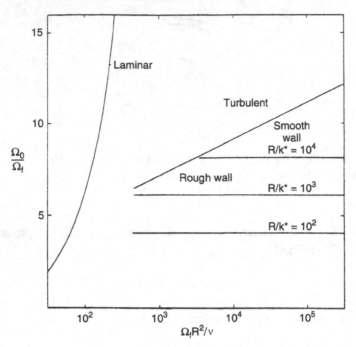

Figure 11.6 Induced centreline angular velocity as a function of Ω_f for a one-dimensional flow of the form $u_\theta\,(r)$.

Expressions (11.8) and (11.9) were first given by Davidson and Hunt (1987). Note that in a turbulent flow u_θ scales linearly with $|\mathbf{B}|$ (with a possible logarithmic correction), whereas in a laminar flow u_θ scales as B^2. These results are summarised in Figure 11.6.

If a different turbulence model is used, then slightly different estimates of the induced swirl are obtained. However, perhaps the more important point to note is that *all* predictions based on integrating (11.5) are substantially out of line with the experimental data, no matter what turbulence model is used (Davidson and Hunt, 1987; Davidson, 1992). The key point is that the force balance (11.5) relies on the time-averaged inertial forces being exactly zero. However, in practice, there are always significant secondary flows, $(u_r, 0, u_z)$, and this secondary motion ensures that the mean inertial forces are finite. Indeed, when Re is large, as it always is, we would expect the inertial forces to greatly exceed the shear stresses, except near the boundaries. Consequently, in the core of the flow, the local force balance should be between F_θ and mean inertia, not between F_θ and azimuthal shear. To obtain realistic predictions of the induced swirl we must embrace the three-dimensional nature of the problem, seek out the sources of secondary motion, and incorporate these into the analysis.

From an industrial perspective there are two distinct cases of particular interest. The first is where we have a long column of liquid, but with the stirring force, F_θ, applied to a relatively short portion of that column. This is relevant to the casting of steel (Figure 11.2a), and in this case the secondary motion arises from differential rotation induced along the length of the column. The second case is where the pool is as deep as it is broad, which is typical of aluminium casting (Figure 11.2b). Here the source of secondary motion is Ekman pumping on the base of the pool. We shall consider each of these in turn, starting with pools which are long and deep, i.e. with steel casting.

11.1.4 The Role of Secondary Flows in Steel Casting

The simplest idealisation of the magnetic stirring of steel is that shown in Figure 11.2(a). Here the fluid occupies a long, cylindrical column of circular cross-section, while the Lorentz force, F_θ, is applied over a relatively short portion of that column and is assumed to be axisymmetric. In such cases a secondary (poloidal) motion in the r–z plane is generated by differential rotation between the forced and unforced regions of the column, as we now show.

Suppose that the mean flow is axisymmetric. Taking the curl of the poloidal components of the Euler equation yields the inviscid azimuthal vorticity equation

$$\mathbf{u} \cdot \nabla(\omega_\theta/r) = \frac{\partial}{\partial z}\left(\frac{\Gamma^2}{r^4}\right), \quad \Gamma = ru_\theta. \qquad (11.10)$$

This tells us that axial gradients in swirl, u_θ, induce azimuthal vorticity, ω_θ, and hence a secondary flow in the r–z plane. (ω_θ is the curl of the poloidal velocity, $(u_r, 0, u_z)$, and so a finite ω_θ implies a finite poloidal recirculation.) Physically, we may picture this generation of azimuthal vorticity, and hence secondary flow, as a spiralling of the poloidal vortex lines by the axial gradients in swirl. That is, the poloidal vortex lines, which represent to the curl of the azimuthal velocity field, u_θ, are spiralled up whenever there is an axial gradient in u_θ, and this provides a constant source of azimuthal vorticity.

An alternative, and perhaps more helpful, interpretation of the secondary motion can be formulated in terms of pressure. The relatively low pressure on the axis of the more rapidly rotating fluid causes the magnetic stirrer to act like a centrifugal pump (Figure 11.7). Fluid is drawn in from the far field, moving along the axis towards the magnetic field. It enters the forced region, is spun up by the Lorentz force and is then thrown to the walls. Finally, the fluid spirals back down the solidification front where, eventually, it loses its excess kinetic energy and angular momentum through wall shear. In the steady state the fluid cannot return until its excess energy is lost, and it

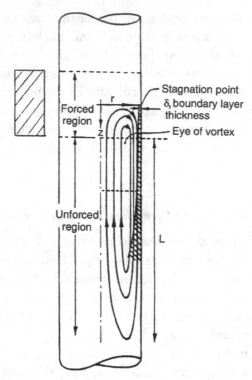

Figure 11.7 The secondary flow driven by the localised stirring of a long steel column. The motion is mirror symmetric about the mid-plane of the forced region, and only the flow in the lower half is shown.

takes a long time for the boundary layer on the outer cylindrical wall to dissipate this energy, essentially because the cross-stream diffusion of energy to the wall is a slow process. Consequently, this centrifugal pumping ensures that a very long portion of the liquid metal column is eventually set into rotation, of axial extent $L \sim u_\theta R^2 / \nu_t$, ν_t being an eddy viscosity (Davidson and Hunt, 1987). Thus, even though F_θ is restricted to a relatively short part of the liquid-metal column, a substantial length of that fluid column is eventually spun up, with the secondary flow transporting angular momentum away from the forced region.

   The picture, therefore, is one of fluid being continuously cycled first through the forced region, where it is spun up, acquiring kinetic energy and angular momentum, and then through the side-wall boundary layers, where that energy and angular momentum are lost. Note that the local force balance in the forced region is between inertia, $\rho\mathbf{u} \cdot \nabla\Gamma$, and the applied Lorentz torque per unit volume,

$$\rho\mathbf{u} \cdot \nabla\Gamma = rF_\theta, \tag{11.11}$$

rather than between F_θ and azimuthal shear. As a consequence, one-dimensional models of the form discussed in §11.1.3 substantially overestimate the induced swirl, typically by a factor of around 500 per cent (Davidson and Hunt, 1987).

An approximate analytical model of this flow has been proposed by Davidson and Hunt, which predicts that the secondary motion is as strong as the primary swirl. Consequently, the force balance $\rho\mathbf{u}\cdot\nabla\Gamma \sim rF_\theta$ yields a characteristic swirl velocity of $u_\theta \sim \sqrt{F_\theta R/\rho}\sim\Omega_f R$. The same model also predicts that the maximum azimuthal velocity occurs, not where F_θ is largest, but rather at the upper and lower edges of the forced region, where F_θ falls to zero. This has been confirmed by experiment. So we conclude that the spatial distribution of swirl does *not* mimic that of F_θ, and that the secondary (poloidal) flow plays a key role in the overall dynamics.

11.1.5 The Role of Ekman Pumping for Non-Ferrous Metals

We now turn to the magnetic stirring of non-ferrous metals, and in particular that of aluminium. Here the liquid-metal pool is typically hemispherical or parabolic in shape, as shown in Figure 11.1(b), and the secondary flow arises from Ekman pumping at the boundaries, as discussed below.

Some hint as to the importance of the secondary motion associated with Ekman pumping can be gleaned from the example discussed in §6.4.2. In this example we looked at the laminar flow of a liquid held between two flat, parallel plates and subject to the force (11.2), i.e. $F_\theta = \frac{1}{2}\sigma B^2\Omega r$. It was found that Ekman-like layers form on the top and bottom surfaces and that these layers induce a secondary, poloidal flow as shown in Figure 11.8. We showed that, outside the Ekman (or Bödewadt) layers, the viscous stresses are negligible and the fluid rotates as a rigid body, the rotation rate being quite different from that predicted by (11.7). In fact, the core angular velocity is

$$\Omega_c = 0.516\Omega_f\left[\frac{\Omega_f w^2}{\nu}\right]^{1/3},\qquad(11.12)$$

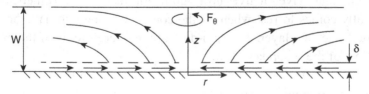

Figure 11.8 Swirling flow and associated Ekman pumping driven between two discs by the force $F_\theta = \frac{1}{2}\sigma B^2\Omega r$. The Ekman pumping in the r–z plane is shown. The motion is mirror symmetric about the mid-plane and so only the bottom half of the flow is illustrated.

where $2w$ is the distance between the plates. Moreover, we saw that the Ekman layers are unaffected by the presence of the forcing (the Lorentz force is negligible by comparison with the viscous forces) and so they look like conventional Bödewadt layers, of the type discussed in §3.8. For example, the thickness of the boundary layers is of the order of $\delta \sim 4(v/\Omega)^{\frac{1}{2}}$.

This simple model problem is discussed at some length in Davidson (1992), where the key features at large Re are shown to be:

  (i) The flow may be divided into a forced, inviscid core, and two viscous boundary layers in which the Lorentz force is negligible.
 (ii) All of the streamlines pass through both regions, collecting kinetic energy and angular momentum in one region, and losing it in the other.
(iii) The applied Lorentz force in the core is balanced by the Coriolis force associated with the core angular velocity.

We shall see shortly that all of these features are characteristic of the magnetic stirring of an aluminium ingot. The main point, however, is (iii). When a secondary flow is present, the Lorentz force is balanced locally by the Coriolis force, not shear, and this is why (11.7) and (11.12) look so different.

Perhaps the key to establishing the distribution of swirl in the magnetic stirring of an aluminium casting lies in the simple, textbook problem of 'spin-down' of a stirred cup of tea, as discussed in §3.8. In this well-known example, the main body of the fluid is predominantly in a state of inviscid rotation. The associated centrifugal force is balanced by a radial pressure gradient, and this pressure gradient is also imposed throughout the boundary layer on the base of the cup. Of course, the swirl in this boundary layer (the Ekman layer) is diminished through viscous drag, and so there is a local imbalance between the radial pressure force and u_θ^2/r. The result is a radial inflow along the bottom of the cup, with the fluid eventually drifting up and out of the boundary layer. In short, we have a kind of Bödewadt layer on the base of the cup. Of course, continuity then requires that the boundary layer be replenished via the side walls and the end result is a form of Ekman pumping, as shown in Figure 11.9. As each fluid particle passes through the Bödewadt boundary layer, it gives up a significant fraction of its kinetic energy and the tea finally comes to rest when all the contents of the cup have been flushed through the Bödewadt layer. The spin-down time, therefore, is of the order of the turnover time of the secondary flow.

In the present context, it is useful to consider a variant of this problem. Suppose now that the tea is continuously stirred. Then it will reach an equilibrium rotation rate in which the rate of work done by the teaspoon exactly balances the rate of dissipation of kinetic energy in the Bödewadt layer. This provides a clue to understanding the magnetic stirring of liquid metal in a confined domain.

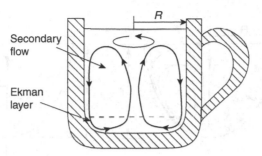

Figure 11.9 Spin-down of a stirred cup of tea.

Let us now return to aluminium casting, where the liquid-metal pool has an aspect ratio of the order of unity, as indicated in Figure 11.2(b). We make no particular assumptions about the shape of the boundary, although we have in mind profiles which are roughly parabolic. Suppose we integrate the time-averaged Navier–Stokes equations around a streamline which is closed in the r–z plane. For an axisymmetric steady-on-average flow, we obtain

$$\oint \mathbf{F} \cdot d\mathbf{x} + \rho \nu_t \oint \nabla^2 \overline{\mathbf{u}} \cdot d\mathbf{x} = 0, \tag{11.13}$$

where $\mathbf{F}$ is the Lorentz force per unit volume and ν_t is an eddy viscosity which, for simplicity, we treat as a constant. Of course, this is an energy balance; it states that all of the energy imparted to a fluid particle by the Lorentz force must be destroyed or diffused away by shear before the particle returns to its original position. But the shear stresses are significant only in the boundary layers. By implication, *all* streamlines must pass through a boundary layer. Of course, Ekman pumping provides the necessary entrainment mechanism. Note also that Bödewadt-like layers can and will form on all surfaces non-parallel to the axis of rotation. The structure of the flow, therefore, is as shown in Figure 11.10(a) (Davidson, 1992). It consists of an interior body of (nearly) inviscid swirl surrounded by Ekman wall jets on the inclined surfaces. Each fluid particle is continually swept first through the core, where it collects kinetic energy and angular momentum, and then through the Ekman layers, where it deposits its energy. The motion is helical, spiralling upward through the core, and downward through the boundary layers.

The fact that all streamlines pass through the Ekman layers has profound implications for the axial distribution of swirl. Let u_b and u_c be characteristic values of the poloidal recirculation $(u_r, 0, u_z)$ in the boundary layer and in the core, respectively. Also, let δ be the boundary layer thickness, R be a characteristic linear dimension of the pool, and $\Gamma = r u_\theta$ be the angular momentum density. Now u_b and u_θ are of similar magnitudes (one drives the other) and so continuity requires

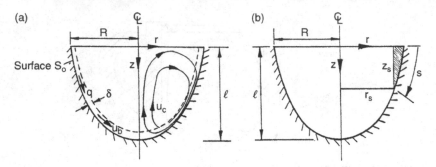

Figure 11.10 (a) Secondary flow induced by forced swirl in a cavity. (b) Coordinate system.

$$u_c \sim u_b(\delta/R) \sim u_\theta(\delta/R). \tag{11.14}$$

That is, the core recirculation is weak by comparison with u_θ. But the core recirculation is related to the core angular momentum, Γ_c, by the inviscid azimuthal vorticity equation (11.10):

$$\mathbf{u} \cdot \nabla(\omega_\theta/r) = \frac{\partial}{\partial z}\left(\frac{\Gamma_c^2}{r^4}\right). \tag{11.15}$$

Combining (11.14) and (11.15) we have (Davidson, 1992),

$$\Gamma_c = \Gamma_c(r)[1 + O(\delta/R)^2]. \tag{11.16}$$

It is extraordinary that, no matter what the spatial distribution of the Lorentz force, the induced swirl in the core is independent of height to second order in δ/R.

Since the flow has a simple, clear structure it is possible to piece together an approximate, quantitative model of the process. We give only a schematic outline here, but more details may be found in Davidson (1992). In the inviscid core the applied Lorentz force is balanced by inertia: $\rho(\mathbf{u} \cdot \nabla)\Gamma_c = rF_\theta$. Since Γ_c is a function only of r, the left-hand side reduces to $\rho u_r \Gamma'_c(r)$, the Coriolis force. Thus we have

$$u_r = rF_\theta/\rho\Gamma'_c(r). \qquad \bullet \tag{11.17}$$

Now all the fluid which moves radially outward is recycled via the boundary layer and so (11.17) may be used to calculate the mass flux in the Ekman layer. In particular, if we apply the continuity equation to the shaded area in Figure 11.10(b), we obtain an estimate of the net mass flux down through the boundary layer,

$$\dot{q} = 2\pi r_s \int_0^\delta u_b dn = 2\pi r_s \int_0^{z_s} u_r dz = \frac{1}{\rho \Gamma'_c(r_s)} \frac{dT}{dr_s}, \tag{11.18}$$

where

$$T = \int_0^{r_s} \left[\int_0^{z_s(r)} 2\pi r^2 F_\theta dz \right] dr. \tag{11.19}$$

Here T is the total magnetic torque applied to the fluid in the region $0 < r < r_s$, where (r_s, z_s) represents the coordinates of the boundary (Figure 11.10(b)). Also, n is a normal coordinate measured from the boundary into the fluid and $u_b(n)$ is the velocity profile in the boundary layer.

Next, we turn our attention to the boundary layer. Equation (11.17) tells us that the ratio of the Lorentz force to inertia in the boundary layer is $rF_\theta / \rho \mathbf{u} \cdot \nabla \Gamma \sim u_c / u_b \sim \delta / R$. Since $\delta \ll R$, we may neglect F_θ in the boundary layer and the azimuthal equation of motion there reduces to a balance between $(\mathbf{u} \cdot \nabla)\Gamma$ and gradients in the azimuthal shear stresses. Integrating this force balance across the boundary layer yields a so-called momentum integral equation, which in this case takes the form (see Davidson, 1992)

$$\frac{\dot{q}}{\pi} \frac{d\Gamma_c}{ds} - \frac{d}{ds} \left\{ \chi \, \Gamma_c \left[\frac{\dot{q}}{\pi} \right] \right\} = -c_f \Gamma_c^2. \tag{11.20}$$

Here $c_f = \tau_\theta / \frac{1}{2} \rho u_\theta^2$ is the dimensionless skin-friction coefficient, which is readily estimated for a turbulent boundary layer, χ is a shape factor related to the velocity profile in the Ekman wall jet (usually taken as 1/6), and s is a curvilinear coordinate measured along the boundary from the surface. Eliminating $\dot{q}$ between (11.18) and (11.20) furnishes

$$\frac{dT}{ds} - \frac{d}{ds} \left\{ \chi \frac{\Gamma_c}{\Gamma'_c(s)} \frac{dT}{ds} \right\} = -\pi c_f \rho \Gamma_c^2. \tag{11.21}$$

This simple equation allows the distribution of the core swirl, $\Gamma_c(r)$, to be calculated whenever the applied Lorentz torque, T, is known. Its predictions have been tested against experiments performed in cones, hemispheres and cylinders, and there is a reasonable correspondence between theory and experiment. However, perhaps the most important conclusions are that (i) there exist Ekman wall jets, which sweep down the solidification front carrying heat and crystal fragments with them, and (ii) Equation (11.16) shows that the induced swirl is quite insensitive to the detailed distribution of the applied Lorentz force. It cares only about the globally averaged magnetic torque.

11.2 Magnetic Damping Using Static Fields

We have seen that the relative movement of a conducting body and a magnetic field can lead to the dissipation of energy. This phenomenon has been used by engineers for over a century to dampen unwanted motion. Indeed, as far back as 1873 we find Maxwell noting, 'A metallic circuit, called a damper, is sometimes placed near a magnet for the express purpose of damping or deadening its vibrations.' Maxwell was talking about a magnetic field moving through a stationary conductor. We are interested in a moving conductor in a stationary field, but of course, this is really the same thing. We have already touched upon magnetic damping in Chapter 6, and here we describe in more detail how magnetic fields are used in certain casting operations to suppress unwanted motion. We shall see that the hallmark of magnetic damping is that the dissipation of energy is subject to the constraint of *conservation of momentum*, and that this constraint is a powerful one.

11.2.1 Metallurgical Applications

We have already seen that a static magnetic field can suppress the motion of an electrically conducting liquid. The mechanism is straightforward: the motion of a liquid across the magnetic field lines induces a current. This leads to Joule dissipation and the resulting rise in thermal energy is accompanied by a fall in magnetic and/or kinetic energy. We are concerned here only with cases where the magnetic Reynolds number is small, so that changes to the applied magnetic field are negligible. In such cases, the rise in Joule dissipation is matched by a fall in kinetic energy. Thus, for example, in an electrically insulated pool, (6.7) gives

$$\frac{d}{dt}\int\left(\frac{1}{2}\rho\mathbf{u}^2\right)dV = -\frac{1}{\sigma}\int \mathbf{J}^2 dV + \text{ viscous dissipation.}$$

In the past two decades this phenomenon has been exploited in a range of metallurgical processes. For example, in the continuous casting of large steel slabs, an intense, static magnetic field (around 0.5 Tesla) is commonly used to suppress motion within the mould (Figure 11.11). Sometimes the motion takes the form of a submerged, turbulent jet which feeds the mould from above; at others it takes the form of large eddies or vortices. In both cases the objective is to keep the free surface of the liquid quiescent, thus avoiding the entrainment of surface debris. In other solidification processes, such as the Bridgeman technique for growing

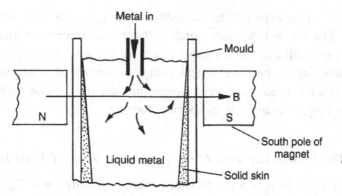

Figure 11.11 Magnetic damping is used to suppress unwanted motion in the continuous casting of steel slabs.

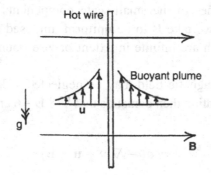

Figure 11.12 Magnetic damping is used to suppress unwanted convection in the hot-wire method of measuring thermal diffusivity.

semi-conductor crystals or the continuous casting of aluminium, it is widely believed that natural convection has a detrimental effect on the metallurgical structure of the solid. Again, the imposition of a static magnetic field is seen as one means of suppressing these unwanted motions.

Magnetic damping is also used in the laboratory measurement of chemical and thermal diffusivities, particularly where solutal or thermal buoyancy can disrupt the measurement technique. For example, in the 'hot-wire' technique for measuring the thermal conductivity of liquid metals the conductivity is determined by monitoring the rate at which heat diffuses into the liquid from a long, thin, vertical wire. This technique relies on conduction being dominant over convection. Yet natural convection is always present to some degree in the form of a buoyant plume. The simplest way of suppressing the unwanted motion is magnetic damping (Figure 11.12).

In this section we examine the magnetic damping of jets, vortices and natural convection. Our aim is to present a unified theoretical framework from which the many disparate published studies may be viewed. We shall see that the hallmark of magnetic damping is that mechanical energy is destroyed while momentum is conserved. It is this need to conserve momentum, despite the dissipation of energy, which gives magnetic damping its special character.

11.2.2 The Need to Conserve Momentum in the Face of Joule Dissipation

We consider flows in which the Reynolds number is large and R_m is low so the magnetic field induced by currents flowing in the liquid is much smaller than the externally imposed field. This covers most laboratory and industrial applications. In view of the large Reynolds number, we may treat the motion as inviscid, except of course when it comes to the small-scale components of turbulence. In the interests of simplicity we take $\mathbf{B}$ to be uniform, imposed in the z-direction, and consider domains which are infinite in extent or else bounded by an electrically insulated surface.

Since the induced magnetic field may be neglected at low R_m, $\mathbf{B}$ is fixed and uniform. Faraday's equation then demands that $\nabla \times \mathbf{E} = \mathbf{0}$ and Ohm's law takes the form

$$\mathbf{J} = \sigma(-\nabla\Phi + \mathbf{u} \times \mathbf{B}), \qquad (11.22)$$

where Φ is the electrostatic potential and $\mathbf{B} = B\hat{\mathbf{e}}_z$ is the uniform, imposed magnetic field. Now we also know, from Ampere's Law, that $\mathbf{J}$ is solenoidal, and so we have

$$\nabla \cdot \mathbf{J} = 0, \qquad \nabla \times \mathbf{J} = \sigma\mathbf{B} \cdot \nabla\mathbf{u}. \qquad (11.23)$$

Equations (11.23) uniquely determine $\mathbf{J}$. (Recall that a vector field is uniquely determined if its divergence and curl are known and suitable boundary conditions are specified.) The key point is that $\mathbf{J}$ is zero if and only if $\mathbf{u}$ is independent of z. Now the Lorentz force per unit mass, $\mathbf{J} \times \mathbf{B}/\rho$, is readily obtained from (11.22),

$$\mathbf{F} = -\frac{\mathbf{u}_\perp}{\tau} + \frac{\sigma(\mathbf{B} \times \nabla\Phi)}{\rho}, \qquad \tau = \rho/\sigma B^2, \qquad (11.24)$$

where $\mathbf{u}_\perp = (u_x, u_y, 0)$ and τ is the Joule damping term. Note that the first term in (11.24) looks like a linear friction term. However, this expression for $\mathbf{F}$ is awkward as it contains the unknown potential Φ. This potential is given by the divergence of Ohm's Law, which yields $\nabla^2\Phi = \mathbf{B} \cdot \boldsymbol{\omega}$, where $\boldsymbol{\omega}$ is the vorticity field. Clearly, when

B and **ω** are mutually perpendicular, the Lorentz force simplifies to $\mathbf{F} = -\mathbf{u}_\perp/\tau$, and so (pressure forces apart) $\mathbf{u}_\perp$ declines on a time scale of τ. This is the phenomenological basis of magnetic damping. Loosely speaking, we may think of rotational motion being damped out provided that its axis of rotation is perpendicular to **B**. The ratio of the damping time, τ, to the characteristic advection time, ℓ/u, gives the interaction parameter,

$$N = \sigma B^2 \ell/\rho u.$$

Typically, N is indicative of the relative sizes of the Lorentz and inertial forces.

We now consider the role of Joule dissipation. This provides an alternative way of quantifying magnetic damping. Combining (11.22) with the inviscid equation of motion,

$$\rho\frac{D\mathbf{u}}{Dt} = -\nabla p + \rho\mathbf{F}, \quad \mathbf{F} = \mathbf{J}\times\mathbf{B}/\rho,$$

yields the energy equation,

$$\frac{dE}{dt} = -\frac{1}{\sigma}\int \mathbf{J}^2 dV = -D, \tag{11.25}$$

where D is the Joule dissipation rate and E is the global kinetic energy. Clearly, E declines until **J** is everywhere zero, which happens only when **u** is independent of z. We can use (11.23) to estimate the rate of decline of energy. Let $\ell_{//}$ and ℓ_{min} be two characteristic length scales for the flow, the first being parallel to **B**. Then (11.23) yields the estimate

$$|\mathbf{J}| \sim \ell_{min}|\nabla\times\mathbf{J}| \sim \ell_{min}|\sigma\mathbf{B}\cdot\nabla\mathbf{u}| \sim \sigma Bu(\ell_{min}/\ell_{//}),$$

and so (11.25) becomes

$$\frac{dE}{dt} = -D \sim -\left(\frac{\ell_{min}}{\ell_{//}}\right)^2 \frac{E}{\tau}. \tag{11.26}$$

This integrates to give

$$E \sim E_0 \exp\left[-\tau^{-1}\int_0^t (\ell_{min}/\ell_{//})^2 dt\right]. \tag{11.27}$$

The implication is that, provided $\ell_{//}$ and ℓ_{min} remain of the same order, the flow will be destroyed on a time scale of τ. Indeed, this might have been anticipated from (11.24).

However, this is not the end of the story. This dissipation is subject to some powerful integral constraints. The key point is that $\mathbf{F}$ cannot create or destroy linear momentum, nor (one component of) angular momentum. For example, since $\mathbf{J}$ is solenoidal, we have $J_i = \nabla \cdot (x_i \mathbf{J})$, and hence Gauss' theorem yields

$$\int \mathbf{F} dV = \rho^{-1} \int \mathbf{J} dV \times \mathbf{B} = 0. \tag{11.28}$$

Thus the Lorentz force cannot itself alter the global linear momentum of the fluid. Similarly, following Davidson (1995), we have

$$(\mathbf{x} \times \mathbf{F}) \cdot \mathbf{B} = \rho^{-1}[(\mathbf{x} \cdot \mathbf{B})\mathbf{J} - (\mathbf{x} \cdot \mathbf{J})\mathbf{B}] \cdot \mathbf{B} = -(B^2/2\rho)\nabla \cdot [\mathbf{x}_\perp^2 \mathbf{J}], \tag{11.29}$$

which integrates to zero over the domain. Evidently, the Lorentz force exerts zero net torque parallel to $\mathbf{B}$, and so cannot create or destroy the corresponding component of angular momentum. (See also the discussion in §6.2.4.) The physical interpretation of (11.29) is that $\mathbf{J}$ may be considered to be composed of many current tubes, and that each of these tubes may, in turn, be considered to be the sum of many infinitesimal current loops, as in the proof of Stokes' theorem. However, the torque on each elementary current loop is $(d\mathbf{m}) \times \mathbf{B}$, where $d\mathbf{m}$ is its dipole moment. Consequently, the global torque, which is the sum of many such terms, can have no component parallel to $\mathbf{B}$.

Now the fact that $\mathbf{F}$ cannot create or destroy linear momentum, nor one component of angular momentum, would not be important if the mechanical forces themselves changed these momenta. However, in certain flows, such as submerged jets, the mechanical (pressure) forces cannot alter the linear momentum of the fluid. In others, such as flow in an axisymmetric container, the pressure cannot alter the axial component of angular momentum. In such cases there is always some component of momentum which is conserved, despite the Joule dissipation. This implies that the flow cannot be destroyed on a time scale of τ, and from (11.27) we may infer that $\ell_{//}/\ell_{\min}$ must increase with time. We might anticipate, therefore, that these flows will exhibit a pronounced anisotropy, with $\ell_{//}$ increasing as the flow evolves, and indeed, this is exactly what happens.

Of course, it has been known for a long time that a strong magnetic field promotes this kind of anisotropy. However, the traditional explanation is rather different from that given above, and so is worth repeating here. If $\nabla \times \mathbf{F}$ is evaluated from (11.22) we obtain

$$\nabla^2(\nabla \times \mathbf{F}) = -\frac{1}{\tau}\frac{\partial^2 \boldsymbol{\omega}}{\partial z^2},$$

and substituting into the inviscid vorticity equation gives

$$\frac{D\boldsymbol{\omega}}{Dt} = \boldsymbol{\omega} \cdot \nabla\mathbf{u} - \frac{1}{\tau}\nabla^{-2}[\partial^2\boldsymbol{\omega}/\partial z^2], \tag{11.30}$$

where the symbolic operator ∇^{-2} can be evaluated using the Biot–Savart law. Phenomenologically, we might consider ∇^{-2} to be replaced by $-\ell_{min}^2$, in which case the Lorentz term promotes a unidirectional diffusion of vorticity along the **B**-lines, with a diffusivity of ℓ_{min}^2/τ. For cases where $\ell_{//} \gg \ell_{min}$, this may be made rigorous (at least in Fourier space) by taking the two-dimensional (*x-y*) Fourier transform of the Lorentz force in (11.30). This leads to a one-dimensional diffusion term in Fourier space, with a diffusivity of $1/(\tau k^2)$. This kind of diffusion argument is a useful one when N is large, so that the non-linear inertial terms are negligible. However, when N is small or moderate, the vortex lines stretch and twist on a time scale of ℓ/u, which is smaller than, or of the order of, τ. In such cases it is difficult to infer much from (11.30) and there is an advantage in returning to the integral arguments given above.

11.2.3 The Magnetic Damping of Submerged Jets

As our first application we turn to the magnetic damping of jets. We are interested primarily in the dissipation of submerged jets.

We start with the slightly artificial problem of the temporal evolution of a long, uniform jet which is dissipated by the sudden application of a magnetic field (Figure 11.13). This provides a useful stepping stone to the more important problem of a submerged jet which evolves in space, rather than in time.

Suppose that we have a unidirectional jet, $\mathbf{u} = u(x,z,t)\hat{\mathbf{e}}_y$, which is initially axisymmetric (Figure 11.13a). At $t = 0$ we (somehow) impose a uniform magnetic field in the z-direction. Current will be induced as shown in Figure 11.13(b), driven

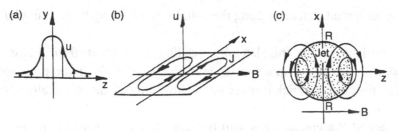

Figure 11.13 Magnetic damping of a jet: (a) initial axisymmetric velocity profile; (b) the induced current; (c) a cross-section through the jet showing current paths. A reverse flow forms at points marked R while the momentum of the jet spreads along the field lines.

parallel to the x-axis by $\mathbf{u} \times \mathbf{B}$, but forced to recirculate back through regions of weak or zero flow by the electrostatic potential. Since the current is two-dimensional we may introduce a streamfunction, ψ, for $\mathbf{J}$ which is related to $\mathbf{u}$ by Ohm's law,

$$\mathbf{J} = \nabla \times [\psi \hat{\mathbf{e}}_y], \quad \nabla^2 \psi = -\sigma B \frac{\partial u}{\partial z}.$$

Our equation of motion,

$$\frac{\partial u}{\partial t} = \frac{B}{\rho} \frac{\partial \psi}{\partial z}, \tag{11.31}$$

then becomes

$$\frac{\partial}{\partial t} \nabla^2 u = -\frac{1}{\tau} \frac{\partial^2 u}{\partial z^2}. \tag{11.32}$$

Evidently, linear momentum is conserved by (11.31), since $\partial \psi / \partial z$ integrates to zero. Now let δ be the thickness of the jet in the x-direction and $\ell_{//}$ be the characteristic length scale for u parallel to $\mathbf{B}$. Then, from conservation of momentum, in conjunction with the energy equation (11.26), we have

$$\frac{dE}{dt} \sim -\left(\frac{\delta}{\ell_{//}}\right)^2 \frac{E}{\tau}, \quad E \sim \rho u^2 \ell_{//} \delta, \quad u \ell_{//} \delta = \text{constant}. \tag{11.33}$$

Clearly, $\ell_{//}/\delta$ must increase with time. If it did not, then E would decline exponentially, which is forbidden by conservation of momentum. For fixed δ, the only possible solution to (11.33) is

$$\ell_{//} \sim \delta (t/\tau)^{1/2}, \quad u \sim u_0 (t/\tau)^{-1/2}. \tag{11.34}$$

Thus the jet spreads laterally along the field lines, evolving from a circular jet into a sheet.

The mechanism for this lateral spreading is evident from Figure 11.13. The induced currents within the jet give rise to a braking force, as expected. However, the current which is recycled to either side of the jet actually accelerates previously stagnant fluid at large $|z|$. Hence the growth in $\ell_{//}$. Notice also that at points marked R a counter-flow will be generated since $\mathbf{F}$ points in the negative y-direction and u is initially zero.

The existence of a counter-flow, as well as the scaling laws (11.34), may be confirmed by exact analysis. For example, taking the Fourier transform of (11.32) leads to an exact solution in terms of hypergeometric functions, as we now show. Let $\hat{u}$ be the transform of u:

$$\hat{u}(k_x, k_z) = 4 \int_0^\infty \int_0^\infty u(x, z) \cos(x k_x) \cos(z k_z) dx dz.$$

Equation (11.32) becomes

$$\frac{\partial \hat{u}}{\partial t} + \cos^2\varphi \frac{\hat{u}}{\tau} = 0, \qquad \cos\varphi = k_z/k,$$

where $k^2 = k_x^2 + k_z^2$. Solving for $\hat{u}$ and taking the inverse transform then yields

$$u(\mathbf{x}, t) = \pi^{-2} \int_0^\infty \int_0^{\pi/2} e^{(\cos^2\varphi)\hat{t}} \cos(x k_x) \cos(z k_z) \hat{u}_0(k) k dk d\varphi,$$

where $\hat{t}$ is the scaled time, t/τ, and $\hat{u}_0(k)$ is the transform of the initial axisymmetric velocity profile, $u_0(r)$. For large $\hat{t}$ this integral is dominated by $\varphi \sim \pi/2$ and can be simplified using the relationship

$$\int_0^\infty e^{-p^2} \cos(\lambda p) dp = (\sqrt{\pi}/2) e^{-\lambda^2/4},$$

to give

$$u(\mathbf{x}, t) = \frac{1}{2\pi\sqrt{\pi\hat{t}}} \int_0^\infty e^{-k^2 z^2/4\hat{t}} \cos(xk) \hat{u}_0(k) k dk. \tag{11.35}$$

It is clear from this integral that, for large $\hat{t}$, u must be of the form

$$u(\mathbf{x}, t) \sim \hat{t}^{-1/2} F(z/\hat{t}^{1/2}, x), \tag{11.36}$$

which confirms the scaling laws (11.34). Of course, the form of F depends on the initial conditions. For example, if we take $u_0(r) = V \exp(-r^2/\delta^2)$, where $r^2 = x^2 + z^2$, then (11.35) yields

$$u(\mathbf{x}, t) = \frac{V}{\sqrt{\pi\hat{t}}} \frac{G(\zeta)}{[1 + z^2/(\delta^2\hat{t})]}, \qquad \zeta = \frac{x^2}{\delta^2 + z^2/\hat{t}}, \tag{11.37}$$

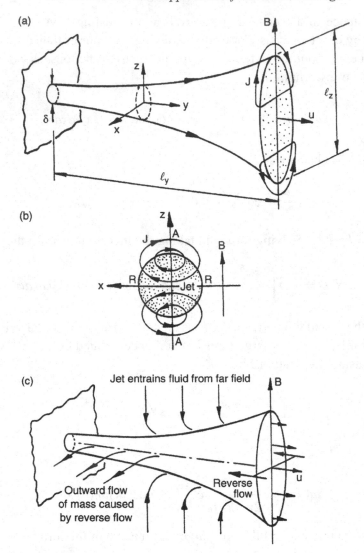

Figure 11.14 MHD jet produced by side-wall injection. (a) The spatial evolution of jet. (b) The current paths. A reverse flow occurs at points marked R. (c) The jet spreads diffusively along the field lines, drawing in fluid from the far field, and the reverse flow produces an outward flow of mass near the wall.

where $G(\zeta)$ is Kummer's hypergeometric function, $G(\zeta) = M(1, 1/2, -\zeta)$. An examination of the shape of $G(\zeta)$ confirms that a reverse flow develops as anticipated in Figure 11.13(c).

Consider now a submerged, steady jet evolving in space, rather than in time. This is illustrated in Figure 11.14(a). It is generated by injecting fluid through a circular aperture in a side-wall and into a uniform magnetic field. We consider the case where

B is weak (N is small) so that the jet inertia is much larger than the Lorentz force. This configuration is particularly relevant to the magnetic damping of jets in castings.

Since N is small, the magnetic field influences the jet only slowly. As a result, the characteristic axial length scale of the jet, ℓ_y, is much greater than ℓ_x and ℓ_z. Now the current must form closed paths. However, each cross-section of the jet looks very much like its neighbouring cross-sections, and so the current must close in the x–z plane, just as it did in the previous example. Figure 11.14(b) illustrates the situation. As before, the induced current recirculates through regions of weak or zero flow. It follows that a reverse flow will form at points marked R, and momentum will diffuse out along the z-axis by precisely the same mechanism as before. Thus the jet cross-section becomes long and elongated. Now if the jet is to spread laterally along the **B**-lines, then continuity of mass requires that there be some entrainment of the surrounding fluid. We would expect, therefore, that the jet draws in fluid from the far field, predominantly at large $|z|$. Conversely, regions of reverse flow on the x-axis will produce an outward flow of mass near the wall (Figure 11.14c).

We now substantiate this picture using the Euler equation. The equation of motion for the jet is very similar to (11.31) and (11.32). In terms of the current stream-function ψ, we have

$$\frac{Du_y}{Dt} = \mathbf{u} \cdot \nabla u_y = \frac{B}{\rho}\frac{\partial \psi}{\partial z} = -\frac{1}{\tau}\nabla^{-2}\left[\frac{\partial^2 u_y}{\partial z^2}\right]. \qquad (11.38)$$

Unlike (11.32), this is non-linear, and so exact solutions are unlikely to be found. However, we may still use conservation of linear momentum, in conjunction with an energy dissipation equation, to obtain scaling laws for the jet. Let M be the momentum flux in the jet, $M = \int u_y^2 dA$. From (11.38) this is conserved. Also, from (11.38) we can construct an energy equation reminiscent of (11.25),

$$\frac{d}{dy}\int\left(\frac{1}{2}u_y^3\right)dA = -\frac{1}{\rho\sigma}\int \mathbf{J}^2 dA. \qquad (11.39)$$

It follows that u_y and $\ell_{//}$ scale as

$$u_y \sim \left[\frac{\tau M^2}{\delta^4 y}\right]^{1/3}, \qquad \ell_{//} \sim \left[\frac{\delta^5 y^2}{\tau^2 M}\right]^{1/3}. \qquad (11.40)$$

(It is readily confirmed that these are the only self-similar scaling laws which satisfy conservation of M as well as the energy equation (11.39); see Davidson, 1995.) Note that the mass flux in the jet increases with y, as shown in Figure 11.14.

It is interesting to compare (11.40) and Figure 11.14 with the two-dimensional jet analysed in §6.2.2. Evidently a two-dimensional jet, where the current paths do not close within the fluid, behaves quite differently from its three-dimensional counterpart. Of course, it is the three-dimensional jet which is the more important in practice.

11.2.4 The Magnetic Damping of Vortices

So far we have considered only cases where the conservation of linear momentum provides the key integral constraint. We now consider examples where conservation of angular momentum is important, i.e. vortices. The discussion is restricted to inviscid fluids.

We start by returning to the model problem of §6.2.4. Suppose we have one or more vortices, of arbitrary shape and orientation, held in a spherical domain. As above, we take R_m to be low and the applied field to be uniform and aligned with the z-axis, $\mathbf{B} = B\hat{\mathbf{e}}_z$. Then, as we saw in §6.2.4, the global magnetic torque is given by

$$\mathbf{T} = \int \mathbf{x} \times (\mathbf{J} \times \mathbf{B})dV = -\frac{1}{4\tau}\int \rho(\mathbf{x} \times \mathbf{u})_\perp dV. \tag{11.41}$$

Let $\mathbf{H}$ be the global angular momentum of the fluid, $\mathbf{H} = \int \mathbf{x} \times \mathbf{u}dV$. (We shall not carry the constant ρ through the calculation.) Then (11.41) gives the inviscid equation of motion,

$$\frac{\partial \mathbf{H}}{\partial t} = -\frac{\mathbf{H}_\perp}{4\tau}, \tag{11.42}$$

and it follows that $\mathbf{H}_{//}$ is conserved while $\mathbf{H}_\perp$ declines exponentially fast (Davidson, 1995):

$$\mathbf{H}_{//} = \text{ constant}; \quad \mathbf{H}_\perp(t) = \mathbf{H}_\perp(0)\exp[-t/4\tau]. \tag{11.43}$$

This applies for any value of the interaction parameter, N, and so is valid even when inertia is dominant and the stretching and twisting of vorticity is more vigorous than the damping. As discussed in §6.2.4, the conservation of $\mathbf{H}_{//}$, combined with the Schwarz inequality, gives us a lower bound on the kinetic energy, $E = \int \frac{1}{2}\mathbf{u}^2 dV$,

$$E \geq H_{//}^2 \left[2\int \mathbf{x}_\perp^2 dV\right]^{-1}. \tag{11.44}$$

This, in conjunction with the energy equation

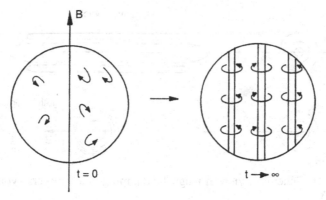

Figure 11.15 Evolution to a two-dimensional state under the influence of an imposed field.

$$\frac{dE}{dt} = -\frac{1}{\rho\sigma}\int \mathbf{J}^2 dV,$$

places constraints on the way in which the flow can evolve. Typically, the energy of the flow decreases through the destruction of $\mathbf{H}_\perp$ until only $\mathbf{H}_{//}$ remains. Since there is a lower bound on E, it follows that the flow must eventually evolve towards a state in which $\mathbf{u}$ is finite, but $\mathbf{J}$ is everywhere zero. From (11.23) it is clear that this asymptotic state must be two-dimensional, consisting of one or more columnar vortices which span the sphere and whose axes are parallel to $\mathbf{B}$, as shown in Figure 11.15.

A natural question to ask is 'How do the vortices evolve into these long, columnar structures?' If the domain is large relative to the initial eddy size, we would expect the initial evolution of the vortices to be independent of the existence of the remote boundaries, and so the tendency for vortices to elongate along the $\mathbf{B}$-lines must be generic, independent of the boundary conditions. This suggests that we dispense with the spherical boundary and consider vortices in an infinite domain. There are two special cases which deserve particular attention. One is where the vorticity is aligned with $\mathbf{B}$, and the other is where $\mathbf{B}$ and $\boldsymbol{\omega}$ are mutually perpendicular. These two cases differ because the electrostatic potential in Ohm's law is governed by $\nabla^2\Phi = \mathbf{B}\cdot\boldsymbol{\omega}$, and so the role of this potential depends crucially on the relative orientation of $\mathbf{B}$ and $\boldsymbol{\omega}$. We start with the simpler case of transverse vortices, where Φ is zero.

In the interests of simplicity we shall consider a two-dimensional vortex whose axis is normal to the imposed magnetic field. Suppose our flow is confined to the (x, z) plane and consists of an isolated vortex whose initial characteristic scale is δ. We shall take the vortex to be initially axisymmetric and subject to a uniform

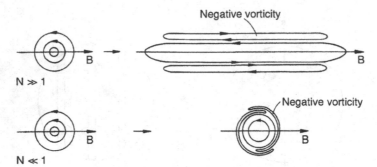

Figure 11.16 Schematic view of magnetic damping of a transverse vortex at low and high N.

magnetic field, **B**, imposed in the z-direction (Figure 11.16). Once again, we shall find that global angular momentum provides one key to determining evolution of the flow.

Since **B** and **ω** are mutually perpendicular, the electrostatic potential is zero, and so (11.24) gives the Lorentz force (per unit mass) and the net magnetic torque as

$$\mathbf{F} = -(u_x/\tau)\hat{\mathbf{e}}_x, \quad T_y = -\tau^{-1}\int zu_x dV = -H_y/2\tau. \tag{11.45}$$

Here H_y is the global angular momentum which may be expressed either in terms of **u** or else in terms of the two-dimensional stream-function, ψ. That is, noting that

$$(\mathbf{x} \times \mathbf{u})_y = 2zu_x - \nabla \cdot (xz\mathbf{u}) = 2\psi - \nabla \cdot (\psi\mathbf{x}),$$

we have

$$H_y = \int (zu_x - xu_z)dV = 2\int zu_x dV = 2\int \psi dV.$$

(In order that all relevant integrals converge, we require that the integral of ψ converge, and this limits our possible choice of initial conditions.) It follows immediately from (11.45) that, even in the low N (non-linear) regime, the angular momentum decays in a remarkably simple manner:

$$H_y(t) = H_y(0)e^{-t/2\tau}. \tag{11.46}$$

This is the two-dimensional counterpart of (11.43). It is tempting to conclude, therefore, that the vortex decays on a time-scale of 2τ. However, this appears to contradict (11.30) which, in the present context, and in the limit of large N, simplifies to

$$\frac{\partial}{\partial t}\nabla^2\omega = -\frac{1}{\tau}\frac{\partial^2\omega}{\partial z^2}. \tag{11.47}$$

In particular, if we Fourier transform (11.47) in the x-direction, this suggests

$$\frac{\partial\hat{\omega}}{\partial t} \sim \frac{1}{\tau k^2}\frac{\partial^2\hat{\omega}}{\partial z^2}, \tag{11.48}$$

which hints at some form of anisotropic diffusion along the field lines. (In fact 11.48 is valid only at late times when the Laplacian is dominated by gradients in the x direction.) If this picture is correct, and we shall see that it is, this diffusion should proceed as

$$\ell_z \sim \delta(t/\tau)^{1/2}. \tag{11.49}$$

We therefore have two conflicting views. On the one hand, (11.46) suggests that the flow is destroyed on a time-scale of 2τ. On the other, (11.49) suggests a continual diffusive evolution of the vortex. We shall now show how these two viewpoints may be reconciled.

Consider the linear case where the magnetic field is relatively intense, so that $N \gg 1$ and nonlinear inertia may be neglected. We introduce the Fourier transform

$$\Psi(k_x, k_z) = 4\int_0^\infty\int_0^\infty \psi(x,z)\,\cos(xk_x)\,\cos(zk_z)dxdz \tag{11.50}$$

and apply this transform to (11.47), rewritten as

$$\frac{\partial}{\partial t}\nabla^2\psi = -\frac{1}{\tau}\frac{\partial^2\psi}{\partial z^2}. \tag{11.51}$$

Evidently,

$$\frac{\partial\Psi}{\partial t} + \cos^2\varphi\,\frac{\Psi}{\tau} = 0\,, \qquad \cos\varphi = k_z/k,$$

where k *is* the magnitude of $\mathbf{k}$. But this is identical to the equation we obtained for an MHD jet in §11.2.3. Thus, without any further work, we may say that at large times,

$$\psi(\mathbf{x}, t) = \frac{1}{2\pi\sqrt{\pi\hat{t}}} \int_0^\infty e^{-k^2 z^2/4\hat{t}} \cos(xk)\Psi_0(k)k\,dk, \qquad (11.52)$$

where $\hat{t} = t/\tau$ and Ψ_0 the transform of ψ at $t = 0$. Evidently, for large $\hat{t}$, this adopts the form

$$\psi(\mathbf{x}, t) \sim \hat{t}^{-1/2} F\left(z/\hat{t}^{1/2}, x\right), \qquad (11.53)$$

where F is determined by the initial conditions. It would appear, therefore, that the arguments leading to (11.49) are substantially correct. An initially axisymmetric vortex progressively distorts into a sheet-like structure, with a longitudinal length scale given by (11.49). Note that (11.53) implies that $u_x \ll u_z$ and $u_z \sim \hat{t}^{-1/2}$. It follows that the energy of the eddy is progressively channelled into the z-component of motion and that the global kinetic energy, $E = \int \frac{1}{2}\mathbf{u}^2 dV$, declines as $E \sim \hat{t}^{-1/2}$.

Let us now consider a specific example. Suppose that the initial eddy structure is described by the Gaussian profile

$$\psi_0(r) = \Phi_0 e^{-r^2/\delta^2}; \quad r^2 = x^2 + z^2. \qquad (11.54)$$

Then (11.52), which is valid for large t, may be integrated to give

$$\psi(\mathbf{x}, t) = \frac{\Phi_0 \delta^2}{\sqrt{\pi\hat{t}}} \frac{\zeta}{x^2} G(\zeta), \qquad \zeta = \frac{x^2}{\delta^2 + z^2/\hat{t}}, \qquad (11.55)$$

where G is Kummer's hypergeometric function, $G(\zeta) = M(1, 1/2, -\zeta)$. Now expressions (11.53) and (11.55) seem to contradict (11.46), which predicts that the angular momentum decays exponentially fast, rather than as a power law. However, (11.55) has an interesting property. For $t \gg \tau$, the global angular momentum, H_y, integrates to zero. It would appear, then, that the structure of the flow at large times is such that the net angular momentum is zero.

The reason for this can be seen from Figure 11.16, which shows the flow for $t \gg \tau$. (The structure of the flow at low N is also shown.) Regions of reverse flow occur either side of the centreline of the vortex. This reverse flow has a magnitude which is just sufficient to cancel the angular momentum of the primary eddy. We conclude, therefore, that the structure of the flow for large t is long and streaky, comprising vortex sheets of alternating sign. In short, the vorticity diffuses along the **B**-lines in accordance with (11.49) while simultaneously adopting a layered structure which has zero net angular momentum, thus satisfying (11.46).

Let us now consider a vortex whose axis is aligned with **B**. For simplicity, we restrict ourselves to axisymmetric vortices, described in terms of cylindrical-polar coordinates (r, θ, z), with **B** parallel to z. We shall neglect viscosity and assume that the initial conditions are such that the integral of the angular momentum converges at $t = 0$. Aspects of this problem have been touched upon in §6.2.3, and here we provide more details.

Suppose we have an isolated region of intense swirl, of characteristic initial scale δ, in an otherwise quiescent fluid. We may uniquely define the instantaneous state of the flow using just two scalar functions: Γ, the angular momentum density, and Ψ, the Stokes stream-function. These are defined through the expressions

$$\mathbf{u} = \mathbf{u}_\theta + \mathbf{u}_p = (\Gamma/r)\hat{\mathbf{e}}_\theta + \nabla \times [(\Psi/r)\hat{\mathbf{e}}_\theta], \tag{11.56}$$

$$\nabla_*^2 \Psi = \frac{\partial^2 \Psi}{\partial z^2} + r \frac{\partial}{\partial r}\left(\frac{1}{r}\frac{\partial \Psi}{\partial r}\right) = -r\omega_\theta. \tag{11.57}$$

Note that the velocity has been divided into azimuthal and poloidal components, which are individually solenoidal. The Lorentz force per unit mass, which is linear in **u**, may be similarly divided, giving

$$\mathbf{F}_p = -\frac{u_r}{\tau}\hat{\mathbf{e}}_r = \frac{1}{r\tau}\frac{\partial \Psi}{\partial z}\hat{\mathbf{e}}_r \quad , \quad F_\theta = -\frac{1}{\tau}\frac{J_r}{\sigma B} = \frac{1}{r\tau}\frac{\partial \varphi}{\partial z}. \tag{11.58}$$

Here φ is the Stokes stream-function for the poloidal current, $\mathbf{J}_p$, normalised by σB. By virtue of Ohm's law (11.23), φ is related to Γ by

$$\nabla_*^2 \varphi = -\partial \Gamma/\partial z. \tag{11.59}$$

The governing equations for Γ and Ψ are the azimuthal components of the momentum and vorticity equations, respectively. In particular, the angular momentum density satisfies

$$\frac{D\Gamma}{Dt} = \frac{1}{\tau}\frac{\partial \varphi}{\partial z}, \tag{11.60}$$

and hence

$$\nabla_*^2 \frac{D\Gamma}{Dt} = -\frac{1}{\tau}\frac{\partial^2 \Gamma}{\partial z^2}. \tag{11.61}$$

Note that (11.60) ensures conservation of the global angular momentum,

$$I_\Gamma = \int \Gamma dV = \text{constant}, \tag{11.62}$$

which is a special case of (11.43) and may be contrasted with the angular momentum of a transverse vortex.

Consider the linear case of large interaction parameter, $N \gg 1$. Then our governing equation for Γ is simply

$$\frac{\partial}{\partial t} \nabla_*^2 \Gamma = -\frac{1}{\tau} \frac{\partial^2 \Gamma}{\partial z^2}, \tag{11.63}$$

whose general form we recognise from (11.32) and (11.51). We expect, therefore, that the angular momentum density will diffuse along the imposed magnetic field at the rate $\ell_{//} \sim \delta(t/\tau)^{1/2}$, in much the same way as vorticity diffuses along the imposed field in a submerged jet and in a transverse vortex. In fact, we can confirm that this is so by using conservation of angular momentum. From (11.26) and (11.62) we have

$$\frac{dE}{dt} \sim -\left(\frac{\delta}{\ell_{//}}\right)^2 \frac{E}{\tau}, \quad E \sim \Gamma^2 \ell_{//}, \quad \Gamma \delta^2 \ell_{//} = \text{constant}, \tag{11.64}$$

which yield the anticipated scaling laws

$$\ell_{//} \sim \delta(t/\tau)^{1/2}, \quad \Gamma \sim \Gamma_0(t/\tau)^{-1/2}. \tag{11.65}$$

These scaling laws may also be confirmed by direct calculation, and in particular by taking the Hankel-cosine transform of (11.63), as shown in Example 6.1 in Chapter 6. By way of a specific example, suppose that, at $t = 0$, we have a spherical blob of swirling fluid, and that our initial condition is the simple Gaussian

$$\Gamma_0(r, z) = \Omega r^2 \exp[-(r^2 + z^2)/\delta^2].$$

Then it is readily confirmed, using the results of Example 6.1, that

$$u_\theta(\hat{t} \to \infty) = \frac{3\sqrt{\pi}}{4} \frac{\Omega \delta}{\hat{t}^{1/2}} \left(\frac{\delta}{r}\right)^4 \zeta^{5/2} H(\zeta), \quad \zeta = \frac{r^2}{\delta^2 + z^2/\hat{t}},$$

where $\hat{t} = t/\tau$ and $H(\zeta)$ is the hypergeometric function $H(\zeta) = M(5/2, 2, -\zeta)$. Curiously, for large ζ, the function H becomes negative, so that the primary vortex is surrounded by an annulus of counter-rotating fluid. This may be attributed to the way in which the induced currents recirculate back through quiescent regions

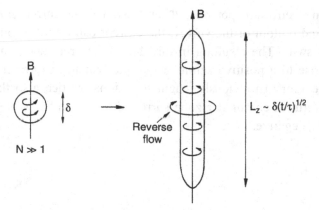

Figure 11.17 Magnetic damping of a parallel vortex at high N. The figure shows schematically the structure of the flow at large times.

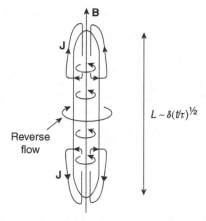

Figure 11.18 Damping of a swirling vortex at high N. The negative radial current above and below the vortex gives rise to a positive azimuthal torque which spins up previously still fluid, while an annulus of reverse flow is generated by the positive radial current.

outside the initial vortex, as discussed below. We conclude, therefore, that the asymptotic structure of a vortex aligned with **B** is as shown schematically in Figure 11.17. It is cigar-like in shape, and quite different in structure to the transverse vortex shown in Figure 11.16. Curiously, though, despite the fact that the two classes of vortices adopt very different spatial structures, their energies both decay as $E \sim (t/\tau)^{-1/2}$.

The mechanism for the propagation of angular momentum along the field lines is shown in Figure 11.18. The term $\mathbf{u}_\theta \times \mathbf{B}$ tends to drive a radial current, J_r. Near the centre of the vortex, where the axial gradient in Γ is small, this is largely counterbalanced by an electrostatic potential, Φ, and so almost no current flows. However, near the top and bottom of the vortex, the current can return through regions of small or zero swirl. The resulting inward flow of current above and below the vortex gives rise to a positive azimuthal torque which, in turn, creates positive angular momentum in previously stagnant regions. Notice also that regions of reverse flow form in an annular zone surrounding the initial vortex where J_r is positive and F_θ negative.

12

Axisymmetric Flows Driven by the Injection of Current

Everything should be made as simple as possible, but not simpler.

A. Einstein

When a current is passed through a liquid metal it causes the metal to pinch in on itself. That is to say, like-signed currents attract each other. If the current is perfectly uniform and there is no free surface, the only effect is to pressurise the liquid. However, often the current is non-uniform; for example, it may spread radially outward from a small electrode placed at the surface of a liquid-metal pool, as in electric-arc welding. In such cases the radial pinch force will also be non-uniform, being largest where the current density is highest (near the electrode). The irrotational pressure force, $-\nabla p$, is then unable to balance the rotational Lorentz force, and motion ensues, with the fluid flowing inward in regions of high current density, and returning through regions of weaker current. Such flows assumed commercial significance towards the end of the nineteenth century when Hall and Héroult developed the first commercial aluminium reduction cells and von Siemens designed the first electric-arc furnace. Shortly thereafter they became the subject of scientific investigation by, amongst others, the American E.F. Northrup. Perhaps the most important modern descendant of the electric-arc furnace is a process known as vacuum-arc remelting, whose flow structure is subtle and unexpected, often verging on the exotic. The fluid mechanics of vacuum-arc remelting provides the focus of this chapter.

12.1 The Need to Purify Metal for Critical Aircraft Parts: Vacuum-Arc Remelting

There are occasions when an ingot cast by conventional means is of inadequate quality, either because of excessive porosity in the ingot or else because it contains unacceptably high levels of pollutants (oxides, refractory material and so on). This is particularly the case in the casting of high-temperature melts, such as titanium or nickel-based alloys, which tend to erode the refractory vessel in which they are melted. Such

351

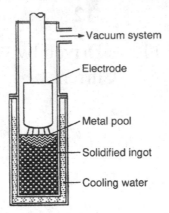

Figure 12.1 Vacuum-arc remelting (VAR).

specialist alloys are widely used in critical aircraft components, particularly in engines, where the slightest imperfection can lead to a fatigue crack and to catastrophic failure. In such situations it is normal to improve the quality of the primary ingot by remelting it in a partial vacuum, burning off the impurities, and then slowly casting a new ingot. This is achieved by a process known as vacuum-arc remelting (VAR).

In effect, VAR looks like a giant version of electric-arc welding, where an arc is struck between an electrode and an adjacent metal surface, with the entire process taking place in a vast vacuum chamber. In VAR the electrode consists of the primary ingot which is to be melted and purified, which typically has a diameter of around one metre. As in electric welding, once an arc is struck the electrode starts to melt and droplets of molten metal fall through the plasma arc to form a liquid-metal pool on a cooled plate below. As this pool slowly solidifies it forms a new, cleaner, ingot (Figure 12.1). The entire process takes place in a partial vacuum and the metallurgical structure of the final ingot depends critically on the temperature distribution and fluid motion within the liquid-metal pool. Of course these, in turn, depend on the buoyancy and Lorentz forces acting on the liquid metal. There is considerable incentive, therefore, to understand the structure of the flow within the pool and how this scales with, say, ingot diameter and current. We shall see that, despite considerable research over the last few decades, this flow is still poorly understood.

The liquid-metal pool in VAR is subject to three primary sources of motion:

(i) the Lorentz force associated with the interaction of the electric current with its self-induced horizontal magnetic field

(ii) buoyancy forces arising from the difference in temperature between the hot plasma above and cooler ingot below

(iii) the Lorentz force induced by the interaction of the electric current with stray, externally imposed magnetic fields, particularly vertical fields.

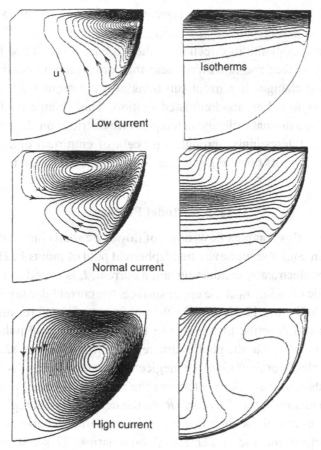

Figure 12.2 Changes in flow (left) and temperature field (right) with current in the liquid pool in a VAR unit. The top panels are typical of a buoyancy-dominated flow (low current), while the bottom panels are characteristic of Lorentz-dominated motion (high current). The normal operating point lies between the two.

In a well-designed VAR unit this latter source of motion is small, and so the flow usually arises from the competing effects of buoyancy and the Lorentz force associated with the horizontal magnetic field, both of which tend to drive motion in a vertical plane (Figure 12.2). However, on those occasions when there is a significant stray vertical magnetic field, perhaps due to the external conductors which carry current to and from the VAR unit, a particularly intense horizontal swirling motion develops in the liquid pool, a motion which is detrimental to the ingot structure. Famously, a very weak stray field of only a few Gauss is sufficient to trigger this intense, undesirable swirl.

In the absence of a stray, external magnetic field there is delicate balance between the Lorentz force associated with the horizontal magnetic field, which tends to drive a poloidal flow which converges at the surface of the pool, and the buoyancy forces associated with the relatively hot upper surface. (The buoyancy-driven motion is opposite in direction to the Lorentz-driven flow, being radially outward at the surface and downward near the cooler lateral boundary.) We shall see that modest changes in current can transform the motion from a buoyancy-dominated flow to a Lorentz-dominated motion. This change in flow regime is accompanied by a dramatic change in temperature distribution (Figure 12.2) and of ingot structure. Interestingly, through a process of empirical optimisation, most commercial VAR units operate just at the verge of this transition in flow regimes.

12.2 A Model Problem

In order to study this complex flow, or set of flows, we shall consider the following model problem. Suppose we have a hemispherical pool of radius R. The boundaries of the pool are electrically conducting and a current, I, is introduced into the pool via an electrode of radius r_0 at the upper surface, the current density being uniform in the electrode (Figure 12.3). We take the pool geometry and resulting flow to be axisymmetric and describe the motion in terms of cylindrical polar coordinates (r, θ, z), with the origin at the pool's surface, as shown in Figure 12.3.

In the simplest configuration we neglect buoyancy forces as well as stray magnetic fields, and we seek to determine the resulting poloidal flow (i.e. flow in the r–z plane) as a function of I, r_0 and R. At the next level of complexity we admit buoyancy forces and allow for a difference in temperature between the relatively hot upper surface and the cooler lateral boundaries. This also tends to drive a poloidal (r–z) flow, though in the opposite direction. Finally, in the most complex configuration, we alloy for an externally imposed vertical magnetic field, which interacts with the radial current in the pool to generate a horizontal (azimuthal) Lorentz force, which then drives a horizontal swirling flow.

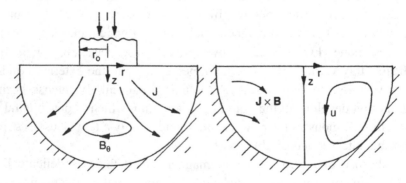

Figure 12.3 Geometry of our model problem.

Consider first the Lorentz-driven flow in the absence of an externally imposed magnetic field. The flow is as shown in Figure 12.3. Like-signed currents attract, and this radially inward force is strongest near the top of the pool, where the current density is largest. So the flow is inward and downward near the upper surface, being recycled via the curved lateral boundary. Let us put this on a firmer mathematical basis. The poloidal current shown in Figure 12.3 gives rise to an azimuthal field, B_θ, with the two related by Ampère's law:

$$2\pi r B_\theta = \mu \int_0^r (2\pi r J_z)dr. \tag{12.1}$$

The interaction of **J** with B_θ gives rise to the Lorentz force,

$$\mathbf{F} = \mathbf{J} \times \mathbf{B}/\rho = -\nabla\left(B_\theta^2/2\rho\mu\right) - [B_\theta^2/(\rho\mu r)]\hat{\mathbf{e}}_r. \tag{12.2}$$

Since the magnetic pressure merely augments the fluid pressure, it does not influence the motion within the pool. We may therefore drop the first term on the right of (12.2) to give

$$\mathbf{F} = -\frac{B_\theta^2}{\rho\mu r}\hat{\mathbf{e}}_r, \tag{12.3}$$

on the understanding that p is augmented by $B_\theta^2/(2\mu)$. This Lorentz force drives a recirculating flow which converges at the surface (where B_θ is largest) and diverges near the base of the pool.

This deceptively simple flow has been the subject of myriad studies, with almost an entire book devoted to it (Bojarevics et al., 1989). Part of the enduring fascination with this flow arises from the fact that it is surprisingly rich. For example, it becomes unsteady and eventually turbulent at a surprisingly low Reynolds number, around Re ~ 10. Moreover, as noted above, it is extremely sensitive to weak, stray magnetic fields of only 1 or 2 Gauss. (The surface geomagnetic field is around ½ Gauss.) In fact, as we shall see, stray magnetic fields of only 1 per cent of the primary azimuthal field, B_θ, are sufficient to completely transform the flow from a poloidal (vertical) motion into a predominantly horizontal, swirling flow. Moreover, even in the absence of a stray magnetic field, the buoyancy forces associated with the hot upper surface are in direct competition with the Lorentz force, tending to drive poloidal motion in the opposite direction, and this competition can produce dramatic changes in flow structure for only modest changes in current.

There is another reason for the fascination with this particular flow. It turns out that there is an exact, non-linear, laminar solution for the special case of a point electrode,

$r_0 \to 0$, feeding current into a semi-infinite domain. This self-similar solution has been quite influential, and some of its features have been used to try and explain certain experimental observations, such as the extreme sensitivity to stray magnetic fields and the associated tendency to develop a strong horizontal swirl. However, as we shall see, this attempt to interpret real events in terms of this idealised solution have been largely misplaced, and indeed the exact solution for a semi-infinite domain tells us very little about the confined flow shown in Figure 12.3.

There are three key differences between the self-similar solution in a semi-infinite domain and a real, confined flow. The first is that the streamlines do not close on themselves in the exact, self-similar solution, but rather enter near the surface at large radius and depart at large z. Consequently, we are free to specify any upstream condition we wish in the self-similar solution, which is clearly not a luxury we have for the case in a real, confined flow. Second, the confined flow develops boundary layers at large Re, and these are very dissipative. Third, in a confined flow the rate of working of the Lorentz force must exactly balance (in the steady state) the rate of viscous dissipation. This is not the case for the idealised self-similar solution, where any difference in the rate of working of the Lorentz and viscous forces simply manifests itself in a difference in the mechanical energy of the incoming and outgoing fluid. In short, a real confined flow is subject to important constraints that the idealised solution is not. We now explore these constraints in a little more detail.

12.3 Integral Constraints and the Work Done by the Lorentz Force

As suggested above, we shall find it convenient to deploy various energy arguments in the sections that follow, so perhaps it is worth saying something about these energy constraints now. The Lorentz force, $\mathbf{J} \times \mathbf{B}$, does work on the fluid. In a confined flow, this causes the kinetic energy to rise until such time as the viscous dissipation matches the rate of working of $\mathbf{J} \times \mathbf{B}$. If, for a given current, we can characterise the relationship between $(\mathbf{J} \times \mathbf{B}) \cdot \mathbf{u}$ and the rate of dissipation of energy, then we should be able to estimate the magnitude of $|\mathbf{u}|$. Thus the key to estimating $|\mathbf{u}|$ often lies in determining the mechanism by which the fluid dissipates the energy injected into the flow. For example, is the dissipation confined to boundary layers or are internal shear layers set up, and what happens to those streamlines which manage to avoid all such dissipative layers? There are two energy equations of importance here, both of which rest on the steady version of the Navier–Stokes equation,

$$\frac{\partial \mathbf{u}}{\partial t} = \mathbf{u} \times \boldsymbol{\omega} - \nabla \left[\frac{p}{\rho} + \frac{u^2}{2} \right] + \nu \nabla^2 \mathbf{u} + \mathbf{F} = 0, \qquad (12.4)$$

where $\mathbf{F}$ is the Lorentz force per unit mass, $\mathbf{J} \times \mathbf{B}/\rho$. The first equation comes from integrating (12.4) around a closed streamline, C, which yields

$$\oint_C \mathbf{F} \cdot d\mathbf{l} = -\nu \oint_C \nabla^2 \mathbf{u} \cdot d\mathbf{l}. \tag{12.5}$$

This represents the balance between the work done by the Lorentz and viscous forces acting on a fluid particle as it passes once around a closed streamline. Specifically, the work done by $\mathbf{F}$ on the fluid particle must be dissipated or diffused out of the particle before it returns to its starting point. The second energy equation comes from taking the product of (12.4) with $\mathbf{u}$ and then integrating the result over the entire domain. Noting that $(\nabla^2 \mathbf{u}) \cdot \mathbf{u} = \nabla \cdot (\mathbf{u} \times \boldsymbol{\omega}) - \boldsymbol{\omega}^2$, we find

$$\int \mathbf{F} \cdot \mathbf{u} \, dV = \nu \int \boldsymbol{\omega}^2 dV. \tag{12.6}$$

This represents a global balance between the rate of working of the Lorentz force and the viscous dissipation. Either (12.5) or (12.6) may be used to estimate the magnitude of $\mathbf{u}$, provided, of course, that F is known. Actually, it is not difficult to show that (12.6) is equivalent to evaluating (12.5) for each streamline in the flow and then summing over all such integrals.

Let us see where integral constraint (12.5) leads for the model problem shown in Figure (12.3), in which $\mathbf{F}$ is given by

$$\mathbf{F} = -\frac{B_\theta^2}{\rho\mu r} \hat{\mathbf{e}}_r.$$

We take C to be the bounding streamline, comprising the upper surface, the axis and the curved outer boundary. Starting with the left-hand integral and integrating in an anti-clockwise direction (in the direction of the flow), we obtain

$$\oint \mathbf{F} \cdot d\mathbf{l} = \left[\int_0^R (B_\theta^2/\rho\mu r)\, dr \right]_{z=0} - \left[\int_0^R (B_\theta^2/\rho\mu r)\, dr \right]_{r^2+z^2=R^2}. \tag{12.7}$$

(The integral of $\mathbf{F}$ along the symmetry axis is zero because $B_\theta = 0$ on $r = 0$.) The first of these integrals is readily evaluated since, from Ampère's law, the surface magnetic field is $2\pi r B_\theta = \mu I (r/r_0)^2$ for $r < r_0$ and $2\pi r B_\theta = \mu I$ for $r > r_0$. This yields

$$\left[\int_0^R (B_\theta^2/\rho\mu r)\, dr \right]_{z=0} = \frac{\mu I^2}{4\pi^2 \rho r_0^2} [1 - (r_0^2/2R^2)]. \tag{12.8}$$

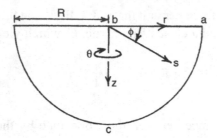

Figure 12.4 Coordinate system used when dealing with a point electrode.

The second integral is, in general, more problematic. However, for the particular case of a point source of current, $r_0 \ll R$, the integral is readily evaluated. In this situation it is useful to introduce the additional (spherical polar) coordinates, s and φ, defined by $s^2 = r^2 + z^2$ and $\cos\varphi = r/s$, as shown in Figure 12.4. It is not difficult to show using (12.1) and (12.3) that, in terms of s and φ, a point source of current leads to

$$\mathbf{J} = \frac{I}{2\pi s^2}\ (\cos\varphi\ \hat{\mathbf{e}}_r + \sin\varphi\ \hat{\mathbf{e}}_z),\quad \mathbf{B} = \frac{\mu I}{2\pi r}(1 - \sin\varphi)\hat{\mathbf{e}}_\theta, \qquad (12.9)$$

and hence

$$\mathbf{F} = -\frac{\mu I^2}{4\pi^2 \rho r^3}(1 - \sin\varphi)^2\hat{\mathbf{e}}_r. \qquad (12.10)$$

This then yields

$$\left[\int_0^R (B_\theta^2/\rho\mu r)dr\right]_{r^2+z^2=R^2} = \frac{\mu I^2}{4\pi^2 \rho R^2}\ [\ln 2 - 1/2]. \qquad (12.11)$$

Combining this with (12.8) gives us, for a point electrode,

$$\oint \mathbf{F}\cdot d\mathbf{l} = \frac{\mu I^2}{4\pi^2 \rho}\left[\frac{1}{r_0^2} - \frac{\ln 2}{R^2}\right]. \qquad (12.12)$$

Of course, for cases where r_0 does not satisfy $r_0 \ll R$, the factor of $\ln 2/R^2$ (but *not* the dominant term, $1/r_0^2$) will need to be modified in (12.12), but this is a detail. The key point is that (12.12), with some correction to the coefficient ln2, represents the work done by $\mathbf{F}$ on a fluid particle as it completes one cycle in the r–z plane. It is the balance between this integral and the viscous dissipation which determines the magnitude of the induced velocity:

$$\oint \mathbf{F} \cdot d\mathbf{l} = \frac{\mu I^2}{4\pi^2 \rho} \left[\frac{1}{r_0^2} - \frac{\ln 2}{R^2} \right] = -\nu \oint \nabla^2 \mathbf{u} \cdot d\mathbf{l}. \tag{12.13}$$

It is interesting to note that $\oint \mathbf{F} \cdot d\mathbf{l}$ tends to infinity as $r_0 \to 0$. This, in turn, suggests that there is something singular about the point electrode problem, and indeed there is. We shall return to this point shortly.

12.4 Structure and Scaling of the Flow

12.4.1 Confined versus Unconfined Domains

We now consider the structure of the flow for the model problem shown in Figure 12.3. In this section we focus on the simplest case where buoyancy forces and stray magnetic fields are neglected. We shall pull in the effects of buoyancy in §12.5 and of stray magnetic fields in §12.6; here we simply ask what flow is driven by force (12.3).

We start with a note of caution. Both electric welding and VAR are characterised by the facts that (i) the electrode has an appreciable size relative to R, (ii) the Reynolds number is high and the flow turbulent and (iii) the presence of the boundary at $s = R$ controls the magnitude of $\mathbf{u}$, since most of the dissipation occurs in the boundary layers. Nevertheless, most studies of this problem have focused on laminar flow driven by a point electrode in a semi-infinite domain. The reason for this concentration on an idealised problem was the discovery by Shercliff (1970), and others, that there exists an exact, self-similar solution of the Navier–Stokes equation for the case of a point electrode on the surface of a semi-infinite domain. Unfortunately, as suggested above, these point electrode, semi-infinite-domain problems can be quite misleading in the context of real, confined flows. There are several key differences. For example, confined domains are subject to (intense) dissipation associated with the boundary layer at $s = R$. This is significant since, as we have seen,

$$\int_V \mathbf{F} \cdot \mathbf{u} dV = \nu \int_V \boldsymbol{\omega}^2 dV. \tag{12.14}$$

That is to say, the global rate of working of the Lorentz force $\mathbf{F}$ must be balanced by viscous dissipation. For confined domains, the right-hand side of (12.14) is dominated by the boundary-layer dissipation and so we might expect the boundary layers to determine the magnitude of $\mathbf{u}$. However, there are no boundary layers in the infinite domain problem, and so we might expect the characteristic velocity in confined and unconfined problems to be rather different.

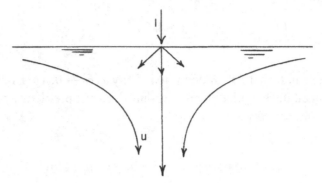

Figure 12.5 Schematic of flow in a semi-infinite domain driven by a point electrode.

A second, related difference is that the streamlines in the semi-infinite problem converge towards the axis but do not close on themselves (Figure 12.5). They are therefore free from integral constraints of the form

$$\oint_c \mathbf{F} \cdot d\mathbf{l} + \nu \oint_c \nabla^2 \mathbf{u} \cdot d\mathbf{l} = 0, \tag{12.15}$$

where C is a closed streamline. Integrals such as (12.15) determine the magnitude of $\mathbf{u}$ in closed-streamline problems, yet are largely irrelevant in cases where the streamlines are open. Thus, for example, any difference between $\int \mathbf{F} \cdot d\mathbf{l}$ and $\nu \int \nabla^2 \mathbf{u} \cdot d\mathbf{l}$ in the semi-infinite problem simply appears as a difference in the mechanical energy of the incoming and outgoing fluid.

A third crucial difference is evident from (12.12). The point electrode problem represents a singular case in which the work done by $\mathbf{F}$ on the fluid becomes infinite:

$$\oint \mathbf{F} \cdot d\mathbf{l} = \frac{\mu I^2}{4\pi^2 \rho r_0^2} \to \infty. \tag{12.16}$$

The implication is that, whenever ν is small, a fluid particle will acquire an infinite amount of kinetic energy as it passes by the electrode. This turns out to be the case, and it is the hallmark of these point electrode, semi-infinite-domain problems that, as ν becomes small (Re is greater than $O(1)$), a singularity appears in the velocity field. Indeed, it is the intriguing appearance of a singularity in $\mathbf{u}$ which has made this point electrode problem such a popular subject of study. It should be emphasised, however, that the appearance of a singularity in $\mathbf{u}$ is simply an artefact of the (unphysical) assumption that r_0 is vanishingly small.

All in all, it would seem that confined and unconfined flows represent quite different problems. Our primary concern here is in confined flows, such as those which occur in VAR. Nevertheless, since the bulk of the literature addresses the semi-infinite-domain, point electrode problem, it would seem prudent to review first the key features of this flow.

12.4.2 Shercliff's Solution for Unconfined Domains

Let us consider a semi-infinite domain and look for a solution of (12.4) in which **F** is given by the point electrode distribution (12.10). It is convenient to introduce the Stokes streamfunction defined by

$$\mathbf{u} = \left(-\frac{1}{r}\frac{\partial \psi}{\partial z}, \; 0, \; \frac{1}{r}\frac{\partial \psi}{\partial r} \right), \tag{12.17}$$

and to take the curl of (12.4), converting it into the vorticity transport equation

$$\mathbf{u} \cdot \nabla \left(\frac{\omega_\theta}{r} \right) = \frac{1}{r}\frac{\partial F}{\partial z} + \nu \left[\nabla^2 \left(\frac{\omega_\theta}{r} \right) + \frac{2}{r}\frac{\partial}{\partial r}\left(\frac{\omega_\theta}{r} \right) \right]. \tag{12.18}$$

We now look for a self-similar solution of (12.18) of the form

$$\psi = -sg(\eta), \qquad \eta = \sin\varphi = z/s, \tag{12.19}$$

where g has the dimensions of m²/s. Substituting into (12.18) and integrating three times we obtain, after some algebra, the governing equation for g:

$$g^2 + 2K(1+\eta)^2\ln(1+\eta) + 2\nu[(1-\eta^2)g' + 2\eta g] = K[a\eta^2 + b\eta + c]. \tag{12.20}$$

Here the primes represent differentiation with respect to η, the constant K is

$$K = \frac{\mu I^2}{4\pi^2\rho}, $$

and a, b and c are constants of integration.

The simplest case to consider is the inviscid one. The constants a, b and c are then determined (in part) by the requirements that (i) u_z is zero on $z=0$ and (ii) u_r is zero on $r=0$. These conditions are equivalent to demanding that $g(0)=g(1)=0$. Inspection of the inviscid version of (12.20) then yields $c=0$ and $a+b=8\ln2$, from which

$$g^2 = K[a\eta(\eta-1) + (8\ln2)\eta - 2(1+\eta)^2\ln(1+\eta)]. \tag{12.21}$$

Of course, the question now is, 'What determines a?' Before answering this question it is instructive to return to (12.8), which gives the integral of $\mathbf{F}$ along the surface of the pool from $r = R$ to $r = 0$. If we let R recede to infinity, we obtain

$$\int_{r=\infty}^{r=0} \mathbf{F} \cdot d\mathbf{l} = \frac{\mu I^2}{4\pi^2 \rho r_0^2} = \frac{K}{r_0^2}. \tag{12.22}$$

This represents the work done on a fluid particle as it moves inward along the surface under the influence of the Lorentz force. Recall that r_0 is the radius of the electrode. For a point source of current this integral diverges. Evidently, in the case of a point electrode, an infinite amount of work is done on each fluid particle as it moves radially inward along the surface. This suggests that something is going to go wrong with our inviscid solution, since we have no mechanism for dissipating the energy created by the Lorentz force, $\mathbf{F}$. In practice, this manifests itself in the following way. We could try to fix a by demanding that u_r be finite on the surface (i.e. the incoming flow has finite energy). In such cases we find that u_z is infinite on the symmetry axis (i.e. the outgoing flow has infinite energy), which is an inevitable consequence of (12.22). The details are simple to check. The requirement that u_r be finite on $z = 0$ demands that $a = 8 \ln 2 - 2$, from which

$$g^2 = K[(8 \ln 2 - 2)\eta^2 + 2\eta - 2(1 + \eta)^2 \ln(1 + \eta)].$$

Near the axis, however, this leads to an axial velocity of

$$u_z = -\frac{K^{1/2}(3 - 4 \ln 2)^{1/2}}{r},$$

which diverges as r tends to zero.

If we now reinstate viscosity into our analysis, it seems plausible that a regular solution of (12.20) will emerge, provided, of course, that the viscous stresses are large enough to combat the tendency for $\mathbf{F}$ to generate an infinite kinetic energy. In practice, this is exactly what occurs. When the (Reynolds-like) parameter $K^{1/2}/\nu$ is less than ~7, regular solutions of (12.20) exist without any singularity in $\mathbf{u}$. For higher values of $K^{1/2}/\nu$, u_z becomes singular on the axis (Bojarevics et al., 1989). Of course, this does not imply that anything special, such as an instability, occurs at the critical value of $K^{1/2}/\nu$. It merely means that our attempt to find a self-similar solution of the form $\psi \sim sg(\eta)$ has failed. Notice that $K^{1/2}/\nu = 7$ corresponds to a relatively low current of around one Amp, which is several orders of magnitude smaller than the currents used in industrial applications.

12.4.3 Confined Flows

Let us now return to flows which are confined to the hemisphere $s < R$ and in which the electrode has a finite radius, r_0, which we now take to be of order R. In VAR the Reynolds number is invariably high. It is natural, therefore, to ask two questions:

(i) What is the structure of the laminar flow when Re is large?
(ii) What is the magnitude of $\mathbf{u}$ when the flow becomes turbulent?

The answer to the first of these questions is surprising: it is likely that there are no stable, laminar flows at moderate-to-high values of Re. The reasons for this are discussed in detail in Kinnear and Davidson (1998) and we will give only a brief summary of the arguments here. Suppose that we have a steady, laminar flow and that $r_0 \sim R$, then the key equation is (12.5),

$$\oint \mathbf{F} \cdot d\mathbf{l} + \nu \oint \nabla^2 \mathbf{u} \cdot d\mathbf{l} = 0. \tag{12.23}$$

This integral constraint is a powerful one. It must be satisfied by *every* closed streamline. When Re is of the order unity (or less) it tells us that $u \sim FR^2/\nu$, where F is a characteristic value of $|\mathbf{F}|$. This gives $u \sim K/\nu R = \mu I^2/(4\pi^2 \rho \nu R)$. Now suppose that Re is large so that boundary layers form on the wall $s = R$. Inside the boundary layer the viscous dissipation is intense while outside it is small. The boundary thickness, δ, is determined by the force balance $(\mathbf{u} \cdot \nabla)\mathbf{u} \sim \nu \nabla^2 \mathbf{u}$, which gives $\delta \sim (\text{Re})^{-1/2} R$. Thus our integral equation applied to a streamline lying close the boundary yields $FR \sim (\nu u R)/\delta^2 \sim u^2$. For a streamline away from the boundary, however, $\nabla^2 \sim R^{-2}$, and so (12.23) demands $FR \sim (\nu u R)/R^2 \sim \nu u/R$. The implication is that the flow in the boundary layer scales as $u_b \sim (FR)^{1/2}$, while that in the core scales according to $u_c \sim FR^2/\nu$, which is much larger than u_b. However, this cannot be so, since the velocity scale in the boundary layer is set by the core velocity. Clearly, something has gone wrong.

The numerical experiments discussed in Kinnear and Davidson (1998) suggest that nature resolves this dilemma in an unexpected way. At surprisingly low Reynolds numbers, of the order of 10, the flow becomes unstable and starts to oscillate. The integral equation (12.23) is then irrelevant. The oscillation consists of a periodic 'bursting' motion in the boundary layer which gives rise to a continual exchange of fluid between the dissipative boundary layer and the less-dissipative core. If we now increase Re a little further, the flow becomes turbulent, which brings us to our second question.

We wish to determine how $|\mathbf{u}|$ scales with I and R in a turbulent flow. Let us apply (12.23) to the time-averaged streamlines of a turbulent flow, with Reynolds stresses replacing the laminar shear stress. Noting that, for a streamline close to the boundary, (12.12) with $r_0 \approx R$ yields

$$\oint \mathbf{F} \cdot d\mathbf{l} \approx (1 - \ln 2)K/R^2,$$

we obtain

$$(1 - \ln 2)\frac{K}{R^2} \sim \frac{\pi R}{2}\frac{\tau_w}{\rho \delta_w}.$$

Here τ_w and δ_w are the wall shear stress and the characteristic length scale for gradients in τ near the wall. Now $\tau_w/\rho \sim (u')^2$ and so our energy balance can be used to estimate the turbulence level in the pool:

$$(1 - \ln 2)\frac{K}{R^2} \sim \frac{\pi R}{2}\frac{(u')^2}{\delta_w}.$$

We now take $u' \sim u/3.5$ and $\delta_w \sim R/10$, where u is a typical mean velocity. (These estimates are characteristic of a confined, turbulent, MHD flow, as observed in, say, induction furnaces.) In this case our energy balance yields

$$u \sim 0.5\frac{K^{1/2}}{R} \sim 0.5\frac{\sqrt{\mu/\rho}\, I}{2\pi R}. \tag{12.24}$$

Velocities compatible with (12.24) are indeed observed in the numerical simulations of such flows.

12.5 The Influence of Buoyancy

So far we have neglected the buoyancy forces acting on the pool. In VAR these are typically as large as the Lorentz force. Indeed, in some cases, they are the dominant forces acting on the liquid. It is useful to start by considering two extremes: one in which buoyancy may be neglected by comparison with $\mathbf{J} \times \mathbf{B}$, and the other in which buoyancy greatly outweighs the Lorentz force. These two extremes are shown in Figure 12.6. Notice that the Lorentz and gravitational forces tend to drive motion in opposite directions.

Davidson and Flood (1994) address the problem of natural convection of liquid metal in an axisymmetric cavity driven by a difference in temperature, ΔT, between the upper surface and the lateral boundary. The maximum velocity in the pool is of the order of

$$u_b \approx 1.3\frac{\alpha}{R}\left(\frac{g\beta R^3 \Delta T}{\alpha^2}\right)^{3/7} = 1.3\frac{\alpha}{R}(\mathrm{Gr})^{3/7}, \tag{12.25}$$

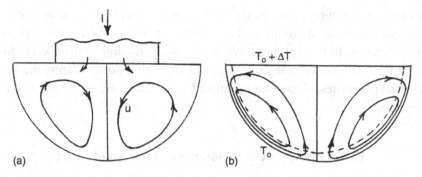

Figure 12.6 Two extremes in vacuum-arc remelting: (a) buoyancy forces are neglected; (b) the Lorentz forces are neglected.

where α is the thermal diffusivity, β the expansion coefficient, and Gr is a Grashof-like number. Compare this with the other extreme where buoyancy is neglected and the flow is driven by $\mathbf{J} \times \mathbf{B}$,

$$u \sim 0.5 \frac{\sqrt{\mu/\rho}\, I}{2\pi R} = 0.5 \frac{K^{1/2}}{R}. \tag{12.26}$$

In cases where the gravitational forces are dominant the fluid diverges at the surface and falls at the outer boundary. When the Lorentz force dominates we have the opposite pattern, with the fluid converging at the surface. We might estimate the point of transition between these two flows by equating (12.25) and (12.26): $\alpha(\text{Gr})^{3/7} \sim K^{1/2}$. Thus, the transition from buoyancy to Lorentz-driven flow should occur when the dimensionless parameter,

$$\chi = \frac{K^{1/2}}{\alpha \text{Gr}^{3/7}} = \frac{\left(\mu I^2/4\pi^2\rho\right)^{1/2}}{\alpha(g\beta R^3 \Delta T/\alpha^2)^{3/7}} = \frac{\sqrt{\mu/\rho}\, I/2\pi}{\sqrt{g\beta R^3 \Delta T}} \text{Gr}^{1/14}, \tag{12.27}$$

exceeds a number of order unity. Note that, if we ignore the weak dependency of χ on $\text{Gr}^{1/14}$, this ratio is independent of thermal diffusivity.

In practice, it is found that the motion resembles a classical buoyancy-driven flow when χ is less than ~0.4. In such cases the Lorentz forces may be neglected when calculating $\mathbf{u}$. Conversely, when χ exceeds ~1.4 the buoyancy forces are unimportant. For intermediate values $0.4 < \chi < 1.4$, the flow may have a complex, multi-cellular structure. This is illustrated in Figure 12.2 where the three sets of panels correspond to $\chi = 0.5, 1.2, 1.5$.

Evidently, modest changes in current can transform the motion from a buoyancy-dominated flow to a Lorentz-dominated one. This change in flow regime is accompanied by a dramatic change in temperature distribution (Figure 12.2) and hence of ingot structure. Interestingly, many commercial VAR units operate just at the verge of transition, at around $\chi \sim 1$, an operating point that has been achieved by empirical optimisation. It is impressive that the steel, titanium and nickel industries have all independently converged to the same operating point, despite the fact that the material properties, electrode current and ingot radius are quite different for the three metals.

12.6 The Apparent Spontaneous Growth of Swirl

There are several industrial processes, including poorly designed VAR units, where current is injected into a liquid-metal pool via its surface, but where the pool is also subject to a weak, stray magnetic field, perhaps associated with remote inductors. In such cases, the weak, stray magnetic field can have an astonishing influence on the motion in the pool, often to the detriment of the process. There is some incentive, therefore, to understand why stray magnetic fields have such a disproportionate influence on the pool motion. It is this question which occupies the remainder of this chapter.

12.6.1 An Extraordinary Experiment

Bojarevics et al. (1989) report an intriguing experiment which exhibits a curious phenomenon, often called 'spontaneous swirl'. In this experiment, current is passed radially downward through an axisymmetric pool of liquid metal, as indicated in Figure 12.7. As we have seen, the interaction of the current density, $\mathbf{J}$, with its associated magnetic field, B_θ, gives rise to a Lorentz force, $\mathbf{F} = \mathbf{J} \times \mathbf{B}/\rho$, which is poloidal (i.e. in the r–z plane). Of course, the resulting motion is also poloidal; at least this is the case at low levels of forcing. At higher levels of current, though, something rather surprising occurs in the experiment. The pool is seen to rotate, and this rotation is much more vigorous than the poloidal motion.

The observed rotation must, in some sense, result from a lack of symmetry in $\mathbf{J}$, or else from a stray magnetic field. That is, a finite azimuthal force, F_θ, is required to maintain the swirl and in particular to overcome the frictional torque exerted on the pool by the boundaries. This additional force is thought to arise from the interaction of $\mathbf{J}$ with a weak, stray magnetic field. In particular, an externally imposed vertical field, B_z, gives rise to the azimuthal force,

$$F_\theta = -\frac{J_r B_z}{\rho} = \frac{B_z}{\rho\mu}\frac{\partial B_\theta}{\partial z}, \tag{12.28}$$

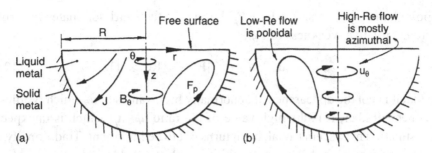

Figure 12.7 The experiment of Bojarevics et al. 1989. (a) Current flows down through the pool producing a poloidal force $\mathbf{F}_p = \mathbf{J}_p \times \mathbf{B}_\theta$. (b) At low levels of current the flow is poloidal, while a higher level of current initiates an intense swirling motion.

in (r, θ, z) coordinates. Nevertheless, it is surprising that a force composed of a large poloidal component plus a weaker azimuthal contribution can give rise to a flow dominated by swirl, i.e. $u_\theta \gg u_r, u_z$. In the experiment the stray field arises in part from the earth's magnetic field (Bojarevics et al., 1989). At the lower levels of current used (~15 Amps) the average magnetic field induced by the current on the pool surface is around 0.6 Gauss, which is comparable with the earth's magnetic field. However, at the highest current levels (~1200 Amps) the average surface value of B_θ is around 40 Gauss, which is a factor of ~100 greater than the earth's magnetic field. The key question, therefore, is why do low levels of azimuthal forcing give rise to disproportionately high levels of swirl? Precisely the same phenomenon is seen at a larger scale in industrial processes such as the vacuum-arc remelting of ingots. Here the stray magnetic field arises from inductors which carry current to and from the apparatus. Unless great effort is made to minimise the stray magnetic field, an intense swirling motion is generated which adversely affects the final product.

In all of these processes the Reynolds number, Re, is high, perhaps $300 \rightarrow 10^4$ in laboratory experiments, and around $10^5 \rightarrow 10^6$ in industrial applications. It is certain, therefore, that these flows are turbulent. Moreover, the phenomenon seems to be particular to high Reynolds numbers, in the sense that there is a value of Re below which disproportionately high levels of swirl are not observed (Bojarevics et al., 1989). There is, however, a second threshold. That is, as we shall see, u_θ dominates the poloidal motion, $\mathbf{u}_p$, only when F_θ exceeds ~$0.01|\mathbf{F}_p|$. (We use subscript p to indicate poloidal components of $\mathbf{u}$ or $\mathbf{F}$.) Below this threshold, the poloidal motion remains dominant, no matter what the value of Re. In particular, if $F_\theta \rightarrow 0$, then there is no swirling motion at all.

Let us summarise the experimental evidence. The term 'spontaneous swirl' is commonly used to describe high-Re flows in which the forcing has both azimuthal and poloidal components, $\mathbf{F} = \mathbf{F}_\theta + \mathbf{F}_p$, but where the swirl dominates the motion despite the relative weakness of $\mathbf{F}_\theta$:

$$u_\theta \gg |\mathbf{u}_p| , \quad F_\theta \ll |\mathbf{F}_p| , \quad (\mathrm{Re} \gg 1). \tag{12.29}$$

Note that it is not the appearance of abnormally high values of u_θ which typifies the experiment. Rather, it is the high value of the ratio $u_\theta/|\mathbf{u}_p|$ which is unexpected. This distinction may seem trivial, but it turns out to be important. Traditionally, this phenomenon was regarded as an instability, with the sudden appearance of swirl marking some instability threshold, rather like the sudden eruption of Taylor vortices in unstable Couette flow. Recently, however, it has been shown that this view is incorrect. We shall see that the magnitude of u_θ is simply governed by the (prescribed) magnitude of F_θ, and that there is nothing mysterious about the level of swirl. In fact, it is an unexpected suppression of $|\mathbf{u}_p|$, rather than a growth in u_θ, which typifies the observations.

12.6.2 But There Is no Spontaneous Growth of Swirl!

We shall show in §12.7 that flows characterised by (12.29) do indeed exist, but that the phrase 'spontaneous swirl' is somewhat of a misnomer. Such flows would be better characterised by the term *poloidal suppression*. That is, the mystery is not that u_θ is unexpectedly large, but that, in the presence of swirl, $|\mathbf{u}_p|$ is disproportionately small. In fact, the magnitude of u_θ is always unambiguously determined by the global torque balance,

$$\rho \int rF_\theta dV = \oint 2\pi r^2 \tau_\theta ds. \tag{12.30}$$

Here s is a curvilinear coordinate measured along the pool boundary and τ_θ is the azimuthal surface shear stress. In a turbulent flow $\tau_\theta = c_f\left(\frac{1}{2}\rho u_\theta^2\right)$ where the skin friction coefficient c_f is, perhaps, of the order 10^{-2}. If R is a typical pool radius this yields the estimate

$$u_\theta \sim \left(RF_\theta/c_f\right)^{1/2} \sim 10\sqrt{RF_\theta}. \tag{12.31}$$

(The same estimate may be obtained by integrating (11.21) along the outer boundary, from the pool surface to the symmetry axis.) Similar estimates may be made for laminar flows, but the details are unimportant. The key point is that u_θ is fixed in magnitude by F_θ.

We shall see that flows of type (12.29) arise from the action of the centrifugal force. That is, there are two driving forces for poloidal motion, the poloidal Lorentz force, $\mathbf{F}_p$, and the centrifugal force $\left(u_\theta^2/r\right)\hat{\mathbf{e}}_r$, and it turns out that these conspire to cancel. Specifically, provided $\mathrm{Re} \gg 1$ and $F_\theta/|\mathbf{F}_p| > 0.01$, the angular momentum of the fluid always distributes itself in such a way that these two forces almost exactly cancel (to within the gradient of a scalar) and that consequently, the poloidal motion is extremely weak. It is this balance between curls of $\mathbf{F}_p$ and $\left(u_\theta^2/r\right)\hat{\mathbf{e}}_r$ which underpins the experimental observations.

12.6.3 Flaws in Traditional Theories Predicting a Spontaneous Growth of Swirl

Traditionally, considerable significance has been attached to the fact that there appears to be a threshold value of Re above which swirl is dominant. This has led some researchers to conclude that the underlying poloidal motion is unstable and that the appearance of swirl is a manifestation of this instability. In particular, it has been popular to study the breakdown of the self-similar poloidal flow associated with the injection of current from a point source located on the surface of a semi-infinite domain (see §12.4.2). Now it happens that at $\mathrm{Re} \sim 7$ this similarity solution breaks down in the sense that the velocity on the axis becomes infinite, as discussed in §12.4.2. However, if just the right amount of swirl is introduced into the far field then, due to the radially inward convection of angular momentum, the singularity on the axis is alleviated. Thus, in semi-infinite domains there exists the possibility of a bifurcation from a non-swirling to a swirling flow, provided, of course, that (some-how) nature provides just the right amount of angular momentum in the far field. This, according to the traditional argument, is the origin of the experimental observations.

However, there are several objections to making the jump from infinite to confined flows. The first point to note is that the appearance of a singularity in the self-similar solution is simply an artefact of the (idealised) assumption of a point source of current, as discussed in §12.4.2, yet the phenomenon of relatively strong swirl is observed in VAR where the electrode spans almost the entire surface of the liquid pool. Second, the infinite domain model relies on angular momentum being injected into the fluid in the far field. In confined domains, where does this angular momentum come from? If swirl exists in a confined flow it must be maintained by an external torque and the magnitude of that torque fixes the magnitude of the swirl in accordance with (12.31). In short, there can be no sudden 'eruption' of swirl due to an increase in Re. Clearly, we must seek an alternative explanation of the observed phenomenon.

12.7 Poloidal Suppression versus Spontaneous Swirl

We now outline the explanation offered in Davidson et al. (1999) for the dominance of swirl in the presence of a stray, vertical magnetic field. Let us assume that the

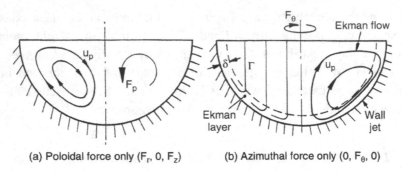

Figure 12.8 Flows driven by poloidal and azimuthal Lorentz forces. (a) Poloidal flow driven by a poloidal force. (b) Angular momentum contours and secondary Ekman flow induced by an azimuthal force.

flow is axisymmetric and the liquid pool occupies a hemisphere of radius R, as shown in Figure 12.7. There are two key points to note. First, it turns out to be easier to generate swirl in a hemispherical pool than poloidal motion, in the following sense. Suppose that we apply a poloidal force $\mathbf{F}_p$ to the pool, measuring the resulting motion shown in Figure 12.8(a). Then (12.24) gives us $|\mathbf{u}_p| \sim \sqrt{R|\mathbf{F}_p|}$. We now remove $\mathbf{F}_p$ and (somehow) apply an azimuthal force F_θ and measure the resulting swirl (Figure 12.8(b)). The resulting flow pattern is discussed in §11.1.5, and consists of angular momentum contours, $\Gamma = ru_\theta$, which are z-independent outside the boundary layers, accompanied by a secondary Ekman flow driven by the side-wall boundary layers. Moreover, from (12.31) we have $u_\theta \sim \left(RF_\theta/c_f\right)^{1/2} \sim 10\sqrt{RF_\theta}$, and so we conclude that

$$u_\theta^2/RF_\theta \approx 100 \; \mathbf{u}_p^2/R|\mathbf{F}_p|.$$

Thus small azimuthal forces, which result from a stray vertical magnetic field, can give rise to a relatively large swirling motion. Indeed, similar azimuthal and poloidal velocities can be generated by azimuthal forces as small as $F_\theta \sim 0.01|\mathbf{F}_p|$.

The second point is more subtle. Suppose that we now allow for both poloidal and azimuthal Lorentz forces, as in our model problem in the presence of a stray, vertical magnetic field. Then, from (12.3) and (12.28), our Lorentz force is

$$\mathbf{F} = -\frac{B_\theta^2}{\rho\mu r}\,\hat{\mathbf{e}}_r + \frac{B_z}{\rho\mu}\frac{\partial B_\theta}{\partial z}\,\hat{\mathbf{e}}_\theta, \tag{12.32}$$

where, for convenience, we take B_z to be constant. An intriguing possibility now presents itself. The azimuthal flow, if large enough, can in principle arrange itself such that the centripetal acceleration, $\left(u_\theta^2/r\right)\hat{\mathbf{e}}_r$, counterbalances the poloidal

Lorentz force to within the gradient of a scalar. In terms of the scaled magnetic field, $\mathbf{h} = \mathbf{B}/\sqrt{\rho\mu}$, this balance is

$$\frac{u_\theta^2}{r}\hat{\mathbf{e}}_r = \frac{h_\theta^2}{r}\hat{\mathbf{e}}_r + \nabla\Phi, \tag{12.33}$$

for some potential Φ, or equivalently,

$$\frac{\partial u_\theta^2}{\partial z} = \frac{\partial h_\theta^2}{\partial z}. \tag{12.34}$$

The significance of (12.34), if it is indeed realised, is the following. The curl of the poloidal equation of motion, subject to the body force (12.32), is the azimuthal vorticity equation,

$$\mathbf{u} \cdot \nabla(\omega_\theta/r) = \frac{\partial}{\partial z}\left[\frac{u_\theta^2}{r^2} - \frac{h_\theta^2}{r^2}\right] + \nu_t \nabla \cdot [r^{-2}\nabla(r\omega_\theta)]. \tag{12.35}$$

Evidently, if the force balance (12.34) can indeed be achieved in the core of the flow (i.e. outside the boundary layers), then that part of the poloidal motion which is driven directly by the Lorentz force will be suppressed. In such a situation, the only residual motion in the r–z plane will be the Ekman pumping associated with the swirling flow. However, this Ekman flow is very weak, except in the boundary layers, and so balance (12.34) is a way of almost totally eliminating the kinetic energy associated with poloidal flow. It also helps keep the kinetic energy associated with u_θ at a modest level. That is, if the only poloidal motion is Ekman pumping, *all* of the streamlines will be flushed through the dissipative Ekman layers where the angular momentum created by F_θ can be efficiently destroyed. (Again, consult §11.1.5 for a discussion of the role of Ekman pumping in confined, swirling flows, and in particular its crucial role in limiting the growth in swirl.) Evidently, if we can find a core solution which satisfies (12.34), then this represents an energetically favourable state.

Of course, for the force balance (12.34) to be realisable we require a sufficiently large swirl. This, in turn, is governed by the azimuthal equation of motion

$$\mathbf{u} \cdot \nabla\Gamma = \frac{\partial}{\partial z}[rh_z h_\theta] + \nu_t \nabla \cdot [r^2\nabla(\Gamma/r^2)], \tag{12.36}$$

where $\Gamma = ru_\theta$ and we have incorporated the azimuthal Lorentz force from (12.32). However, we have already noted that a relatively large azimuthal motion can be generated by a modest azimuthal force, in the sense that u_θ^2/RF_θ is two orders of magnitude larger than $\mathbf{u}_p^2/R|\mathbf{F}_p|$, i.e. $u_\theta^2/RF_\theta \approx 100\ \mathbf{u}_p^2/R|\mathbf{F}_p|$. So we might hope to

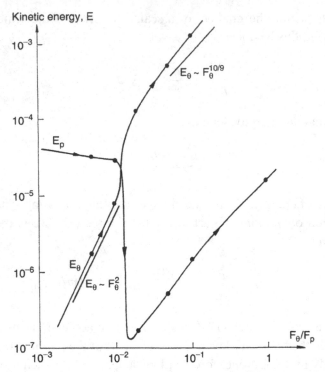

Figure 12.9 Variation of kinetic energy of the swirling (E_θ) and poloidal (E_p) components of motion as the ratio $F_\theta/|\mathbf{F}_p|$ is increased.

access the low energy state associated with (12.34) even for relatively weak levels of azimuthal forcing.

In Davidson et al. (1999) it is shown that Equation (12.34) is readily satisfied even for very weak vertical fields and that consequently the flow does indeed adopt this low energy state. In fact, when F_θ exceeds the modest threshold of $F_\theta \sim 0.01|\mathbf{F}_p|$, the flow is dominated by swirl, despite the weakness of the azimuthal forcing.

This is illustrated by the sequence of numerical experiments shown in Figure 12.9. These relate to the flow of liquid steel in a hemispherical pool of radius $R = 0.1$ m. The flow was taken to be turbulent and the Reynolds stresses were estimated using the $\kappa - \varepsilon$ turbulence closure model. The computations allowed for a uniform axial magnetic field $h_z = -h_0$, where h_0 lies in the range $0 \rightarrow 1$ m/s. It is convenient to introduce the symbol $\hat{h}$ to represent the maximum value of h_θ, so that $h_0/\hat{h}$ is a measure of relative size of $F_\theta/|\mathbf{F}_p|$. With $\hat{h}$ fixed at 1 m/s, the flow was calculated for a range of values of $h_0/\hat{h}$, corresponding to 0, 0.005, 0.01, 0.02, 0.05, 0.1 and 1.0. This represents the full range from $F_\theta = 0$ to $F_\theta \sim |\mathbf{F}_p|$. The resulting variation of the kinetic energies E_θ and E_p (defined as the integrals of $u_\theta^2/2$ and

$\mathbf{u}_p^2/2)$ is shown in Figure 12.9. Note that, when the azimuthal force reaches a value of $F_\theta \sim 0.01 \, |\mathbf{F}_p|$, the swirl and poloidal motions have similar intensities. As F_θ is further increased, the energy of the poloidal motion collapses, dropping by a factor of 100. This represents the transition to the regime governed by (12.34), in which the curl of the poloidal Lorentz force is balanced by the curl of the centrifugal force.

In summary, the experimental observations of Bojarevics et al. (1989) are not associated with an unexpected growth in swirl, but rather with an unexpected suppression of the poloidal motion.

13

MHD Instabilities in Aluminium Reduction Cells

Electricity is of two kinds, positive and negative. The difference is,
I presume, that one comes a little more expensive, but is more durable;
the other is a cheaper thing, but the moths get in.

Stephen Leacock

The amount of electrical energy required to reduce alumina to aluminium in electrolysis cells is staggering. In North America, for example, around 5 per cent of all generated electricity is used to produce aluminium. Worldwide, 50 million tonnes of aluminium are produced annually by electrolysis and this requires around 0.7×10^{12} kWh. At a notional cost of 5 cents per kWh, the corresponding electricity bill is around \$35 billion per year! Yet much of this energy, approximately 40 per cent, is wasted in the form of I^2R heating of the electrolyte used to dissolve the alumina, energy which is subsequently lost to the environment. Needless to say, strenuous efforts have been made to reduce these losses, mostly centred around minimising the depth of electrolyte in the cells. However, the aluminium industry is faced with a fundamental problem. When the depth of electrolyte is reduced below some critical threshold, the electrolysis cell becomes unstable. It is this instability, which is driven by MHD forces, which is the subject of this chapter.

13.1 The Prohibitive Cost of Reducing Alumina to Aluminium

13.1.1 Early Attempts to Produce Aluminium by Electrolysis

It is not an easy matter to produce aluminium from mineral deposits. The first serious attempt to isolate elemental aluminium was that of Humphrey Davy, Faraday's mentor at the Royal Institution. (In fact, Davy's preferred spelling – aluminum – is still used today in North America.) In 1809 he passed an electric current through fused compounds of aluminium and into a substrate of iron. Although an alloy of aluminium and iron resulted in place of the pure aluminium

374

he sought, Davy had at least managed to prove that aluminium oxide was indeed reducible.

Oersted, and later Wöhler, set aside electricity and concentrated on chemical means of isolating aluminium. By 1827 Wöhler was able to produce small quantities of aluminium powder by displacing the metal from its chloride using potassium. Later, in the 1850s, potassium was replaced by sodium, which was cheaper, and aluminium fluoride was substituted for the more volatile chloride. Wöhler's laboratory technique had at last become commercially viable and the industrial production of aluminium began. However, those chemical processes were all swept aside by the revolution in electrical technology initiated by Faraday. In particular, the development of the dynamo made it possible to produce aluminium by electrolysis.

The electrolytic route was first proposed by Robert Bunsen in 1854, but it was not until 1886 that a continuous commercial process was developed. It was a twenty-two year old college student from Ohio, Charles Martin Hall, and the Frenchman Paul Héroult, who made this breakthrough: the Frenchman as a result of good fortune (which he had the wit to pursue), and the American as a result of systematic enquiry. Hall and Héroult realised that molten cryolite, a mineral composed of fluorine, sodium and aluminium, readily dissolves alumina and that a current passed through the solution will decompose the alumina, leaving the cryolite unchanged. Full commercial production of aluminium began on Thanksgiving Day 1888 in Pittsburgh in a company founded by Hall. An example of an early Hall–Héroult reduction cell is shown in Figure 13.1. Remarkably, over a century later, the process is virtually unchanged.

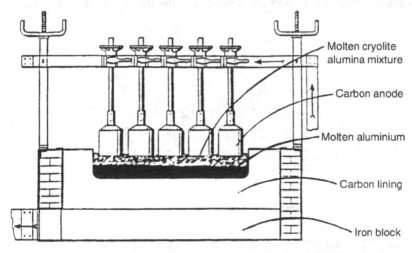

Figure 13.1 A schematic of a 1920s aluminium reduction cell (side view).

13.1.2 An Instability in Modern Reduction Cells and Its Financial Consequences

Today most aluminium is produced by electrolysis (a modest amount is recycled from scrap) and the cells which are used look remarkably similar to those envisaged by Hall and Héroult. A schematic of the end view of a modern cell is shown in Figure 13.2. (A side view, by contrast, would typically show 8 pairs of anode blocks arranged along the length of the cell.) A large vertical current, perhaps 300kA or more, flows downward from the carbon anode blocks, passing first through the electrolytic layer, where it reduces the alumina, and then through a liquid aluminium pool before finally being collected at the carbon cathode at the base of the cell. The liquid layers are broad and shallow, perhaps 4 m × 10 m in plan, yet only a few centimetres deep. The liquids are maintained at a temperature of around 950°C (aluminium melts at 660°C) and alumina is continually fed into the top of the cell while liquid aluminium is tapped off at the base. In a typical smelter, 100 or more reduction cells will sit side-by-side, with the electrical current passed from cell to cell.

The liquid aluminium is an excellent electrical conductor, the carbon a moderate one and the electrolyte (cryolite) a very poor conductor. Consequently, much of the electrical energy consumed by the cell is lost in Ohmic heating of the cryolite, typically around 40 per cent. In fact, these losses are vast, and so there is considerable incentive to lower the resistance of the electrolyte layer by reducing its thickness. (Note the difference in electrolyte thickness in Figures 13.1 and 13.2.) In most modern cells the depth of the electrolyte below the anode blocks is restricted to 4.5–5 cm, while the aluminium pool is usually 20–30 cm deep. It is estimated that, worldwide, every millimetre of this thin electrolyte layer results in I^2R losses of around 6 billion kWh/year, the electrical power being converted to

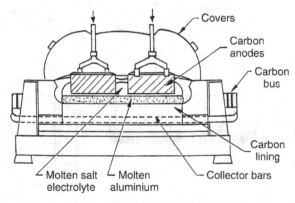

Figure 13.2 A modern reduction cell (end view). In side view there are typically 8 pairs of anode blocks.

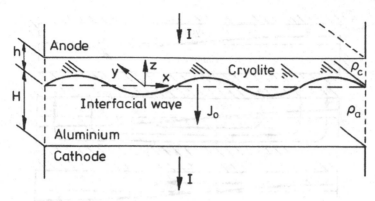

Figure 13.3 Interfacial waves in a reduction cell.

heat which is then lost to the environment. (A small fraction of the Ohmic heating within a cell is captured as thermal energy in the liquid aluminium, which is periodically tapped off. However, this is only a tiny percentage of the input energy to the cell.) At a notional cost of 5 cents per kWh, we might estimate that each millimetre of electrolyte results in lost profits of $300 million per year. So, if reduction cells could be made to operate with a cryolite depth of, say, 4 cm instead of 4.5 cm, that would save over a billion dollars a year!

The energy problem aside, this process works reasonably well. However, there is one fundamental problem. It turns out that unwanted disturbances are readily triggered at the electrolyte-aluminium interface (Figure 13.3). In effect, these are long-wavelength, interfacial gravity waves, modified by the intense magnetic and electric fields which pervade the cell. Under certain conditions these disturbances are observed to grow, disrupting the operation of the cell, possibly even resulting in a short circuit if the wave amplitude is large enough for the liquid aluminium to make contact with the anode blocks. These instabilities have been the subject of much research over the last few decades, since they represent the greatest single impediment to increasing the energy efficiency of reduction cells. In particular, the cryolite depth, h, must be maintained above a certain critical value to ensure stability, which imposes a massive energy penalty. There is much incentive, therefore, to understand the nature of this instability and seek ways of neutralising it.

13.2 Attempts to Model Unstable Interfacial Waves in Reduction Cells

To some extent, the mechanism of the instability is clear. Tilting the interface causes a perturbation in current density, $\mathbf{j}$. Excess current is drawn into the aluminium at those points where the thickness of the highly resistive cryolite is reduced, and less current is drawn at points where the electrolyte depth is increased.

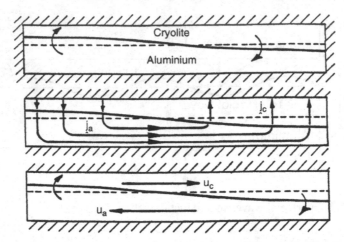

Figure 13.4 Perturbations in current caused by a movement of the interface. In the long-wavelength limit the perturbation current, **j**, is predominantly vertical in the cryolite and horizontal in the aluminium. The 'sloshing' motion in the two liquids is predominantly horizontal.

Since the carbon cathode is much more resistive than the aluminium, these perturbations in vertical current feed into the aluminium but do not penetrate the carbon cathode. In the long-wavelength approximation ($kh \rightarrow 0$, k being the wavenumber) the perturbed current in the electrolyte is purely vertical while that in the aluminium is horizontal (to leading order in kh). This perturbation in current is shown schematically in Figure 13.4 for the simplest of wave shapes.

Now the change in current causes a perturbation in the Lorentz force, $\delta \mathbf{F} = \mathbf{j} \times \mathbf{B}_0 + \mathbf{J}_0 \times \mathbf{b}$, where $\mathbf{J}_0$ is the unperturbed (downward) current density in the cell, $\mathbf{B}_0$ the ambient magnetic field in the absence of disturbances, and **b** the perturbation in magnetic field associated with the current perturbation **j**. (The ambient magnetic field, $\mathbf{B}_0$, arises from the passage of current down through each cell, as well as the transfer of current from cell to cell.) In the long-wavelength limit, the dominant contribution to $\delta \mathbf{F}$ can be shown to be $\mathbf{j} \times \mathbf{B}_z$, where $\mathbf{B}_z$ is the vertical component of the ambient magnetic field in the cell (see §13.4 below). The key question, therefore, is whether or not this change in Lorentz force amplifies the initial motion.

After many years of research, this issue was finally resolved by Sneyd and Wang (1994) and Bojorevics and Romerio (1994). Subsequently their analyses were generalised by Davidson and Lindsay (1998). These authors all simplify the geometry to that of a closed, rectangular domain (i.e. a shoebox), as shown in Figure 13.4. Sneyd and Wang start by noting that, in the absence of a magnetic field, the interface may support an infinite number of conventional standing waves. The normal modes associated with these form an orthogonal set of functions, so that one can represent an arbitrary disturbance of the interface as the superposition of many such gravitational modes. When the Lorentz force is absent, these modes are decoupled. However, when the Lorentz force is taken into account, certain gravitational modes are coupled. That is, the redistribution of current caused by one mode gives rise to a Lorentz force which, when Fourier-decomposed, can excite many other modes. This leads to a coupled set of equations of the form

$$\ddot{\mathbf{x}} + \mathbf{\Omega}\mathbf{x} = \hat{\varepsilon}\mathbf{K}\mathbf{x}, \quad \hat{\varepsilon} = J_0 B_z / \rho_a H. \tag{13.1}$$

Here $\mathbf{x}$ is a column vector which represents the amplitudes of the gravitational modes, $\mathbf{\Omega}$ is diagonal with elements equal to the square of the conventional gravitational frequencies, H and ρ_a are the depth and density of the aluminium, and $\mathbf{K}$ is the interaction matrix which arises from the perturbed Lorentz force $\mathbf{j} \times \mathbf{B}_z$. Now it turns out that $\mathbf{K}$ is skew-symmetric and so complex eigenvalues, and hence instabilities, are guaranteed when $\hat{\varepsilon}$ is large enough. Unfortunately however, (13.1) sheds little light on the all-important *instability mechanism*. Consequently, before going on to describe the instability in detail, we shall discuss a simple mechanical analogue which highlights the basic instability mechanism. This is due to Davidson and Lindsay (1998) and relies on the fact that, in the long-wavelength limit, the motion in the aluminium is purely horizontal (Figure 13.4).

13.3 A Simple Mechanical Analogue for the Instability

Suppose we replace the liquid aluminium with a thin, rigid, aluminium plate attached to the centre of the anode by a light, rigid strut. The strut is pivoted at its top and so the plate is free to swing as a compound pendulum about two horizontal axes, x and y (see Figure 13.5). Let the plate have thickness H, lateral dimensions L_x, L_y and density ρ_a. The gap h between the plate and the anode is filled with an electrolyte of negligible inertia and very poor electrical conductivity. A uniform current density, J_0, passes vertically downward into the plate and is

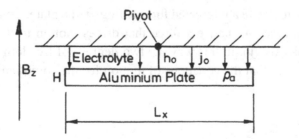

Figure 13.5 The compound pendulum shown here contains all of the essential physics of the reduction cell instability.

tapped off at the centre of the plate. Finally, suppose that there is an externally imposed, uniform, vertical magnetic field, B_z.

Evidently, we have replaced one mechanical system (the cell), which has an infinite number of degrees of freedom, with another which has only two degrees of freedom. However, electrically the two geometries are alike. Moreover, the nature of the motion in the two cases is not dissimilar. In both systems we have horizontal movement of the aluminium associated with a tilting of the electrolyte-aluminium interface. In a cell this takes the form of a 'sloshing' back and forth of the aluminium as the interface tilts first one way and then the other (Figure 13.4).

Let θ_x and θ_y be the angles of rotation of the plate measured about x and y axes. Then a local change in the electrolyte resistance resulting from a local change in fluid depth causes a perturbation in cryolite current density, $j_c(x, y, t)$, and this perturbation is readily expressed in terms of θ_x and θ_y:

$$\mathbf{j}_c = [J_0(\theta_y x - \theta_x y)/h]\hat{\mathbf{e}}_z.$$

Of course, this perturbation in vertical current feeds into the aluminium plate, causing changes in the horizontal current flow in the plate. It is not difficult to show that, as a result of the rotations θ_x and θ_y, the perturbation of the current I in the aluminium plate is given by (see Davidson and Lindsay, 1998)

$$\delta I_x = \frac{J_0 L_y \theta_y}{2h}\left[(L_x/2)^2 - x^2\right],$$

$$\delta I_y = -\frac{J_0 L_x \theta_x}{2h}\left[(L_y/2)^2 - y^2\right].$$

The change in the net Lorentz force acting on the plate (i.e. the volume integral of $\mathbf{j} \times \mathbf{B}_z$) can be calculated from these expressions and the equations of motion for the compound pendulum then follow. Note that a tilting of the plate about the y axis induces both motion and a horizontal current flow in the x direction. This then interacts with $\mathbf{B}_z$ to create a y-directed force acting on the plate. Thus motion in one direction, say x, tends to induce a force that drives motion in the perpendicular direction (in this case, y). In this way the two degrees of freedom of the plate are coupled. It turns out that the equations of motion for the compound pendulum are

$$\ddot{\gamma}_x + \varpi_x^2 \gamma_x = -(J_0 B_z/\rho_a H)\gamma_y = -\hat{\varepsilon}\gamma_y, \tag{13.2a}$$

$$\ddot{\gamma}_y + \varpi_y^2 \gamma_y = (J_0 B_z/\rho_a H)\gamma_x = \hat{\varepsilon}\gamma_x, \tag{13.2b}$$

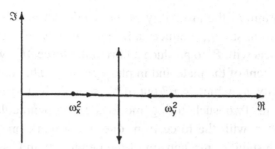

Figure 13.6 Variation of ϖ^2 with $J_0 B_z$ in the complex plane for the compound pendulum.

where $\gamma_x = \theta_x / L_x^2$, $\gamma_y = \theta_y / L_y^2$ and ϖ_x, ϖ_y are the conventional gravitational frequencies of the pendulum. Note the cross-coupling of θ_x and θ_y, as well as the similarity between (13.1) and (13.2). If we look for solutions of (13.2) of the form $\gamma \sim \exp(j\varpi t)$ then we find constant-amplitude oscillations for small values of $\hat{\varepsilon} = J_0 B_z / \rho_a H$ and exponentially growing oscillations (unstable solutions) for large $\hat{\varepsilon}$. The transition occurs at $\hat{\varepsilon} = \frac{1}{2} |\varpi_x^2 - \varpi_y^2|$.

Figure 13.6 shows the movement of ϖ^2 in the complex plane as $\hat{\varepsilon}$ is increased. The square of the two natural frequencies, ϖ_1^2 and ϖ_2^2, move along the real axis until they meet. At this point, they move off into the complex plane and an instability develops. The important points to note are:

(1) The tendency for instability depends only on the magnitude of $J_0 B_z / \rho_a H$ and on the natural gravitational frequencies ϖ_x and ϖ_y.
(2) To minimise the danger of an instability, it is necessary to keep $J_0 B_z / \rho_a H$ low and the gravitational frequencies well apart. The closer the natural frequencies are, the lower the threshold value of $J_0 B_z$ at which an instability appears. (Circular and square plates are unstable for vanishingly small values of $J_0 B_z$.)
(3) The system is unstable whenever

$$\frac{J_0 B_z}{\rho_a H} \geq \frac{1}{2} |\varpi_x^2 - \varpi_y^2|.$$

(4) When $\hat{\varepsilon}$ is large, the unstable normal mode corresponds to a rotating, tilted plate (Davidson and Lindsay, 1998).

Very similar behaviour is seen in reduction cells governed by (13.1). In particular, the sensitivity of reduction cells to the destabilising influence of $J_0 B_z$ depends on the initial separation of the gravitational frequencies. The closer the gravitational frequencies, the lower the stability threshold. Moreover, unstable waves frequently correspond to a rotating, tilted interface.

The physical origin of the instability of the pendulum is now clear. Tilting the plate in one direction, say θ_x, produces a horizontal flow of current in the aluminium which interacts with B_z to produce a horizontal force δF_x, which is perpendicular to the movement of the plate and in phase with θ_x. This tilting also produces a horizontal velocity, u_y, which is $\pi/2$ out of phase with the force δF_x and mutually perpendicular to it. Two such tilting motions in perpendicular directions can reinforce each other, with the force from one doing work on the motion of the other. This is the instability mechanism of the pendulum and essentially the same thing happens in a reduction cell.

We may think of the Lorentz force as playing two roles. In the first instance it modifies the gravitational frequencies, pulling them together on the real axis. Once these frequencies coincide, so that the plate oscillates at the same frequency in two directions, the Lorentz force adopts a second role in which it supplies energy to the pendulum. Unstable motion then follows.

A simple energy argument shows why, whenever the plate oscillates with a single frequency in two perpendicular directions, an instability is inevitable. From the expressions for δI_x and δI_y we can calculate the net Lorentz force acting on the plate, $\delta \mathbf{F}$, and hence the rate of work done by that force, $\delta \mathbf{F} \cdot \mathbf{u}$, $\mathbf{u}$ being the velocity of the plate. After a little algebra we find

$$\delta \mathbf{F} \cdot \mathbf{u} = J_0 B_z L_x L_y \left[L_y^2 \theta_x \dot{\theta}_y - L_x^2 \theta_y \dot{\theta}_x \right] / 12. \tag{13.3}$$

Now suppose that θ_x and θ_y both oscillate at frequency ϖ, but are $\pi/2$ out of phase. Then the time-averaged value of $\delta \mathbf{F} \cdot \mathbf{u}$ is non-zero, implying unstable motion.

Note that this *instability mechanism* is independent of the action of gravity. That is, provided the plate can oscillate at the same frequency in two directions, it will become unstable. (Circular and square plates are unstable at arbitrarily small values of B_z.) The *stability threshold*, on the other hand, does depend on gravity, in that it is dependent on the initial separation of the gravitational frequencies.

We now return to reduction cells. We shall see in §13.5.1 that, in the absence of friction, interfacial waves are described by the wave-like differential equation,

$$\frac{\partial^2 \mathbf{q}}{\partial t^2} - c^2 \nabla^2 \mathbf{q} = \varpi_B^2 [\hat{\mathbf{e}}_z \times \mathbf{q}]_P, \qquad \mathbf{q} \cdot \mathbf{n} = 0, \tag{13.4}$$

where $\mathbf{q}$ is the horizontal mass flux in the cryolite. Here $\mathbf{n}$ is the unit normal to the horizontal perimeter of the fluid layers, $\varpi_B^2 = J_0 B_z / (\bar{\rho} h H)$ is a measure of the destabilising MHD effects, $c^2 = (\rho_a - \rho_c) g / \bar{\rho}$ is the hydrodynamic interfacial wave speed, ρ_c and ρ_a are the densities of the cryolite and aluminium, $\bar{\rho} = (\rho_c / h) + (\rho_a / H)$ is a weighted density average, and h and H are the depths of the cryolite and aluminium layers, respectively. The subscript P on the term $\hat{\mathbf{e}}_z \times \mathbf{q}$ implies that we take only the irrotational (i.e. potential) component of $\hat{\mathbf{e}}_z \times \mathbf{q}$ in accordance with a Helmholtz decomposition, as described in §13.5.1. Equation (13.4) is valid quite generally and makes no assumption regarding the existence or shape of the lateral boundaries. Also, note that all of the electromagnetic effects are captured by the single parameter ϖ_B^2.

The independent physical parameters which control solutions of (13.4) are evidently ϖ_B^2, c^2 and the characteristic horizontal length scales of the motion, which are introduced through the boundary condition $\mathbf{q} \cdot \mathbf{n} = 0$. Dimensional analysis then tells us that the stability characteristics of (13.4) are determined by the product of $k_B = \varpi_B/c$ with the characteristic horizontal length scale of the motion, L. Thus the key dimensionless group that characterises the onset of instability is

$$\varepsilon = k_B^2 L^2 = \frac{\varpi_B^2 L^2}{c^2} = \frac{J_0 B_z L^2}{(\rho_a - \rho_c)ghH}. \tag{13.5}$$

Note the crucial appearance of h in (13.5). There are four special cases of interest, which we shall discuss in detail in §13.6.

(i) When $\varpi_B = 0$, we recover the standard hydrodynamic wave equation for inviscid interfacial disturbances,

$$\frac{\partial^2 \mathbf{q}}{\partial t^2} - c^2 \nabla^2 \mathbf{q} = \mathbf{0}.$$

(ii) If ϖ_B is non-zero, and we look for solutions in the channel $0 < x < L$, then we find unstable travelling waves.

(iii) For a closed circular domain, (13.4) yields unstable standing waves for vanishingly small values of ϖ_B. (Remember, (13.4) does not include friction and a circular compound pendulum is unstable for vanishingly small $J_0 B_z$.) Moreover, the unstable normal mode corresponds to a rotating, tilted interface, just like that of the pendulum.

(iv) If we place (13.4) in a rectangular domain, we recover the matrix equation of Sneyd & Wang (1994),

$$\ddot{\mathbf{x}} + \mathbf{\Omega}\mathbf{x} = \hat{\varepsilon}\mathbf{K}\mathbf{x}, \quad \hat{\varepsilon} = J_0 B_z/\rho_a H,$$

where the coupling matrix, $\mathbf{K}$, is skew-symmetric and $\mathbf{\Omega}$ is diagonal with elements equal to the square of the conventional gravitational wave frequencies. (Note the similarity to (13.2).)

We now provide a schematic derivation of (13.4). The enthusiastic or the sceptical will find a more detailed discussion in Davidson and Lindsay (1998).

13.4 Simplifying Assumptions and a Model Problem

A model of the cell is shown in Figure 13.3. The undisturbed depths of cryolite and aluminium are h and H, and the unperturbed current flow is purely vertical and has magnitude J_0. We use a Cartesian coordinate system, (x, y, z), where the positive direction of z is upward and the origin for z lies at the undisturbed interface. On occasion, we shall refer to cells which are rectangular in plan view, and these are given lateral dimensions L_x and L_y. However, much of the analysis can be applied to any shape of cell.

We take the characteristic time scale for the wave motion to be much greater than the diffusion time of the magnetic field. That is, we make the pseudo-static approximation $\mu\sigma uh \ll 1$, where μ is the permeability of free space, σ is the conductivity, and u is a typical velocity. Thus, each time the interface moves, the current immediately relaxes to a new equilibrium distribution. Ohm's law is then

$$\mathbf{J} = \sigma\mathbf{E} = -\sigma\nabla\Phi \ , \quad \nabla^2\Phi = 0.$$

We are concerned only with linear stability, and so we consider infinitesimal perturbations of the interface of the form $z_s = \eta$, $\eta \ll h, H$. The corresponding distributions of $\mathbf{J}$ and $\mathbf{B}$ are

$$\mathbf{J} = \mathbf{J}_0 + \mathbf{j} = -J_0\hat{\mathbf{e}}_z - \sigma\nabla\varphi, \mathbf{B} = \mathbf{B}_0 + \mathbf{b},$$

where φ is the perturbation in the electrostatic potential and the boundary conditions on $\mathbf{J}$ and φ arise from the ranking of the electrical conductivities. That is, we have

$$\sigma_a \gg \sigma_{carbon} \gg \sigma_c,$$

where the subscripts 'a' and 'c' refer to the aluminium and cryolite. It is not difficult to show that this requires $\varphi_c = 0$ on $z = h$ and $\partial\varphi_a/\partial z = 0$ on $z = -H$. The first of the boundary conditions states that the anode potential is fixed, while the second ensures that $\mathbf{j}$ does not penetrate into the carbon cathode blocks.

We shall assume that the fluid is inviscid, that surface tension can be ignored, and that there is no background motion in the unperturbed state. The first of these assumptions means that our equations of motion cannot mimic the damping of high-wavenumber perturbations which occurs in practice. To compensate for this, we simply ignore those modes whose wavelengths are shorter than a certain (small but arbitrary) value. The last of these three assumptions (i.e. $\mathbf{u}_0 = \mathbf{0}$) greatly simplifies the stability analysis. However, this simplification does severely limit the allowable distributions of $\mathbf{B}_0$. That is, to ensure that the perturbation occurs about an equilibrium configuration, we must satisfy $\nabla \times (\mathbf{J}_0 \times \mathbf{B}_0) = \mathbf{0}$. Given our assumed distributions of $\mathbf{J}_0$, this demands that $\mathbf{B}_0$ be of the form

$$\mathbf{B}_0 = \Big(B_x(x,y), B_y(x,y), B_z\Big), \tag{13.6}$$

where B_z is spatially uniform. We shall assume that all three components of $\mathbf{B}_0$ are of the same order of magnitude. From Ampère's law, $\nabla \times \mathbf{B} = \mu \mathbf{J}$, this implies that $B_x \sim B_y \sim B_z \sim \mu J_0 L$ where L is a typical lateral dimension.

Our final assumption relates to the aspect ratio of the liquid layers. We shall assume that $kh \ll 1$, where k is a typical wavenumber. In effect, we use the shallow-water approximation. This leads directly to a number of simplifying features. In particular, as a result of the shallow-water approximation, and to leading order in kh, it may be shown that

(a) $\mathbf{j}$ is vertical in the cryolite.
(b) $\mathbf{j}$ is horizontal in the aluminium and is uniformly distributed across that layer.
(c) The perturbed Lorentz force acting on the cryolite may be neglected.
(d) The velocity in each layer is uniform in z and horizontal.
(e) The dominant contribution to the perturbed Lorentz force in the aluminium is
   $\mathbf{j} \times (B_z \hat{\mathbf{e}}_z)$.

In fact, it is not difficult to see how these simplifications arise. Consider the situation shown in Figure 13.4, where the disturbance has a long wavelength. Approximations (a) and (b) are purely geometric and are a consequence of the ranking of the electrical conductivities. That is, the dominant resistance to the flow of current is the thin sheet of cryolite, so that the current passes directly downward through this layer (condition (a)). The aluminium, which is a very good conductor, is almost an equipotential surface, so that spatial variations of J_z in the cryolite (due to undulations of the interface) lead to a 'shorting' of the perturbed current through the aluminium. This 'shorted' current is almost purely horizontal (condition (b)). The neglect of the perturbed Lorentz force in the cryolite (condition (c)) stems from the fact that $\mathbf{j}_c \ll \mathbf{j}_a$, which in turn arises from the aspect ratio $kh \ll 1$. Moreover, the uniformity of the velocity in the two fluid layers (condition (d)) follows from the fact that the Lorentz force in the aluminium is independent of depth.

This leaves only simplification (e) to justify, and here there is some subtlety in the argument. Using subscripts H and V to indicate horizontal and vertical components of $\mathbf{J}$ and $\mathbf{B}$, it seems reasonable to neglect $\mathbf{j}_H \times \mathbf{B}_H$ and $\mathbf{j}_V \times \mathbf{B}_H$ because the former is vertical and so merely perturbs the vertical pressure gradient, while the latter is much smaller than $\mathbf{j}_H \times \mathbf{B}_V$, by virtue of (b). Finally, the neglect of $\mathbf{J}_0 \times \mathbf{b}$ relies on the fact that $|\mathbf{b}|$ is of order $|\mu \mathbf{j}_H H|$, while $|\mathbf{B}_0|$ is of order $\mu J_0 L$, so that $|\mathbf{J}_0 \times \mathbf{b}|$ is of order H/L smaller than $|\mathbf{j} \times \mathbf{B}_V|$.

We now construct a shallow-water model of the instability based on these assumptions, which will lead us back to (13.4). This model is more general than the 'mode-by-mode' description of (13.1) in that it makes no assumption regarding the existence or shape of the lateral boundaries.

13.5 A Shallow-Water Wave Equation for the Model Problem

13.5.1 The Shallow-Water Wave Equations

We start with conventional shallow-water theory. It is not difficult to show that, to second order in kH, the pressure in each layer is hydrostatic. As a consequence, we may apply the conventional shallow-water equation to each layer in turn. This is a two-dimensional equation for the horizontal motion,

$$\rho \frac{D\mathbf{u}_H}{Dt} + \rho g \nabla H_a = -\nabla P_0 + \mathbf{F}_H,$$

where $H_a(x, y, t)$ is the aluminium depth, P_0 is the interfacial pressure, and $\mathbf{F}_H$ is the horizontal body force in each layer. The unfamiliar term on the left arises from the horizontal gradient in pressure. For example, the pressure at the base of the aluminium layer is $P_0 + \rho_a g H_a$, so that the horizontal pressure force at the base of this layer is $-\nabla P_0 - \rho_a g \nabla H_a$. Similarly, the pressure at the top of the cryolite is $P_0 + \rho_c g H_c$, and so the horizontal pressure force just under the anode is $-\nabla P_0 + \rho_c g \nabla H_c$, or equivalently $-\nabla P_0 - \rho_c g \nabla H_a$.

Note that, since $\mathbf{F}_H$ is independent of z (to leading order in kH), our shallow-water equation is a strictly two-dimensional equation of motion. We now linearise our equation of motion about a base state of zero background motion. Taking $H_a = H + \eta(x, y, t)$, we obtain

$$\rho \frac{\partial \mathbf{u}_H}{\partial t} + \rho g \nabla \eta = -\nabla P_0 + \mathbf{F},$$

where, in the interests of brevity, we have dropped the subscript H on $\mathbf{F}$. Although $\mathbf{u}_H$ is a two-dimensional velocity field, vertical movement of the interface means that the two-dimensional divergences of $\mathbf{u}_{Hc}$ and $\mathbf{u}_{Ha}$ are both non-zero. In fact, standard shallow-water theory tells us that continuity demands

$$\nabla \cdot (H\mathbf{u}_a) = -\nabla \cdot (h\mathbf{u}_c) = -\frac{\partial \eta}{\partial t}.$$

(Again, we have dropped the subscript H for convenience.)

Next, we replace $\mathbf{u}_a$ and $\mathbf{u}_c$ by the volume fluxes $\mathbf{q}_a = H\mathbf{u}_a$ and $\mathbf{q}_c = -h\mathbf{u}_c$. Also, by virtue of condition (c) above, we may take $\mathbf{F}_c = \mathbf{0}$ (to leading order in kH). This is valid because, as we have seen, the current perturbation in the cryolite is an order of magnitude smaller than that in the aluminium. The governing equations become

$$\frac{\rho_c}{h} \frac{\partial \mathbf{q}_c}{\partial t} - \rho_c g \nabla \eta = \nabla P_0, \tag{13.7}$$

$$\frac{\rho_a}{H} \frac{\partial \mathbf{q}_a}{\partial t} + \rho_a g \nabla \eta = -\nabla P_0 + \mathbf{F}_a, \tag{13.8}$$

$$\nabla \cdot \mathbf{q}_c = \nabla \cdot \mathbf{q}_a = -\frac{\partial \eta}{\partial t}. \tag{13.9}$$

We now perform a so-called Helmholtz decomposition on $\mathbf{q}$ and write $\mathbf{q} = \mathbf{q}_R + \mathbf{q}_P$. That is, we divide $\mathbf{q}$ into a solenoidal, rotational part and an irrotational component of finite divergence, as discussed in Davidson and Lindsay (1998). Such a decomposition is unique if appropriate boundary conditions are prescribed. The boundary conditions on $\mathbf{q}_a$ and $\mathbf{q}_c$ are that $\mathbf{q} \cdot \mathbf{n}$ vanishes on the lateral boundary, S. An appropriate decomposition is therefore

$$\nabla \times \mathbf{q}_P = \mathbf{0}, \qquad \nabla \cdot \mathbf{q}_P = -\frac{\partial \eta}{\partial t}, \qquad \mathbf{q}_P \cdot \mathbf{n} = 0 \quad \text{on } S, \tag{13.10}$$

$$\nabla \times \mathbf{q}_R = \nabla \times \mathbf{q}, \qquad \nabla \cdot \mathbf{q}_R = 0, \qquad \mathbf{q}_R \cdot \mathbf{n} = 0 \quad \text{on } S. \tag{13.11}$$

Now (13.7) tells us that $\mathbf{q}_R$ is zero in the electrolyte, while (13.10) demands that $\mathbf{q}_P$ be the same in both layers: $\mathbf{q}_c = \mathbf{q}_P$; $\mathbf{q}_a = \mathbf{q}_P + \mathbf{q}_R$. We now rewrite (13.7) and (13.8) in terms of $\mathbf{q}_P$ and $\mathbf{q}_R$, eliminate P_0 by adding the equations, and use (13.10) to express η in terms of $\mathbf{q}_P$. The resulting equation of motion is

$$\bar{\rho} \frac{\partial^2 \mathbf{q}_P}{\partial t^2} - \Delta \rho g \nabla^2 \mathbf{q}_P = \left[\frac{\partial \mathbf{F}_a}{\partial t} \right]_P = \frac{\partial \mathbf{F}_P}{\partial t}, \tag{13.12}$$

where $\bar{\rho} = \rho_c/h + \rho_a/H$ and $\Delta \rho = \rho_a - \rho_c$. The subscript P on the bracket implies that we take only the irrotational component of the corresponding term; that is, like $\mathbf{q}$, $\mathbf{F}_a$ is subject to a Helmholtz decomposition, $\mathbf{F}_a = \mathbf{F}_P + \mathbf{F}_R$. Note that, in deriving (13.12), we have taken $\mathbf{F}_R = (\rho_a/H)\partial \mathbf{q}_R/\partial t$, so that (13.11) demands $\mathbf{F}_R \cdot \mathbf{n} = 0$. Thus it is implicit in (13.12) that $\mathbf{F}_P \cdot \mathbf{n} = \mathbf{F}_a \cdot \mathbf{n}$. Note also that, when the Lorentz force is zero, we recover the conventional equation for interfacial waves in the shallow-water limit:

$$\frac{\partial^2 \mathbf{q}_P}{\partial t^2} - c^2 \nabla^2 \mathbf{q}_P = \mathbf{0}, \qquad c^2 = \Delta \rho g / \bar{\rho}. \tag{13.13}$$

We now evaluate $\mathbf{j}_a$, and hence $\mathbf{F}_a$, using the long-wavelength approximation. In the cryolite we have, to leading order in kh, $\partial^2 \Phi / \partial z^2 = 0$, from which

$$\Phi_c = \Phi_0 z / (h - \eta) + O(kh),$$
$$\mathbf{j}_c = -(J_0 \eta / h)\, \hat{\mathbf{e}}_z + O(kh).$$

This current passes into the aluminium, and so the linearised boundary conditions on $\mathbf{j}_{za}$ are

$$\mathbf{j}_{za} = -(J_0 \eta / h)\hat{\mathbf{e}}_z \quad \text{on} \quad z = 0,$$
$$\mathbf{j}_{za} = \mathbf{0} \qquad\qquad \text{on} \quad z = -H.$$

It is readily confirmed that the conditions of zero divergence and zero curl, as well as the boundary conditions given above, are satisfied by

$$\mathbf{j}_a = \mathbf{j}_H(x, y,\ t) - (1 + z/H)(J_0 \eta / h)\, \hat{\mathbf{e}}_z, \tag{13.14}$$

where $\mathbf{j}_H$ is the horizontal component of the current density in the aluminium, which satisfies

$$\nabla \times \mathbf{j}_H = \mathbf{0}, \quad \nabla \cdot \mathbf{j}_H = \frac{J_0 \eta}{Hh}, \quad \mathbf{j}_H \cdot \mathbf{n} = 0 \quad \text{on} \quad S. \tag{13.15}$$

Comparing Equation (13.15) with (13.10) we find the remarkably useful result

$$\frac{\partial \mathbf{j}_H}{\partial t} = -\frac{J_0}{hH}\, \mathbf{q}_P. \tag{13.16}$$

This is the key relationship which allows us to express the Lorentz force in terms of the fluid motion and therefore deserves some special attention. The physical basis for (13.16) is contained in Figure 13.4. When the interface tilts, there is a horizontal flow of current from the high to the low side of the interface. Simultaneously, there is a horizontal rush of the aluminium in the opposite direction. It is this coupling which lies at the heart of the instability, and which is expressed by (13.16).

We now invoke condition (e) of §13.4, which states that the leading order term in the Lorentz force arises from the interaction of $\mathbf{j}_H$ with the background component of B_z, $\mathbf{F}_a = \mathbf{j}_H \times \mathbf{B}_z$. Substituting for $\mathbf{F}_a$ in (13.12), and introducing

$$\varpi_B^2 = \frac{J_0 B_z}{\bar{\rho} h H},$$ (13.17)

we find, after a little algebra,

$$\frac{\partial^2 \mathbf{q}_P}{\partial t^2} - c^2 \nabla^2 \mathbf{q}_P = \varpi_B^2 [\hat{\mathbf{e}}_z \times \mathbf{q}_P]_P, \quad \mathbf{q}_P \cdot \mathbf{n} = 0.$$ (13.18)

Finally, to obtain the most compact version of our wave equation, it is convenient to introduce potentials for $\mathbf{q}_P$ and $\mathbf{F}_P = \mathbf{F}_a - \mathbf{F}_R$:

$$\mathbf{q}_P = \nabla \varphi_P, \quad \frac{\partial \mathbf{F}_P}{\partial t} = \bar{\rho} \varpi_B^2 \nabla \Psi.$$

Noting that $\mathbf{F}_a = \mathbf{j}_H \times \mathbf{B}_z$ is solenoidal, and hence $\mathbf{F}_P$ is also solenoidal, (13.18) becomes

$$\frac{\partial^2 \varphi_P}{\partial t^2} - c^2 \nabla^2 \varphi_P = c^2 k_B^2 \Psi, \quad \nabla^2 \Psi = 0,$$ (13.19)

where

$$k_B^2 = \frac{\varpi_B^2}{c^2} = \frac{J_0 B_z}{\Delta \rho g h H}.$$ (13.20)

The corresponding boundary conditions on φ_P and Ψ are

$$\nabla \varphi_P \cdot \mathbf{n} = 0, \quad \nabla \Psi \cdot \mathbf{n} = (\nabla \varphi_P \times \mathbf{n})_z.$$ (13.21)

Note that the boundary condition on Ψ comes directly from (13.16) and from the requirement that $\mathbf{F}_P \cdot \mathbf{n} = \mathbf{F}_a \cdot \mathbf{n}$.

13.5.2 Key Dimensionless Groups

At last we are in a position to investigate cell stability. Solving (13.19) subject to boundary conditions (13.21) will determine the stability of the interface. Note that (13.19) is valid for any shape of domain, since we have made no assumptions about the lateral boundaries. We shall see that (13.19) can support both standing waves and travelling waves, and that both may go unstable.

Consider now a rectangular domain of size L_x, L_y, with $L_x > L_y$. We can make (13.19) dimensionless by re-scaling t according to $\hat{t} = k_B c t$ and $\mathbf{x}$ as $\hat{\mathbf{x}} = k_B \mathbf{x}$. In scaled units, (13.19) becomes

$$\ddot{\varphi}_P - \nabla^2 \varphi_P = \Psi, \qquad \nabla^2 \Psi = 0.$$

Evidently, the behaviour of interfacial waves in a rectangular domain is uniquely controlled by the (scaled) boundary shape, i.e. by $\hat{L}_x = k_B L_x$ and $\hat{L}_y = k_B L_y$. It follows that the stability threshold in a rectangular domain is uniquely determined by just two dimensionless parameters,

$$\varepsilon = k_B^2 L_x L_y = \frac{J_0 B_z L_x L_y}{\Delta \rho g h H}, \qquad r = L_x / L_y. \tag{13.22}$$

Here ε is a dimensionless version of $\hat{\varepsilon}$ introduced in (13.1). It follows from (13.22) that, in the absence of friction, the onset of instability in a rectangular cell will be characterised by a relationship of the form

$$\left[\frac{J_0 L_x L_y B_z}{\Delta \rho g h H} \right]_{\text{critical}} = \left[\frac{I B_z}{\Delta \rho g h H} \right]_{\text{critical}} = F(L_x / L_y), \tag{13.23}$$

where I is the net current passing through the cell and F is some function which is determined by the solution of (13.19). In the presence of friction this generalises to

$$\left[\frac{I B_z}{\Delta \rho g h H} \right]_{\text{critical}} = F(L_x / L_y, \text{ friction}),$$

(Davidson, 2000). Evidently, the instability is exacerbated by a large cell current, I, by an ambient magnetic field, B_z, and by a small cryolite depth, h. However, there is great incentive to make I large, to maximise productivity, and h as small as possible, to minimise Ohmic losses. Consequently, the traditional approach to cell design has been to minimise B_z through a careful arrangement of the currents entering and leaving the cell.

13.6 Solutions of the Wave Equation

Our shallow-water equation supports both travelling waves and standing waves. We shall show that both may become unstable.

13.6.1 Travelling Waves

Consider an infinitely long channel of width L, say $0 < x < L$. The easiest way of identifying travelling waves is to write both Ψ and φ_P in the form

$$\varphi_P = \hat{\varphi}(x) \exp[j(\varpi t - k_y y)]$$

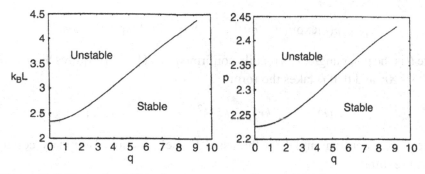

Figure 13.7 Neutral stability curves corresponding to (13.24).

and define a second wavenumber, k_x, through the expression $k_x^2 = (\varpi^2/c^2) - k_y^2$. Then (13.19) gives the eigenvalue problem

$$\hat{\varphi}'' + k_x^2 \hat{\varphi} = -k_B^2 \hat{\Psi}, \qquad \frac{\partial \hat{\varphi}}{\partial x} = 0 \quad \text{on} \quad x = 0, L,$$

$$\hat{\Psi}'' - k_y^2 \hat{\Psi} = 0, \qquad \frac{\partial \hat{\Psi}}{\partial x} = jk_y \hat{\varphi} \quad \text{on} \quad x = 0, L.$$

After a little algebra this yields a dispersion relationship for k_x in the form

$$2(k_B L)^4 \left[\cosh q \cos p - 1 + \frac{1}{2}(p/q - q/p)\sinh q \sin p \right]$$
$$+ (p^2 + q^2)^2 (p/q)\sinh q \sin p = 0 \tag{13.24}$$

where $p = k_x L$ and $q = k_y L$. When the Lorentz forces are zero ($k_B = 0$), this gives $k_x = m\pi/L$, which represents conventional travelling waves in a channel. For a finite value of k_B, and for an arbitrary wavenumber, k_y, we can always find a solution of (13.24) for which k_x is real. This represents stable travelling waves. However (13.24) also supports unstable waves. That is, for real values of k_B and k_y, we can find complex values of k_x which satisfy (13.24). This leads to complex frequencies and therefore to unstable motion. Figure 13.7 shows the neutral stability curve for waves in the range $0 < q < 10$.

13.6.2 Standing Waves in Circular Domains

We now consider waves in a closed, circular domain. This is of interest as it demonstrates the instability in a particularly simple way. Suppose the fluids occupy the domain $0 < r < R$, and consider solutions of the form

$$\varphi_P = \hat{\varphi}(r)\exp[j(\theta - \varpi t)] \; , \qquad \Psi = \hat{\Psi}(r)\exp[j(\theta - \varpi t)],$$

where θ is the polar angle. It is readily confirmed that (13.19) requires $\hat{\Psi}$ to be linear in r, $\hat{\Psi} = Ar$, and that $\hat{\varphi}$ takes the form

$$\hat{\varphi}(r) = CJ_1(kr) - (k_B^2/k^2)Ar \; , \qquad k = \varpi/c,$$

where C is a constant and J_1 is the usual Bessel function. Boundary conditions (13.21) require,

$$\hat{\varphi}'(R) = 0 \; , \qquad \hat{\varphi}(R) = jRA,$$

which yields the dispersion relation

$$k_B^2 J_2(kR) = jk^2 J_1'(kR).$$

This requires that k is complex, and so the waves are unstable for all non-zero k_B. The key point, though, is that the interface near marginal stability is of the form

$$\eta \sim J_1(kr)\sin(\theta - \varpi t),$$

which represents a rotating, tilted interface. This is precisely what is expected from the compound pendulum analogue of §13.3.

13.6.3 Standing Waves in Rectangular Domains

We now turn to rectangular domains, which of course are the most important for real cells. Here it is convenient to rewrite (13.19) in matrix form. This can be achieved by expanding $\mathbf{q}_P$ in a set of orthogonal cosine functions, ψ_i, of the form

$$\mathbf{q}_P = \sum \mathbf{q}_i = \sum k_i^{-1} x_i(t) \nabla \psi_i,$$

where

$$\psi_i = \psi_{mn} \sim \cos(m\pi x/L_x)\cos(n\pi y/L_y) \; ,$$
$$k_i^2 = k_{mn}^2 = (m\pi/L_x)^2 + (n\pi/L_y)^2 \; ,$$

and $x_i(t)$ are the amplitudes of the modes. Of course, ψ_i are just the gravitational modes in the absence of Lorentz forces and k_i are the corresponding wavenumbers. We now take the dot-product of (13.18) and $\nabla \psi_i$ and integrate over the domain, V. The end result is

$$\ddot{x}_i(t) + \varpi_{gi}^2\, x_i = \varpi_B^2 \sum K_{ij}x_j, \tag{13.25}$$

where ϖ_{gi} are the conventional gravitational frequencies, the interaction matrix K_{ij} has elements

$$K_{ij} = k_i^{-1}k_j^{-1} \int \left(\nabla\psi_j \times \nabla\psi_i \right)_z dx dy, \tag{13.26}$$

and we have normalised the modes ψ_i such that

$$\int \psi_i^2 dV = 1. \tag{13.27}$$

Note that K_{ij} is skew-symmetric, $K_{ij} = -K_{ji}$, and so has all of its diagonal elements equal to zero. Finally we truncate x_i at some suitably small wavenumber and rewrite (13.25) in matrix form:

$$\ddot{\mathbf{x}} + \Omega_g\mathbf{x} = \varpi_B^2\mathbf{K}\mathbf{x}. \tag{13.28}$$

Let us now consider some of the more general properties of (13.28). Consider first the case where ϖ_B is much greater than the gravitational frequencies of the truncated system. In this case, (13.28) gives

$$\frac{d^4\mathbf{x}}{dt^4} = -\varpi_B^4\mathbf{S}_1\mathbf{x}, \quad \mathbf{S}_1 = -\mathbf{K}\mathbf{K}. \tag{13.29}$$

The equivalent eigenvalue problem is

$$\mathbf{S}_1\mathbf{x} = -(\varpi/\varpi_B)^4\mathbf{x} = \lambda\mathbf{x}. \tag{13.30}$$

Now $\mathbf{S}_1$ is real, symmetric and has positive diagonal elements. It follows that the eigenvalues of (13.30) are real and at least some of them are positive. We conclude, therefore, that for large ϖ_B at least some frequencies of our truncated system are complex and the flow unstable.

Let us now return to the general eigenvalue problem represented by (13.28):

$$\left(\Omega_g - \varpi_B^2\mathbf{K} \right)\mathbf{x} = \lambda\mathbf{x}, \quad \lambda = \varpi^2. \tag{13.31}$$

Suppose that $\mathbf{x}$ is truncated after N modes and that the diagonal elements of Ω_g are arranged in order of increasing frequency, from ϖ_{g1}^2 to ϖ_{gN}^2. Then we may show that in the truncated system the eigenvalues, λ, have the following general properties:

(a) $\varpi_{g1}^2 \leq \text{Re}(\lambda) \leq \varpi_{gN}^2$;

(b) $\sum \lambda_i = \sum \varpi_{gi}^2$;

(c) λ_i/ϖ_B^2 are zero or pure complex if $\varpi_B^2 \gg \varpi_{gN}^2$.

These three properties are sufficient to define the general behaviour of the eigen-values as a function of ϖ_B. The first property follows from the skew-symmetry of **K**. That is, if $\bar{x}_i$ is the complex conjugate of x_i, then

$$\sum_i \left(\varpi_{gi}^2 - \lambda\right)|x_i|^2 = \varpi_B^2 \sum_i \sum_j K_{ij} x_j \bar{x}_i.$$

If we normalise the eigenvectors to have unit magnitude and take the complex conjugate of the transpose of this equation, we obtain

$$\text{Re}(\lambda) = \sum \varpi_{gi}^2 \ |x_i|^2.$$

Condition (a) then follows. Condition (b), on the other hand, arises from the fact that the sum of the eigenvalues equals the trace of $\mathbf{\Omega}_g - \varpi_B^2 \mathbf{K}$, while condition (c) is a standard result for skew-symmetric matrices.

The situation is therefore clear. As ϖ_B is increased, the eigenvalues move along the real axis but remain within the limits $\varpi_{g1}^2 < \lambda < \varpi_{gN}^2$. At some critical value of ϖ_B, two or more eigenvalues become complex (an inevitable consequence of condition (c)) and do so in the form of complex conjugate pairs (condition (b)). However, the real part of the complex eigenvalues remain bounded by the least and largest gravitational frequency of the truncated set of modes (condition (a)).

We now present a simple numerical example which illustrates the phenomenon. We shall show that, frequently, it is not the pair of modes with the closest gravita-tional frequencies which go unstable first. Moreover, the modes which go unstable at the lowest value of $J_0 B_z$ need not be the most dangerous. Often the highest growth rates are observed in the pairs of modes which are the second or third to go unstable. Of course, it is the modes with the highest growth rates which are most likely to survive the friction which is inevitably present in any real flow. We start by rewriting (13.31) in dimensionless form. We use $k_1 = \pi/L_y$ as a characteristic (inverse) length scale and introduce the dimensionless eigenvalues $\hat{\lambda} = \varpi^2/(ck_1)^2$. It is also convenient to reintroduce

$$\varepsilon = k_B^2 L_x L_y = \frac{J_0 B_z L_x L_y}{\Delta \rho g h H}$$

as our dimensionless control parameter.

Table 13.1 *Development of unstable modes in a rectangular cell of aspect ratio* $L_y / L_x = 0.3$.

Instability	Modes	Comments
First....	$(3,0) + (0,1)$	Restabilises
Second....	$(2,0) + \frac{1}{2}[(3,0) + (0,1)]$	–
Third....	$(1,1) + (2,1)$	Furthest to the right
Fourth....	$(1,0) + \frac{1}{2}[(3,0) + (0,1)]$	Furthest to the left

Consider the case $L_y / L_x = 0.3$, which is typical of a real cell. The trajectories of the dimensionless eigenvalues, $\hat{\lambda}$, in the complex plane are shown in Figure 13.8. Three ranges of ε are shown, corresponding to (a) $\varepsilon < 8.12$, (b) $\varepsilon < 10.1$ and (c) $\varepsilon < 13.5$. Figure 13.8(a) shows that, by $\varepsilon = 8.12$, one pair of eigenvalues has coalesced and moved into the complex plane. In fact, these complex eigenvalues first appear at $\varepsilon = 3.90$, through the interaction of the (3, 0) and (0, 1) modes. (We classify the eigenvalues in terms of their mode number (m, n) at $\varepsilon = 0$.) By $\varepsilon = 10.1$, the complex eigenvalues have returned to the real axis and a new pair of unstable frequencies have appeared. This arises from an interaction of a (2, 0) mode with one of the pair of previously unstable eigenvalues. By $\varepsilon = 13.5$, two additional unstable pairs have appeared. One arises from the interaction of (1, 1) and (2, 1) modes, and the other through the interaction of the (1, 0) mode with the second of the pair of previously unstable modes. The behaviour is summarised in Table 13.1.

This simple example exhibits two interesting features. First, it is not the modes with the closest gravitational frequencies ϖ_{gi} which go unstable. In fact, the closest gravitational frequencies are the (0, 1) and (1, 1) modes, yet at no time do they combine to produce an instability. Second, by $\varepsilon = 13.5$, the largest growth rate is exhibited not by the first instability, but by the fourth one. Given that any real flow has dissipation, it is this last instability which is the most likely to survive in practice.

So the idea that the modes with the closest gravitational frequencies are the most dangerous is a little too simplistic. In this regard it is important to note that **K** is very sparse. Indeed only around one in five mode-pairs are coupled. In general, then, relatively few modes exchange energy. It is not difficult to show that an instability cannot develop from these uncoupled modes, so it is only the separation of the coupled modes which is important. Thus a stability criterion based on keeping all gravitational modes apart is overly conservative. This point is of considerable practical importance.

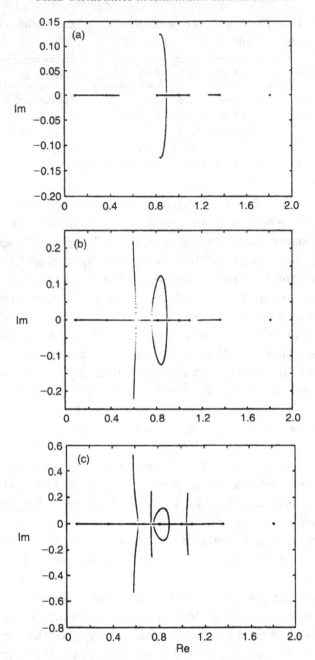

Figure 13.8 Instabilities in a rectangular cell of aspect ratio $L_y/L_x = 0.3$. The three panels show the trajectory of the dimensionless eigenvalues in the complex plane corresponding to varying values of ε: (a) $\varepsilon < 8.12$, (b) $\varepsilon < 10.1$, (c) $\varepsilon < 13.5$.

13.7 Implications for Cell Design and Potential Routes to Saving Energy

There are many idealisations embedded in our stability analysis and so it would be imprudent to consider it as an accurate working model of a real cell. For example, we have ignored friction as well as the fact that the anode is not monolithic, but rather consists of pairs of discrete carbon blocks. Nevertheless, our analysis does capture the key instability mechanism and so the broad conclusions of the model should remain valid. In particular, we might expect (13.23),

$$\left[\frac{J_0 L_x L_y B_z}{\Delta\rho g h H}\right]_{\text{critical}} = \left[\frac{I B_z}{\Delta\rho g h H}\right]_{\text{critical}} = F\left(L_x/L_y, \text{friction}\right),$$

to characterise the onset of instability.

The industry trend over the last 3 decades has been to make the cell current, I, as large as possible, to maximise productivity, and h as small as possible, to minimise Ohmic losses in the cell. Clearly, both of these trends exacerbate the MHD instability. (In last decade the current in successive generations of cells has risen from 300kA to 500kA, while the depth h has been maintained at around 4.5–5.0 cm.) To counter the growing risk of instability, the traditional approach has been to

(i) choose the cell aspect ratio, L_x/L_y, to ensure that the natural frequencies of the dominant interfacial waves are well-separated
(ii) minimise the ambient vertical field, B_z, through a careful arrangement of the currents paths entering and leaving the cell
(iii) carefully monitor the cryolite depth, h, in order to detect the onset of an instability.

The current practice is to continuously monitor the cryolite depth, h, using measurements of the anode–cathode voltage drop in individual anode blocks, and if an instability is detected in the form of a fluctuating voltage, the depth h is increased somewhat to stabilise the cell.

However, if we wish to eliminate the instability completely, then more drastic action is required, and several options are proposed in Davidson (2000). One possibility is to introduce vertical baffle plates into the liquid aluminium, whose function is to break up the long-wavelength sloshing motions. (Recently, at least one patent along these lines has been applied for.) There are other, related, forms of passive control, all designed to absorb wave energy and so stabilise the cell. One involves introducing an irregular surface on the carbon cathode block, which increases the frictional drag on the liquid-metal layer. Another strategy is to try and absorb wave energy at the periphery of the cell so as to weaken the waves when they reflect from the lateral boundaries, thus diminishing the ability of reflected waves to establish unstable standing waves. In this respect it is important to note that there are so-called edge-channels which run around the perimeter of nearly all cells and which are open to the atmosphere (Figure 13.2). If wide enough, these edge-channels can act as a pressure release mechanism for waves impacting on the lateral boundary, diverting some of the wave energy into hydraulic surges within the edge-channel.

A more ambitious control strategy is to continuously monitor the position of the interface and slowly tilt the anode block assembly in sympathy with any unstable wave, so as to keep the electrolyte thickness roughly uniform across the cell. This has been termed the 'anode-tilt' strategy (Davidson, 2000). If implemented correctly, it will prevent the build up of current perturbations and so remove the driving force for the instability. Such an approach builds on the fact that present cells already have the capacity to raise or lower individual anode blocks in response to voltage measurements. The approach is also helped by the observation that: (i) the wave motion is extremely slow (with periods measured in minutes and growth rates in many tens of minutes), (ii) the unstable liquid interface tilts by only a fraction of a degree, and (iii) the required vertical movement in anode position is small (a millimetre or so). However, there are also practical problems associated with this kind of active control, and such a scheme has yet to be successfully implemented.

Whatever control strategy is adopted, there is a pressing need to reduce the cryolite depth and increase the energy efficiency of reduction cells. As noted at the start of this chapter, if electrolysis cells could be made to operate with a cryolite depth of 4 cm, instead of 4.5 cm, that could save over a billion dollars a year. As the pressure to reduce energy consumption and move towards a more sustainable environment increases, it seems inevitable that a new generation of Hall–Héroult cells will eventually emerge.

Exercises

13.1 Many ways have been proposed for eliminating the MHD instability in reduction cells, as discussed above. An additional strategy is to use sloping cathode blocks to continuously drain the aluminium and so avoid the build up of a thick aluminium layer. Why do you think this has not yet been implemented, despite considerable research?

13.2 Find the normal mode shape for the oscillations of the compound pendulum shown in Figure 13.5. Confirm that for large $\hat{\varepsilon}$ the mode consists of a rotating, tilted plate.

13.3 Give a simple physical explanation, based on the dynamics of the compound pendulum, as to why travelling wave instabilities in a channel are inevitable, as described by (13.24).

Part IV
Applications in Physics

If you want to become a physicist, you must do three things – first, study mathematics, second, study more mathematics, and third, do the same.
Arnold Sommerfeld

Perhaps one of the earliest applications of MHD in geophysics was Faraday's suggestion that movement in the oceans might be responsible for fluctuations in the earth's magnetic field. Its origins in astrophysics, on the other hand, probably dates back to the end of the nineteenth century, when astrophysicists realised that the sun played host to a magnetic field. Shortly thereafter Larmor speculated that the motion of a conducting fluid across the field lines embedded within a rotating body (say our sun) might induce precisely those currents required to support the original field. Of course, it took many decades to add substance to this tentative proposal and a century later there are still many unanswered questions in solar and planetary dynamo theory. In the meantime, MHD has become a crucial ingredient of many aspects of astrophysics, from the formation of stars to the triggering of solar flares and the generation of turbulence within accretion discs. There is also a sizable community trying to understand how planetary dynamos work, and yet another trying to fathom why it is so hard to confine a fusion plasma in a stable configuration using magnetic fields.

Of course, we cannot possibly do justice to these many interesting and diverse topics here, each of which could provide, and indeed has provided more than enough material to fill an entire monograph. So the purpose of the following chapters is much more modest: to give a mere hint as to some of the issues involved and to provide a first stepping stone to more advanced study.

14

The Geodynamo

We may imagine, as Gilbert did, the Earth to be wholly or in part a magnet, such as a magnet of steel, or we may conceive it to be an electromagnet, with or without a core susceptible of induced magnetism. In the present state of our knowledge this second hypothesis seems to be the more probable and indeed we now have reasons for believing in terrestrial currents. ... The question which occurs now is this: can the magnetic phenomena at the earth's surface, and above it, be produced by an internal distribution of closed currents occupying a certain limited space below the surface.

Kelvin, 1867

We might follow Kelvin and add a second question: 'What maintains those terrestrial currents?'

14.1 Why Do We Need a Dynamo Theory for the Earth?

Dynamo theory is the name given to the process of magnetic field generation by the inductive action of a moving, conducting fluid, i.e. the conversion of mechanical energy into magnetic energy through the stretching and twisting of the magnetic field lines. It is generally agreed that this is the source of the earth's magnetic field, since the temperature of the earth's interior is well above the Curie point at which ferro-magnetic material loses its permanent magnetism. Moreover, the earth's magnetic field cannot be the static relic of some primordial field trapped within its interior. Such a field would have decayed long ago. To see why this is so, suppose that there is negligible motion in the earth's core. The product of $\mathbf{B}$ with Faraday's law yields, in the absence of core motion, the energy equation

$$\frac{\partial}{\partial t}[\mathbf{B}^2/2\mu] = -\nabla \cdot [\mathbf{E} \times \mathbf{B}/\mu] - \mathbf{J}^2/\sigma. \qquad (14.1)$$

Integrating over all space and noting that the Poynting flux, $\mathbf{E} \times \mathbf{B}/\mu$, integrates to zero, we obtain

$$\frac{d}{dt}\int_{V_\infty} (\mathbf{B}^2/2\mu)dV = \frac{dE_B}{dt} = -\int_{V_C} (\mathbf{J}^2/\sigma)dV, \tag{14.2}$$

where E_B is the total magnetic energy and V_C is the electrically conducting core of the earth. As we might have anticipated, the magnetic energy decays at a rate fixed by the Ohmic dissipation in the core.

The rate of decay of the magnetic field can be determined by normal mode analysis of the diffusion equation, $\partial\mathbf{B}/\partial t = \lambda\nabla^2\mathbf{B}$, applied to a sphere and coupled to an external (irrotational) magnetic field lying outside the sphere. The resulting eigenvalue problem yields, for the slowest decaying mode, a characteristic decay time of $t_d = R_C^2/(\lambda\pi^2)$, where R_C is the outer radius of the conducting core. Given that $R_C = 3.48 \times 10^6$m and $\lambda \approx 0.7$ m^3/s, the decay time is $t_d \approx 5 \times 10^4$ years. However, the earth's dipole field has been in existence for at least 10^9 years, and so there must be some additional mechanism at work which maintains E_B against these Ohmic losses.

Historically there have been many attempts to explain the origins of the earth's magnetic field, other than MHD. Now all abandoned, these include a magnetic crust, the Hall effect, thermoelectric effects, rotating electrostatic charges, and even, as a last act of desperation, a proposed modification to Maxwell's equations. The electrostatic argument arises from the fact that the earth's surface is negatively charged. In fact, this charge is so great that near the earth's surface there exists, on average, an atmospheric electric field of ~100V/m, directed radially inward. This surface charge is maintained by lightning storms that are charging the earth, relative to the upper atmosphere, with an average current of 1800 Amperes!

One by one these tentative theories have all been abandoned, often because they fail on an order of magnitude basis. It was Larmor who, in 1919, first suggested a self-excited fluid dynamo (in the solar context) in his paper: 'How could a rotating body like the sun become a magnet?' With the discovery of the liquid core of the earth, Larmor's ideas became relevant to the earth, and the notion of a self-excited fluid dynamo operating in the core made rapid progress through the work of Elsasser, Bullard and Parker. By the late 1950s many of the basic physical ingredients of modern dynamo theory were in place (see, in particular, Parker, 1955), though it took several more decades to add mathematical substance to these intuitive, physical ideas. In any event, there now exists a substantial body of theoretical work, including a number of idealised fluid dynamos and a myriad of successful numerical experiments. The problem, however, is that the sparsity of observational evidence means it is unclear how the various idealised theories and

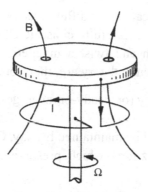

Figure 14.1 The homopolar disc dynamo.

numerical experiments relate to events in the earth's core, or indeed the cores of other planets.

Mechanical analogues of a self-excited fluid dynamo are readily constructed. A simple, and common, example is the homopolar disc dynamo, shown in Figure 14.1. Here a solid metal disc rotates with a steady angular velocity Ω and a twisted wire (which we shall assume is constructed from the same material as the disc) provides a current path between the rim and axis of the disc, as shown. It is readily confirmed that, provided Ω is large enough, any small magnetic field which exists at $t = 0$ will grow exponentially in time. First we note that rotation of the disc results in an EMF of $\Omega\Phi/2\pi$ between the axis and the edge of the disc, Φ being the magnetic flux which links the disc. (This may be confirmed by integrating Ohm's law radially across the disc and assuming that the flux through the disc is axisymmetric.) This EMF then drives a current, I, which evolves according to

$$L\frac{dI}{dt} + RI = \text{EMF} = \frac{\Omega M I}{2\pi},$$ (14.3)

where M is the mutual inductance of the current loop and the rim of the disc, and L and R are the self-inductance and resistance of the total circuit. Evidently I, and hence **B**, grows exponentially whenever Ω exceeds $2\pi R/M$. This increase in magnetic energy is accompanied by a corresponding rise in the torque, T, required to drive the disc, the source of the magnetic energy being the mechanical power, $T\Omega$. In any real situation, however, this exponential rise in T cannot be maintained for long, and eventually the applied torque will fall below that needed to maintain a constant Ω. At this point the disc will slow down, eventually reaching the critical level of $\Omega = 2\pi R/M$. T, Ω and **B** then remain steady. Note that, if R is dominated by the resistance of the wire, the criterion $\Omega > 2\pi R/M$ can be rewritten as $\Omega r^2_{\text{wire}}/\lambda > O(1)$, where r_{wire} is the radius of the wire. This is the first hint that planetary dynamos require a large magnetic Reynolds number.

Now this kind of mechanical device differs greatly from a fluid dynamo because the current is directed along a carefully constructed path. Nevertheless, a highly conducting fluid can stretch and twist a magnetic field so as to intensify E_B. The central questions in dynamo theory are, therefore,

(i) Can we construct a steady (or steady-on-average) velocity field which leads to dynamo action?
(ii) Can such a velocity field be maintained by, say, Coriolis and buoyancy forces in the face of the Lorentz force which, presumably, tends to suppress the fluid motion?

Question (i) is known as the *kinematic dynamo problem* and it was resolved many decades ago. The second question, that of a dynamically consistent dynamo, is altogether much harder. It is now agreed that the answer to question (ii) is also 'yes', if only because dynamically self-consistent dynamos are routinely produced in computer simulations. However, it should be said from the outset that there is, as yet, no generally accepted picture as to the precise mechanisms which underpin the geodynamo. The many successful numerical experiments, for example, operate in a regime very far from that encountered in a planetary core, being overly viscous by a factor of 10^9 and very weakly driven, unlike convection in the core which is virtually inviscid and very strongly driven. So, while nearly all theoreticians agree that strongly helical convection is probably the key to a successful planetary dynamo, the mechanisms by which that helicity is established are still debated.

14.2 Sources of Convection, Reversals and Key Dimensionless Groups

14.2.1 The Structure of the Earth and Sources of Convection

The earth, which is close to spherical, has an outer radius of 6370 km and comprises a metal core, a rocky mantle and a thin, outer crust (Figure 14.2). The mantle which surrounds the earth's core has a poor electrical conductivity, except perhaps close to the core-mantle boundary (CMB), and so plays no direct role in the geodynamo. The core itself is divided into two parts. The liquid outer core has a radius of approximately $R_C = 3480$ km, while the inner core is solid and has a radius of $R_i = 1220$ km. The composition of the core can be inferred, in part, from seismology and in part from estimates of the chemical makeup of those iron meteorites which helped form the earth. The inner core is thought to be composed of an iron-nickel alloy, being mostly iron with around an 8 per cent nickel content. The outer core is also predominantly composed of iron and nickel, though seismology shows that, on average, it is around 15 per cent lighter than the inner core, with an abrupt drop in density of 5 per cent at the inner core boundary (ICB). Since iron and nickel have

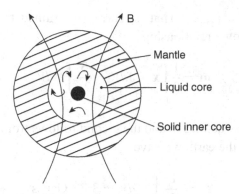

Figure 14.2 Schematic of the structure of the earth.

similar densities, there must be a weak admixture of lighter elements in the outer core, such as sulphur, oxygen or silicon.

The maximum temperature of the core is similar to that on the surface of the sun, being around 6000K at the ICB. Despite this high temperature, the inner core is slowly growing by solidification. At first sight it is odd that freezing occurs near the centre of the earth, where it is hottest, rather than at the mantle. However, it is the high pressure at the centre that causes freezing there. It is probable that the core was initially completely molten and that the solid inner core first nucleated around 10^9 years ago (although its age is frequently disputed), and has been slowly growing ever since at a rate of ~1 mm/year. This slow solidification of the core helps drive convection in two distinct ways. First, there is latent heat release which supplements the conventional thermal buoyancy. Second, when iron mixed with impurities solidifies, it is the relatively pure iron that freezes first. Thus the iron which freezes onto the inner core is relatively free of the lighter impurities and leaves behind a layer of fluid adjacent to the ICB that is rich in lighter elements. This light fluid then floats out towards the mantle, giving rise to compositional convection.

In short, there are two primary sources of motion in the outer core – thermal and compositional convection – and it is generally agreed that these are the main sources of energy that drive the geodynamo. In numerical simulations of planetary cores, this radial buoyancy flux is often found to be concentrated in and around the equatorial plane, where it takes the form of random meandering plumes floating out towards the mantle, and also near the rotation axis, where it comprises buoyant upwellings.

14.2.2 Field Structure and Reversals

The external magnetic field of the earth is dipole-like, with a magnetic axis roughly aligned with the rotation axis. (There is an 11° mismatch.) The dipole moment has a strength of $|\mathbf{m}| = 7.9 \times 10^{22}$ Am^2, from which we can estimate the average axial field strength in the core, $\overline{B}_z$. That is, currents confined to a spherical volume V satisfy the well-known relationship

$$\mathbf{m} = \frac{1}{2}\int_V \mathbf{x} \times \mathbf{J}dV = \frac{3}{2\mu}\int_V \mathbf{B}dV, \tag{14.4}$$

where the first equality defines $\mathbf{m}$ and the second follows from the Biot–Savart law (Jackson, 1998). For the earth we have

$$\overline{B}_z = \frac{1}{V_C}\int_{V_C} B_z dV = 3.75 \text{ Gauss.} \tag{14.5}$$

Although the external magnetic field is dipole-like, the dominant structure of the field in the core is the subject of much discussion and little agreement. One of the central components of the early dynamo theories of Elsasser and Parker was the idea that, as a result of angular momentum conservation, differential rotation would be common in the core of a planet. (The idea is that, in some statistical sense, angular momentum conservation means that fluid spins less rapidly as it moves out towards the mantle, and more rapidly as it returns towards the solid inner core, although this is harder to prove than it sounds – see §14.5.3.) Such differential rotation operating at a high magnetic Reynolds number would sweep out an east-west field from the north-south dipole, as described in §5.5.3 (and shown in Figure 5.8). As we noted in Chapter 5, this process can lead to an east-west field of order $R_m\overline{B}_z$, where $R_m = u\ell/\lambda$ is based on a typical velocity and length scale representative of the differential rotation. Since R_m is usually assumed to be reasonably large in planetary dynamos, this suggests that the dominant field in the core may be an east-west field, and indeed some of the early numerical simulations exhibited a dominant east-west field (Figure 14.3). However, most of the current numerical experiments display only modest differential rotation, largely because the dipole field resists being azimuthally sheared and tends to clamp the core flow into a state close to that of rigid-body rotation. So most of the (admittedly imperfect) numerical simulations exhibit an east-west field that is no greater than the dipole field. Of course, it is entirely possible that this picture will change as the simulations become more planet-like, so there is still much debate as to the dominant field within the core of the earth.

The study of paleomagnetism shows that the earth's dipole field reverses in what appears to be a random manner. These reversals are relatively rapid, typically $10^3 \rightarrow 10^4$ years in duration, whereas the time between reversals is of the order of 10^6 years. The reversal timescale is comparable with the magnetic diffusion time for the solid inner core, $t_d = R_i^2/(\lambda\pi^2) \sim 6900$ years, and considerably longer than

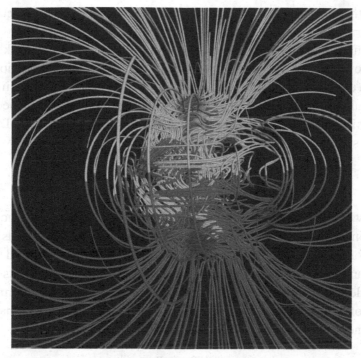

Figure 14.3 An early numerical simulation of the geodynamo showing the magnetic field structure. Note the intense azimuthal field wrapped around the inner core. (Courtesy of Gary Glatzmaier.)

the convective turn-over time in the core (which is around 300 years). This suggests that the rate-determining step for magnetic reversals might be the time taken for the field to diffuse out of the solid inner core. The implication is that the magnetic inertia of the inner core helps stabilise the terrestrial field. Similar reversals are often seen in (the imperfect) numerical simulations of planetary dynamos, and the underlying mechanisms are still being debated. Perhaps the oldest and simplest explanation rest on the fact that the core motion is highly chaotic because of the non-linear coupling of the buoyancy, velocity and magnetic fields. This randomness can result in occasional transient declines in convection which might cause the dynamo to temporarily shut down. Magnetic fields suppress convection, so as the field in the core collapses, the convection can re-establish itself and dynamo action restart. In this picture a reversal occurs when the resurgent field happens to have the opposite polarity to the original one, whereas those cases where the resurgent field has the same polarity as the original are classified as excursions. However, this picture of reversals is too simplistic, since the reversal time is then the core diffusion time, which is much too long. As discussed in Exercise 14.9 at the end of the chapter, the observed fast reversal time implies a significant, if transient, change in the core convection pattern, which in turn may drive a rapid readjustment of the internal field towards a transient, multipolar structure.

In summary, five questions any theory of the geodynamo might seek to answer are; (i) 'How does the field regenerate itself?', (ii) 'Is differential rotation, and the associated east-west magnetic field, an important part of this cycle?', (iii) 'Why is the external field dipole-like?', (iv) 'Why is the dipole (almost) aligned with the rotation axis?' and (v) 'What causes a reversal in the dipole field?'. To date, we have only speculative answers to these questions.

14.2.3 Key Dimensionless Groups

Let us now turn to the various dimensionless groups that characterise motion in the core, starting with R_m. The characteristic value of the magnetic Reynolds number is different for the large and small-scale motion. A typical large-scale fluid velocity in the core can be estimated from the rate at which the magnetic field drifts across the core-mantle boundary, and this suggests $|\mathbf{u}| \sim 0.2$ mm/s. Also, the most recent estimates of electrical conductivity give $\lambda \approx 0.7$ m^2/s, so we might estimate a large-scale magnetic Reynolds number to be

$$R_m = \frac{|\mathbf{u}|(R_C - R_i)}{\lambda} \sim 600.$$

On the other hand, an estimate of the minimum length scale for motion in the core might be obtained by demanding that $R_m \sim 1$ at the small scales, since any motion below this scale will be subject to strong Ohmic dissipation. Taking $|\mathbf{u}| \sim 0.2$ mm/s and $u\ell_{min}/\lambda = O(1)$, this suggests, albeit tentatively, the estimate $\ell_{min} \sim 4$ km.

Let us now consider the relative magnitude of the forces operating in the earth's core. The angular velocity of the earth is $\Omega = 7.28 \times 10^{-5}$s^{-1}, which yields a Rossby number of

$$\mathrm{Ro} = \frac{|\mathbf{u}|}{\Omega(R_C - R_i)} \sim 10^{-6}.$$

Evidently inertia, $\mathbf{u} \cdot \nabla \mathbf{u}$, is irrelevant in the core, even at the smallest scale of the motion, ℓ_{min}. So we are obliged to neglect $\mathbf{u} \cdot \nabla \mathbf{u}$ in any theory of the geodynamo. Of course, the absence of $\mathbf{u} \cdot \nabla \mathbf{u}$ means that the Kolmogorov picture of turbulence, with its cascades driven by inertia, does not apply, and so any turbulence in the core will be a consequence of the non-linear coupling between $\mathbf{u}$, $\mathbf{B}$ and buoyancy. Note that the low value of Ro also means that inertial waves are likely to be commonplace and so, although we are obliged to drop $\mathbf{u} \cdot \nabla \mathbf{u}$, we cannot neglect $\partial \mathbf{u}/\partial t$ in the governing equations.

Table 14.1 *Estimated properties of the core. (a) Properties which can be measured or inferred. (b) Time scales in the core where, for illustrative purposes, we take* λ_{wave} *= 20 km as the inertial-wave wavelength. (c) Various dimensionless groups.*

(a) Measured or inferred properties		
Outer core radius	R_C	3480 km
Inner core radius	R_i	1220 km
Angular velocity	Ω	7.28×10^{-5} s^{-1}
Mean density of fluid	ρ	10.9×10^3 kg/m^3
Magnetic diffusivity	λ	0.7 m^2/s
Mean axial field in core	$\overline{B}_z$	3.7 Gauss
Characteristic velocity	$|\mathbf{u}|$	~ 0.2 mm/s
Minimum length-scale for motion	ℓ_{min}	~ 4 km ?
(b) Time scales		
Magnetic diffusion time	$t_d = R_C^2/(\lambda\pi^2)$	5×10^4 years
Convective time scale	$(R_C - R_i)/|\mathbf{u}|$	300 years
Time for inertial waves to traverse the core	$\pi(R_C - R_i)/\Omega\lambda_{wave}$	8 weeks
(c) Dimensionless groups		
Magnetic Reynolds number	$R_m = \frac{|\mathbf{u}|(R_C-R_i)}{\lambda}$	$O(600)$
Rossby number	$\mathrm{Ro} = \frac{|\mathbf{u}|}{\Omega(R_C-R_i)}$	$O(10^{-6})$
Magnetic Prandtl number	$\mathrm{Pr}_m = \nu/\lambda$	$O(10^{-6})$
Ekman number	$\mathrm{Ek} = \frac{\nu}{\Omega R_C^2}$	$O(10^{-15})$
Elsasser number	$\Lambda = \frac{\sigma B^2}{\rho\Omega}$	$O(1)$

We turn now to the viscous forces. The kinematic viscosity of liquid iron is $\nu \approx 10^{-6}$m^2/s, which is less viscous than air at room temperature. So the Ekman and magnetic Prandtl numbers are

$$\mathrm{Ek} = \frac{\nu}{\Omega R_C^2} \sim 10^{-15}, \quad \mathrm{Pr}_m = \frac{\nu}{\lambda} \sim 10^{-6},$$

as noted in Table 14.1. These particularly small values of Ek and Pr_m are difficult to achieve in numerical simulations because of the relatively small length scales they create. So, in the current generation of dynamo simulations one is usually restricted to $\mathrm{Ek} > 10^{-6}$, and consequently the viscous stresses are overestimated by a thought-provoking factor of 10^9. However, perhaps the primary significance of our estimates of Pr_m and Ek are that (i) $\mathrm{Pr}_m \ll 1$ requires that viscous dissipation be negligible by comparison with Ohmic dissipation and (ii) $\mathrm{Ek} \sim 10^{-15}$ demands that the viscous stresses be tiny in the core, so they are unlikely to play any significant role in the dynamics. This suggests that any dynamo model that relies on the presence of viscous stresses is unlikely to be relevant to the core of the earth, or indeed of any other planet. We shall return to this point in §14.4.

Given that both $\mathbf{u} \cdot \nabla \mathbf{u}$ and the viscous stresses are much too small to play any significant role, we might ask what *is* the primary force balance in the core? The mean density in the liquid core is $\rho = 10.9 \times 10^3 \text{ kg/m}^3$, from which we can calculate the magnitude of the *Elsasser number*,

$$\Lambda = \sigma B^2 / \rho \Omega. \tag{14.6}$$

This dimensionless group is usually taken as a measure of the ratio of the Lorentz to Coriolis forces. If we take $B = \bar{B}_z \sim 4\text{G}$, then we find $\Lambda \sim 0.2$, although the typical value of $|\mathbf{B}|$ in the core is likely to be larger than $\bar{B}_z$, especially if differential rotation sweeps out an east-west field. In any event, $\Lambda = O(1)$ is often interpreted as indicative of an approximate balance between Lorentz and Coriolis forces. Since buoyancy is the driving force for motion in the core, it is usually assumed that the Coriolis and buoyancy forces are also of the same order of magnitude in a planetary dynamo, and since dynamo action saturates once the Lorentz forces act back on the flow, it seems plausible that the Lorentz, Coriolis and buoyancy forces are all of similar magnitudes in a steady-on-average dynamo. (Actually, this is somewhat naïve because in a flow that is approximately geostrophic, with relatively weak gradients in the axial direction, the bulk of the Coriolis force is irrotational and so balanced by pressure gradients – see §14.3.)

14.3 A Comparison with the Other Planets

One of the main stumbling blocks to progress in geodynamo theory is the sparsity of observational data about events in the earth's core, particularly the structure of the velocity and magnetic fields. It seems prudent, therefore, to examine the behaviour of the other planets in the hope that this might expose trends in the observational data that an examination of the earth alone cannot reveal. In particular, it is natural to examine how the intensity and structure of planetary magnetic fields vary with the core size, core composition and rotation rate of the planets.

14.3.1 The Properties of the Other Planets

The planets in our solar system are divided up into three groups. Mercury, Venus, the earth and Mars have rocky mantles and liquid iron cores and are known collectively as the terrestrial planets (Figure 14.4). Jupiter and Saturn, on the other hand, are referred to as the gas giants (for obvious reasons) and are primarily composed of hydrogen and helium, with the extreme pressures in their interiors

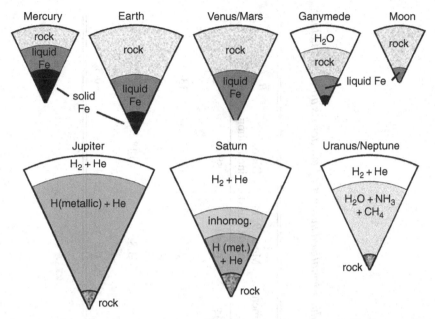

Figure 14.4 Approximate structure of the planets. (Courtesy of Uli Christensen.)

giving rise to metallic hydrogen. The third group, Uranus and Neptune, are called the ice giants. These are composed mostly of water, ammonia and methane, with a rocky inner core consisting of silicates, and an atmosphere rich in hydrogen and helium. The properties of these planets are listed in Table 14.2, where the mean axial magnetic fields in the conducting cores have been calculated from the observed dipole moments, **m**, using (14.4).

Mercury's external field is strongly dipolar though it is very weak (a tiny fraction of a Gauss) with the associated Elsasser number for the core being small, $\Lambda = \sigma B^2 / \rho \Omega \sim 10^{-4}$. The reason why Mercury's field is so weak is uncertain, although it has been attributed to the core convection being strongly suppressed by stable stratification, with convection confined to the deepest parts of the core. Whatever the reason for its weak field, what sets Mercury apart from the dynamos operating in the gas giants and the earth is the very low value of the Elsasser number, and so in some sense Mercury's dynamo could be anomalous.

Neither Venus nor Mars currently have dynamos, which is a little surprising as their sizes and internal structures are similar to that of the Earth. Venus has no detectable magnetic field while Mars exhibits some remnant magnetism, though this is largely confined to the southern hemisphere. This suggests that Venus has yet to develop a dynamo, while Mars once had a dynamo, though this has long since shut down. It is thought that an absence of dynamo action in Venus may be because

Table 14.2 *Properties of the planets. (Note the relatively uniform value of the normalised magnetic field across all the planetary dipoles. This is discussed in Exercise 14.8.) (Source: Davidson, 2013.)*

Planet	Rotation period (days)	Mean radius of planet (10^3 km)	Core radius (10^3 km)	Dipole moment ($10^{22}\,\mathrm{Am^2}$)	Mean B_z in core (Gauss)	Dipole inclination	$\dfrac{\bar{B}_z/\sqrt{\rho\mu}}{\Omega R_C} \times 10^6$
Mercury	58.6	2.44	1.8	0.004	0.014	5°	5.5
Venus	243	6.05	3.2	0	–	–	–
Earth	1	6.37	3.48	7.9	3.7	11°	13
Mars	1.03	3.39	1.8	0	–	–	–
Jupiter	0.413	69.9	55	150,000	18	9.6°	5.2
Saturn	0.440	58.2	29	4600	3.8	<0.1°	2.3
Uranus	0.718	25.3	~18	390	~1.3	59°	~1.0
Neptune	0.671	24.6	~20	200	~0.5	47°	~0.3

the convection is too weak, for one of two reasons. First, Venus has not nucleated a solid inner core, so there is no compositional buoyancy to supplement the thermal buoyancy. Second, the mantle convection pattern in Venus is quite different to that of the earth. There are no tectonic plates in Venus, but rather a single plate that acts like a rigid lid. This is likely to suppress the heat flux through the mantle and core and so reduce the thermal power available to drive convection in the core. The demise of Mars' dynamo is also attributed to a decline in core convection, one suggestion being that the pattern of mantle convection may have changed from one of plate tectonics to a Venus-like rigid lid, suppressing the core convection in the process.

Jupiter and Saturn both have strong magnetic fields, which are thought to be maintained by convective dynamos. The metallic hydrogen in their cores has an electrical conductivity somewhat less than that of iron, but given the size of the gas giants, it is likely that R_m is nevertheless very large. However, the crucial difference between the gas giants and the terrestrial planets, as far as dynamo action is concerned, is the absence of a rocky mantle. Rather, in the gas giants there is a gradual transition from metallic to molecular hydrogen. In the case of Jupiter, the metallic hydrogen constitutes the bulk of the planet (Figure 14.4), and so dynamo action is likely to be widely distributed, while in Saturn the metallic hydrogen is confined to the inner part of the planet and is enveloped by a stably stratified, helium-rich transition layer. One particularly striking feature of Saturn's external magnetic field is that it is almost perfectly axisymmetric, with a near perfect alignment of the rotation and magnetic axes. Since Cowling's theorem (which we will discuss in §14.5.4) forbids an axisymmetric dynamo, the internal field must be non-axisymmetric, with the higher harmonics filtered by the outer parts of the conducting core.

Finally, we turn to the ice giants, which possess magnetic fields quite unlike those of the other planets. The external fields discussed so far are all strongly dipolar, with the magnetic axis more or less aligned with the rotation axis. However, the external fields of the ice giants have a highly complex spatial structure, and to the extent that there is a dipole component, this is more aligned with the equator than the poles (Table 14.2). It seems likely, therefore, that the origin of the magnetic fields in the ice giants is quite different to those of the inner planets, and so we shall not dwell on these anomalous cases. Rather, we shall compare the observable features of the geodynamo with the dynamos operating in Mercury, Jupiter and Saturn.

14.3.2 Trends in the Strengths of the Planetary Dipoles: Scaling Laws

There are three observable properties of these planets which are important: (i) the size of the planet, from which we can infer a conducting core size, R_C; (ii) the planetary rotation rate, Ω; and (iii) the dipole field strength, $\mathbf{m}$, from which we can determine the mean axial field strength, $\overline{B}_z$, using (14.4). Other important parameters are the mean rate of working of the buoyancy force per unit mass in the core, P, and the magnetic diffusivity, λ. The first of these is related to the radial convective heat flux (per unit area), $\dot{q}$, according to $P = (g\beta/\rho c_p)\dot{q}$, where β is the thermal expansion coefficient, g is the local gravitational acceleration, and c_p the specific heat (Christensen, 2010). Unfortunately, $\dot{q}$, and hence P, cannot be readily estimated for some of the planets and so P is not always a well-constrained parameter. Estimates of $\dot{q}$ are available for the earth, however, (see Exercise 14.1) and these show that some suitable measure of the driving force for convection, say a Rayleigh number, is around 10^4 times the critical value at which non-magnetic convection would first appear. In this sense the convection is strongly driven and likely to be turbulent. In any event, it is natural to look for a relationship between R_C, Ω, $\overline{B}_z$, P and λ, which comes down to constructing dimensionless groups from these variables and then establishing scaling laws between these groups, either by theory or by observation. Such scaling laws, which are rather speculative, are discussed in §14.11 and Exercises 14.1–14.8.

An early suggestion was that $\Lambda = \sigma B^2/\rho\Omega \sim 1$ for all steady planetary dynamos, which is based on the idea that the Lorentz force, which is taken to be of order $|\mathbf{J} \times \mathbf{B}| \sim \sigma u B^2$, is of the same order of magnitude as the Coriolis force, $2\rho\mathbf{u} \times \mathbf{\Omega}$. However, as noted in §14.2, this is a little too simplistic, as the flow in a rapidly rotating fluid, such as in the core of the earth, is invariably strongly anisotropic, with typical flow structures being highly elongated along the rotation axis (see §3.9). This is important because the curl and divergence of the Coriolis force are

$$\nabla \times (\mathbf{u} \times \mathbf{\Omega}) = (\mathbf{\Omega} \cdot \nabla)\mathbf{u}, \quad \nabla \cdot (\mathbf{u} \times \mathbf{\Omega}) = \mathbf{\Omega} \cdot \nabla \times \mathbf{u}.$$

So when the flow is strongly anisotropic, the curl of the Coriolis force is much smaller than its divergence. Consequently, if we perform a Helmholtz decomposition on the Coriolis force, dividing it into irrotational and solenoidal-rotational parts, the dominant term is the former. However, the irrotational part of the Coriolis force is simply balanced by pressure gradients (the so-called quasi-geostrophic force balance) and so the Lorentz force, if order one, should be balanced against the much weaker rotational component. When this more careful force balance is performed we obtain $\Lambda \sim \ell_\perp/\ell_{//} \ll 1$, where subscripts $//$ and $\perp$ indicate characteristic length scales parallel and perpendicular to the rotation axis. A second problem with the proposal $\Lambda \sim 1$ is that the estimate $|\mathbf{J} \times \mathbf{B}| \sim \sigma u B^2$ assumes R_m is not much larger than one. So, one way or another, the suggestion that $\Lambda \sim 1$ seems to lack theoretical support, and there is certainly some spread in the values of Λ in Table 14.3.

More recent scaling theories for planets have focussed on a different measure of B,

Table 14.3 *Estimates of the Elsasser number,* Λ, *and* $\Pi_B = \overline{B}_z/\Omega R_C \sqrt{\rho\mu}$ *based on the mean axial field in the planetary cores.* (*We use the estimates of* $\lambda \sim 0.7 \ m^2/s$, $\rho \sim 10^4 \ kg/m^3$ *for the terrestrial planets, and* $\lambda \sim 30 \ m^2/s$, $\rho \sim 10^3 \ kg/m^3$ *for the gas giants. Included for comparison is a fully convective, low-mass, M-dwarf star, which has an 800 Gauss dipole field of the* α^2 *type.*)

Planet (or star)	Rotation period (days)	Elsasser number $\Lambda = \sigma \overline{B}_z^2/\rho\Omega$	$\dfrac{\overline{B}_z/\sqrt{\rho\mu}}{\Omega R_C}$	$\sqrt{\dfrac{P}{\Omega^2\lambda}}$	$\sqrt{\dfrac{P}{\Omega^2\lambda}}\sqrt{\dfrac{V_P}{\Omega R_C}}$
Mercury	58.6	2×10^{-4}	5.5×10^{-6}	–	–
Earth	1	0.2	13×10^{-6}	0.012	8.1×10^{-5}
Jupiter	0.413	0.5	5.2×10^{-6}	0.011	4.9×10^{-5}
Saturn	0.440	0.02	2.3×10^{-6}	0.007	3.4×10^{-5}
V374 Pegasi	0.44	$\sim 10^4$	17×10^{-6}	–	–

$$\Pi_B = \frac{\overline{B}_z/\sqrt{\rho\mu}}{\Omega R_C}, \quad \text{or else} \quad \Pi_B = \frac{B_{rms}/\sqrt{\rho\mu}}{\Omega R_C}, \qquad (14.7a, b)$$

where B_{rms} is often taken to be $\sim 10\overline{B}_z$. It is interesting to note that this second dimensionless measure of $\overline{B}_z$ is relatively constant across the planets (Table 14.3). Order-of-magnitude force balances then give (see Davidson, 2014, Exercise 14.6, or §14.11),

$$\frac{B_{rms}/\sqrt{\rho\mu}}{\Omega R_C} \sim \sqrt{\frac{P}{\Omega^2\lambda}}\sqrt{\frac{\ell_\perp}{R_C}}, \qquad (14.8)$$

where $\sqrt{P/\Omega^2\lambda} \sim 10^{-2}$ for the earth, Jupiter and Saturn. (P is unknown for Mercury.) The question is, then: 'What sets the scale $\ell_\perp$?'. Three options are discussed in Exercise 14.8. Taking $u\ell_\perp/\lambda \sim 1$ returns us, rather surprisingly, to $\Lambda \sim 1$, while assuming $V_P\ell_\perp/\lambda \sim 1$, where $V_P = (PR_C)^{1/3}$, leads to $B_{rms}/\sqrt{\rho\mu} \sim V_P$, which has been suggested by Christensen (2011), but for different reasons. There is only modest observational support for the first option, while the second, which asserts that B is independent of Ω, has some support, but lacks theoretical under-pinning. A third option, which is based on the idea that the columnar vortices in planetary cores are created by inertial waves (see §14.8.2), is the hypothesis that $\ell_\perp$ is the minimum length scale at which inertial waves can propagate to form columnar vortices, which is $u/\Omega\ell_\perp \approx 0.2$. This leads to $u \sim V_P$ and $\ell_\perp/R_C \sim V_P/\Omega R_C$, which is a good fit to the observational data for the earth, Jupiter and Saturn (final column in Table 14.3).

14.4 Tentative Constraints on Planetary Dynamo Theories

We have already hinted that the current generation of numerical simulations of planetary dynamos cannot get close to planet-like conditions, and we shall see that this is also true of laboratory experiments. (It is not possible to squeeze the core of a planet into the laboratory without making major compromises on dominant force balances and key dimensionless groups.) Given the sparsity of observable data, and the inability of numerical or physical experiments to probe the appropriate para- meter regime, it is important to find other ways of constraining potential theories of the geodynamo.

Perhaps the first point to note is that, despite having very different sizes, rotation rates and internal structures, the magnetic fields in Mercury, Jupiter, Saturn and the earth are all remarkably similar (Table 14.3). Specifically, they are all strongly dipolar with a magnetic axis more or less aligned with the rotation axis. Moreover, their dimensionless field strengths, as measure by Π_B, are surprisingly similar. The implication is that the dynamo mechanisms operating in all four planets are, in some zero-order sense, similar. This, in turn, suggests that the dynamical mechan- isms behind dynamo action are simple and robust, and not contingent on, say, very particular mechanical boundary conditions.

The second important observation comes from the discussion of force balances in §14.2. Here we noted that the viscous stresses are extremely small, the implica- tion being that viscosity cannot play any significant role in core dynamics. Now it might be argued that allowing viscous boundary layers (Ekman layers) to form on the mantle is a way of getting around this constraint, but this too is problematic. On the one hand, the conventional thickness of laminar Ekman layers (as described in §3.8), combined with the estimate of $Ek \sim 10^{-15}$ for the earth, would require such layers to be a tiny fraction of a metre thick, which makes no sense from a physical point of view.[1] On the other hand, there is no mantle in either of the gas giants on which such layers can form, and we have just suggested that similar dynamos operate in the terrestrial planets and gas giants. In short, planetary dynamo theories should probably not be reliant on the existence of viscous stresses or viscous boundary layers.

The third point relates to our estimate of the Rossby number in the core of the earth (Table 14.1). This is small, even at the smallest scales of motion, and so the non-linear term $\mathbf{u} \cdot \nabla \mathbf{u}$ is necessarily negligible in the core and cannot play a role in any theory of the geodynamo. (It does, however, play a role in some of the

---

[1] One might have hoped to sidestep this problem by invoking the concept of an eddy viscosity. However, balancing the Coriolis force by a Reynolds stress in the Ekman layer requires that the thickness of that layer scales as $\delta \sim u/\Omega \sim 3\text{m}$, which is still much too small.

numerical experiments!) This is another reason why we cannot readily invoke the idea of an eddy viscosity to replace v.

Our fourth, and final, observation concerns the convective heat flux per unit area out through the core of the earth, $\dot{q}$. Typical estimates of $\dot{q}$ show that a suitable measure of the convective driving force in the core (say, a Rayleigh number) is around 10^4 times the critical value at which non-magnetic convection would first appear (see, for example, Christensen, 2011). Consequently the convection in the core is strongly driven and likely to be highly turbulent. In the terminology adopted by those working on rotating convection, the flow is classified as highly super-critical. The numerical simulations, by contrast, are only weakly driven and highly viscous.

Taken together, these various observations tentatively suggest, but certainly do not demand, that we seek an explanation of the geodynamo which

(i) is reliant on only simple, robust dynamical processes associated with a rapidly rotating fluid (so that a similar dynamo can operate across a range of rapidly rotating planets with different properties)

(ii) is insensitive to mechanical boundary conditions and in particular to the existence of a mantle

(iii) does not rely on viscous or turbulent Reynolds stresses and in particular does not rely on the formation of Ekman layers

(iv) has negligible non-linear inertia, $\mathbf{u} \cdot \nabla \mathbf{u}$

(v) embraces the fact that the convection is highly super-critical (strongly driven), and therefore likely to be strongly chaotic.

The first of these suggests that waves, either inertial waves (as discussed in §3.9.2) or magnetostrophic waves (see §7.4), may play an important role.

Constructing a theory of the geodynamo, or indeed any planetary dynamo, comes in two parts. The first step, which is purely kinematic, asks the question, 'Can we construct a steady or steady-on-average velocity field which leads to dynamo action?'. This is now well developed and there are many candidate velocity fields. The second, and much harder, task is to identify which of these candidate kinematic dynamos can operate in a dynamically self-consistent manner within the interior of a planet. Perhaps the guidelines listed above become relevant at this point, but unfortunately there is still very little agreement as to how these can be satisfied in a coherent model. While we have some notion of what happens in the overly viscous, weakly driven numerical experiments, the gap between these simulations and planetary cores is much too large to allow us to transfer this understanding to natural dynamos.

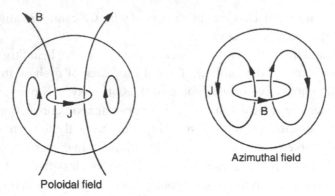

Poloidal field

Azimuthal field

Figure 14.5 Azimuthal and poloidal magnetic fields and their associated currents.
Note that the azimuthal field is confined to the conducting core.

14.5 Elementary Kinematic Theory: Phenomena, Theorems and Dynamo Types

14.5.1 A Survey: Six Important Kinematic Results

Let us start with the simpler kinematic theory and with six elementary, though important, ideas in planetary dynamo theory. We adopt a frame of reference rotating with the earth and use cylindrical polar coordinates, (r, θ, z), centred on the core of the earth. When using such coordinates it is natural to divide a vector field, say $\mathbf{B}$, into its azimuthal, $\mathbf{B}_\theta = (0, B_\theta, 0)$, and poloidal, $\mathbf{B}_p = (B_r, 0, B_z)$, components. For axisymmetric fields $\mathbf{B}_\theta$ and $\mathbf{B}_p$ are separately solenoidal, though this in not true of non-axisymmetric fields. Also, with axisymmetric fields the current that supports the poloidal field, $\mathbf{B}_p$, is purely azimuthal, while $\mathbf{B}_\theta$ is associated with a poloidal current density. This is illustrated in Figure 14.5. Note that Ampère's law in the form $\oint \mathbf{B} \cdot d\mathbf{r} = \mu \int \mathbf{J} \cdot d\mathbf{S}$ tells us that the azimuthal field cannot extend past the core–mantle boundary.

There are six basic ideas which are central to most planetary dynamo theory and which we seek to introduce here.

(1) The magnetic Reynolds number must be large.
(2) A combination of natural convection plus angular momentum conservation will tend to drive differential rotation (departures from rigid-body rotation) in planetary cores. Such differential rotation, if it is strong enough, will spiral out an azimuthal field from the observed dipole. This is called the Ω-effect and it is shown schematically in Figure 14.6. In some theories of the geodynamo this east-west field is the dominant one. However, the dipole field tends to resist being sheared in the θ-direction and often clamps the flow into a state close to that of rigid-body rotation. In such cases there is little or no Ω-effect.

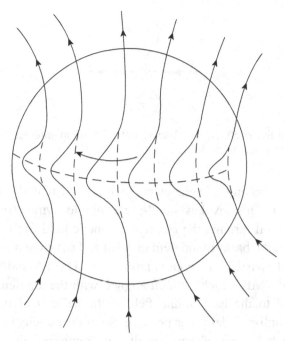

Figure 14.6 Differential rotation sweeps out an east-west field from the dipole field.

(3) Cowling's (1934) theorem states that an axisymmetric dynamo is not possible. This theorem is just one of a number of anti-dynamo theorems which all assert that a dynamo will fail if the flow and magnetic field are constrained to be overly symmetric.

(4) An axisymmetric dipole field is supported by azimuthal currents, as shown in Figure 14.5. More generally, an evolution equation for the axial component of a planetary field, integrated over the core, shows that this field is sustained by the volume integral of the azimuthal EMF, $(\mathbf{u} \times \mathbf{B})_\theta$. This is true whether or not the $\mathbf{u}$ and $\mathbf{B}$ fields are axisymmetric. However, it happens that contributions to this EMF that come from the axisymmetric components of $\mathbf{u}$ or $\mathbf{B}$ integrate to zero over the core. This is an example of Cowling's theorem (see item 3 above). Since departures from axial symmetry in $\mathbf{u}$ are thought to occur at a scale much less than R_C, the dipole field is probably supported by the cumulative effect of many small, local EMFs, $(\mathbf{u} \times \mathbf{B})_\theta$, each generated by a small-scale, non-axisymmetric disturbance. To be effective, these local EMFs must be additive (of the same sign), rather than of random sign. Thus a crucial component of any planetary dynamo theory is the need to ensure that the small-scale azimuthal EMFs are additive across each hemisphere in the core.

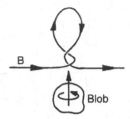

Figure 14.7 In the α-effect, small-scale helical motion interacts with the local mean field to produce twisted flux loops.

(5) An important mechanism in planetary dynamos, due to Parker (1955), is the so-called α-effect. This invokes small-scale, non-axisymmetric, helical disturbances that spiral through the ambient magnetic field. As these disturbances punch through the background field they lift and twist the magnetic field lines, creating small, twisted flux loops (Figure 14.7). This lift-and-twist mechanism drives a local EMF, which is often aligned with the ambient magnetic field, being parallel to the background field if the helicity of the disturbance is negative and anti-parallel if it is positive. So, if the helicity of the disturbances in some extended region of space are all of the same sign, these local EMFs can combine, like a bank of batteries, to drive a global current in the core. This large-scale current then induces a second, large-scale magnetic field which adds to the original one.

(6) Differential rotation, which converts a dipole field into an azimuthal one, is not in itself enough to drive a dynamo, and to complete the regenerative cycle $\mathbf{B}_p \to \mathbf{B}_\theta \to \mathbf{B}_p$ some theories invoke Parker's α-effect to convert $\mathbf{B}_\theta$ back into a poloidal field. These are known as $\alpha - \Omega$ dynamos. Yet other theories regard any differential rotation as incidental to dynamo action and focus instead on small-scale helical disturbances, often in the form of long, thin convection cells aligned with the rotation axis. These helical convection cells generate $\mathbf{B}_\theta$ from the dipole field via an α-effect, and then convert $\mathbf{B}_\theta$ back into a poloidal field using a second α-effect. These are called α^2 dynamos. At the time of writing, most numerical dynamos are of this second type.

We shall outline these six key ideas one by one in the sections that follow. Of course, the most important of these, the key new concept, is Parker's lift-and-twist mechanism (the α-effect), and we shall spend some time discussing this in §14.5.6. However, let us start by reminding ourselves that in a dynamo the magnetic Reynolds number must be large.

14.5.2 A Large Magnetic Reynolds Number Is Required

Let us return to (14.2), only this time we allow for motion in the core. The product of **B** with Faraday's law gives

$$\frac{\partial}{\partial t}\left[\frac{\mathbf{B}^2}{2\mu}\right] = -\nabla \cdot [(\mathbf{E} \times \mathbf{B})/\mu] - \mathbf{J} \cdot \mathbf{E}, \tag{14.9}$$

and on substituting for **E** using Ohm's law, and integrating over all space, we find

$$\frac{d}{dt}\int_{V_\infty}(\mathbf{B}^2/2\mu)dV = -\int_{V_C}[\mathbf{u} \cdot (\mathbf{J} \times \mathbf{B})]dV - \int_{V_C}(\mathbf{J}^2/\sigma)dV, \tag{14.10}$$

which might be compared with (14.2). The first integral on the right is the rate of working of the Lorentz force and represents an exchange of energy between the magnetic and velocity fields. The second integral is, of course, the Ohmic dissipation, *D*. In order to maintain a dynamo, we require

$$\dot{W} = -\int_{V_C}[\mathbf{u} \cdot (\mathbf{J} \times \mathbf{B})]dV \geq \int_{V_C}(\mathbf{J}^2/\sigma)dV, \tag{14.11}$$

i.e. the rate of conversion of kinetic energy into magnetic energy, $\dot{W}$, must exceed the Ohmic dissipation, *D*.

The ratio of $\dot{W}$ to *D* is clearly of the order of $R_m = u\ell/\lambda$ for some suitably defined *u* and ℓ. Evidently the magnetic Reynolds number must be reasonably large to achieve a dynamo and indeed there is a minimum value of R_m below which a dynamo is not possible. We can identify such a minimum by placing bounds on the integrals $\dot{W}$ and *D*. For example, from the Schwarz inequality we have

$$\mu^2 \dot{W}^2 \leq u^2_{\max}\left[\int_{V_C}(\nabla \times \mathbf{B}) \times \mathbf{B}dV\right]^2 \leq u^2_{\max}\int_{V_\infty}\mathbf{B}^2dV\int_{V_C}(\nabla \times \mathbf{B})^2dV, \tag{14.12}$$

where $u_{\max}$ is the maximum velocity in the core. Moreover, for current confined to a sphere, the calculus of variations gives

$$\int_{V_C}(\nabla \times \mathbf{B})^2dV \geq \frac{\pi^2}{R_C^2}\int_{V_\infty}\mathbf{B}^2dV, \tag{14.13}$$

from which

$$|\dot{W}| \le \frac{u_{\max} R_C}{\pi\lambda} D. \qquad (14.14)$$

Clearly, a necessary condition for a dynamo is

$$R_m = \frac{u_{\max} R_C}{\lambda} \ge \pi. \qquad (14.15)$$

Many tighter bounds on R_m exist (see, for example, Moffatt, 1978). In practice, it is found that dynamo action in a sphere is not possible unless R_m exceeds a few multiples of 10. In numerical simulations of the geodynamo, for example, magnetic Reynolds numbers in excess of 50 or 60 are usually required for a dynamo.

Similar limits on R_m exist in other geometries. For example, the kinematic dynamo of Ponomarenko consists of a helical pipe flow of diameter d embedded in an otherwise stationary conducting medium. The velocity field is $\mathbf{u} = (0, \ \Omega r, \ V)$ in $(r, \ \theta, \ z)$ coordinates, Ω and V being constants, and the helical pitch most favourable to dynamo action corresponds to $V = 0.66\Omega d$. Defining $Rm = u_{\max} d/\lambda$ we obtain a dynamo provided that $R_m > 35$, i.e. a few multiples of 10. Note that the resulting magnetic field is asymmetric, despite the axial symmetry of the base flow, as demanded by Cowling's theorem. Note also that the flow is helical, which is a recurring theme in this chapter.

14.5.3 Differential Rotation in the Core and the Ω-Effect

We now consider the possible role played by differential rotation in the core, which was an important component of the early theories of Elsasser and Parker. This differential rotation is thought to have its origins in the buoyant upwellings which occur above and below the inner core, within the so-called *tangent cylinder*; an imaginary cylinder which is coaxial with the rotation axis, circumscribes the inner core, and spans the outer core, as shown in Figure 14.8.

At its simplest level, the Ω-effect operates as follows. Thermal or compositional buoyancy within the tangent cylinder will tend to set up a recirculating flow in the core. In an inertial frame of reference fluid parcels tend to conserve their angular momentum as they move (pressure and Lorentz forces apart), and so we might expect $(\mathbf{x} \times \mathbf{u})_z = ru_\theta$ to be conserved in an inertial frame. Thus, as buoyant fluid moves out to greater r, driven by the buoyant upwellings above and below the inner core, u_θ will tend to fall as r increases. So we might expect the fluid near the inner core to spin slightly faster than that near the mantle. In a frame of reference rotating at Ω, the fluid surrounding the inner core will then have a positive value of u_θ. If this is correct, then an east-west field will be swept out from the dipole field, as shown

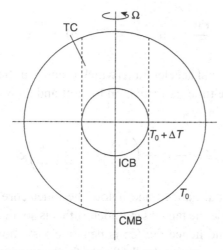

Figure 14.8 The tangent cylinder circumscribes the inner core. (Courtesy of B. Sreenivasan.)

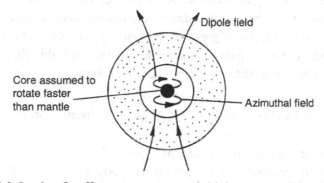

Figure 14.9 In the Ω–effect, an east-west field is generated by differential rotation.

in Figure 14.9. Moreover, since the solid inner core is coupled to the surrounding fluid by the dipole field, and this reacts back when it is sheared, we might expect the inner core to also spin slightly faster than the mantle, being dragged around by the fluid. Note that the resulting azimuthal field is anti-symmetric about the equator, being negative in the north and positive in the south for a dipole pointing northward.

The main problem with this argument is that it ignores pressure and Lorentz forces, and so an alternative explanation for the Ω-effect is often given. Let us return to our rotating frame of reference. If we continue to neglect Lorentz forces we have, in the Boussinesq approximation,

$$\frac{\partial \mathbf{u}}{\partial t} = 2\mathbf{u} \times \mathbf{\Omega} - \nabla(p/\rho) + \frac{\delta\rho}{\rho}\mathbf{g}, \tag{14.16}$$

where $\mathbf{g}$ is the gravitational acceleration (which is anti-parallel to $\mathbf{x}$) and ρ the mean density. Neglecting the time derivative on the left and taking the curl we have, in cylindrical polar coordinates,

$$2\Omega\frac{\partial u_\theta}{\partial z} = \frac{\partial}{\partial r}\left(\frac{\delta\rho}{\rho}\right)g_z - \frac{\partial}{\partial z}\left(\frac{\delta\rho}{\rho}\right)g_r. \tag{14.17}$$

We might expect the fluid above and below the inner core to be relatively light compared to that outside the tangent cylinder. If this is so, then $\partial(\delta\rho)/\partial r > 0$ within the tangent cylinder and hence $\partial u_\theta/\partial z$ is negative just above the inner core and positive just below. Once again, the fluid near the inner core is predicted to rotate faster than that near the mantle and an east-west field is swept out from the dipole field, as shown in Figure 14.9.

However, even this argument is overly simplistic, with the spatial distribution of $\delta\rho$ assumed. So let us return to the idea of angular momentum conservation, only now we allow for the effects of pressure and Lorentz forces and stay in the rotating frame of reference. Once again we start with (14.16), only this time we retain the time derivative and form the angular momentum equation

$$\frac{\partial}{\partial t}\mathbf{x} \times \mathbf{u} = 2\mathbf{x} \times (\mathbf{u} \times \mathbf{\Omega}) + \nabla \times (p\mathbf{x}/\rho) + (\text{Lorentz torque}). \tag{14.18}$$

We now integrate this over the cylindrical volume V_1 shown in Figure 14.10, which is coaxial with the rotation axis and symmetric about the equator. Focusing on the z-component of (14.18), we find that the pressure term drops out, because it cannot generate an axial torque on the surface of the cylinder, while the Coriolis torque transforms according to $2\left(\mathbf{x} \times (\mathbf{u} \times \mathbf{\Omega})\right)_z = -\Omega\nabla \cdot (r^2\mathbf{u})$. The end result is

$$\frac{d}{dt}\int_{V_1} (\mathbf{x} \times \mathbf{u})_z dV = -\Omega\oint_{S_1} r^2\mathbf{u} \cdot d\mathbf{S} + (\text{Lorentz torque}), \tag{14.19}$$

where S_1 is the surface of V_1. Using mass conservation applied to V_1, this can be rewritten in the more convenient form

$$\frac{d}{dt}\int_{V_1} (\mathbf{x} \times \mathbf{u})_z dV = \Omega\int_{S_T+S_B} (R^2 - r^2)\,\mathbf{u} \cdot d\mathbf{S} + (\text{Lorentz torque}), \tag{14.20}$$

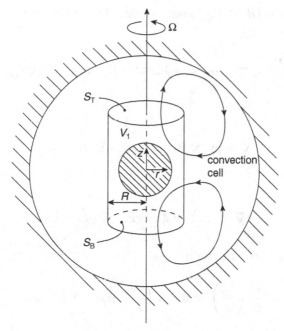

Figure 14.10 As a result of convection and angular momentum conservation, the angular velocity of the fluid near the inner core is larger than that near the CMB.

where S_T and S_B are the top and base of the cylinder V_1 and R is its radius. The effect of convection out through S_T and S_B is clearly to increase $(\mathbf{x} \times \mathbf{u})_z$ within the cylinder V_1, which to some extent will be opposed by a Lorentz torque associated with the twisting of the dipole field. Now in the limit that V_1 spans the core from south to north, the first integral on the right of (14.20) is zero. On the other hand, it is non-zero and positive when V_1 is relatively short. It follows that the fluid near the inner core is preferentially spun up by the convection, but at the expense of the fluid near the mantle, and once again we end up with the fluid near the inner core rotating faster than that near the mantle.

Crucially, however, we have neglected the back reaction of the dipole field, which resists shearing in the θ-direction and tends to suppress any differential rotation. So perhaps the most we can conclude is that, provided the dipole field is not too strong, there are tentative grounds for believing that differential rotation will accompany convection in the core, with excess rotation near the inner core. However, a strong dipole field will tend to suppress differential rotation, and indeed such a partial suppression is commonly observed in the numerical simulations.

The effect of this differential rotation on the dipole field is most easily seen if we assume axial symmetry. As usual, we split $\mathbf{B}$ and $\mathbf{u}$ into azimuthal, $\mathbf{B}_\theta = (0, B_\theta, 0)$, and poloidal, $\mathbf{B}_p = (B_r, 0, B_z)$, components. The azimuthal component of the induction equation is then

$$\frac{\partial \mathbf{B}_\theta}{\partial t} = \nabla \times [\mathbf{u}_p \times \mathbf{B}_\theta] + \nabla \times [\mathbf{u}_\theta \times \mathbf{B}_p] + \lambda \nabla^2 \mathbf{B}_\theta. \qquad (14.21)$$

Noting that

$$\nabla \times [\mathbf{u}_p \times \mathbf{B}_\theta] = -r \mathbf{u}_p \cdot \nabla (B_\theta/r) \hat{\mathbf{e}}_\theta,$$

and

$$\nabla \times [\mathbf{u}_\theta \times \mathbf{B}_p] = r \mathbf{B}_p \cdot \nabla (u_\theta/r) \hat{\mathbf{e}}_\theta,$$

this reduces to the scalar equation

$$\frac{D}{Dt}\left(\frac{B_\theta}{r}\right) = \mathbf{B}_p \cdot \nabla\left(\frac{u_\theta}{r}\right) + \lambda r^{-2} \nabla_*^2 (rB_\theta), \qquad (14.22)$$

where ∇_*^2 is the usual Stokes operator

$$\nabla_*^2 = \frac{\partial^2}{\partial z^2} + r \frac{\partial}{\partial r} \frac{1}{r} \frac{\partial}{\partial r}. \qquad (14.23)$$

The interaction of a dipole field B_z with differential rotation, $\partial (u_\theta/r)/\partial z$, is now clear. If B_z is positive (i.e. the dipole field points to the north), and u_θ/r is a maximum near the inner core, then $\mathbf{B}_p \cdot \nabla (u_\theta/r)$ will be negative in the north and positive in the south, thus generating an azimuthal field which is anti-symmetric about the equator and negative in the north (Figure 14.11).

In the steady state (14.22) suggests that $B_\theta \sim (u_\theta \ell/\lambda) B_z$. So if R_m based on u_θ is large, we would expect B_θ to be the dominant field in the core. However, the existence of significant differential rotation in planetary dynamos, and hence the status of B_θ, has proved to be controversial. Seismic studies, which seek to detect differential rotation of the inner core (rather than the surrounding fluid), vary from those whose findings are inconclusive (because the expected value of u_θ/R_i is so small), through to those that report effectively zero differential rotation. The temptation, then, is to assume that there is no significant differential rotation in the fluid. However, such a conclusion is thrown into doubt by the fact that the inner core may be gravitationally locked to density fluctuations in the mantle, so that a local excess rotation in the fluid may not be transferred to the solid inner core.

Of the numerical simulations that yield dynamo action, a few show significant differential rotation and an associated strong east-west field, whilst many others

Figure 14.11 Generation of an east-west field by differential rotation. This is called the Ω-effect.

show only weak differential rotation and an east-west field no greater than $|\mathbf{B}_p|$. The implication is that, in most of the simulations, any latent differential rotation is largely suppressed by the poloidal magnetic field, which threads through the core and reacts back on the fluid when azimuthally sheared. Of course, since none of these numerical simulations is close to an earth-like parameter regime, the question of significant differential rotation, and by implication a strong east-west field, remains an open one.

14.5.4 An Axisymmetric Dynamo Is Not Possible: Cowling's Theorem

Cowling's theorem states that an axisymmetric dynamo is not possible. There are two ways in which the theorem may be understood. The simplest is called Cowling's neutral-point argument.

Suppose that we seek a steady, axisymmetric dynamo in which $\mathbf{B}$ is poloidal, $\mathbf{B}_p = (B_r, 0, B_z)$, $\mathbf{J}$ is azimuthal, and $\mathbf{u}$ is also poloidal. Since the dynamo is steady, Faraday's law demands $\nabla \times \mathbf{E} = 0$ and Ohm's law reduces to $\mathbf{J} = \sigma(-\nabla V + \mathbf{u} \times \mathbf{B})$. The electrostatic potential is governed by the divergence of Ohm's law,

$$\nabla^2 V = \nabla \cdot (\mathbf{u} \times \mathbf{B}) = \mathbf{B} \cdot \boldsymbol{\omega} - \mu \mathbf{u} \cdot \mathbf{J},$$

and since $\boldsymbol{\omega}$ and $\mathbf{J}$ are both azimuthal, this gives $V = 0$. It follows that, for this simple, steady configuration, $\mathbf{J} = \sigma \mathbf{u} \times \mathbf{B}$. Now in an axisymmetric poloidal field there is always at least one *neutral ring*, N, where $|\mathbf{B}| = 0$ and the $\mathbf{B}$-lines are locally closed in the r–z plane (Figure 14.12). Evidently $\mathbf{J} = 0$ at the neutral ring. But Ampère's circuital law, $\oint \mathbf{B} \cdot d\mathbf{r} = \mu \int \mathbf{J} \cdot d\mathbf{S}$, applied to a field line surrounding N demands that $\mathbf{J}$ be non-zero near N, which seemingly contradicts $\mathbf{J} = \sigma \mathbf{u} \times \mathbf{B}$ since $\mathbf{B}$ is zero at the neutral point. The implication is that such a configuration is impossible, which is Cowling's neutral-point theorem. Actually, this is a little glib. A formal proof of the neutral-point theorem is a little more complicated, as one has

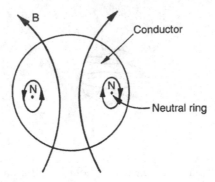

Figure 14.12 The neutral ring, N, in an axisymmetric poloidal field.

to consider the behaviour of $|\mathbf{B}|$ and $|\mathbf{J}|$ in the limit of the $\mathbf{B}$-line in Ampère's law shrinking onto the point N. However, this more careful analysis does indeed reveal the expected contradiction (see, for example, Moffatt, 1978).

A less constrained, if more cumbersome, proof of Cowling's theorem proceeds as follows. Suppose that we allow for both poloidal and azimuthal velocity and magnetic fields, which may be steady or unsteady, but which are axisymmetric:

$$\mathbf{B}(r,\ z,\ t) = \mathbf{B}_P + \mathbf{B}_\theta, \quad \mathbf{u}(r,\ z,\ t) = \mathbf{u}_P + \mathbf{u}_\theta.$$

The induction equation can similarly be divided into poloidal and azimuthal parts,

$$\frac{\partial \mathbf{B}_p}{\partial t} = \nabla \times [\mathbf{u}_p \times \mathbf{B}_p] + \lambda \nabla^2 \mathbf{B}_p, \tag{14.24}$$

$$\frac{\partial \mathbf{B}_\theta}{\partial t} = \nabla \times [\mathbf{u}_p \times \mathbf{B}_\theta] + \nabla \times [\mathbf{u}_\theta \times \mathbf{B}_p] + \lambda \nabla^2 \mathbf{B}_\theta. \tag{14.25}$$

Since $\mathbf{B}_P$ is solenoidal we can introduce a vector potential $\mathbf{A}_\theta$ defined by $\mathbf{B}_P = \nabla \times \mathbf{A}_\theta = \nabla \times [(\chi/r)\ \hat{\mathbf{e}}_\theta]$, where χ is the flux function for $\mathbf{B}_P$. We now uncurl (14.24) to give

$$\frac{\partial \mathbf{A}_\theta}{\partial t} = \mathbf{u}_p \times \mathbf{B}_p + \lambda \nabla^2 \mathbf{A}_\theta, \tag{14.26}$$

which yields

$$\frac{D\chi}{Dt} = \lambda \nabla_*^2 \chi, \tag{14.27}$$

where ∇_*^2 is the Stokes operator (14.23). The induction equation for $\mathbf{B}_\theta$, on the other hand, simplifies to (14.22):

$$\frac{D}{Dt}\left(\frac{B_\theta}{r}\right) = \mathbf{B}_p \cdot \nabla\left(\frac{u_\theta}{r}\right) + \lambda r^{-2}\nabla_*^2(rB_\theta). \tag{14.28}$$

We recognise the Ω-effect in (14.28), whereby a finite $\mathbf{B}_P$, interacting with gradients in the azimuthal swirl, $\nabla(u_\theta/r)$, can generate an azimuthal field, $\mathbf{B}_\theta$. The problem, however, lies in (14.27). This tells us that χ will eventually diffuse to zero (there is no source term), at which point $\mathbf{B}_P$ vanishes, as does the source term in (14.28). Thus, as $\mathbf{B}_P$ decays, so does $\mathbf{B}_\theta$. Evidently, an axisymmetric dynamo is not possible.

Cowling's theorem was the first of many similar anti-dynamo theorems, all of which state that if too many symmetries are imposed on the flow and the magnetic field then a dynamo cannot be sustained. Historically, Cowling's theorem played an important role because it made the likelihood of a self-excited fluid dynamo seem less plausible. However, Parker's heuristic, but insightful, analysis in 1955 changed all of that, as we shall see in §14.5.6.

14.5.5 An Evolution Equation for the Axial Field

We now seek an evolution equation for the axial field in the core. En route we shall discover the key to circumventing Cowling's theorem. We start with (14.4),

$$\mathbf{m} = \frac{1}{2}\int_{V_C} \mathbf{x} \times \mathbf{J}dV = \frac{3}{2\mu}\int_{V_C} \mathbf{B}dV,$$

which can be rewritten as

$$\mathbf{m} = \frac{3}{2\mu}\int_{V_C} \mathbf{B}dV = \frac{\sigma}{2}\int_{V_C} \mathbf{x} \times (\mathbf{E} + \mathbf{u} \times \mathbf{B})dV. \tag{14.29}$$

Since $\nabla \times \left[\left(\frac{1}{2}\mathbf{x}^2\right)\mathbf{E}\right] = \left(\frac{1}{2}\mathbf{x}^2\right)\nabla \times \mathbf{E} + \mathbf{x} \times \mathbf{E}$, our equation for $\mathbf{m}$ becomes

$$\frac{3}{2\mu}\int_{V_C} \mathbf{B}dV = \frac{\sigma}{2}\int_{V_C} \mathbf{x} \times (\mathbf{u} \times \mathbf{B})dV - \frac{\sigma}{4}\int_{V_C} (R_C^2 - \mathbf{x}^2)\frac{\partial\mathbf{B}}{\partial t}dV, \tag{14.30}$$

which may be rearranged to give the evolution equation

$$\frac{d}{dt}\int_{V_C} (R_C^2 - \mathbf{x}^2)\mathbf{B}dV = 2\int_{V_C} \mathbf{x} \times (\mathbf{u} \times \mathbf{B})dV - 6\lambda\int_{V_C} \mathbf{B}dV. \qquad (14.31)$$

Of particular interest is the axial component of this equation, which reduces to an evolution equation for the axial field:

$$\frac{d}{dt}\int_{V_C} (R_C^2 - \mathbf{x}^2)B_z dV = 2\int_{V_C} r(\mathbf{u} \times \mathbf{B})_\theta\, dV - 6\lambda\int_{V_C} B_z dV. \qquad (14.32)$$

Evidently, the axial field is maintained by the volume integral of $r(\mathbf{u} \times \mathbf{B})_\theta$, and in the steady state, when $\mathbf{E}$ is irrotational, this reduces to

$$3\lambda\int_{V_C} B_z dV = \int_{V_C} r(J_\theta/\sigma)dV = \int_{V_C} r(\mathbf{u} \times \mathbf{B})_\theta dV. \qquad (14.33)$$

This is to be expected since, by virtue of Ampère's law, an axial magnetic field requires azimuthal currents to support it, and these are driven by the azimuthal EMF, $(\mathbf{u} \times \mathbf{B})_\theta$. However, the first integral on the right of (14.32) is necessarily zero when the velocity and magnetic fields are axisymmetric. This can be seen as follows. If $\mathbf{B}$ is axisymmetric we can write $\mathbf{B} = \mathbf{B}_p + \mathbf{B}_\theta$ where $\mathbf{B}_p = (B_r,\ 0,\ B_z)$ is solenoidal and so can be expressed in terms of a vector potential: $\mathbf{B}_p = \nabla \times \mathbf{A}_\theta$, $\nabla \cdot \mathbf{A}_\theta = 0$. We then have $r(\mathbf{u} \times \mathbf{B})_\theta = -\mathbf{u}_p \cdot \nabla(rA_\theta)$, and hence

$$\int_{V_C} r(\mathbf{u} \times \mathbf{B})_\theta\, dV = -\int_{V_C} \mathbf{u}_p \cdot \nabla(rA_\theta)dV = -\oint_{S_C}(rA_\theta)\mathbf{u}_p \cdot d\mathbf{S} = 0, \qquad (14.34)$$

since $\nabla \cdot \mathbf{u}_p = 0$ for an axisymmetric velocity field. Evidently, sustained dynamo action in a sphere is not possible if $\mathbf{u}$ and $\mathbf{B}$ are both axisymmetric. Of course, this is an illustration of Cowling's theorem.

Let us now suppose that $\mathbf{B}$ and $\mathbf{u}$ are not axisymmetric, and write $\mathbf{B} = \overline{\mathbf{B}}(r,z) + \mathbf{b}$ and $\mathbf{u} = \overline{\mathbf{u}}(r,\ z) + \mathbf{v}$, where $\overline{\mathbf{B}}$ and $\overline{\mathbf{u}}$ are azimuthal averages of $\mathbf{B}$ and $\mathbf{u}$, and $\mathbf{b}$ and $\mathbf{v}$ have zero azimuthal means. Then the cross terms $\mathbf{v} \times \overline{\mathbf{B}}$ and $\overline{\mathbf{u}} \times \mathbf{b}$ integrate to zero and (14.32) becomes

$$\frac{d}{dt}\int_{V_C} (R_C^2 - \mathbf{x}^2)\overline{B}_z dV = 2\int_{V_C} r(\mathbf{v} \times \mathbf{b})_\theta dV - 6\lambda\int_{V_C} \overline{B}_z dV. \qquad (14.35)$$

We conclude that the axial field is maintained by an azimuthal EMF generated by non-axisymmetric fluctuations. This is the key to unlocking the dynamics of planetary dynamos. The dipole field is supported by an azimuthal EMF associated with non-axisymmetric fluctuations, and we shall see shortly that this EMF arises from the interaction of helical disturbances with the east-west field, i.e. Parker's α-effect. It is the helical lifting and twisting of the azimuthal field that generates the local EMF $(\mathbf{v} \times \mathbf{b})_\theta$. Crucially, if the helicity is uniformly of one sign in the north and of another sign in the south, say left-handed spirals in the north and right-handed spirals in the south, then the local EMFs $(\mathbf{v} \times \mathbf{b})_\theta$ are additive in each hemisphere, giving rise to global currents which then support the dipole field.

There is another important message contained in (14.35). Introducing angled brackets to represent an azimuthal average, (14.35) tells us that, provided the small-scale EMFs are additive over scales of order R_C, then

$$\langle \mathbf{v} \times \mathbf{b} \rangle_\theta \sim \frac{\lambda}{u R_C} u \overline{B}_z \ll u \overline{B}_z. \tag{14.36}$$

The implication is that, in a steady dynamo, the perturbation field $\mathbf{b}$ is much weaker than the mean dipole field $\overline{B}_z$. We shall see later that, in general, we also expect $\mathbf{b}$ to be much weaker than the mean azimuthal field.

The discussion above suggests that an important part of the dynamo story is the local EMF induced by small, helical disturbances lifting and twisting the background field. We shall now describe Parker's lift-and-twist mechanism.

14.5.6 A Glimpse at Parker's Helical Dynamo Mechanism

We now turn to the all-important lift-and-twist mechanism of Parker, now known as the α-effect. As before, we divide $\mathbf{B}$ and $\mathbf{u}$ into axisymmetric and non-axisymmetric parts: $\mathbf{B} = \overline{\mathbf{B}}(r, z) + \mathbf{b}$ and $\mathbf{u} = \overline{\mathbf{u}}(r, z) + \mathbf{v}$. Our starting point is to return to the integral equation (14.35) which, in the steady state, demands

$$\int_{V_C} \overline{B}_z dV = \frac{1}{3\lambda} \int_{V_C} r(\mathbf{v} \times \mathbf{b})_\theta \, dV. \tag{14.37}$$

This tells us that the dipole field is supported by the EMF associated with multiple non-axisymmetric disturbances, and it turns out that Parker's α-effect, operating on the east-west field, is the source of this EMF. Let us see if we can understand how this lift-and-twist mechanism works, and in particular why it induces an EMF aligned with the background magnetic field. To focus thoughts, we shall examine

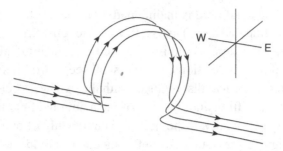

Figure 14.13 One of Parker's helical disturbances lifts and twists the east-west magnetic field lines. (Adapted from Cowling, 1957.)

the consequences of a single, helical disturbance spiralling through the east-west field, as shown in Figure 14.13.

We now need to decide what sort of disturbance we should take in Figure 14.13. To that end we recall from §3.9 that turbulence in a rapidly rotating fluid tends to be dominated by long, thin columnar vortices aligned with the rotation axis, sometimes in the form of wave packets composed of inertial waves (Figure 3.33). Such a columnar wave packet, dispersing from a localised disturbance, is shown in Figure 14.14. As discussed in §3.9, the flow in such wave packets has maximum helicity, with **v** parallel to $\nabla \times$ **v**. Moreover, upward-propagating wave packets carry negative helicity and downward-propagating packets positive helicity (Figure 3.33). Now it so happens that the more strongly forced of the numerical planetary dynamos yield a flow structure somewhat similar to that shown in Figure 3.33. Specifically, outside the tangent cylinder the flow is dominated by long, thin columnar vortices aligned with the rotation axis and filled with high levels of helicity. Moreover, in the same numerical simulations these columnar vortices are observed to be the source of the α-effect, lifting and twisting the background field lines as they spiral up and down through the core. It seems natural, therefore, to take as our single, localised disturbance a columnar vortex (or wave packet) of maximum helicity and aligned with the rotation axis. In the interests of simplicity, we examine the extreme case in which the vortex (or wave packet) is highly elongated and so the helical flow may be considered to be locally independent of z.

We now calculate the EMF generated by such a columnar vortex spiralling through a background east-west field. The arguments follow those of Davidson (2014). To keep the calculation simple, we shall take the magnetic Reynolds number to be less than one, so that we can use the low-R_m approximation. This may seem odd since R_m based on the core size is large, but in fact the more strongly driven numerical simulations suggest that the columnar vortices are so thin and

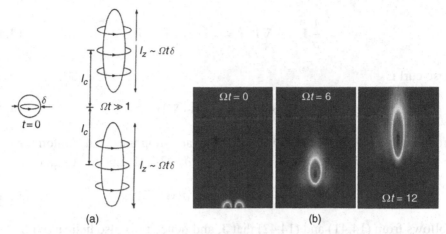

Figure 14.14 (a) An initial blob of vorticity converts itself into a pair of columnar eddies (transient Taylor columns) via inertial wave propagation, Ro ≪ 1. This kind of spontaneous formation of columnar wave packets is the hallmark of low-Ro motion. (b) The energy distribution at different times in the r–z plane ($z > 0$) calculated for an initial condition consisting of a single Gaussian eddy located at the origin. The times from left to right are $\Omega t = 0,6,12$.

elongated that R_m based on their radius is not much larger than unity. In any event, we shall generalise the analysis to larger R_m later.

We shall assume that the horizontal scale of our disturbance is sufficiently small, say around 10 km, that we may take the local background field to be uniform. We adopt local Cartesian coordinates, with z aligned with the rotation axis and the background field pointing in the x-direction, $\mathbf{B} = \bar{B}\hat{\mathbf{e}}_x$. Our velocity field then takes the form

$$\mathbf{v} = \mathbf{v}(x,y), \quad \boldsymbol{\omega} = \nabla \times \mathbf{v} = \delta^{-1}\mathbf{v}, \tag{14.38}$$

where the constant δ may be positive or negative and $|\delta|$ is the characteristic transverse scale of the columnar disturbance. Clearly, if we envisage our columnar disturbance to be an inertial wave packet, then δ will be negative for an upward-traveling wave packet and positive for a downward-propagating packet. Introducing angled brackets to represent a horizontal spatial average over the cross-section of the vortex, it is readily confirmed that (14.38) requires

$$\langle v_z^2 \rangle = \langle v_x^2 + v_y^2 \rangle, \quad \langle v_x v_z \rangle = \langle v_y v_z \rangle = 0. \tag{14.39}$$

Let us now consider the current induced by our helical disturbance. In the low-R_m approximation the induced field, $\mathbf{b}$, is governed by (6.4) in the form

$$\frac{1}{\sigma}\mathbf{J} = -\nabla V + \mathbf{v} \times \overline{\mathbf{B}} = \lambda \nabla \times \mathbf{b}, \tag{14.40}$$

whose curl is

$$\lambda \nabla \times \nabla \times \mathbf{b} = (\overline{\mathbf{B}} \cdot \nabla)\mathbf{v}. \tag{14.41}$$

Introducing a vector potential for $\mathbf{v}$, $\mathbf{v} = \nabla \times \mathbf{c}$, adopting the usual Coulomb gauge, $\nabla \cdot \mathbf{c} = 0$, and noting that (14.38) requires $\mathbf{c} = \delta \mathbf{v}$, (14.41) uncurls to give

$$\lambda \nabla \times \mathbf{b} = (\overline{\mathbf{B}} \cdot \nabla)\mathbf{c} = \delta(\overline{\mathbf{B}} \cdot \nabla)\mathbf{v}. \tag{14.42}$$

It follows from (14.41) and (14.42) that $\mathbf{J}$, and hence $\mathbf{b}$, is also helical, with

$$\mathbf{b} = \mathbf{b}(x, y), \qquad \mu \mathbf{J} = \nabla \times \mathbf{b} = \delta^{-1}\mathbf{b}. \tag{14.43}$$

The spatially averaged EMF, $\langle \mathbf{v} \times \mathbf{b} \rangle$, induced by the columnar vortex is now readily calculated. First we note that the divergence of (14.40) yields

$$\nabla^2 V = \overline{\mathbf{B}} \cdot \nabla \times \mathbf{v} = -\overline{\mathbf{B}} \cdot \nabla^2 \mathbf{c} = -\nabla^2(\overline{\mathbf{B}} \cdot \mathbf{c}), \tag{14.44}$$

which, for a localised disturbance, gives $V = -\overline{\mathbf{B}} \cdot \mathbf{c}$. Ohm's law now becomes

$$\lambda \nabla \times \mathbf{b} = \lambda \delta^{-1}\mathbf{b} = \nabla(\overline{\mathbf{B}} \cdot \mathbf{c}) + \mathbf{v} \times \overline{\mathbf{B}}, \tag{14.45}$$

from which

$$\lambda \delta^{-1}\mathbf{v} \times \mathbf{b} = \mathbf{v} \times (\mathbf{v} \times \overline{\mathbf{B}}) + \mathbf{v} \times \nabla(\overline{\mathbf{B}} \cdot \mathbf{c}) = \mathbf{v} \times (\mathbf{v} \times \overline{\mathbf{B}})$$
$$+ (\overline{\mathbf{B}} \cdot \mathbf{v})\mathbf{v} - \nabla \times ((\overline{\mathbf{B}} \cdot \mathbf{c})\mathbf{v}).$$

We now take a spatial average over the disturbance, which eliminates the curl on the right:

$$\langle \mathbf{v} \times \mathbf{b} \rangle = -\frac{\delta}{\lambda}\langle \mathbf{v}^2\overline{\mathbf{B}} - 2(\overline{\mathbf{B}} \cdot \mathbf{v})\mathbf{v} \rangle. \tag{14.46}$$

In terms of the components of $\mathbf{v}$ this becomes

$$\langle \mathbf{v} \times \mathbf{b} \rangle = -\frac{\delta \overline{B}}{\lambda}\left\langle \left(v_z^2 + (v_y^2 - v_x^2)\right)\hat{\mathbf{e}}_x - 2v_x v_y \hat{\mathbf{e}}_y \right\rangle, \tag{14.47}$$

or, using (14.39),

$$\langle \mathbf{v} \times \mathbf{b} \rangle = -\frac{2\delta\overline{B}}{\lambda} \langle v_y^2 \hat{\mathbf{e}}_x - v_x v_y \hat{\mathbf{e}}_y \rangle. \tag{14.48}$$

If the disturbance happens to be axisymmetric, as in Figure 14.14, that symmetry allows us to rewrite (14.47) as

$$\langle \mathbf{v} \times \mathbf{b} \rangle = -\frac{\delta}{\lambda} \langle v_z^2 \rangle \overline{\mathbf{B}} = -\frac{\delta}{2\lambda} \langle v^2 \rangle \overline{\mathbf{B}}. \tag{14.49}$$

Alternatively, if we make the weaker assumption of symmetry about the mean field only, then

$$\langle \mathbf{v} \times \mathbf{b} \rangle = -\frac{2\delta}{\lambda} \langle v_y^2 \rangle \overline{\mathbf{B}}. \tag{14.50}$$

Either way, the mean induced EMF is aligned with the background magnetic field.

This is the key result: the helical disturbance has produced a mean EMF aligned with $\overline{\mathbf{B}}$. In particular, since $h = \mathbf{v} \cdot \boldsymbol{\omega} = \delta^{-1} v^2$, (14.50) tells us that $\langle \mathbf{v} \times \mathbf{b} \rangle$ is anti-parallel to $\overline{\mathbf{B}}$ if the helicity is positive (downward-propagating wave packets), and parallel to $\overline{\mathbf{B}}$ if the helicity is negative (upward-propagating wave packets). This is illustrated in Figure 14.15 for the case of positive helicity. To emphasise the point, we can rewrite (14.49) as

$$\langle \mathbf{v} \times \mathbf{b} \rangle = -\frac{\delta^2}{2\lambda} \langle h \rangle \overline{\mathbf{B}}. \tag{14.51}$$

It has become conventional to write expressions like (14.51) in the form

$$\langle \mathbf{v} \times \mathbf{b} \rangle = \alpha \overline{\mathbf{B}}, \tag{14.52}$$

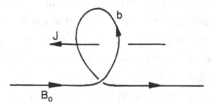

Figure 14.15 Helical disturbances with positive helicity generate an EMF which is anti-parallel to **B**. In the case of wave packets, this corresponds to downward-propagating waves (either inertial waves or magnetostrophic waves). Conversely, helical disturbances with negative helicity tend to generate an EMF which is parallel to **B**. This corresponds to upward-propagating wave packets.

which is the origin of the phrase α-effect. However, by convention, the induction process encapsulated by (14.52) need not be limited to low R_m, and so estimates of α may differ considerably from the value $\alpha = -\delta^2 \langle h \rangle / 2\lambda$ suggested by (14.51).

Perhaps some comments are in order here. First, for simplicity, we have assumed that the magnetic Reynolds number is low. However, we shall see that it is possible to generalise the analysis to large R_m. Second, although we have in mind here columnar vortices or inertial wave packets, magnetostrophic waves have very similar properties, to the extent that (i) they have maximum helicity and (ii) that helicity is negative for upward-propagating waves packets and positive for downward-propagating waves. Third, this locally induced EMF would be of little interest if a neighbouring disturbance had helicity of the opposite sign, as the two EMFs would then tend to cancel. This emphasises the need to somehow coordinate the disturbances so that they are of uniform helicity in each hemisphere, say left-handed spirals in the north and right-handed spirals in the south. One of the central questions in planetary dynamo theory is how that coordination can be achieved in a simple, robust manner.

14.5.7 Different Classes of Planetary Dynamo

Turbulence in a rapidly rotating fluid tends to be dominated by a random sea of long-lived, columnar vortices roughly aligned with the rotation axis (Davidson, 2013). Most numerical simulations of the geodynamo show precisely that. In the weakly driven simulations the columnar vortices are relatively wide, almost steady, and arranged around the solid inner core as a regular array of alternating cyclones and anti-cyclones, spanning the outer core from mantle to mantle. Since these simulations are highly viscous, Ekman layers form on the mantle, and the resulting Ekman pumping imparts helicity to the core flow, driving an α-effect. This, in turn, produces a laminar dynamo which has all the regularity of a Swiss watch. However, in the more strongly driven simulations the columnar vortices are much thinner and decidedly transient (turbulent), though they still tend to appear in cyclone–anticyclone pairs and are concentrated outside the tangent cylinder and hence arranged around the solid inner core. As with their weakly driven counterparts, these columnar vortices are observed to be highly helical and to drive an α-effect, and through this a dynamo. Crucially, though, some of these more transient vortices do not intersect with the mantle (Figure 14.16), and so Ekman pumping is no longer the sole source of helicity in such cases (Olson, Christensen and Glatzmaier, 1999).

Of course, we do not know what happens in the core of a real planet, and it is worth bearing in mind that convection in the earth's core is much more strongly driven than even the most ambitious of the numerical simulations. It seems

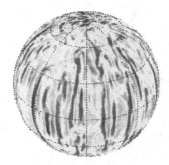

Figure 14.16 The radial velocity near the mantle in a numerical simulation of the geodynamo. (Adapted from Christensen and Wicht, 2007.)

plausible, nevertheless, that the core flow is dominated by columnar vortices, as the simulations suggest, and that these are, in part, interior objects (not attached to the mantle) and are helical, driving an α-effect, and hence a dynamo. An important question, then, is, 'What maintains the helicity in the columnar vortices?'

Two distinct classes of dynamo models have emerged from the phenomenology of Parker's lift-and-twist mechanism: the $\alpha - \Omega$ dynamo and the α^2 dynamo. In the $\alpha - \Omega$ class of dynamos significant large-scale differential rotation is assumed to exist and the dynamo is maintained by the following cycle of events:

(1) The Ω-effect sweeps out B_θ from the dipole field and this east-west field is the dominant one in the core.
(2) Random, small-scale helical disturbances interact with B_θ to produce a multitude of small-scale EMFs, $(\mathbf{v} \times \mathbf{b})_\theta$, and when azimuthally averaged these combine to give a global EMF, as described in §14.5.5.
(3) This global EMF, $\langle \mathbf{v} \times \mathbf{b} \rangle_\theta$, drives azimuthal currents and these have the same sign in both hemispheres and are such as to support the original dipole field.

Note that step (2) requires that the helicity be of uniform (or nearly uniform) sign in each hemisphere, so that the local EMFs are additive.

One of the key constraints of such a dynamo is point (3) above: the azimuthal currents driven by $\langle \mathbf{v} \times \mathbf{b} \rangle_\theta$ must support the original dipole field. Let us explore this constraint in more detail. Suppose that, for the sake of argument, the dipole points to the north. Then the Ω-effect will sweep out an azimuthal field that is negative in the north and positive in the south, as shown in Figure 14.17. However, in order to support the original dipole field we require azimuthal currents that are positive in both hemispheres. Given that the orientation of $\langle \mathbf{v} \times \mathbf{b} \rangle_\theta$ relative to $\overline{\mathbf{B}}_\theta$ has the opposite sign to that of the local helicity, this tells us that such a dynamo requires a helicity distribution that is predominantly positive in the north and negative in the south. This kind of $\alpha - \Omega$ dynamo is discussed at length in, for

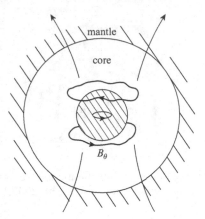

Figure 14.17 Differential rotation produces an azimuthal field which is anti-symmetric about the equator.

example, Moffatt (1978), and one of its hallmarks is that the dynamo is typically oscillatory, reversing at regular intervals. This is highly reminiscent of the solar dynamo, though not of planetary dynamos. In any event, most of the (admittedly imperfect) numerical simulations of planetary dynamos do not show a strong Ω-effect outside the tangent cylinder, where the α-effect resides, and so do not produce $\alpha - \Omega$ dynamos. (There are, however, signs of an Ω-effect inside the tangent cylinder, as discussed below, although this does not contribute to dynamo action in these particular simulations.)

Because of the uncertainty over the existence of an Ω-effect in planetary dynamos, the focus in recent years has turned to α^2 dynamos. Here any large-scale differential rotation which exists is thought to be largely incidental to dynamo action. Rather, attention is focussed on the α-effect, driven by the thin columnar vortices outside the tangent cylinder. Within such columns the flow is helical and B_θ is generated from $\mathbf{B}_p$ through an α-effect operating on B_r, while B_z is regenerated from B_θ through a second α-effect, reminiscent of that in the $\alpha - \Omega$ dynamo. A typical example of this kind of cycle is shown in Figure 14.18 where, to focus thoughts, the dipole points to the north and, in line with many of the numerical simulations, the helicity outside the tangent cylinder is taken to be negative in the north and positive in the south (the opposite to that needed for an $\alpha - \Omega$ dynamo).

This kind of α^2 dynamo operates almost entirely outside the tangent cylinder and is thought to work as follows. The columnar vortices interact most strongly with the transverse (radial and azimuthal) components of the magnetic field, lifting and twisting the transverse field lines. As discussed above, this produces an EMF which is aligned with the transverse magnetic field when h is negative,

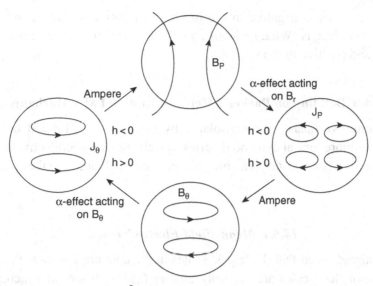

Figure 14.18 Cartoon of an α^2 dynamo based on the helicity distribution observed outside the tangent cylinder in most numerical simulations.

and anti-parallel to the transverse field when h is positive. Thus the helical convection interacts with the radial component of the dipole magnetic field to produce a radial EMF which is positive in both the north and the south. (Note that, as B_r changes sign, so does h.) This then drives a poloidal current density which has a quadruple structure as shown in Figure 14.18, being outward in regions of a strong radial EMF and returning near the equator where the radial EMF is weaker. The associated azimuthal magnetic field then follows from Ampère's law, and is positive in the north and negative in the south, as observed outside the tangent cylinder in most numerical simulations. (Note, however, that B_θ is sometimes observed to adopt the opposite signs within the tangent cylinder. Since the azimuthal field shown in Figure 14.18, driven by an α^2 dynamo, is opposite in sign to that shown in Figure 14.17, the observed reversal of B_θ within the tangent cylinder is probably due to a local Ω-effect, driven by large-scale upwellings.) Returning to Figure 14.18, the α-effect now operates on the east-west field and, given the asymmetric helicity distribution, this drives an azimuthal current which is positive in both the north and the south, which is exactly what is required to support the original dipole field. Thus the dynamo cycle is completed.

Figure 14.18 is a zero-order model of what is observed in many of the recent numerical dynamos (at the time of writing), where dynamo action largely resides outside the tangent cylinder, and is driven by the intensely helical flow within the columnar vortices. Of course, for this to work it is vital that we have a helicity distribution which is negative in the north and positive in the south.[2] The key question, therefore, is 'What maintains a helicity distribution which is negative in the north and positive in the south?'

14.6 Building on Parker's Helical Lift-and-Twist Mechanism

Parker's intuitive ideas, as encapsulated by Figure 14.15, formed the basis of a number of more formal dynamo theories, which are known collectively as mean-field models, or as *mean-field electrodynamics*. We shall describe these now, albeit briefly.

14.6.1 Mean-Field Electrodynamics

We have already seen that the largest scales in the core are $\ell \sim R_C - R_i \sim 10^3$ km, while the smallest scales are probably around $\ell_{min} \sim 10$ km, at which point R_m approaches $R_m \sim 1$ and Ohmic dissipation becomes large. (An independent estimate of the size of the small-scale motion is given in Exercise 14.4 at the end of this chapter.) In order to formalise the idea of $\alpha - \Omega$ and α^2 dynamos, it is convenient to suppose there is a separation of scales between the large-scale axisymmetric fields, $\bar{\mathbf{u}}$ and $\overline{\mathbf{B}}$, and the small-scale helical disturbances, $\mathbf{v}$ and $\mathbf{b}$, that generate the α-effect. One can then adopt a form of Reynolds averaging in order to assess the influence of the small scales on the large. So let us suppose that the velocity and magnetic fields are divided into a large-scale, slowly varying part, $\bar{\mathbf{u}}$ and $\overline{\mathbf{B}}$, and a small-scale component, $\mathbf{v}$ and $\mathbf{b}$. It is convenient to use azimuthal averaging to distinguish between the two components, and so we divide $\mathbf{B}$ and $\mathbf{u}$ into axisymmetric and non-axisymmetric parts:

$$\mathbf{B} = \overline{\mathbf{B}}(r, z, t) + \mathbf{b}, \quad \mathbf{u} = \bar{\mathbf{u}}(r, z, t) + \mathbf{v},$$

where $\overline{\mathbf{v}} = \overline{\mathbf{b}} = \mathbf{0}$. The induction equation may also be separated into mean and fluctuating components, to give

$$\frac{\partial \overline{\mathbf{B}}}{\partial t} = \nabla \times [\bar{\mathbf{u}} \times \overline{\mathbf{B}}] + \nabla \times \langle \mathbf{v} \times \mathbf{b} \rangle + \lambda \nabla^2 \overline{\mathbf{B}}, \qquad (14.53)$$

---

[2] In principle, α^2 dynamos can operate when the helicity takes the same sign in both hemispheres, in which case the east-west field is symmetric about the equatorial plane. In practice, however, the distributions of both h and B_θ are almost invariably anti-symmetric about the equator in convection-driven dynamos.

$$\frac{\partial \mathbf{b}}{\partial t} = \nabla \times [\overline{\mathbf{u}} \times \mathbf{b}] + \nabla \times [\mathbf{v} \times \overline{\mathbf{B}}] + \nabla \times [\mathbf{v} \times \mathbf{b} - \langle \mathbf{v} \times \mathbf{b} \rangle] + \lambda \nabla^2 \mathbf{b}. \quad (14.54)$$

The second of these equations is linear in $\mathbf{b}$ and has the source term $\nabla \times [\mathbf{v} \times \overline{\mathbf{B}}]$. The linearity of (14.54) ensures that $\mathbf{b}$ and $\overline{\mathbf{B}}$ are linearly related, and hence $\langle \mathbf{v} \times \mathbf{b} \rangle$ is also linearly related to $\overline{\mathbf{B}}$. Since $\overline{\mathbf{B}}$ is locally uniform on the scale of $\mathbf{b}$, $\langle \mathbf{v} \times \mathbf{b} \rangle$ will depend primarily on the local value of $\overline{\mathbf{B}}$ and so we may write $\langle \mathbf{v} \times \mathbf{b} \rangle_i = \alpha_{ij}\overline{B}_j + \beta_{ijk}\partial\overline{B}_j/\partial x_k$ for some tensors α_{ij} and β_{ijk}. For the particularly simple case of $\alpha_{ij} = \alpha\delta_{ij}$ and $\beta_{ijk} = \beta\varepsilon_{ijk}$, as in (14.51), the mean part of the induction equation becomes

$$\frac{\partial \overline{\mathbf{B}}}{\partial t} = \nabla \times [\overline{\mathbf{u}} \times \overline{\mathbf{B}}] + \nabla \times [\alpha\overline{\mathbf{B}}] + \lambda_e\nabla^2\overline{\mathbf{B}}, \quad \lambda_e = \lambda + \beta,$$

where α is a pseudo-scalar with the dimensions of a velocity and β is taken to be a constant of order $\alpha\ell$. In this approximation, the poloidal and azimuthal components of the averaged induction equation change from (14.24) and (14.25) to

$$\frac{\partial \mathbf{B}_p}{\partial t} = \nabla \times [\mathbf{u}_p \times \mathbf{B}_p] + \nabla \times [\alpha\mathbf{B}_\theta] + \lambda_e\nabla^2\mathbf{B}_p, \quad (14.55)$$

$$\frac{\partial \mathbf{B}_\theta}{\partial t} = \nabla \times [\mathbf{u}_p \times \mathbf{B}_\theta] + \nabla \times [\mathbf{u}_\theta \times \mathbf{B}_p] + \nabla \times [\alpha\mathbf{B}_p] + \lambda_e\nabla^2\mathbf{B}_\theta, \quad (14.56)$$

where we have omitted the overbars for clarity. The evolution equations for χ, the flux function for $\mathbf{B}_p$, and the azimuthal magnetic field, B_θ, are now

$$\frac{D\chi}{Dt} = \alpha r B_\theta + \lambda_e\nabla_*^2\chi, \quad (14.57)$$

$$\frac{D}{Dt}\left(\frac{B_\theta}{r}\right) = \mathbf{B}_p \cdot \nabla\left(\frac{u_\theta}{r}\right) + r^{-1}\left(\nabla \times [\alpha\mathbf{B}_p]\right)_\theta + \lambda_e r^{-2}\nabla_*^2(rB_\theta), \quad (14.58)$$

which might be compared with (14.27) and (14.28).

There is now direct feedback from $\mathbf{B}_\theta$ to $\mathbf{B}_p$ through (14.57), which completes the regenerative cycle $\mathbf{B}_p \to \mathbf{B}_\theta \to \mathbf{B}_p$ in the $\alpha - \Omega$ dynamo. Moreover, $\mathbf{B}_\theta$ is linked to $\mathbf{B}_p$ through both an Ω-effect, $\mathbf{B}_p \cdot \nabla(u_\theta/r)$, and the α-effect, $\nabla \times [\alpha\mathbf{B}_p]$, in (14.58). Thus there is the possibility of a dynamo cycle $\mathbf{B}_p \to \mathbf{B}_\theta \to \mathbf{B}_p$ operating through the α-effect alone, i.e. an α^2 dynamo. Integration of (14.57) and (14.58), with suitable distributions of α, do indeed produce dynamo action, both of the $\alpha - \Omega$ and α^2 type.

The various solutions and their properties are discussed in detail in Moffatt (1978). The $\alpha - \Omega$ dynamos tend to be oscillatory, like the sun's magnetic field, and dynamo action is possible only if the so-called dynamo number, $|\alpha| u_\theta R_C^2/\lambda^2$, is large enough, say greater than 50 or so (depending on the exact definition). On the other hand, α^2 dynamos tend to be non-oscillatory and require $|\alpha| R_C/\lambda$ to be larger than 10 or 20, depending of course on the prescription of α. In both cases α is usually taken to be anti-symmetric about the equator.

14.6.2 A More Careful Look at the α-Effect

Clearly, mean-field dynamos require some prescription of α as an input, which raises the question of how α can be estimated. So let us see if we can generalise the analysis of §14.5.6. First, it is important to recall that α is a pseudo-scalar and, like **B**, changes sign under the coordinate transformation $\mathbf{x} \rightarrow -\mathbf{x}$. That is to say, both α and **B** change sign if we move from a right-handed to a left-handed frame of reference. Now we might expect α to depend on $h = \mathbf{v} \cdot \nabla \times \mathbf{v}$, v_α, ℓ_α and λ, where v_α and ℓ_α are representative velocity and length scales of the small-scale dynamics and h the corresponding helicity. Note that h is the only pseudo-scalar in this list, so that α must be proportional to h. Dimensional analysis then demands

$$\alpha = \frac{\overline{h}\ell_\alpha}{v_\alpha} F(v_\alpha \ell_\alpha/\lambda), \tag{14.59}$$

for some dimensionless function F. The two obvious cases to consider are $v_\alpha \ell_\alpha/\lambda \gg 1$ and $v_\alpha \ell_\alpha/\lambda \leq 1$.[3] At large values of $v_\alpha \ell_\alpha/\lambda$ we might expect λ not to be a relevant parameter, and so we have

$$\alpha \sim -\overline{h}\ell_\alpha/v_\alpha, \qquad v_\alpha \ell_\alpha/\lambda \gg 1. \tag{14.60}$$

Note the minus sign, which is needed to be consistent with Figure 14.15. Estimates like (14.60) are commonly made in stellar dynamo models. For low R_m, $v_\alpha \ell_\alpha/\lambda \leq 1$, the induction equation tells us that $|\mathbf{b}| \sim v_\alpha |\overline{\mathbf{B}}| \ell_\alpha/\lambda$, and so the function F must be linear and we obtain

$$\alpha \sim -\overline{h}\ell_\alpha^2/\lambda, \qquad v_\alpha \ell_\alpha/\lambda \leq 1, \tag{14.61}$$

which might be compared with (14.51). In fact, if the small scales can be considered to be statistically homogeneous, at least in a local sense, then we have the exact result (Davidson, 2013)

---

[3] Recall that the low magnetic Reynolds number approximation usually works well up to $R_m \sim 1$.

$$\alpha_{ij} = -\lambda^{-1} [\ \langle \mathbf{c} \cdot \mathbf{v} \rangle \delta_{ij} - \langle c_i v_j + c_j v_i \rangle], \qquad v_\alpha\, \ell_\alpha / \lambda \ll 1, \qquad (14.62)$$

where $\mathbf{c}$ is the vector potential for $\mathbf{v}$. In this case it is the pseudo-scalar $\mathbf{c} \cdot \mathbf{v}$, rather than h, which plays the key role. However, for disturbances with maximum helicity, i.e. $\mathbf{v} = \pm \ell_\alpha \nabla \times \mathbf{v}$, we have $\mathbf{c} \cdot \mathbf{v} = \ell_\alpha^2 h$ and so (14.62) scales in the same way as (14.61). Estimates like (14.61) are commonly used in cartoons of planetary dynamos.

We can make more progress in estimating α if we embrace the fact that the small-scale motion that generates the α-effect takes the form of long, thin columnar eddies aligned with the rotation axis. In particular, it is natural to try and generalise the simple calculation of §14.5.6. Just such an analysis is given in Davidson and Ranjan (2015), which we now describe.

It is assumed that the small-scale motion is columnar, perhaps a collection of convection rolls, or else a sea of inertial or magnetostrophic wave packets. Such wave packets would be generated in regions well populated by buoyant anomalies (such as near the equatorial plane) and then propagate up and down the rotation axis towards the mantle. It is also assumed that a typical transverse dimension of the columnar eddies, δ, is much less than R_C, so that $\overline{\mathbf{B}}$ varies slowly on the scale of δ and may be treated as locally uniform. Since the analysis is local, and the global boundary conditions play no role, it is convenient to adopt local Cartesian coordinates with z aligned with the rotation axis. We then make the following additional assumptions:

(i) Axial gradients in $\mathbf{v}$ are very small and so we may write $\mathbf{v} \approx \mathbf{v}(x, y)$.
(ii) The fluctuations in velocity have maximal helicity, as would be the case for inertial or magnetostrophic wave packets.
(iii) The fluctuations are statistically homogeneous, at least locally.
(iv) Following (14.36), it is assumed that $|\mathbf{b}|$ is much smaller than the local mean field $|\overline{\mathbf{B}}|$, a situation known as *first-order smoothing* in mean-field theory.

In Davidson and Ranjan (2015) the validity of these various assumptions is tested for the particular case of a sea of inertial wave packets generated by random buoyant anomalies located near the equator. It turns out that they are a reasonable approximation. Given these various assumptions, the starting point for the analysis is Ohm's law written in the form

$$\frac{\partial \mathbf{A}}{\partial t} = \mathbf{u} \times \mathbf{B} - \nabla \Phi - \lambda \nabla \times \mathbf{B}, \qquad (14.63)$$

where $\mathbf{A}$ is the vector potential for $\mathbf{B}$ and we have substituted for $\mathbf{E}$ using Faraday's law in the form of (2.46). Writing $\mathbf{B} = \overline{\mathbf{B}} + \mathbf{b}$, $\mathbf{u} = \bar{\mathbf{u}} + \mathbf{v}$, $\mathbf{A} = \bar{\mathbf{A}} + \mathbf{a}$ and $\Phi = \overline{\Phi} + \varphi$, subtracting the averaged version of (14.63) from the un-averaged version and applying first-order smoothing, we obtain

$$\frac{\partial \mathbf{a}}{\partial t} = \mathbf{v} \times \overline{\mathbf{B}} - \nabla \varphi - \lambda \nabla \times \mathbf{b}. \tag{14.64}$$

Next we introduce a vector potential for the velocity field, $\mathbf{v} = \nabla \times \mathbf{c}$, and as usual we adopt the Coulomb gauge for $\mathbf{c}$, $\nabla \cdot \mathbf{c} = 0$. Then the potential φ is governed by the divergence of (14.64),

$$\nabla^2 \varphi = \nabla \cdot (\mathbf{v} \times \overline{\mathbf{B}}) = \overline{\mathbf{B}} \cdot \boldsymbol{\omega} = -\nabla^2 (\overline{\mathbf{B}} \cdot \mathbf{c}), \tag{14.65}$$

where $\boldsymbol{\omega} = \nabla \times \mathbf{v}$ is the vorticity. It follows that $\varphi = -\overline{\mathbf{B}} \cdot \mathbf{c} + \hat{\varphi}, \quad \nabla^2 \hat{\varphi} = 0$.

The next step is to invoke assumption (ii) in the form $\mathbf{c} = \delta \mathbf{v} = \delta^2 \boldsymbol{\omega}$, where δ is a (positive or negative) constant having the dimensions of length. The first thing to note is that, if $\mathbf{v}$ has maximal helicity, then so usually does $\mathbf{b}$. That is, using $\mathbf{c} = \delta \mathbf{v}$, (14.64) may rewritten as

$$[\partial / \partial t - \lambda \nabla^2](\mathbf{a} - \delta\,\mathbf{b}) = 0, \tag{14.66}$$

and so even if $\mathbf{b}$ starts out as non-helical, it progressively becomes so. So let us take $\mathbf{a} = \delta \mathbf{b}$, which in any event holds for both inertial and magnetostrophic waves. In this case our governing equation (14.64) becomes

$$\frac{\partial \mathbf{b}}{\partial t} + \frac{\lambda}{\delta^2} \mathbf{b} = \frac{1}{\delta}[\mathbf{v} \times \overline{\mathbf{B}} - \nabla \varphi], \tag{14.67}$$

from which

$$\mathbf{v} \times \mathbf{b} + \frac{\delta^2}{\lambda} \mathbf{v} \times \frac{\partial \mathbf{b}}{\partial t} = \frac{\delta}{\lambda}[\mathbf{v} \times (\mathbf{v} \times \overline{\mathbf{B}}) + \nabla \times (\varphi \mathbf{v}) - \varphi \nabla \times \mathbf{v}]. \tag{14.68}$$

If we now assume statistical homogeneity we find

$$\langle \varphi \boldsymbol{\omega} \rangle = -\langle (\overline{\mathbf{B}} \cdot \mathbf{c}) \boldsymbol{\omega} \rangle = -\langle (\overline{\mathbf{B}} \cdot \mathbf{v}) \mathbf{v} \rangle, \tag{14.69}$$

and so (14.68) averages to give

$$\langle \mathbf{v} \times \mathbf{b} \rangle + \frac{\delta^2}{\lambda} \langle \mathbf{v} \times \frac{\partial \mathbf{b}}{\partial t} \rangle = -\frac{\delta}{\lambda} \langle v^2 \overline{\mathbf{B}} - 2(\mathbf{v} \cdot \overline{\mathbf{B}}) \mathbf{v} \rangle, \tag{14.70}$$

which is a generalisation of (14.46). The final step is to combine assumptions (i) and (ii) in the form $\mathbf{v} = \mathbf{v}(x, y) = \delta \boldsymbol{\omega}$, and again invoke statistical homogeneity.

(Note that δ may be positive or negative, depending on the sign of the helicity.) After a little algebra we find that $\mathbf{v} = \mathbf{v}(x,y) = \delta \boldsymbol{\omega}$ demands

$$\langle v_x v_z \rangle = \langle v_y v_z \rangle = 0 , \qquad \langle v_z^2 \rangle = \langle v_x^2 + v_y^2 \rangle, \qquad (14.71)$$

as in (14.39), and so (14.70) yields

$$\langle \mathbf{v} \times \mathbf{b} \rangle + \frac{\delta^2}{\lambda} \left\langle \mathbf{v} \times \frac{\partial \mathbf{b}}{\partial t} \right\rangle = -\frac{2\delta}{\lambda} \langle v_\perp^2 \overline{\mathbf{B}}_\perp - (\mathbf{v}_\perp \cdot \overline{\mathbf{B}}_\perp) \mathbf{v}_\perp \rangle, \qquad (14.72)$$

where $\overline{\mathbf{B}}_\perp$ and $\mathbf{v}_\perp$ are the components of $\overline{\mathbf{B}}$ and $\mathbf{v}$ which are perpendicular to the rotation axis. Note that there is no mean EMF in the axial direction.

Expression (14.72) simplifies further if we make additional assumptions, assumptions which are less likely to be good approximations in the core of the earth. For example, if two-dimensional isotropy holds in the x–y plane then $\langle v_x v_y \rangle = 0$ and $\langle v_x^2 \rangle = \langle v_y^2 \rangle$, and so (14.72) yields a generalisation of (14.49) and (14.51),

$$\langle \mathbf{v} \times \mathbf{b} \rangle + \frac{\delta^2}{\lambda} \left\langle \mathbf{v} \times \frac{\partial \mathbf{b}}{\partial t} \right\rangle = -\frac{\delta \langle v^2 \rangle \, \overline{\mathbf{B}}_\perp}{2\lambda} = -\frac{\delta^2 \langle h \rangle \, \overline{\mathbf{B}}_\perp}{2\lambda}, \qquad (14.73)$$

where $h = \mathbf{v} \cdot \boldsymbol{\omega}$. (Such two-dimensional isotropy is unlikely to be a good approximation in the core of the earth where the presence of the field $\overline{\mathbf{B}}_\perp$ will almost certainly produce significant anisotropy in the transverse plane, as discussed in §14.8.3.)

Alternatively, if the time dependence of $\mathbf{b}$, as measured by a frequency ϖ, is such that $R_m = \varpi \delta^2 / \lambda < 1$, and two-dimensional isotropy happens to apply, then we have the particularly simple approximation

$$\langle \mathbf{v} \times \mathbf{b} \rangle = -\frac{\delta^2 \langle h \rangle \, \overline{\mathbf{B}}_\perp}{2\lambda}, \qquad (14.74)$$

which is (14.51).

Given these various estimates of α, it is possible to integrate (14.57) and (14.58) and search for self-excited dynamos, and indeed a variety of $\alpha - \Omega$ and α^2 solutions have been found. Although this formalism of an α-effect rests on an assumed separation of scales, which possibly does not formally exist in the core, it does at least provide a convenient conceptual framework within which to classify and rationalise the many dynamo mechanisms which have been unearthed in the numerical simulations.

Estimate (14.73) is also useful as it may be combined with integral relationships of the form of (14.35) to provide a kinematic criterion for dynamo action, as we now show. The first step is to generalise (14.35) to include the mean azimuthal field.

14.6.3 Exact Integrals Relating the Large-Scale Field to the Small-Scale EMF

In §14.5.5 we derived the integral relationship

$$\frac{d}{dt}\int_{V_C}\frac{1}{2}\left(R_C^2 - \mathbf{x}^2\right)\overline{B}_z dV = \int_{V_C} r\langle \mathbf{v}\times\mathbf{b}\rangle_\theta dV - 3\lambda\int_{V_C}\overline{B}_z dV, \tag{14.75}$$

which in the steady-state gives

$$\int_{V_C}\overline{B}_z dV = \frac{1}{3\lambda}\int_{V_C} r\langle \mathbf{v}\times\mathbf{b}\rangle_\theta dV = \frac{1}{3\lambda}\int_{V_C} r\left(\overline{J}_\theta/\sigma\right)dV. \tag{14.76}$$

This confirms that the large-scale dipole field is maintained by an azimuthal EMF associated with non-axisymmetric disturbances. It also yields the estimate

$$\langle \mathbf{v}\times\mathbf{b}\rangle_\theta \sim \frac{\lambda}{uR_C}u\overline{B}_z \ll u\overline{B}_z, \tag{14.77}$$

which suggests that the perturbation field $\mathbf{b}$ is much weaker than the mean dipole field $\overline{B}_z$.

It is natural to seek an equivalent equation for the large-scale azimuthal field. This is readily obtained by integrating the azimuthal average of the poloidal components of (14.63)

$$\frac{\partial\mathbf{A}}{\partial t} = \mathbf{u}\times\mathbf{B} - \nabla\Phi - \frac{1}{\sigma}\mathbf{J}, \tag{14.78}$$

rewritten as

$$\frac{\partial(\mathbf{x}\times\mathbf{A})}{\partial t} = \mathbf{x}\times(\mathbf{u}\times\mathbf{B}) + \nabla\times(\Phi\mathbf{x}) - \frac{1}{\sigma}\mathbf{x}\times\mathbf{J}. \tag{14.79}$$

However, we must integrate over one hemisphere only, as $\overline{B}_\theta$ is antisymmetric about the equator. To this end, we follow a procedure similar to that which led to (14.75), but ignore terms such as

$$\frac{d}{dt}\int_{V_C}\frac{1}{2}\left(R_C^2 - \mathbf{x}^2\right)\left(\overline{B}_\theta/r\right)dV,$$

on the assumption that the magnetic field may be considered as steady when averaged over the scale of the core. The first step is to take the cross product of $\mathbf{x}/r$ with the azimuthal average of the poloidal components of (14.78), and then integrate this over the northern hemisphere, V_N. Next we note that $\overline{B}_\theta$ is zero at the core-mantle boundary and on the equator, and that, by virtue of Ampère's law,

$$\frac{1}{r\sigma}\left(\mathbf{x}\times\overline{\mathbf{J}}\right)_\theta = -\lambda\nabla\cdot[\overline{B}_\theta\mathbf{x}/r] + \lambda\overline{B}_\theta/r. \tag{14.80}$$

So the volume integral of $\left(\mathbf{x}\times\overline{\mathbf{J}}\right)_\theta/r\sigma$, which arises from the last term on the right of (14.79), converts to a volume integral of $\lambda\overline{B}_\theta/r$. The end result is (Davidson and Ranjan, 2015),

$$\int_{V_N}\frac{1}{2}\left(R_C^2-\mathbf{x}^2\right)\left[\overline{\mathbf{u}}_p\cdot\nabla\left(\frac{\overline{B}_\theta}{r}\right)-\overline{\mathbf{B}}_p\cdot\nabla\left(\frac{\overline{u}_\theta}{r}\right)\right]dV - \int_{z=0}\frac{1}{2}\left(R_C^2-r^2\right)\frac{\overline{B}_z\overline{u}_\theta}{r}dA$$
$$= \int_{V_N}\frac{1}{r}[z\langle\mathbf{v}\times\mathbf{b}\rangle_r - r\langle\mathbf{v}\times\mathbf{b}\rangle_z]dV - 2\pi\oint\overline{\Phi}\mathbf{x}\cdot d\mathbf{r} - \lambda\int_{V_N}\left(\overline{B}_\theta/r\right)dV$$

$$\tag{14.81}$$

where we have applied Stokes' theorem to the integral involving $[\nabla\times(\overline{\Phi}\mathbf{x})]_\theta$ to convert it into a line integral of $\overline{\Phi}$. (This line integral is anticlockwise around the perimeter of a quarter circle in the r–z plane defined by the equator, the core-mantle boundary and the z-axis.) We recognise both the mean advection of $\overline{B}_\theta/r$ and the Ω-effect on the left of (14.81), consistent with their appearance in (14.58), while on the right we have the mean poloidal EMF induced by the small-scale fluctuations, as well as an integral measure of the Ohmic dissipation.

Most of the numerical simulations suggest that the Ω-effect does not contribute significantly to dynamo action outside the tangent cylinder, and of course there is no dynamo effect associated with simple advection of the mean azimuthal field. So we might tentatively drop the mean flow terms on the left of (14.81). Recalling that columnar motion ensures that $\langle\mathbf{v}\times\mathbf{b}\rangle_z \ll \langle\mathbf{v}\times\mathbf{b}\rangle_r$, as captured by (14.72), we obtain the much simpler expression

$$\int_{V_N}\frac{z}{r}\langle\mathbf{v}\times\mathbf{b}\rangle_r dV - 2\pi\oint\overline{\Phi}\mathbf{x}\cdot d\mathbf{r} \approx \lambda\int_{V_N}\left(\overline{B}_\theta/r\right)dV. \tag{14.82}$$

The same equation holds for the southern hemisphere, but with a different contour for the line integral of $\overline{\Phi}$. Now the large-scale azimuthal magnetic field relies on the presence of mean poloidal currents, as shown in Figure 14.18, and (14.82) clearly demonstrates that in an α^2 dynamo these currents are maintained by a mean radial EMF associated with the non-axisymmetric fluctuations.

The appearance of the mean electrostatic potential in (14.82), despite its absence in (14.76), is important and can be understood from Equation (14.78) and Figure 14.18. The mean EMF $\langle \mathbf{v} \times \mathbf{b} \rangle_r$ drives current radially outward and, because the mantle is non-conducting, this poloidal current returns in regions of low mean EMF, in particular near the equator. It is the electrostatic potential that forces the mean poloidal current to recirculate near the equatorial plane. Thus the role of the line integral in (14.82) is to oppose, to some extent, the mean EMF $\langle \mathbf{v} \times \mathbf{b} \rangle_r$. However, we can still equate, approximately, the first and last terms in (14.82), and so we obtain the estimate

$$\langle \mathbf{v} \times \mathbf{b} \rangle_r \sim \frac{\lambda}{u R_C} u |\overline{B}_\theta| \ll u |\overline{B}_\theta|. \tag{14.83}$$

Evidently, in an α^2 dynamo, the fluctuating field, $\mathbf{b}$, is much weaker than the mean east-west field, $\overline{B}_\theta$, as well as the mean axial field, $\overline{B}_z$.

14.6.4 Putting the Pieces Together: A Kinematic Criterion for Dynamo Action

We are now in a position to derive a kinematic criterion for dynamo action of the α^2 type. Between them, (14.76) and (14.82) dictate the relationship between the large-scale magnetic field and the mean EMF induced by non-axisymmetric fluctuations in a steady-on-average α^2 dynamo:

$$\int_{V_C} \overline{B}_z dV = \frac{1}{3\lambda} \int_{V_C} r \langle \mathbf{v} \times \mathbf{b} \rangle_\theta dV, \quad \int_{V_N} (\overline{B}_\theta/r) dV \sim \frac{1}{\lambda} \int_{V_N} \frac{z}{r} \langle \mathbf{v} \times \mathbf{b} \rangle_r dV. \tag{14.84}$$

When combined with our estimate of the α-effect in §14.6.2, this provides the basis for an α^2 dynamo. (Note that the second equation in (14.84) applies equally to the northern and southern hemispheres.) Consider, for example, the somewhat artificial case where we assume local isotropy in planes normal to the rotation axis. (We shall correct for this shortly.) Then (14.73) might be rewritten to include the β-effect as

$$\langle \mathbf{v} \times \mathbf{b} \rangle \sim -\frac{\lambda^{-1}\delta^2 \langle h \rangle \, \overline{\mathbf{B}}_\perp}{1 + (\delta^2 \varpi/\lambda)^2} - \beta \nabla \times \overline{\mathbf{B}}, \tag{14.85}$$

where $\beta \sim \alpha \delta$, ϖ is the wave frequency and we have averaged over a wave period. (We are assuming here that the columnar vortices that drive the α-effect are in fact helical wave packets.) This may be combined with (14.84) to give

$$\overline{B}_z \sim -\frac{R_C \delta^2 \langle h \rangle}{\lambda \lambda_e [1 + (\delta^2 \varpi / \lambda)^2]} \overline{B}_\theta, \quad \overline{B}_\theta \sim \mp \frac{R_C \delta^2 \langle h \rangle}{\lambda \lambda_e [1 + (\delta^2 \varpi / \lambda)^2]} \overline{B}_r, \quad (14.86)$$

where $\lambda_e = \lambda + \beta$ and the upper (lower) sign in (14.86) applies in the north (south). Evidently, a steady-on-average dynamo with a dipole pointing to the north requires that $\langle h \rangle$ and $\overline{B}_\theta$ be of opposite signs in both the north and south. If the helicity outside the tangent cylinder is negative in the north and positive in the south, then $\overline{B}_\theta$ outside the tangent cylinder must be positive in the north and negative in the south, as observed in most numerical simulations and as shown in Figure 14.18. ($\overline{B}_\theta$ inside the tangent cylinder is often of the opposite sign, presumably due to a local Ω-effect.) All in all, the situation is very like that shown in Figure 14.18.

One of the weaknesses of (14.86) is that, as a result of the transverse magnetic field, $\overline{\mathbf{B}}_\perp$, the columnar vortices are likely to be more sheet-like than tubular, elongated in the direction of $\overline{\mathbf{B}}_\perp$. This anisotropy in the transverse plane is not captured by (14.85), and so a more refined analysis is offered in Davidson and Ranjan (2015). If γ represents the characteristic aspect ratio of the columnar vortices in the transverse $(r - \theta)$ plane, with $\gamma > 1$, then the mean EMF is estimated as

$$\langle \mathbf{v} \times \mathbf{b} \rangle \sim -\frac{\delta^2 \langle h \rangle \, \overline{\mathbf{B}}_\perp}{\lambda \gamma^2} - \beta \nabla \times \overline{\mathbf{B}}, \quad (14.87)$$

where δ is the smaller of the two transverse length scales, $\delta^2 \varpi / \lambda$ is assumed to be at most of order one, $\beta \sim \delta^2 \langle \mathbf{v}^2 \rangle / \lambda$, and $\lambda_e \sim \lambda + (\delta v / \lambda)^2 \lambda \sim \lambda$. Estimate (14.86) is then replaced by

$$\overline{B}_z \sim -\frac{R_C \delta^2 \langle h \rangle}{\lambda_e \lambda \gamma^2} \overline{B}_\theta, \quad \overline{B}_\theta \sim \mp \frac{R_C \delta^2 \langle h \rangle}{\lambda_e \lambda \gamma^2} \overline{B}_r. \quad (14.88)$$

We conclude that such an α^2 dynamo is possible provided that $|\alpha| R_C / \lambda_e = O(1)$, which in practice becomes $|\alpha| R_C / \lambda_e \sim 10$. Equivalently, we require

$$\frac{R_C \langle \mathbf{v}^2 \rangle^{1/2}}{\lambda} \frac{|\delta| \langle \mathbf{v}^2 \rangle^{1/2}}{\lambda} \sim 10 \gamma^2. \quad (14.89)$$

Given that $R_C u / \lambda \sim 900$ and $\delta u / \lambda \sim 1$ (see Exercise 14.4), this suggests an anisotropy factor of $\gamma \sim 10$.

14.7 The Numerical Simulations of Planetary Dynamos

The numerical simulation of planetary dynamos is a considerable challenge, for at least three reasons. First, the buoyant forcing in the simulations is usually constrained to be much weaker than in the core of the earth, and so the simulations are invariably underpowered by some considerable margin. Second, the tiny viscosity in the core, which gives rise to the extremely low values of the Ekman and magnetic Prandtl numbers shown in Table 14.1, leads to small length-scales that cannot be resolved in the simulations. The third problem relates to the time step, Δt. This is controlled by the Courant condition, and in particular by the need to resolve the passage of (fast) inertial waves through each cell in the grid, which in turn demands that $\Delta t < O(\Omega^{-1})$. So, in order to enable time steps on a geological time scale, the numerical dynamos invariably under-rotate.

Despite the fact that the simulations are only weakly driven, are much too viscous, and significantly under-rotate, a number of the numerical dynamos exhibit earth-like features. For example, many have a quasi-steady dipole aligned with the rotation axis, a slow westward drift of the field across the mantle, and even occasional reversals. Given that these numerical experiments must be predicting much that is unrepresentative of core motion, yet can produce plausible-looking dynamos, perhaps the most important question to ask is, 'What is it that the simulations are getting right?' That is to say, what hydrodynamic features do numerical dynamos and planetary cores have in common that enables at least some of the simulations to produce earth-like magnetic fields? Extended reviews of these intriguing but flawed simulations may be found in Jones (2011) and Christensen (2011).

The forcing in these numerical experiments is typically measured using the Rayleigh-like number $Ra = g_0\beta\Delta TR_C/\Omega\nu$, where g_0 is the gravity at the core surface, β the thermal expansion coefficient, and ΔT the superadiabatic temperature difference across the core. The current simulations may be classified into one of two categories: weakly driven and moderately driven dynamos (Figure 14.19). In the former, the Rayleigh-like number $Ra = g_0\beta\Delta TR_C/\Omega\nu$ is only a few multiples of the critical value at which non-magnetic convection first sets in, say 10 times critical, and the flow is laminar or mildly chaotic. In the moderately driven simulations, which are somewhat more earth-like, the Rayleigh number significantly exceeds the critical value, perhaps by a factor of $50\rightarrow100$. Here the flow is turbulent, rather than mildly chaotic. To place things in perspective, in the geodynamo Ra may be as high as $10^3\rightarrow10^4$ times the critical value (Christensen, 2011), way beyond anything the numerical simulations can currently reach.

In the weakly driven dynamos the magnetic Reynolds number, $R_m = u_{rms}(R_c - R_i)/\lambda$, is usually less than 100, while the moderately driven

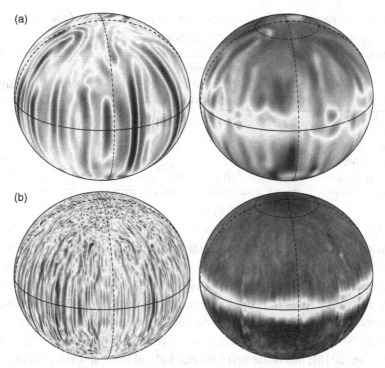

Figure 14.19 A comparison of weakly and moderately driven dynamos. The images on the left show the radial velocity field near the mantle, and those on the right the magnitude of the radial magnetic field near the mantle. (a) A weakly driven dynamo, 11 times the critical Rayleigh number. (b) A moderately driven dynamo, 46 times critical. (From Sreenivasan, 2010.)

flows can reach values of several hundred, which is not unlike the earth. In both classes of simulations Pr_m and $Ek = \nu/\Omega R_C^2$ are usually given values of $Pr_m = O(1)$ and $Ek \sim 10^{-6}$, whereas we have $Pr_m \sim 10^{-6}$ and $Ek \sim 10^{-15}$ in the core of the earth. Table 14.4 provides a comparison of the dimensionless parameters in the numerical simulations and in the earth's core. Perhaps the most striking observation is that the numerical experiments are overly viscous by a factor of at least 10^9, as measured by the Ekman number. So, instead of being largely irrelevant, as in a planet, the viscous stresses in the simulations are often dominant.

The Rossby numbers in the simulations are also a matter of concern. In the liquid core we have $Ro = |\mathbf{u}|/\Omega\ell \sim 10^{-6}$ based on large-scale estimates of $|\mathbf{u}|$ and ℓ,[4] perhaps rising to around $Ro \sim 10^{-2}$ for the smallest scales present in the core. This means that the inertial term $\mathbf{u} \cdot \nabla\mathbf{u}$ is negligible at all scales. By way of contrast, $Ro \sim 10^{-3} \to 10^{-2}$ in most simulations, again based on large-scale estimates of $|\mathbf{u}|$

---

[4] An independent estimate of the large-scale Rossby number is given in Exercise 14.3.

Table 14.4 *A comparison of dynamo parameters in the core of the earth and in the numerical simulations. (Adapted from Christensen, 2011.)*

Parameter	Ekman number	Magnetic Prandtl number	Modified Rayleigh number	Rossby number
Definition	$\mathrm{Ek} = \frac{\nu}{\Omega R_C^2}$	$\mathrm{Pr}_m = \frac{\nu}{\lambda}$	$\mathrm{Ra} = \frac{g_0 \beta \Delta T R_c}{\Omega \nu}$	$\mathrm{Ro} = \frac{|\mathbf{u}|}{\Omega(R_C - R_i)}$
Estimated values in the liquid core	10^{-15}	10^{-6}	$(10^3 \rightarrow 10^4)\mathrm{Ra}_{crit}$	10^{-6}
Value in simulations	$10^{-3} \rightarrow 10^{-6}$	$0.1 \rightarrow 10$	$(10 \rightarrow 100)\mathrm{Ra}_{crit}$	$10^{-3} \rightarrow 10^{-2}$

and ℓ. This suggests that $\mathrm{Ro} \sim 0.05 \rightarrow 1$ at the small scales in such simulations, which means that, in some of the numerical experiments, inertial waves will not be sustained at the scale of the thin columnar vortices. Interestingly, there is considerable evidence that, as the small-scale Rossby number rises above a value somewhat less than unity, say $\mathrm{Ro} \sim 0.2$, the large-scale dipole field is lost in the numerical experiments, to be replaced by a random, disorganised magnetic field (Roberts and King, 2013). Since inertial waves cease to exist when Ro exceeds ~ 0.3 (Bin Baqui and Davidson, 2015), this is the first hint that helical inertial wave packets may play an important role in some of the numerical dynamos, helping to sustain a dominant dipole field.

The results of the numerical simulations vary considerably depending on the boundary conditions used and the parameter regime adopted. Nevertheless, some general observations may be made. When describing numerical dynamos it is useful to distinguish between events inside and outside the tangent cylinder. In the weakly driven simulations, which we shall discuss first, the dynamo is located outside the tangent cylinder. The flow there is organised into columnar vortices, aligned with the rotation axis. Within the columnar vortices the flow is strongly helical and the sign of the rotation alternates from column to column, being cyclonic ($u_\theta > 0$) in one and anti-cyclonic ($u_\theta < 0$) in the next. However, the helicity density, $h = \mathbf{u} \cdot \nabla \times \mathbf{u}$, is uniformly of one sign in each hemisphere, but of opposite signs in the north and south. Outside the tangent cylinder we find $h < 0$ in the north and $h > 0$ in the south, as shown in Figure 14.18. With an appropriate choice of simulation parameters and boundary conditions the poloidal field can be made strongly dipolar, while the azimuthal field is more or less antisymmetric about the equator. Within the tangent cylinder B_θ is negative in the north and positive in the south (for a dipole pointing to the north), consistent with the Ω-effect illustrated in Figure 14.17. However, the dominant azimuthal field in the weakly driven simulations lies outside the tangent cylinder, where the signs of B_θ are reversed, with B_θ positive in the north and negative in the south, as shown in Figure 14.18. These dynamos are therefor of the α^2 type, with dynamo action taking place outside the tangent cylinder and large-scale differential rotation playing no significant role. Since $B_\theta > 0$ and $h < 0$ in the north, and $B_\theta < 0$ and $h > 0$ in the south, we have $hB_\theta < 0$ in both hemispheres, consistent with (14.86).

In the moderately driven dynamos the motion is more turbulent with a wider range of scales. The flow outside the tangent cylinder is not unlike that for a weakly driven dynamo, consisting of thin columnar convection cells, though the cells are thinner, more sinuous, and less likely to intersect the mantle. These moderately driven dynamos are mostly of the α^2 type, rather like their weakly driven partners. The main difference is that the flow within the tangent cylinder is now quite vigorous, consisting of upwellings near the polar axis. These upwellings give rise to strong differential rotation, and hence to a significant Ω-effect. Consequently, the azimuthal field is now concentrated within the tangent cylinder, being negative in the north and positive in the south (assuming the dipole points to the north). Crucially, though, this Ω-effect does not contribute significantly to dynamo action in these moderately forced simulations. Of course, an important question is what happens if the driving force is increased yet further. It seems likely that an Ω-effect will persist, generating a strong azimuthal field inside the tangent cylinder, but whether or not this will lead to something resembling a classical $\alpha - \Omega$ dynamo remains unclear.

14.8 Speculative Dynamo Cartoons Based on the Numerical Simulations

14.8.1 Searching for the Source of the North-South Asymmetry in Helicity

Let us now combine the observational evidence of §14.4 with the intriguing results of the numerical simulations discussed above. The (admittedly imperfect) numerical experiments suggest that the geodynamo is of the α^2 type, located outside the tangent cylinder, and driven by the strongly helical flow within thin, columnar vortices. Such a dynamo can operate only if the helicity is predominantly of opposite signs in each hemisphere, and the simulations display negative helicity in the north and positive helicity in the south (outside the tangent cylinder), as shown in Figure 14.18.

If we are to transfer this knowledge to planetary cores, we are obliged to find a simple, robust, mechanism of producing helical columnar vortices whose helicity is of one sign in the north and another in the south, ideally negative in the north and positive in the south. Moreover, from the discussion in §14.4, we would like that mechanism to be

(i) insensitive to mechanical boundary conditions and in particular to the exis-
     tence of a mantle, so that similar dynamos can operate in the earth and the gas
     giants,
(ii) not reliant on viscous or Reynolds stresses and in particular not reliant on the
     formation of Ekman layers, partly because the viscous and Reynolds stresses
     are so tiny, and partly because there is no mantle in the gas giants on which
     such layers can form.

In the more weakly forced numerical dynamos, which are also extremely
viscous, the columnar vortices intersect the mantle, and so the dominant source
of the core helicity is Ekman pumping. However, in the more strongly driven
simulations, Ekman pumping provides only part of the helicity (Olson, Christensen
and Glatzmaier, 1999), and in any event Ekman pumping is an unlikely candidate
for a planetary core, as suggested above. So we are faced with finding an alternative
mechanism for the production and subsequent spatial segregation of helicity in the
core. Moreover, such a mechanism should be simple and robust, so that it can
operate equally in the terrestrial planets and gas giants.

One traditional approach to this problem is to consider the role of the pseudo-
scalar $\mathbf{\Omega}\cdot\mathbf{g}$ in determining the north-south asymmetry. The idea is that, in the
absence of Ekman pumping, the required north-south asymmetry must, in some
sense, be the result of buoyant convection within the interior of a rapidly rotating
fluid. It is natural, therefore, to ask if the pseudo-scalar $\mathbf{\Omega}\cdot\mathbf{g}$ plays a role in
determining the asymmetric helicity distribution. To show that it can, we take the
dot product of $\mathbf{u} \times \mathbf{\Omega}$ with the curl of the Boussinesq equation

$$\frac{\partial \mathbf{u}}{\partial t} = 2\mathbf{u} \times \mathbf{\Omega} - \nabla(p/\rho) + b\mathbf{g} + \mathbf{J} \times \mathbf{B}/\rho,$$

where $b = \rho'/\rho$, ρ' being the density perturbation. Ignoring the Lorentz force, as well
as the time derivative on the left, we find

$$2\Omega^2(\mathbf{u}_\perp \cdot \nabla \times \mathbf{u}_\perp) = \mathbf{\Omega} \cdot \mathbf{u} \times (\nabla b \times \mathbf{g}) = \mathbf{\Omega} \cdot \nabla b(\mathbf{u} \cdot \mathbf{g}) - \mathbf{\Omega} \cdot \mathbf{g}(\mathbf{u} \cdot \nabla b),$$

where the subscript $\perp$ indicates planes perpendicular to the rotation axis. This does
indeed establish a local relationship between $\mathbf{\Omega}\cdot\mathbf{g}$ and one component of helicity.
However, in a rapidly rotating fluid which is permeated with columnar structures,
one might have expected the helicity to be dominated by its other components, such as
$\mathbf{u}_{//}\cdot \nabla \times \mathbf{u}_\perp$, rather than by $\mathbf{u}_\perp \cdot \nabla \times \mathbf{u}_\perp$ (Moffatt, 1978). Moreover, this mechanism
will tend to induce positive helicity in the north and negative in the south, when
$\mathbf{u}. \nabla b > 0$, which is the opposite of that observed outside the tangent cylinder in the

numerical simulations. So perhaps we need to look for an alternative way of establishing the required north-south asymmetry in helicity.

We have already hinted that helical wave packets, composed of either inertial waves or magnetostrophic waves, are promising candidates for the production and spatial segregation of helicity. Such waves are ubiquitous in a rapidly rotating fluid, have maximum helicity, are generated by buoyant anomalies, and possess the intriguing property that upward-propagating wave packets carry with them negative helicity and downward-propagating waves positive helicity. (See §3.9.2 and §7.4 for a discussion of inertial and magnetostrophic waves, and in particular Figure 3.33 for a demonstration of the spatial segregation of helicity by inertial wave packets.) Moreover, this mechanism of helicity generation, in which a single, localised source of buoyancy can generate helicity throughout a large volume via the initiation and subsequent dispersion of helical wave packets, is potentially more efficient than the local mechanism associated with $\mathbf{\Omega} \cdot \mathbf{g}$. So let us take a moment to consider the possible role of such waves.

The primary sources of motion in the core are buoyant anomalies which float slowly outward from the inner core towards the mantle. Since the buoyant anomalies migrate slowly, taking around 300 years to reach the mantle, they are obliged to continually emit low-frequency inertial wave packets in the absence of a significant mean field, or magnetostrophic wave packets if there is a substantial local mean field. Moreover, we have already shown that such low-frequency inertial waves packets will spread along the rotation axis, spontaneously elongating to form columnar, helical vortices (see Figures 3.33 and 14.14). Now the perfect asymmetry in the helicity of upward- and downward-propagating waves can be formally proved only for monochromatic inertial (or magnetostrophic) waves. However, it so happens that this property carries over to wave packets composed of a spread of wavenumbers, as shown in Davidson and Ranjan, 2015, and as discussed in Example 14.1 below. So, if we can identify an energy source for waves (in the form of buoyant anomalies) which is located in or around the equator, that energy source will produce a stream of wave packets, each wave packet resembling a columnar vortex, and each wave packet carrying negative helicity northward or positive helicity to the south. Provided the waves are destroyed by Ohmic dissipation before reaching the mantle, and so do not reflect, the resulting helicity distribution is exactly that required by an α^2 dynamo operating outside the tangent cylinder.

Rather remarkably, many (though certainly not all) of the numerical dynamos indicate that the buoyancy flux outside the tangent cylinder is indeed biased towards the equatorial plane (see, for example, Olson, Christensen and Glatzmaier, 1999, or Sakuraba and Roberts, 2009), although the results depend somewhat on whether a fixed heat flux boundary condition is adopted at the mantle, rather than the more common choice of a fixed temperature difference. (A fixed heat flux boundary condition is the correct one for the earth, as the thermal resistance to heat flowing out to the earth's crust is dominated by the mantle.) This equatorial buoyancy flux takes the form of a random sea of meandering plumes floating radially outward to the mantle. If these plumes excite wave packets, as they must, and those waves dissipate before reflecting at the mantle, we have all the necessary ingredients for an α^2 dynamo of the type seen in the more strongly forced numerical simulations.

14.8.2 A Speculative Weak-Field Cartoon

Let us now explore in a little more detail the possibility of an α^2 dynamo driven by helical wave packets. We start by considering the weak-field case, where the dynamic influence of the ambient magnetic field is ignored and so the buoyant anomalies floating out along the equatorial plane generate inertial wave packets, rather than magnetostrophic waves. Our discussion follows that of Davidson (2014) and Davidson and Ranjan (2015).

Consider an isolated blob of buoyant material of scale δ sitting in the equatorial plane. The fluid is taken to be Boussinesq, with $b = \rho'/\rho$, and we adopt local Cartesian coordinates with z pointing northward, x radially outward and y in the azimuthal direction. The governing equation at low Ro is then

$$\frac{\partial \mathbf{u}}{\partial t} = 2\mathbf{u} \times \mathbf{\Omega} - \nabla(p/\rho) + b\mathbf{g}, \quad \mathbf{g} = -g\hat{\mathbf{e}}_x, \qquad (14.90)$$

and the vorticity equation is evidently

$$\frac{\partial \boldsymbol{\omega}}{\partial t} = 2(\mathbf{\Omega} \cdot \nabla)\mathbf{u} + \nabla b \times \mathbf{g}. \qquad (14.91)$$

Since ρ' is governed by an advection-diffusion equation it evolves on a slow time scale set by $\mathbf{u}$. Inertial waves, by contrast, evolve on the fast time scale of Ω^{-1}. Thus, at low Ro, we may take ρ' as quasi-steady as far as the initiation of inertial waves is concerned. Treating ρ' as independent of time, the operator $\nabla \times (\partial/\partial t)$ applied to (14.91) gives

$$\frac{\partial^2}{\partial t^2}\left(\nabla^2 \mathbf{u}\right) + (2\mathbf{\Omega} \cdot \nabla)^2 \mathbf{u} = (2\mathbf{\Omega} \cdot \nabla)(\mathbf{g} \times \nabla b). \qquad (14.92)$$

Evidently the buoyancy acts as a source of low-frequency inertial waves. These waves necessarily propagate in the $\pm\mathbf{\Omega}$ directions at a speed of $\Omega\delta$, carrying negative helicity to the north and positive helicity to the south.

Consider the somewhat artificial problem in which the buoyant blob is suddenly introduced at $\mathbf{x} = 0$ at $t = 0$. Then, after a time t, we find wave fronts located at $z \sim \pm\delta\Omega t$ above and below the blob, and within the cylindrical region defined by $z \sim \pm\delta\Omega t$ we have low-frequency inertial waves which have originated from the buoyant blob. The situation is somewhat reminiscent of the transient Taylor columns generated by an impulsively moved penny, as shown in Figure 3.34. We might therefore expect these waves to create transient Taylor columns similar to those shown in Figure 3.34, and this is confirmed by (14.91) which, when the wave frequency is low, demands $(\mathbf{\Omega} \cdot \nabla)\mathbf{u} \approx 0$ outside the buoyant blob. In short, quasi-two-dimensional columnar vortices spontaneously emerge from the buoyant blob.

We can determine the structure of these columnar vortices (transient Taylor columns) by considering the vertical jump conditions across the buoyant blob. Since ρ' is quasi-steady and the inertial waves are of low frequency, (14.91) within the buoyant blob reduces to

$$2(\mathbf{\Omega} \cdot \nabla)\mathbf{u} + \nabla b \times \mathbf{g} \approx 0, \tag{14.93}$$

whose curl is

$$2(\mathbf{\Omega} \cdot \nabla)\boldsymbol{\omega} \approx \mathbf{g}\nabla^2 b - (\mathbf{g} \cdot \nabla)\nabla b. \tag{14.94}$$

From (14.94) we find that the integrated vertical jump condition in vorticity across the blob is $\Delta\omega_z \approx 0$, while (14.93) tells us that the vertical jump conditions in velocity are

$$\Delta u_z \approx -\frac{g}{2\Omega}\int(\partial b/\partial y)dz \tag{14.95}$$

and $\Delta u_x \approx 0$, $\Delta u_y \approx 0$. From the jump condition for ω_z we see that a cyclonic (or anticyclonic) columnar vortex below the buoyant blob must corresponds to a cyclonic (or anticyclonic) vortex above the blob. Also, for a Gaussian-like blob, (14.95) tells us that Δu_z is positive for $y < 0$ and negative for $y > 0$. Hence u_z, which is antisymmetric about the plane $z = 0$, diverges from the plane $z = 0$ for $y < 0$ and converges to $z = 0$ for $y > 0$. Finally we recall that upward-propagating inertial waves have negative helicity, while downward-propagating waves have positive helicity. Combining all of this information we conclude that the inertial wave dispersion pattern consists of a pair of cyclonic and anticyclonic columnar

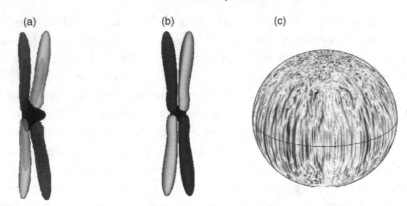

Figure 14.20 (a) A buoyant blob floats radially outward while emitting low-frequency inertial waves. Ro = 0.1. The shading indicates the direction of u_z in the wave packets; light grey for $u_z < 0$ and dark grey for $u_z > 0$. The dispersion pattern consists of four columnar wave packets: a cyclonic/anticyclonic pair above the blob and a matching pair below. (b) As for (a), but with Ro = 0.01. (c) Alternating cyclonic/anticyclonic pairs in a numerical dynamo. (From Davidson and Ranjan, 2015).

vortices above the blob, matched to a pair of cyclonic and anticyclonic vortices below the blob. The cyclones above and below are located at the same value of y (the same azimuthal angle in the earth), as are the anticyclones, and the anticyclones are located at negative y (smaller azimuthal angle), and the cyclones at positive y (larger azimuthal angle).

This dispersion pattern is illustrated in Figure 14.20, which shows a numerical simulation of the flow generated by a buoyant blob of initially Gaussian profile. The buoyant blob slowly migrates horizontally under the influence of gravity while generating low-frequency inertial waves. The images are taken from Davidson and Ranjan (2015) and correspond to a time of $\Omega t = 8$, with a Rossby number of Ro = 0.1 in (a) and Ro = 0.01 in (b). The shading indicates the direction of u_z (light grey for $u_z < 0$, dark grey for $u_z > 0$) and the dispersion pattern is exactly as anticipated above. This might be compared with the typical flow pattern seen in numerical dynamo simulations, which consists of thin, alternating cyclones and anticyclones aligned with the rotation axis, as shown in Figure 14.20c. (Note that the dispersion pattern from a buoyant blob is quite different when it lies within the tangent cylinder, where **g** and Ω are aligned. This is discussed in Exercise 14.11 at the end of this chapter.)

Figure 14.21, which is also taken from Davidson and Ranjan (2015), shows the flow resulting from a random distribution of buoyant blobs concentrated in a horizontal layer and drifting slowly in the x direction. The images show the distribution of axial velocity and, once again, we see that the inertial waves disperse

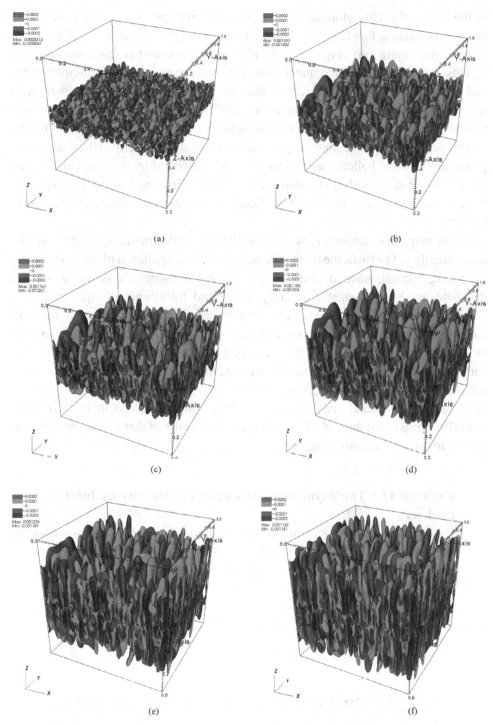

Figure 14.21 Dispersion of wave energy from a sea of random buoyant blobs. The panels show iso-surfaces of u_z (light grey for $u_z < 0$, dark grey for $u_z > 0$) at $\Omega t = 2, 4, 6, 8, 10, 12$. (From Davidson and Ranjan, 2015).

in the form of axially elongated wave packets, arranged as cyclone-anticyclone pairs. It is striking that the overall flow pattern is highly reminiscent of that seen in dynamo simulations (Figure 14.20c). Equally important is the observation in Davidson and Ranjan (2015) that the wave packets carry negative helicity upward and positive helicity downward and that the helicity is close to maximal, with $|h| \approx |\mathbf{u}||\boldsymbol{\omega}|$. (See also the discussion in Example 14.1 below). Thus we have a robust mechanism not only for the generation of helicity by buoyant blobs, but also for the spatial segregation of regions of positive and negative helicity. The mean EMF generated by these helical wave packets is well described by the α-effect estimate of (14.74), as discussed in Davidson and Ranjan (2015), and so the requirements for a weak-field α^2-dynamo driven by these helical waves are captured by criterion (14.89).

In summary, one cartoon of a weak-field dynamo is that random, buoyant plumes float radially out towards the mantle in and around the equatorial plane, continually triggering inertial wave packets as they float outwards. These wave packets dissipate as they propagate, possibly attenuated by Ohmic dissipation before reaching the mantle. However, they are likely to live long enough to carry a significant negative helicity to the north and positive helicity to the south, exactly as required by an α^2 dynamo. Of course, one can conceive of alternative dynamo cartoons, and the sparsity of observational data makes it hard to choose between the various candidates. In any event, the scaling laws for α^2 planetary dynamos driven by helical wave packet are discussed in Davidson (2014) and in Exercise 14.8 (labelled option 3) at the end of this chapter. The results of that scaling analysis, as shown in Tables 14.3, are rather encouraging.

Example 14.1 The Asymmetric Dispersion of Helicity by Inertial Wave Packets

Consider the dispersion pattern shown in Figure 14.21, where inertial wave packets are launched up and down the rotation axis from a quasi-steady slab of buoyant anomalies, which we take to sit on or near the equatorial plane. Show that the governing equation

$$\frac{\partial \mathbf{u}}{\partial t} = 2\mathbf{u} \times \boldsymbol{\Omega} - \nabla(p/\rho) + b\mathbf{g}$$

can be rewritten as

$$\frac{\partial \mathbf{u}}{\partial t} = 2\boldsymbol{\Omega} \cdot \nabla \mathbf{c} - \nabla(p^*/\rho) + b\mathbf{g}, \qquad \nabla^2(p^*/\rho) = \nabla \cdot (b\mathbf{g}),$$

where $\mathbf{c}$ is the usual vector potential for velocity. The Poisson equation for p^* can be inverted and, given that $b\mathbf{g}$ is quasi-steady (on the inertial-wave time scale), show that the p^* is also quasi-steady and falls off rapidly with distance from the buoyant cloud.

Now take the curl of the momentum equation and show that the helicity is governed by

$$\frac{\partial}{\partial t}(\mathbf{u} \cdot \boldsymbol{\omega}) = 2\Omega\nabla \cdot (u_z\mathbf{u} + c_z\boldsymbol{\omega}) - \nabla \cdot \left(p^*\boldsymbol{\omega}/\rho + \mathbf{u} \times (b\mathbf{g})\right) + 2\boldsymbol{\omega} \cdot (b\mathbf{g}).$$

We recognise the last term on the right as a source of helicity, which is confined to the buoyant cloud, and the second divergence on the right as a flux which redistributes helicity within the cloud, but quickly falls to zero outside the cloud. The first term on the right, therefore, represents the flux which is exclusively responsible for the far-field dispersion of helicity from the cloud.

Consider a horizontal surface, S_N, that lies well above the buoyant cloud, but within the upward-propagating inertial wave packets. Integrate the helicity equation over all space above that plane to show

$$\frac{d}{dt}\int \mathbf{u} \cdot \boldsymbol{\omega}\,dV = -2\Omega\int_{S_N} (u_z^2 + c_z\omega_z)\,dS.$$

If, instead, the surface were to lie below the cloud (call it S_S), and we integrate over all space below that surface, show that we get exactly the same result, but without the negative sign on the right.

Finally show that, if we have statistical homogeneity in horizontal planes, and we ignore axial gradients by comparison with transverse gradients, then $\langle c_z\omega_z\rangle = \langle u_x^2 + u_y^2\rangle$ where the angled brackets represent a horizontal average in the x–y plane. It follows that the net upward flux of helicity through S_N is strictly negative, while the net downward flux through S_S is positive, which is consistent with Figure 14.21 and what we know about monochromatic inertial waves. Thus the spatial segregation of positive and negative helicity above and below the buoyant cloud is guaranteed. (A similar argument applies to the dispersion of magnetostrophic waves from a localised buoyant cloud.) ■

14.8.3 A Speculative Strong-Field Cartoon

One problem with the discussion above is that it ignores the dynamic influence of the mean magnetic field, particularly the transverse field component $\overline{\mathbf{B}}_\perp$. As inertial wave packets generated near the equatorial plane propagate towards the mantle, they encounter an ever-stronger azimuthal and radial magnetic field. This has two immediate consequences. First, the wave packets elongate along the transverse field lines, $\overline{\mathbf{B}}_\perp$, evolving into sheet-like structures, as shown in Figure 14.22. Second, the inertial waves launched near the equator gradually transform into magnetostrophic waves, whose dispersion characteristics are very different to those of inertial waves (see §7.4). The dynamos operating in planetary cores probably belong to the strong-field regime, so any helical-wave cartoon for

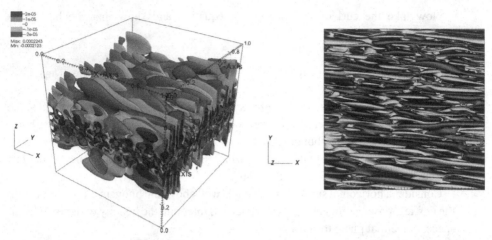

Figure 14.22 Repeat of the calculation shown in Figure 14.21, but including the Lorentz force which elongates the wave packets along the magnetic field lines ($\Omega t = 10$). The magnetic field is in the x direction and the shading indicates the direction of u_z in the wave packets, with light grey for $u_z < 0$ and dark grey for $u_z > 0$. (From Davidson and Ranjan, 2015.)

an α^2-planetary dynamo must incorporate the influence of the transverse field on the wave dispersion.

In some sense, the zero-order cartoon does not change by much, since magnetostrophic waves share the property with inertial waves that negative helicity is transported upward and positive helicity downward (see §7.4). So the spatial segregation of helicity, which is so crucial to an α^2-dynamo, can still occur. The two main consequences of the development of magnetostrophic waves are that (i) anisotropy in the transverse plane makes the calculation of the α-effect more delicate; and (ii) the group velocity of magnetostrophic waves is much slower than that of inertial waves. For example, in Davidson and Ranjan (2015) it is estimated that the wave packets launched from the equatorial plane start out with a group velocity of order 1 m/s (that of inertial waves), but rapidly slow down to a group velocity of ~0.4 mm/s as the waves transform into magnetostrophic waves. It is also estimated that the waves dissipate before reaching the mantle and so do not reflect, as required by an α^2-dynamo.

14.9 Dynamics of the Large Scales: the Taylor Constraint

We now turn from small-scale dynamics (the α-effect) to large-scale dynamics, and in particular to an angular momentum constraint which imposes restrictions on dynamo action. This is called the Taylor constraint, after Taylor (1963), and arises because viscous stresses and non-linear inertia, $\mathbf{u} \cdot \nabla \mathbf{u}$, are negligible in the core.

Let us consider the angular momentum balance of a cylindrical control volume, V, like the one shown in Figure 14.10. The cylinder has radius R, is coaxial with the rotation axis, and spans the core, being bounded top and bottom by the mantle. Our governing equation of motion is

$$\frac{\partial \mathbf{u}}{\partial t} = 2\mathbf{u} \times \mathbf{\Omega} - \nabla(p/\rho) + \frac{\rho'}{\rho}\mathbf{g} + \rho^{-1}\mathbf{J} \times \mathbf{B}, \tag{14.96}$$

where we have neglected the viscous term because of the smallness of the Ekman number. Now (14.96) yields

$$\frac{\partial}{\partial t}(\mathbf{x} \times \mathbf{u}) = 2\mathbf{x} \times (\mathbf{u} \times \mathbf{\Omega}) + \nabla \times (p\mathbf{x}/\rho) + \rho^{-1}\mathbf{x} \times (\mathbf{J} \times \mathbf{B}), \tag{14.97}$$

the z component of which we integrate over the cylindrical control volume V. The pressure term drops out because it cannot exert an axial torque on the surface of V. The Coriolis term also integrates to zero because

$$2\int_V \left(\mathbf{x} \times (\mathbf{u} \times \mathbf{\Omega})\right)_z dV = -\Omega\int_V \nabla \cdot (r^2 \mathbf{u})dV = -R^2\Omega\oint_S \mathbf{u} \cdot d\mathbf{S} = 0, \tag{14.98}$$

and so our axial angular momentum balance reduces to

$$\frac{d}{dt}\int_V \rho r u_\theta dV = \int_V r(\mathbf{J} \times \mathbf{B})_\theta dV. \tag{14.99}$$

Now it is natural to look for the possibility of a steady dynamo, in which case (14.99) demands that, in terms of Maxwell stresses acting on the cylindrical surface of V,

$$\int_V r(\mathbf{J} \times \mathbf{B})_\theta dV = R\int_{r=R}(B_r B_\theta/\mu)dS = 0. \tag{14.100}$$

This is known as *Taylor's constraint*, and fields that satisfy (14.100) are known as a *Taylor state*. The physical origin of (14.100) is clear: since we are obliged to neglect non-linear inertial and viscous forces in the core, there is nothing left to balance the axial Lorentz torque acting on V. So if B_r has the form of a simple dipole field then the Taylor constraint requires that B_θ oscillates with z in such a way as to satisfy (14.100) through the cancelation of regions of positive and negative magnetic torque. Note that (14.100) must be satisfied for all values of R within the core, and so this constraint may be applied to a cylindrical annulus or thickness δ_r that spans the core from north to south.

Of course, it is improbable that the instantaneous magnetic field within the interior of a planet, whose velocity field is inevitably chaotic, will exactly satisfy this constraint at any one instant, though presumably it remains close (in some statistical sense) to a Taylor state. The key question, therefore, is what natural mechanism allows the core flow, and its associated magnetic field, to relax towards a Taylor state, albeit subject to stochastic fluctuations. Here it is important to note that the geostrophic component of azimuthal motion, $\bar{u}_\theta(r)$, can alter the distribution of B_θ through radial gradients in $\bar{u}_\theta/r$, i.e. through the Ω-effect. So we have the possibility that, if there is an excess magnetic torque acting on any one cylindrical annulus, then $\bar{u}_\theta(r)$ increases as a result of that torque, and in such a way that $\bar{u}_\theta/r$ sweeps out (from B_r) an additional contribution to B_θ that reduces the net Lorentz torque acting on that annulus. In short, we have a natural mechanism for pushing the magnetic field towards a Taylor state.

It practice it seems likely that the core fluctuates about a Taylor state, with the various fluid annuli executing damped torsional oscillations. To picture how these oscillations arise it is important to first note that the situation is not as constrained as it might at first seem. The net torque arising from a closed system of currents interacting with its self field is necessarily zero. So, in a global sense, the Taylor constraint is automatically satisfied, and the problem is more one of a local torque imbalance between adjacent annuli. That is to say, if there is an excess magnetic torque acting on one cylindrical annulus, there will be equal and opposite torques on adjacent annuli. So it seems likely that torsional oscillations develop between adjacent annuli, with the radial magnetic field acting as a magnetic spring, linking the annuli. Such oscillations are governed by (14.99) and damped by Ohmic dissipation. So we might envisage the core being constantly subject to torsional oscillations which are both maintained and dissipated by the magnetic field. In this way the core can constantly fluctuate around a Taylor state without having to satisfy (14.100) exactly at any one instant.

14.10 Laboratory Dynamo Experiments

The last two decades have witnessed not only the rise of numerical dynamo experiments, but also laboratory dynamo experiments. Perhaps the first electrically homogeneous laboratory dynamo experiment was that of Lowes and Wilkinson (1963), though it was much more mechanical than fluid. It consisted of two solid cylindrical rotors constructed from a ferromagnetic iron alloy and imbedded within a solid block of the same material. The cylinders had a diameter of ~7 cm and spun around two mutually perpendicular axes, with the gap between the rotors and the surrounding metal block filled with a thin layer of mercury. When spun at several hundred rpm, corresponding to a magnetic Reynolds number of around 200, the experiment produced spontaneous dynamo action, generating a magnetic field of around 10 Gauss. While mildly suggestive of a geophysical dynamo, we had to wait a further 36 years before dynamo action in a conducting fluid was demonstrated in the laboratory. This long awaited breakthrough came at the turn of the century, and like the proverbial London bus, the much anticipated arrival came not once but twice, in quick succession.

14.10.1 Two Classic Experiments

In early 2000 the German newspapers announced that there had been a closely fought race to produce the first liquid-metal dynamo experiment. The race had been between German engineers at the nuclear centre in Karlsruhe and Latvian physicists in Riga, both groups having reported successful dynamos in late 1999. In both experiments liquid sodium (around 2 m³) was pumped in large volumes through a network of (non-magnetic) stainless steel channels which incorporated guide vanes to induce swirl in certain regions. Admittedly, the mean flow in both experiments was constrained to follow highly optimised paths whose efficacy had been carefully checked ahead of time against kinematic dynamo calculations. Nevertheless, the ability to realise spontaneous dynamo action in a fluid was a major accomplishment.

The choice of liquid sodium for these (and many subsequent) dynamo experiments is natural. There are three liquid metals commonly used in MHD experiments because of their low melting point: mercury, gallium and sodium. However mercury is extremely heavy and has only a modest electrical conductivity, while gallium is expensive and so not suitable for particularly large experiments. Sodium, on the other hand, is light ($\sim 900 \text{kg/m}^3$), melts at 98°C, and has a high electrical conductivity ($\lambda \approx 0.08 \text{ m}^2/\text{s}$). Of course there is a significant fire hazard in the event of a sodium leak, but MHD experimentalists have long since learned to live with this ever-present threat.

The Karlsruhe experiment is shown in Figure 14.23. It consisted of 52 stainless steel ducts of circular cross-section, 0.98 m long and 0.21 m in diameter, and arranged into a compact cylindrical assembly of overall diameter 1.85 m. In the outer region of each duct swirl was induced by guide vanes, and so the motion there was helical. The flow in adjacent channels alternated between upward and downward motion, though the guide vanes were arranged to ensure that the helicity was uniformly of one sign across all of the channels. The wall thickness of the ducts was 1 mm and the effective diffusivity of the combined sodium and stainless steel walls was around $\lambda \approx 0.1 \text{ m}^2/\text{s}$, the steel being less conducting than the sodium.

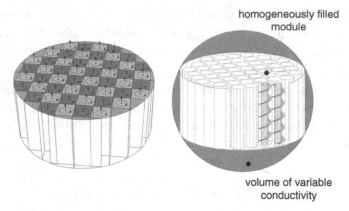

homogeneously filled
module

volume of variable
conductivity

Figure 14.23 The Karlsruhe dynamo experiment. (Courtesy of Robert Stieglitz.)

The channels were interconnected at the top and bottom of the assembly and three pumps were used to drive the flow, each capable of delivering a flow rate of $0.04\ \mathrm{m}^3/\mathrm{s}$. The temperature of the sodium was carefully controlled to be close to 120°C and steady dynamo action was realised when the magnetic Reynolds number based on the radius of a single channel reached a value of order unity. The resulting magnetic field saturated at around 200–300 Gauss and was maintained for over an hour, taking the form of a dipole whose axis was perpendicular to that of the ducts.

The Riga experiment, by contrast, was a modification of the Ponomarenko dynamo (see §14.5.2). It consisted of three concentric cylindrical ducts, each 3 m long, as shown in Figure 14.24. In the central duct the flow spiralled downward, mimicking the Ponomarenko configuration, whereas the flow returned without swirl in the intermediate annulus. The outer annulus was filled with quiescent sodium whose purpose was to increase the magnetic diffusion time of the overall assembly. The temperature of the sodium varied between experiments, from 160 to 300°C, and a monumental flow rate of around $0.7\ \mathrm{m}^3/\mathrm{s}$ was maintained throughout. An oscillatory dynamo of several hundred Gauss was realised when the magnetic Reynolds number, based on the diameter of the inner cylinder, exceeded a value of around 40. (In theory, the critical magnetic Reynolds number for the Ponomarenko dynamo is around 35, as discussed in §14.5.2.)

These two experiments were triumphs of engineering, and had the immense psychological benefit of removing any doubt that dynamo action could be realised in the laboratory. Both are reviewed in Gailitis et al. (2002). Their weakness, of course, is that the fluid in both cases was constrained to follow a prescribed path which is more or less known in advance to produce a kinematic dynamo. The obvious next step was to remove some of this constraint and seek dynamo action within an extended body of liquid metal which is free from internal walls or

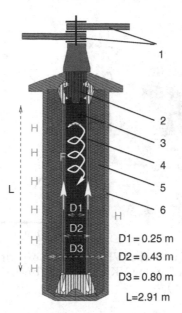

D1 = 0.25 m
D2 = 0.43 m
D3 = 0.80 m
L = 2.91 m

Figure 14.24 The Riga dynamo experiment. (Courtesy of Gunter Gerbeth.)

guide vanes. However, this next step proved to be particularly demanding, and has met with only limited success.

14.10.2 More Recent Experiments

The successes in Karlsruhe and Riga inspired a new generation of less constrained laboratory dynamo experiments. None of these can pretend to mimic a planetary dynamo, as one cannot squeeze the core of a planet into a laboratory without making major concessions over key force balances and controlling dimensionless groups. Nevertheless, they constitute interesting experiments in their own right. Of these second-generation experiments only one has so far generated a magnetic field, and even then this was achieved only after the incorporation of soft iron rotors. In the interests of brevity, we shall restrict attention below to just two of these experiments. One is French, and was originally devised by Stephan Fauve in Lyon and then executed in the sodium facility at Cadarache by researchers from ENS-Paris, ENS-Lyon and CEA-Saclay. The other is American, and is based in the Physics Department at Maryland. Both use sodium as the working fluid.

The 'French washing-machine', or von Kármán Sodium (VKS) experiment, to give it its more technical name, is shown in Figure 14.25(a). Housed in the sodium facility at Cadarache, it consists of a 0.5 m-long cylindrical vessel with axially aligned, counter-rotating impellors at either end, rather like a chemical engineer's

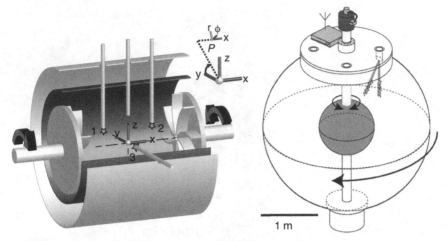

Figure 14.25 (a) The Cadarache (VKS) dynamo experiment. (b) The Maryland experiment. (Courtesy of (a) the VKS team and (b) Dan Lathrop, from the PhD thesis of D. Zimmerman.)

mixing vessel. The impellors drive a turbulent swirling flow whose mean velocity distribution consists of (i) a swirling flow of opposite signs either side of the mid-plane of the cylinder and (ii) a poloidal flow that is radially outward at the surfaces of the discs and radially inward near the mid-plane of the cylinder. The mean flow was intended to drive an Ω–effect through the left-right differential rotation, as well as induce field line stretching at the hyperbolic point in the poloidal flow at the centre of the cylinder. Although initial trials proved unsuccessful, magnetic field generation was realised once ferromagnetic blades were attached to the impellors. Of course, a weakness of the experiment as now configured lies in the need for ferromagnetic material, which renders the configuration distinctly non-astrophysical. Nevertheless, this constitutes the first example of a laboratory dynamo in which the fluid is free to follow its own path, which is a significant achievement.

The Maryland configuration resembles a similar, though much smaller, experiment in Grenoble. It consists of concentric spherical surfaces which may be rotated independently (so-called 'spherical Couette flow'), as shown in Figure 14.25(b). Once again the fluid is sodium and the vessel walls constructed from stainless steel. The inner and outer diameters are 1 m and 3 m, respectively, and the inner sphere may rotate at up to 15Hz and the outer one up to 4Hz. Geometrically, the configuration is clearly more planet-like than the VKS experiment, though the driving force for motion is differential rotation rather than the buoyancy that drives flow in the core of a planet. Crucially, the Rossby number in the experiment is greater than unity, and so much larger than in the earth's core, where Ro $\sim 10^{-6}$. If a low value of

Ro turns out to be central to dynamo action in planetary cores, which it may well do, then such an experiment will not be able to mimic a planetary dynamo, though of course it may achieve dynamo action by other means. In any event, at the time of writing, the experiment has yet to yield a dynamo. Both the VKS and Maryland experiments are reviewed in some detail in Roberts and King (2013).

14.11 Scaling Laws for Planetary Dynamos (Reprise)

We close this chapter by returning to an important question first raised in §14.3.2: 'How does the dipole field strength of a planet scale with the planetary rotation rate, core size and the rate of working of the buoyancy forces?'

It should be noted from the outset that this is a difficult and controversial topic, made all the more delicate by the fact that much of the evidence for scaling laws comes from the highly viscous and weakly forced numerical dynamos, whereas the dominant force balances and resulting scaling laws are likely to be very different in the planets. As a consequence, there is little agreement as to which of the various published scaling laws actually apply to the planets. Many of the scaling laws which have been proposed are explored in the exercises at the end of this chapter. Here we merely note the main results.

The various scaling laws all start with an assumption about the dominant force balance in the core. Consider a Boussinesq fluid in which β is the expansion coefficient, T' the temperature perturbation, and $P = -\beta T' \mathbf{u} \cdot \mathbf{g}$ the rate of working of the thermal buoyancy force per unit mass. P can be expressed in terms of the time-averaged convective heat flux per unit area, $\dot{q} = \rho c_p \overline{T' \mathbf{u}}$, according to $P = (g\beta/\rho c_p)\dot{q}$, and since rough estimates of $\dot{q}$ are available for the earth ($\dot{q} \approx 0.02 \mathrm{Wm}^{-2}$), Jupiter ($\dot{q} \approx 5.4 \mathrm{Wm}^{-2}$) and Saturn ($\dot{q} \sim 2.0 \mathrm{Wm}^{-2}$), but not Mercury, we can estimate P for three of the planetary dynamos.

As discussed in §14.3.2, the bulk of the Coriolis force is irrotational and balanced by a pressure gradient. Nevertheless, it seems plausible that the curl of the Coriolis, buoyancy and Lorentz forces are all of similar magnitudes in a saturated dynamo. Moreover, most models assume that there are only two characteristic length scales in this force balance, $\ell_\perp$ and $\ell_{//} \sim R_C$, the scales perpendicular and parallel to the rotation axis. This leads to the force balance (see Exercise 14.1)

$$\frac{B^2/\rho\mu}{\ell_\perp^2} \sim \frac{P}{u\ell_\perp} \sim \frac{\Omega u}{R_C}. \tag{14.101}$$

The scaling laws which result from this balance are established in Exercise 14.2, and their consequences discussed in Exercises 14.3→14.5. These laws are usually expressed in terms of the dimensionless groups

$$\Pi_B = \frac{B_{rms}/\sqrt{\rho\mu}}{\Omega R_C}, \qquad \text{Ro} = \frac{u}{\Omega R_C}, \qquad \Pi_p = \frac{P}{\Omega^3 R_C^2}, \qquad (14.102)$$

where B_{rms} is a typical core field strength, often estimated to be around $B_{rms} \sim 10\overline{B}_z$. Several versions of these scaling laws exist and they differ in the way that $\ell_\perp$ is estimated. For example, scaling laws aimed at the viscous numerical dynamos, as distinct from planets, usually specify $\ell_\perp/R_C \sim \text{Ek}^{1/3}$ on the grounds that the viscous forces are of order unity. Other estimates of $\ell_\perp$ are commonly used. One particular scaling, $\Pi_B \sim \Pi_P^{1/3}$, has attracted considerable attention as it asserts that the magnetic energy is independent of rotation rate. Indeed, in Christensen (2011) it is argued that the scaling $\Pi_B \sim \Pi_P^{1/3}$ applies to both the planets and fully convective stars.

It seems likely, however, that the columnar vortices in the earth's core are more sheet-like than tubular, and so require two perpendicular length scales to characterise their cross-section. The resulting force balance is then more subtle, as discussed in Exercise 14.6. The net effect is that (14.101) is replaced by

$$\frac{B^2}{\rho\mu\ell_\perp^2}\frac{\lambda}{uR_C} \sim \frac{P}{u\ell_\perp} \sim \frac{\Omega u}{R_C}, \qquad (14.103)$$

provided that the low-R_m approximation may be applied on the scale of $\ell_\perp$ (which limits us to planets and excludes stars). The resulting scaling laws are then

$$\frac{B_{rms}/\sqrt{\rho\mu}}{\Omega R_C} \sim \sqrt{\frac{P}{\Omega^2\lambda}}\sqrt{\frac{\ell_\perp}{R_C}}, \quad \frac{u^2}{V_P^2} \sim \frac{V_P}{\Omega\ell_\perp}, \qquad (14.104)$$

where the velocity V_P is characteristic of the forcing and is defined as $V_P = (PR_C)^{1/3}$. The key question now is how to estimate $\ell_\perp$, and here we must distinguish between the viscous numerical dynamos and planets. For the numerical dynamos we might take $\ell_\perp/R_C \sim \text{Ek}^{1/3}$, and this converts (14.104) into

$$\Pi_B \sim \Pi_P^{7/18}\text{Pr}_m^{1/6}(V_P R_C/\lambda)^{1/3}, \qquad (14.105)$$

$$\text{Ro} \sim \Pi_P^{4/9}\text{Pr}_m^{-1/6}(V_P R_C/\lambda)^{1/6}. \qquad (14.106)$$

The quantity $(V_P R_C/\lambda)^{1/3}$ is almost always of order unity in the numerical dynamos, lying in the range 5→15 (Davidson, 2014). For such limited data sets, these viscous scaling laws reduce to the deceptively simple results $\Pi_B \sim \Pi_P^{7/18}\text{Pr}_m^{1/6}$ and $\text{Ro} \sim \Pi_P^{4/9}\text{Pr}_m^{-1/6}$. These are surprisingly close to the scaling laws

Table 14.5 *Estimates of* $\Pi_B = B_{rms}/\Omega R_C\sqrt{\rho\mu}$, $\sqrt{P/\Omega^2\lambda}\sqrt{V_P/\Omega R_C}$, Λ *and* $V_P^4/\Omega^2\lambda^2$, *with* Π_B *and* Λ *based on the mean axial field in the planetary cores and the assumption that* $B_{rms} \sim 10\overline{B}_z$. *(We use the estimates of* $\lambda \sim 0.7\ m^2/s$, $\rho \sim 10^4 kg/m^3$ *for the earth, and* $\lambda \sim 30\ m^2/s$, $\rho \sim 10^3 kg/m^3$ *for the gas giants.)*

Planet	$\Pi_B = \frac{B_{rms}/\sqrt{\rho\mu}}{\Omega R_C}$ Observed	$\Pi_B \sim \sqrt{\frac{P}{\Omega^2\lambda}}\sqrt{\frac{V_P}{\Omega R_C}}$ Prediction 14.107	$\Pi_B \sim \sqrt{\frac{\lambda}{\Omega R_C^2}}$ Prediction $\Lambda = 1$	$\Lambda = \frac{\sigma B_{rms}^2}{\rho\Omega}$ Observed	$\Lambda \sim \frac{V_P^4}{\Omega^2\lambda^2}$ Prediction 14.107
Earth	13×10^{-5}	8.1×10^{-5}	2.8×10^{-5}	22	8.3
Jupiter	5.2×10^{-5}	4.9×10^{-5}	0.7×10^{-5}	49	43
Saturn	2.3×10^{-5}	3.4×10^{-5}	1.5×10^{-5}	2.3	5.4

$\Pi_B \sim \Pi_P^{0.31}Pr_m^{0.16}$ and $Ro \sim \Pi_P^{0.44}Pr_m^{-0.13}$, which have been derived empirically by several authors from a suit of numerical dynamos (see Exercise 14.7).

Of course, viscosity is unlikely to be important in a planetary core, and so for planets $\ell_\perp$ must be determined in a different way. Several options are discussed in Exercise 14.8. The simplest approach is to assume that $u\ell_\perp/\lambda \sim 1$, which, combined with (14.104), leads back to the classical criterion $\Lambda \sim 1$. Another approach is based on the idea that the columnar vortices in planetary cores are created by inertial wave packets (see §14.8.2). So an alternative hypothesis is that $\ell_\perp$ is the minimum length scale at which inertial waves can propagate to form columnar vortices, which is set by $u/\Omega\ell_\perp \approx 0.2$. This, in conjunction with (14.104), yields $u \sim V_P$ and hence $\ell_\perp/R_C \sim V_P/\Omega R_C$, from which we obtain

$$\frac{B_{rms}/\sqrt{\rho\mu}}{\Omega R_C} \sim \sqrt{\frac{P}{\Omega^2\lambda}}\sqrt{\frac{V_P}{\Omega R_C}}, \quad \Lambda = \frac{\sigma B_{rms}^2}{\rho\Omega} \sim \frac{V_P^4}{\Omega^2\lambda^2}. \quad (14.107)$$

Table 14.5, in which we use the estimate $B_{rms} \sim 10\overline{B}_z$, shows that the speculative laws (14.107) are in good agreement with the observational data for the earth, Jupiter and Saturn, and probably a better fit than the traditional assumption of $\Lambda \sim 1$ and $\Pi_B \sim R_\lambda^{-1/2}$.

Note that (14.104) does not apply to stars. In Exercise 14.10 it is suggested that for fully convective, rotationally dominated stars (14.107) be replaced by $\Pi_B \sim \Pi_P^{1/2}$ and $Ro \sim \Pi_P^{1/3}$.

Exercises

The exercises in this chapter are somewhat different from those in the rest of the book. They take the form of a series of extended exercises with commentary and relate to a variety of tentative scaling laws which have been proposed for planetary dynamos. This remains a controversial topic and so the scaling laws discussed should be regarded as speculative, whose validity remains largely unestablished.

14.1 *The zero-order force balance in the core*

Consider the momentum equation written for a Boussinesq fluid in the earth's core,

$$\frac{\partial \mathbf{u}}{\partial t} = 2\mathbf{u} \times \mathbf{\Omega} - \nabla(p/\rho) - \beta T'\mathbf{g} + \rho^{-1}\mathbf{J} \times \mathbf{B},$$

where $\mathbf{g}$ is the gravitational acceleration, β the expansion coefficient and T' the temperature perturbation. Note that, since the Ekman and Rossby numbers are very small in the core, we are obliged to drop the viscous and non-linear inertial terms. Let $\ell_{//}$ and $\ell_{\perp}$ be characteristic length scales parallel and perpendicular to the rotation axis, with $\ell_{//} \gg \ell_{\perp}$ and $\ell_{//} \sim R_C$. In a quasi-geostrophic flow much of the Coriolis force is irrotational and so balanced by a pressure gradient, with the relatively small rotational part balanced by the buoyancy and Lorentz forces. Show that, on the assumption that the curl of the Coriolis, buoyancy and Lorentz forces are of the same order,

$$\mathbf{\Omega} \cdot \nabla \mathbf{u} \sim \nabla \times (\beta T'\mathbf{g}) \sim \nabla \times (\mathbf{B} \cdot \nabla \mathbf{B}/\rho\mu),$$

we have the force balance

$$\frac{\Omega u}{\ell_{//}} \sim \frac{g\beta T'}{\ell_{\perp}} \sim \frac{B^2}{\rho\mu\ell_{\perp}^2}.$$

It is convenient to rewrite the buoyancy force in terms of the time-averaged rate of production of energy per unit mass, $P = -\beta\overline{T'\mathbf{u}} \cdot \mathbf{g}$. This, in turn, can be expressed in terms of the time-averaged convective heat flux per unit area, $\dot{\mathbf{q}} = \rho c_p\overline{T'\mathbf{u}}$, according to

$$P = \frac{g\beta}{\rho c_p}\dot{q},$$

where $\dot{q}$ is the radial convective heat flux through the core and $g = |\mathbf{g}|$. The terrestrial convective heat flux is, perhaps, 3TWatts (though estimates vary), giving $\dot{q} \approx 0.02 \mathrm{Wm}^{-2}$. Doubling this to allow for the additional rate of working of the compositional convection gives an effective value of P of $P \approx 5.1 \times 10^{-13} \, \mathrm{m}^2 \mathrm{s}^{-3}$. Show that, in terms of P, our tentative force balance becomes (Davidson, 2013)

$$\frac{B^2/\rho\mu}{\ell_\perp^2} \sim \frac{P}{u\ell_\perp} \sim \frac{\Omega u}{R_C}.$$

14.2 *Somewhat simplistic scaling laws for planetary dynamos*
The tentative force balance in Exercise 14.1 involves six parameters: Ω, R_C and P, which we may consider as given features of a planet, and $B/\sqrt{\rho\mu}$, u and $\ell_\perp$, which we may take as dependent parameters. Between them, these six parameters contain the dimensions of length and time. Use the Buckingham Pi theorem to show that four dimensionless groups can be constructed from these parameters:

$$\Pi_P = \frac{P}{\Omega^3 R_C^2}, \qquad \Pi_B = \frac{B/\sqrt{\rho\mu}}{\Omega R_C}, \qquad \mathrm{Ro} = \frac{u}{\Omega R_C}, \qquad \frac{\ell_\perp}{R_C}.$$

Of these, the first dimensionless group may be considered as given, or prescribed, in a particular planet and the other three as functions of Π_P. Note that, using the estimate of P above, we have $\Pi_P \sim 1.1 \times 10^{-13}$ for the earth. Now use the force balance in Exercise 14.1 to confirm the tentative scaling laws

$$\Pi_B = \frac{B/\sqrt{\rho\mu}}{\Omega R_C} \sim \Pi_P^{1/4} \; (\ell_\perp/R_C)^{3/4},$$

$$\mathrm{Ro} = \frac{u}{\Omega R_C} \sim \Pi_P^{1/2} \; (\ell_\perp/R_C)^{-1/2}.$$

14.3 *An independent, if speculative, estimate of the Rossby number for the earth*
Eliminate $\ell_\perp$ from the scaling laws above to show that

$$\mathrm{Ro} \sim \Pi_B^{-2/3} \, \Pi_P^{2/3}.$$

Now use the estimates $P \approx 5.1 \times 10^{-13} \, \mathrm{m}^2 \mathrm{s}^{-3}$ (from Exercise 14.1) and $\Pi_B \approx 1.3 \times 10^{-5}$ (based on a core field strength of 3.7 Gauss) to give

$Ro \sim 4 \times 10^{-6}$. This is close to the estimate given in Table 14.1, based on the westward drift velocity of $u \sim 0.2$ mm/s.

14.4 *An independent, if speculative, estimate of the perpendicular length scale in the earth*
Use the estimates of Π_B and P given above, along with the scaling

$$\Pi_B \sim \Pi_P^{1/4} \ (\ell_\perp / R_C)^{3/4},$$

to obtain the estimate $\ell_\perp \sim 6.5 \times 10^{-3} R_C \sim 22$ km. Note that this is not much larger than the minimum length scale of 4 km given in Table 14.1, set by the condition $u\ell_{min}/\lambda \sim 1$.

14.5 *Comparing the simplistic scaling analysis with viscous numerical dynamos*
To close the system in Exercise 14.2 we need an independent estimate of $\ell_\perp$. In many of the (overly viscous) numerical dynamos this is set by the viscous stresses, which demand $\ell_\perp / R_C \sim Ek^{1/3}$, and hence

$$\Pi_B \sim \Pi_P^{1/4} Ek^{1/4}, \quad Ro \sim \Pi_P^{1/2} Ek^{-1/6}.$$

Of course, this kind of viscous dominance is most unlikely to hold in planets. (The determination of $\ell_\perp$ in planetary dynamos remains an open question.) Show that these viscous scaling laws for numerical dynamos can be rewritten as

$$\Pi_B \sim \Pi_P^{1/3} Pr_m^{1/4} \left(P^{1/9} R_C^{4/9} / \lambda^{1/3} \right)^{-3/4},$$

$$Ro \sim \Pi_P^{4/9} Pr_m^{-1/6} \left(P^{1/9} R_C^{4/9} / \lambda^{1/3} \right)^{1/2}.$$

In the current generation of numerical dynamos the quantity $P^{1/9} R_C^{4/9} / \lambda^{1/3}$ is almost always of order unity, lying in the range $5 \rightarrow 15$ in most numerical simulations. (See the discussion in Davidson, 2014.) For such limited data sets, these viscous scaling laws reduce to the deceptively simpler relationships $\Pi_B \sim \Pi_P^{1/3} Pr_m^{1/4}$ and $Ro \sim \Pi_P^{4/9} Pr_m^{-1/6}$. These are reasonably close to the scaling laws $\Pi_B \sim \Pi_P^{0.31} Pr_m^{0.16}$ and $Ro \sim \Pi_P^{0.44} Pr_m^{-0.13}$, which have been derived empirically by several authors through a statistical analysis of a suit of numerical dynamos. (See, for example, Christensen, 2011.)

The scaling $\Pi_B \sim \Pi_P^{1/3}$ has attracted considerable attention in the literature as it suggests that the strength of the dipole field, or equivalently the magnetic energy, is independent of rotation rate. In Christensen (2011) it is argued that the scaling $\Pi_B \sim \Pi_P^{1/3}$ is not limited to the viscous numerical dynamos, but also applies to both the planets and fully convective M-dwarf stars, the central assertion being that the magnetic energy is determined by the buoyant forcing only, and independent of rotation.

14.6 *A refined scaling analysis, embracing anisotropy in planes normal to the rotation axis*

A weakness of the scaling analysis given in the exercises above is that it is far from clear how the resulting scaling laws can be made to satisfy the kinematic dynamo criterion (14.89). Moreover, because of the radial and azimuthal magnetic field components in the core, it is likely that the columnar eddies are strongly anisotropic in planes normal to the rotation axis, as suggested in Figure 14.22. In particular, energy propagation along the transverse components of the magnetic field lines, either by low-R_m diffusion or by waves, will cause the columnar eddies to have cross-sections which are more sheet-like than tubular, characterised by two perpendicular length scales rather than by one. So let us replace $\ell_\perp$ in the examples above by two transverse length scales, ℓ_B and $\ell_\perp$ ($\ell_B > \ell_\perp$), where ℓ_B measures the characteristic length scale parallel to the local component of magnetic field in the transverse plane, while $\ell_\perp$ is the smaller of the two transverse length scales. Moreover, in order to simplify the analysis, we shall assume that the low-R_m approximation may be applied on the scale of $\ell_\perp$, which limits us to the terrestrial planets and the numerical dynamos, possibly includes the gas giants, but definitely excludes stars. Show that, if **B** is dominated by the large-scale field, then the curl of the Lorentz force now takes the form

$$\nabla \times \nabla \times (\rho^{-1}\mathbf{J} \times \mathbf{B}) \sim \rho^{-1}\mathbf{B} \cdot \nabla(\nabla \times \mathbf{J}) \sim \rho^{-1}\mathbf{B} \cdot \nabla(\sigma\mathbf{B} \cdot \nabla\mathbf{u}),$$

from which

$$\nabla \times (\rho^{-1}\mathbf{J} \times \mathbf{B}) \sim \frac{\sigma B^2 u \ell_\perp}{\rho \ell_B^2}.$$

Now use the *kinematic* criterion for dynamo action (14.89), rewritten as

$$\frac{R_C u}{\lambda} \frac{\ell_\perp u}{\lambda} \sim \left(\frac{\ell_B}{\ell_\perp}\right)^2,$$

to show that

$$\nabla \times \left(\rho^{-1}\mathbf{J} \times \mathbf{B}\right) \sim \frac{B^2}{\rho\mu\ell_\perp^2} \frac{\lambda}{uR_C}.$$

Hence, confirm that the force balance given in Exercise 14.1 must be modified to read

$$\frac{B^2}{\rho\mu\ell_\perp^2} \frac{\lambda}{uR_C} \sim \frac{P}{u\ell_\perp} \sim \frac{\Omega u}{R_C},$$

and so the scaling laws of Exercise 14.2 change to (Davidson, 2014)

$$\Pi_B \sim \Pi_P^{1/2} R_\lambda^{1/2} \left(\ell_\perp/R_C\right)^{1/2},$$

$$\mathrm{Ro} \sim \Pi_P^{1/2} (\ell_\perp/R_C)^{-1/2},$$

where $R_\lambda = \Omega R_C^2/\lambda$. Show that these scalings dictate that the Elsasser number, Λ, is related to the small-scale magnetic Reynolds number, $u\ell_\perp/\lambda$, and intermediate length, ℓ_B, by

$$\Lambda = \frac{\sigma B^2}{\rho\Omega} \sim \left(\frac{u\ell_\perp}{\lambda}\right)^2 \sim \frac{\ell_B^2}{R_C\ell_\perp}.$$

Finally show that the refined scaling laws may be rewritten in terms of the magnetic and kinetic energy density as

$$\frac{B^2/\rho\mu}{V_P^2} \sim \frac{V_P\ell_\perp}{\lambda}, \quad \frac{u^2}{V_P^2} \sim \frac{V_P}{\Omega\ell_\perp},$$

where the velocity scale V_P is characteristic of the forcing and is defined as $V_P = (PR_C)^{1/3}$. (Using the estimates of P given in Exercise 14.8, rough estimates of V_P for the earth, Jupiter and Saturn are 1 cm s^{-1}, 20 cm s^{-1} and 10 cm s^{-1}, respectively.)

We may now reinterpret the assertion by Christensen (2011) that the magnetic energy is independent of rotation rate as a statement that $\ell_\perp$ is independent of Ω. Moreover, the ratio of magnetic energy to kinetic energy is evidently predicted to be

$$\frac{B^2/\rho\mu}{u^2} \sim \frac{\Omega\ell_\perp^2}{\lambda} = \frac{u\ell_\perp}{\lambda}\frac{\Omega\ell_\perp}{u}.$$

Given that we require $u/\Omega\ell_\perp < 0.2$ in order to maintain a dipole field (see §14.7), this is consistent with the observation that the magnetic energy is larger than the kinetic energy.

14.7 *Comparing the refined, anisotropic scaling analysis with viscous numerical dynamos*

It is still unclear what fixes the unknown length scale $\ell_\perp$ in planetary cores. However, in the overly viscous numerical dynamos (but not in planets), it is likely that $\ell_\perp$ is set by viscous forces, which demands $\ell_\perp/R_C \sim \text{Ek}^{1/3}$. Show that, in such cases, the scaling laws of Exercise 14.6 become

$$\Pi_B \sim \Pi_P^{1/2} R_\lambda^{1/2}\, \text{Ek}^{1/6}, \quad \text{Ro} \sim \Pi_P^{1/2} \text{Ek}^{-1/6},$$

or equivalently

$$\Pi_B \sim \Pi_P^{7/18} \text{Pr}_m^{1/6}(V_P R_C/\lambda)^{1/3},$$

$$\text{Ro} \sim \Pi_P^{4/9} \text{Pr}_m^{-1/6}(V_P R_C/\lambda)^{1/6}.$$

As noted in Exercise 14.5, the quantity $(V_P R_C/\lambda)^{1/3}$ is almost always of order unity in the numerical dynamos, usually lying in the range $5 \rightarrow 15$. For such limited data sets, these viscous scaling laws reduce to $\Pi_B \sim \Pi_P^{7/18} \text{Pr}_m^{1/6}$ and $\text{Ro} \sim \Pi_P^{4/9} \text{Pr}_m^{-1/6}$, which is surprisingly close to the scalings $\Pi_B \sim \Pi_P^{0.31} \text{Pr}_m^{0.16}$ and $\text{Ro} \sim \Pi_P^{0.44} \text{Pr}_m^{-0.13}$ which have been derived empirically from a suit of numerical dynamos, as discussed in Exercise 14.5.

14.8 *Searching for a dynamo scaling law for the planets*

Perhaps the most striking feature of Tables 14.2 and 14.3 is that Π_B has a similar numerical value for both the terrestrial planets and the gas giants. If we base Π_B on the *rms* field strength in the core, B_{rms}, and guess $B_{rms} \sim 10\overline{B}_z$ (which corresponds to ~ 40 Gauss in the case of the earth), then Table 14.3 suggests $\Pi_B \sim 10^{-4}$. Of course, the question is then, 'Why should Π_B be similar for all the planets and what sets its value?' Perhaps a comparison of the earth with the other planets offers a clue here. Show that our refined prediction for the magnetic field strength can be rewritten as (Davidson, 2014),

$$\frac{B_{rms}/\sqrt{\rho\mu}}{\Omega R_C} \sim \sqrt{\frac{P}{\Omega^2\lambda}}\sqrt{\frac{\ell_\perp}{R_C}}.$$

Also, use the data in Exercise 14.2 to show that, for the earth, $P/\Omega^2\lambda \sim 1.4 \times 10^{-4}$. For Jupiter and Saturn, on the other hand, measurements of luminosity suggest $\dot{q} \approx 5.4\text{Wm}^{-2}$ and $\dot{q} \sim 2.0\text{Wm}^{-2}$, respectively. This gives $P \approx 1.2 \times 10^{-10}\text{m}^2\text{s}^{-3}$ and $P/\Omega^2\lambda \sim 1.3 \times 10^{-4}$ for Jupiter, and $P \approx 4.3 \times 10^{-11}\text{m}^2\text{s}^{-3}$ and $P/\Omega^2\lambda \sim 0.52 \times 10^{-4}$ for Saturn. (The convective heat flux in Mercury's core is unknown.) These various estimates of $P/\Omega^2\lambda$ are surprisingly similar: $\sqrt{P/\Omega^2\lambda} \sim 0.007 \rightarrow 0.012$. So, if we assume that $\sqrt{\ell_\perp/R_C}$ does not vary by too much from planet to planet, one rather tentative explanation for the relative uniformity of Π_B in Table 14.3 is that it just so happens that the values of $\sqrt{P/\Omega^2\lambda}$ do not differ greatly across the planets.

It remains unclear what fixes the unknown length scale $\ell_\perp$ in planetary cores. One guess for the terrestrial planets might be to equate $\ell_\perp$ to the minimum length scale in the core, which is probably set by $u\ell_{min}/\lambda \sim 1$. Show that the scaling laws of Exercise 14.6 then reduce to

$$\Lambda = \frac{\sigma B^2}{\rho\Omega} \sim 1, \quad \text{Ro} \sim \Pi_P R_\lambda, \quad \Pi_B \sim R_\lambda^{-1/2}, \quad \ell_B \sim \sqrt{R_C\ell_{min}},$$

the first of which is, of course, very familiar. However, the hypothesis that $u\ell_\perp/\lambda \sim 1$ lacks much direct support, other than the observation that is $\Lambda \sim 1$ for the earth and Saturn.

An alternative guess might be to assume $V_P\ell_\perp/\lambda \sim 1$. Show that the scaling laws of Exercise 14.6 then take us back to Christensen's (2011) scaling $\Pi_B \sim \Pi_P^{1/3}$, which does has some empirical support. Of course, one is then obliged to ask, 'Why $V_P\ell_\perp/\lambda \sim 1$?'

A third approach, which is based on the idea that the columnar vortices in planetary cores are created by inertial waves (see §14.8.2), is the hypothesis that $\ell_\perp$ is the minimum length scale at which inertial waves can propagate to form columnar vortices, i.e. $u/\Omega\ell_\perp \approx 0.2$ (Bin Baqui and Davidson, 2015). Show that the refined scaling of Exercise 14.6 then yields $u \sim V_P$, and hence $\ell_\perp/R_C \sim V_p/\Omega R_C$, from which we predict

$$\Pi_B = \frac{B_{rms}/\sqrt{\rho\mu}}{\Omega R_C} \sim \sqrt{\frac{P}{\Omega^2\lambda}}\sqrt{\frac{V_P}{\Omega R_C}}, \quad \Lambda = \frac{\sigma B_{rms}^2}{\rho\Omega} \sim \frac{V_P^4}{\Omega^2\lambda^2}.$$

Table 14.5 shows that these predictions are a reasonable fit to the observational data for the earth, Jupiter and Saturn.

14.9 *Geomagnetic reversals*

Consider the integral equation (14.35) and associated α-effect estimate (14.51):

$$\frac{d}{dt}\int_{V_C} (R_C^2 - \mathbf{x}^2)\overline{B}_z dV = 2\int_{V_C} r(\mathbf{v} \times \mathbf{b})_\theta dV - 6\lambda \int_{V_C} \overline{B}_z dV, \quad \langle \mathbf{v} \times \mathbf{b} \rangle = -\frac{\delta^2}{2\lambda}\langle h \rangle\overline{\mathbf{B}}.$$

The simple idea that reversals are passively triggered by a temporary decline in convection is inconsistent with this integral equation, as the dipole field would then reverse on the diffusive time scale of $R_C^2/(\lambda\pi^2) \sim 5 \times 10^4$ yrs. By contrast, reversals occur on the fast time scale of $10^3 \rightarrow 10^4$ yrs, which is comparable with the time taken for the dipole field to diffuse out of the solid inner core, $R_i^2/(\lambda\pi^2)$. So a field reversal must be actively driven by a change in convection pattern. Suppose the core convection were to change so that the buoyant plumes drifting out to the mantle became stronger near the poles than at the equator. The helical waves triggered by these plumes would then be directed primarily from the mantle to the equatorial plane, reversing the sign of helicity outside the tangent cylinder. Show that the α^2 dynamo of Figure 14.18 would then become an anti-dynamo, with the α-effect integral in the equation above acting to suppress B_z. Show also that this reversed α-effect will create a multicellular east-west field, which in turn creates a multipolar poloidal field.

14.10 *Searching for a dynamo scaling law for fully convective stars*

Exercises 14.6$\rightarrow$14.8 assume that the low-R_m approximation may be applied on the scale of $\ell_\perp$, which limits us to the planets and excludes stars. For fully convective, rotationally dominated stars it is, perhaps, better to return to the scaling laws of Exercise 14.2,

$$\Pi_B \sim \Pi_P^{1/4} \ (\ell_\perp/R_C)^{3/4}, \quad \text{Ro} \sim \Pi_P^{1/2} \ (\ell_\perp/R_C)^{-1/2}.$$

Combine these with the hypothesis of Exercise 14.8, that the width of the columnar eddies is set by the condition $u/\Omega\ell_\perp \approx 0.2$, to give the *non-equipartition* scaling laws

$$u \sim V_P = (PR_C)^{1/3}, \quad \text{Ro} \sim \Pi_P^{1/3}, \quad \Pi_B \sim \Pi_P^{1/2}.$$

Pegasi-V374 has radius $R \approx 2 \times 10^5$ km and its luminosity gives $\dot{q} \approx 4\text{MWm}^{-2}$, from which P $\sim 10^{-4}\text{m}^2\text{s}^{-3}$. Estimate Π_B using $\Pi_B \sim \Pi_P^{1/2}$ and the equipartition scaling $\Pi_B \sim \Pi_P^{1/3}$. Compare the two estimates with the measured value in Table 14.3.

14.11 *Dispersion of inertial waves from a buoyant blob with* **g** *and* **Ω** *aligned*
The dispersion pattern from a buoyant blob is different from that shown in Figure 14.20 when **g** and **Ω** are aligned. Repeat the calculation which leads up to (14.95), but for the case where **g** is antiparallel to **Ω**, as would be the case inside the tangent cylinder and in the north. Consider the case where the blob is buoyant and has a Gaussian profile. Use the new jump conditions, along with the fact that upward- (downward-) propagating wave packets have negative (positive) helicity, to show that the dispersion pattern for inertial waves is quite different to that on the equatorial plane, i.e. that shown in Figure 14.20. In particular, show that the dispersion pattern now consists of an axially propagating wave packet with anticyclonic rotation and positive axial velocity above the centre of the blob, matched to a downward-propagating cyclonic wave packet below the centre of the blob in which the fluid velocity is also upward. (Recall that

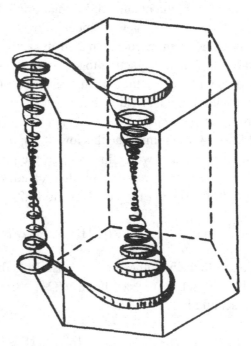

Figure 14.26 Convection pattern for rotating Rayleigh–Bénard convection when the planform is hexagonal. (From Veronis, 1959.)

the fluid and group velocities need not be in the same direction, and in this case the two axial velocities, group and fluid, are aligned above the blob but anti-parallel below.) Show also that these central wave packets are surrounded by an annular region, located above and below the outer edge of the blob, in which the upward-propagating wave packet now has cyclonic rotation but a downward fluid velocity, and the downward-propagating wave packet possess anticyclonic rotation and a downward fluid velocity.

Perhaps it is not surprising that this arrangement of helical vortices is highly reminiscent of that seen in rotating, Rayleigh–Bénard convection (Figure 14.26). Of course, the main difference is that the vortices do not propagate as wave packets in the Rayleigh–Bénard problem, but none-theless snapshots of the two flows are rather similar.

15

Stellar Magnetism

> One lot cogitates on the way of religion,
> Another ponders on the path of mystical certainty;
> But I fear one day the cry will go up,
> 'Oh you fools, neither this nor that is the way!'
> *Omar Khayyam*

Perhaps sunspots were the earliest manifestation of solar MHD to capture the imagination of scientists and philosophers. The capricious nature of sunspots had been a source of speculation since the first observations in ancient China, but considered debate in the West probably dates back to the early seventeenth century and to the development of the telescope by Galileo. Indeed it was Galileo's *Letters on Solar Spots*, published in Rome in 1613, which precipitated the clash between Galileo and the church, a clash that culminated in Galileo's arrest in 1633. Thus the battle between science and religion began, a skirmish which had still not abated by 1860 when Huxley and Bishop Wilberforce debated Darwin's *Origin of the Species*.

Records of sunspot appearances have been kept more or less continuously since Galileo's time. By 1843 it was realised that the appearance of spots followed an eleven-year cycle (although there was a curious dearth of sunspots during the reign of the Roi Soleil in France, known as the Maunder minimum). The reason for the eleven-year cycle remained a mystery for some time, but it was clear by the middle of the nineteenth century that there was an electromagnetic aspect to the problem. As Maxwell noted in 1873 when discussing terrestrial magnetic storms, 'It has been found that there is an epoch of maximum disturbance every eleven years, and that this coincides with the epoch of maximum number of sunspots in the sun.' However, it was not until the development of MHD that scientists began to understand some of the observations.

In this chapter we provide a brief introduction to stellar magnetism, focussed on our own star. We begin in §15.1 with the sun itself, discussing both its interior (the solar dynamo and sunspot formation) and its atmosphere (coronal flux loops and

Table 15.1 *Physical characteristics of the sun.*

Age	~4.6 × 10^9 yrs
Radius, $R_\odot$	6.96 × 10^5 km
Mass, $M_\odot$	1.99 × 10^{30}kg
Surface temperature	~5800 K
Equatorial surface rotation period (synodic)	26.3 days

solar flares). The sun's corona eventually evolves into the solar wind which then spirals outward through our solar system. Satellite measurements of fluctuations in the solar wind have recently become a topic of considerable interest, in large part because they provide a test bed for various theories of MHD turbulence, and so we discuss the solar wind in §15.2. We close in §15.3 with an introduction to accretion discs, those vast, disc-like spirals of gas which form around young and dying stars. Of particular current interest are the protoplanetary discs that form around young stars and provide the breeding ground for planets. The discussion throughout is largely qualitative, intended as a stepping stone to further study.

There are many excellent textbooks and monographs available to those who wish to pursue these topics in more depth. For example, Pringle and King (2007) provide an introduction to astrophysical fluid mechanics, while Mestel (1999) gives a definitive account of stellar magnetism. Regards the sun, Cravens (1997) and Priest (2014) offer detailed overviews, with strong sections on the solar wind and solar flares, respectively, while Hughes, Rosner and Weiss (2007) provide a thoughtful discussion of the solar convection zone and the solar dynamo. As for accretion discs, their elementary properties are discussed in the introductory text by Carroll and Ostlie (1996), while the standard disc model is described in detail in Frank, King and Raine (2002).

15.1 The Dynamic Sun

15.1.1 The Sun's Interior and Atmosphere

The sun is composed mostly of hydrogen and helium, being 71 per cent hydrogen and 27 per cent helium by mass fraction. It has a radius of $R_\odot = 6.96 \times 10^5$ km, a mass of $M_\odot = 1.99 \times 10^{30}$ kg and a surface temperature of $T_\odot \approx 5800$ K (see Table 15.1). With a (synodic) equatorial rotation period of ~26 days, the sun is a rather modest rotator. The interior of the sun is therefore not rotationally constrained to the same extent as the earth's core, with Rossby numbers in the convection zone of the order of Ro ~ 0.2 to 0.9, rather than the tiny value of Ro ~ 10^{-6} found in the liquid core of the earth.

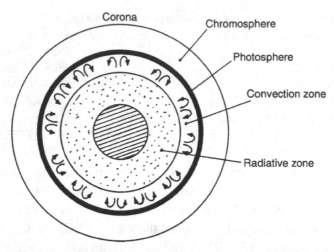

Figure 15.1 The structure of the sun. The layers are not to scale, with the depth of the photosphere and chromosphere exaggerated for clarity. (The chromosphere is less than 1 per cent of the solar radius.)

The north-south field strength measured near the poles is around 10 Gauss, whereas the peak field within sunspots is of the order of 3000 Gauss, which is thought to be indicative of the subsurface east-west field strength. The historical records of sunspots provide crucial clues as to the behaviour of the sun, not just in the solar atmosphere (solar flares are more common at sunspot maximum), but also deep within the interior, where the sunspot flux tubes are thought to originate. Magnetic fields play a particularly important role in the solar atmosphere, triggering solar flares and heating the corona to such a high temperature that it expands radially outward from the sun, spiralling through the solar system in the form of a tenuous plasma wind. The average distance from the sun to the earth is $1\text{AU} = 1.50 \times 10^8$ km, and it takes two or three days for the solar wind to reach us from the sun, whereas light traverses the same distance in around 8 minutes.

The sun's interior is conventionally divided into three zones. The *core* has a radius of around $R_\odot/4$ and is the seat of nearly all the fusion power (Figure 15.1). It has a high density and temperature and contains around 50 per cent of the sun's mass. This is surrounded by the so-called *radiative zone*, which is hydrodynamically stable and extends up to a radius of $0.71R_\odot = 4.9 \times 10^5$ km. The gas in the radiative zone is quiescent and radiant heat diffuses outward along the radial temperature gradient. The outer region is called the *convection zone* and is $\sim 2 \times 10^5$ km deep. It is convectively unstable and transports heat to the surface of the sun (the photosphere) by turbulent convection. Here, convection cells transport heat and mass across the annulus on a time scale of weeks. Although the convection zone contains only 2 per cent of the solar mass, it is the source of nearly all MHD activity, including the initiation of coronal flux tubes and sunspots, as well as the location of the solar dynamo.

The existence of the solar convection zone is evident in the granular appearance of the photosphere. In photographs it looks rather like a gravel path and is reminiscent of multicellular Bénard convection. The granules (polygonal convection cells) are continually evolving on a time scale of minutes and have diameters of around 700 km and a typical velocity measured in fractions of a km/s. They are bright at the centre, where hot plasma is rising to the surface, and darker at the cell boundaries, where cooler plasma falls. Because they are influenced by surface radiation, these granules are not necessarily representative of the scale of the internal motion deep within the convection zone.

The radiative zone rotates more or less as a rigid body. However the convection zone does not, with the equator rotating faster than the poles at the photosphere. (The synodic surface rotation period at the equator is 26.3 days, while that in the polar regions is around 36 days.) The distribution and cause of the differential rotation within the convection zone has been the subject of considerable debate and several surprises. Initially it was assumed that the angular velocity was primarily a function of distance from the rotation axis, in line with the Taylor–Proudman theorem (though recall that the Rossby number in the convection zone is not much smaller than unity). This long-standing believe was finally laid to rest by helioseismological measurements in the late 1980s, which showed that the angular velocity in the convection zone is primarily a function of latitude only. The radiative zone, by comparison, rotates as a rigid body at a rate intermediate between the equatorial and polar surface rotation rates. The transition from differential rotation in the convection zone to rigid-body rotation in the radiation zone turns out to be surprisingly abrupt, occurring in a thin, highly sheared layer known as the *tachocline*. This intense shear layer is perhaps 20,000 km deep in the equatorial regions and is a prolific generator of east-west field through the Ω–effect. Many of the more important MHD processes in the sun have their origins in the tachocline, and it is interesting to speculate as to whether or not a similar layer forms in all magnetically active stars of the same class.

The solar atmosphere is also divided up into three regions. The surface of the sun is called the photosphere. This is a thin, transparent layer of relatively dense material about 500 km deep. It has a mean temperature of ~5800 K, with the temperature falling from around 6500 K at the base of the photosphere down to ~4400 K at the top. Above this lies the hotter, lighter chromosphere, which is around 2000 km deep. The temperature of the chromosphere rises from 4400 K near its base to around 20000 K at its top, and as the gas heats up, so its density falls, eventually reaching around 10^{-5} times that of the density in the photosphere. Thus the solar atmosphere is extremely tenuous by the time we reach the top of the chromosphere.

The outermost layer of the solar atmosphere is called the corona, which has no clear upper boundary and extends in the form of the solar wind out to the planets. There is a sharp rise in temperature in passing from the chromosphere to the corona, from 20000 K at the top of the chromosphere to around 10^6K at the base of the corona. This unexpected rise in temperature is possibly due to the random yet frequent reconnection of magnetic flux loops (so-called nano-flares), though this is still debated. These flux tubes arch up from photosphere into the corona and then back down to the photosphere. There is a continual jostling of the field lines in the photosphere, causing them to become stretched and entangled in the corona. Every so often these field lines snap back to a lower energy state by magnetic reconnection, releasing magnetic energy which subsequently reappears as heat.

15.1.2 Is There a Solar Dynamo?

The natural decay time for a magnetic dipole field in the sun is around 10^{10} years, which is about the age of the sun itself, i.e. ~5×10^9 years. This is not inconsistent with the notion that the solar field is the relic of some galactic field which was trapped in the solar gas at the time of the sun's formation. However, the sun's magnetism is constantly varying in a manner that cannot readily be explained by some frozen-in primordial field. Sunspot activity could be attributable to transient, small-scale processes, but the periodic (22-year) variation of the sun's global field suggests that theories based on a frozen-in relic are implausible. (There have been attempts to explain the 22-year oscillation in the east-west field in terms of a periodic torsional oscillation of the subsurface differential rotation, but such cartoons have largely been abandoned, as discussed below.) It seems likely, therefore, that the explanation of solar magnetism lies in dynamo action within the convective zone. Note that while dynamo theory is invoked to explain the unexpected persistence of the earth's magnetic field, it is invoked in the solar context to explain the rapid evolution of the sun's field.

The dynamo theories which have been developed in the context of the earth and the sun are, however, very different. In the core of the earth, velocities are measured in fractions of a millimetre per second, and as a result R_m is rather modest, a few hundred at most. In the convection zone of the sun, on the other hand, velocity fluctuations are some fraction of a km/s, giving $R_m \sim 10^8$. While concerns in geodynamo theory often centre on finding turbulent motions which have an R_m high enough to induce significant field line stretching, in the solar dynamo the problem is of the opposite nature. R_m is so high that the molecular diffusivity becomes very weak, and so extremely large gradients in the magnetic field must develop in order to allow the flux tube reconnections which are needed to explain the observed behaviour and are essential for dynamo action.

15.1.3 Sunspots and the 11-Year Solar Cycle

At the base of the convection zone, intense east-west flux tubes are generated within the tachocline via the Ω-effect. These are subject to magnetic buoyancy, with the pressure inside a flux tube lower than that outside, the difference being the magnetic pressure, $B^2/2\mu$. Occasionally these buoyant flux tubes become unstable and break loose from the tachocline, floating up through the convective zone. Sunspots are a manifestation of these buoyant flux tubes after they erupt into the solar atmosphere (Figure 15.2). These dark spots usually appear in pairs and are the foot-points of the flux tube in the solar surface, where the intense local magnetic field (1000G→4000G) suppresses fluid motion and cools the surface. The spots are typically 10^4 km in diameter (much larger than a granule) and they appear mainly near the equatorial plane. Often they occur in groups (an *active region*) and this gives rise to an increased brightness, called *photospheric faculae*.

The intensity of sunspot activity fluctuates on a regular 11-year cycle. At sunspot minimum there may be no sunspots, while at the maximum there are typically around a hundred. After their rapid initial formation, sunspot pairs may survive for some time, disintegrating over a period of days or weeks, the so-called 'following spot' vanishing first. The fragments of the flux tube which caused the spot are then convected around by the photospheric flows accumulating along granular boundaries.

The area of the photosphere covered by sunspots varies during the 11-year cycle. At the minimum point new spots appear first at latitudes of $\sim\pm30°$. The number of active zones then increases, gathering towards the equator, until finally the sunspot activity dies away. The magnetic field in an active region is

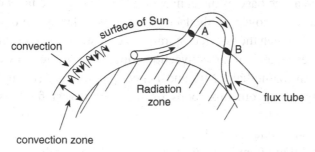

Figure 15.2 A buoyant flux tube rises up through the convection zone and erupts through the surface of the sun. Sunspots form at A and B where the magnetic field suppresses the turbulence.

predominantly azimuthal (east-west field lines), so that the sunspot pairs are more or less aligned with a line of latitude. They rotate with the surface of the sun, the leading spot being slightly closer to the equator than the following spot. As one might expect, leading and following spots are observed to have opposite polarities, i.e. opposite signs of **B** · *d***S**, where *d***S** is part of the solar surface. However, all pairs in one hemisphere have the same orientation (e.g. **B** directed outward in the leading spot and **B** directed inward in the following spot) and this orientation reverses as we move from one hemisphere to the other. This suggests that the subsurface azimuthal field is unidirectional in each hemisphere and antisymmetric about the equator: a picture which is consistent with an azimuthal field being swept out from a dipole field by differential rotation. Crucially, however, the field orientation in the sunspot pairs reverses from one 11-year cycle to the next, suggesting a periodic variation of the subsurface azimuthal field every 22 years. If this azimuthal field is generated from a dipole field by differential rotation, then this, in turn, suggests a periodicity in the dipole field, or else a periodic reversal in the differential rotation which sweeps out the azimuthal field. If the latter were true there would be no need to invoke a solar dynamo to explain the 22-year cycle. However, observations suggest that it is the first explanation which is correct. The sun's dipole field reverses at sunspot maximum (90° out of phase with the east-west field), strongly suggesting that the solar magnetic field is maintained by dynamo action in the convection zone.

15.1.4 *The Location of the Solar Dynamo and Dynamo Cartoons*

We have already noted that the equator rotates faster than the poles at the photosphere, while the radiative zone rotates as a rigid body at a rate somewhere between the equatorial and polar surface rotation rates. The transition from one angular velocity distribution to the other takes place in a thin, highly sheared, stably stratified region at the base of the convection zone, i.e. in the tachocline.

Even before the discovery of the tachocline, most solar dynamo models were of the $\alpha - \Omega$ type, in which the Ω-effect spirals out east-west flux tubes from the dipole field in the lower regions of the convection zone. (The α-effect, by contrast, is the generation of an azimuthal EMF, and hence a poloidal field, by small helical disturbances spiralling out flux loops from the east-west field, as discussed in Chapter 14.) The ratio of the azimuthal to dipolar fields associated with the Ω-effect is, from Equation (14.22), of the order of $R_\Omega \sim |d\Omega/dr|\ell^3/\lambda$, where $d\Omega/dr$ is a typical gradient in angular velocity and ℓ a suitable length scale. Since R_Ω is large in the sun, the east-west field in the tachocline is particularly strong, of the order of $10^4 \to 10^5$ Gauss, which is much greater than the nominal dipole field strength of 10G. As this east-west field is spiralled up it is subject to instabilities driven by radial gradients in field strength, as discussed below. It turns out that those flux tubes which have a modest field strength are held in place by the intrinsic stable stratification of the tachocline, whereas those with excessive fields, perhaps 10^5 Gauss or greater, can break free. Once freed, the flux tubes float upward though the action of magnetic buoyancy. That is, if a flux tube is in approximate equilibrium with its surroundings, the internal gas pressure of the tube, p_i, is less than the external pressure, $p_e \approx p_i + B^2/2\mu$, and so the flux tube is buoyant with the plasma density inside the tube less than that outside. Thus, every now and then an unstable magnetic flux tube breaks free from the tachocline to arch upward into the convection zone, emerging a few weeks later in the photosphere (Figure 15.2).

However, not all of the flux tubes which escape the tachocline make it to the photosphere. As isolated flux tubes rise through the action of magnetic buoyancy, they undergo a dramatic expansion, and flux conservation tells us that there is a corresponding fall-off in field strength. In many cases the rising flux tubes fragment and lose their identity. Indeed, numerical simulations suggest that only those flux tubes with a distinctly spiralled internal magnetic structure remain intact as they rise up through the convection zone[1].

The magnetic destabilisation of the tachocline, and the resulting flux tube eruptions, are discussed in Hughes, Rosner and Weiss (2007). A simple model problem is sufficient to convey many of the key ideas. Suppose we have a horizontal magnetic field, $B_0(z)$, which sits in an otherwise stable, compressible atmosphere and is a smooth function of height z. As we shall see, such a field can drive an instability, provided that $B_0(z)$ decreases sufficiently rapidly with height, the mechanism being essentially the same as that which causes an isolated flux tube to float upward. For simplicity we shall take gravity, $\mathbf{g} = -g\hat{\mathbf{e}}_z$, to be uniform, any perturbation of the plasma properties to be adiabatic and the plasma to be electrically ideal.

In the absence of a magnetic field it is well known (see, for example, Davidson, 2013) that such an atmosphere is stable to adiabatic perturbations, provided that

$$\frac{d\rho_0}{dz} + \frac{\rho_0 g}{c_0^2} < 0, \tag{15.1}$$

where $c_0 = \sqrt{\gamma p_0/\rho_0}$ is the adiabatic speed of sound, γ the ratio of specific heats, and the term involving c_0 is usually referred to as the adiabatic lapse rate. In such cases it is conventional to introduce the generalised Väisälä–Brunt frequency, N, defined through

---

[1] This spiralled magnetic field is also evident in solar prominences. The significance of a spiralled magnetic field for the stability of flux tubes is discussed in §16.3.4 in the context of fusion plasmas.

$$N^2 = \frac{g}{\gamma}\frac{d}{dz}\ln\left(\frac{p_0}{\rho_0^\gamma}\right), \tag{15.2}$$

which, in the absence of a magnetic field, satisfies

$$N^2 = \frac{g}{\rho_0}\left|\frac{d\rho_0}{dz} + \frac{\rho_0 g}{c_0^2}\right|. \tag{15.3}$$

(By way of comparison, in an incompressible Boussinesq fluid the Väisälä–Brunt frequency is defined through the expression $N^2 = (g/\rho_0)|d\rho_0/dz|$.)

Let us now consider the influence of the horizontal field, $B_0(z)$, on our otherwise stable atmosphere. If the unperturbed field is in equilibrium with its surroundings we have

$$\frac{d}{dz}\left(p_0 + \frac{B_0^2}{2\mu}\right) = -\rho_0 g. \tag{15.4}$$

For simplicity we consider perturbations in which the wavevector, $\mathbf{k}$, satisfies $\mathbf{B}_0 \cdot \mathbf{k} = 0$, so that there is no bending of the field lines. (These are known as interchange modes.) Suppose that a perturbation raises a small fluid element by an amount δz. Since we are assuming that the perturbations are adiabatic and that the plasma is ideal, the displaced fluid element satisfies two conditions: (i) $\delta p = c_0^2 \delta\rho$, which comes from $p \sim \rho^\gamma$, and (ii) $\delta B/B_0 = \delta\rho/\rho_0$, which is a consequence of $\mathbf{B}_0 \cdot \mathbf{k} = 0$ and the compressible analogue of (5.6),

$$D(\mathbf{B}/\rho)/Dt = (\mathbf{B}/\rho)\cdot\nabla\mathbf{u}.$$

Moreover, if we assume that the displacement is sufficiently slow for the displaced fluid parcel to maintain approximate equilibrium with its surroundings, we have

$$\delta\left(p + \frac{B^2}{2\mu}\right) = d\left(p_0 + \frac{B_0^2}{2\mu}\right) = -\rho_0 g(\delta z). \tag{15.5}$$

Note that, in these expressions, δ represents a *Lagrangian perturbation* which follows a fluid particle as it moves, whereas d represents an increment in any background variable, e.g. $d\rho_0 = (d\rho_0/dz)dz$.

An instability now arises if the density change in the rising fluid element is less than the corresponding change in background density: $\delta\rho < d\rho_0$. It is left as an exercise for the reader to show that (15.5), along with $\delta p = c_0^2\delta\rho$ and $\delta B/B_0 = \delta\rho/\rho_0$, demands that the equilibrium be unstable if, at any point,

$$\frac{d\rho_0}{dz} + \frac{\rho_0 g}{c_0^2 + a_0^2} > 0 \ , \qquad a_0^2 = \frac{B_0^2}{\rho_0 \mu} . \tag{15.6}$$

Evidently, this is a simple generalisation of the non-magnetic case, as given by (15.1). In cases where the non-magnetic atmosphere is stable, as in the tachocline, this criterion can be rewritten as (Hughes, Rosner and Weiss, 2007)

$$-\frac{g \ a_0^2}{c_0^2} \frac{d}{dz} \ln\left(\frac{B_0}{\rho_0}\right) > N^2 , \tag{15.7}$$

where N is the generalised Väisälä–Brunt frequency defined by (15.2).

Clearly, if B_0/ρ_0 decreases with height sufficiently rapidly, the magnetic field can destabilise an otherwise stable atmosphere. This is thought to be the key flux release mechanism in the tachocline, which is strongly stable in the absence of a magnetic field. (Actually, it turns out that certain three-dimensional perturbations, which weakly bend the field lines, are more dangerous than the interchange modes, with $\ln(B_0/\rho_0)$ in (15.7) being replaced by $\ln B_0$ in the corresponding stability criterion. These three-dimensional modes are discussed in Hughes, Rosner and Weiss, 2007.)

Three-dimensional, non-linear numerical simulations suggest that such instabilities do indeed give rise to isolated flux tubes which arch up from the zone of background magnetic field. This is consistent with the idea that particularly intense regions of azimuthal flux in the tachocline become unstable, with the erupting flux tubes being anchored in the tachocline while the more buoyant portions of the flux tubes rise up through the convection zone to form sunspots in the photosphere (Figure 15.2).

To summarise, provided the azimuthal magnetic field in the tachocline is not too intense, it can be stored there because of the strong, stable stratification. However, when the east-west field becomes too intense, say 10^5 Gauss or greater, the tachocline becomes locally unstable, giving rise to buoyant flux tubes which rise up through the convection zone. Many of these tubes disintegrate as they rise, their flux being dispersed across the convection zone, while a few remain coherent and eventually reach the photosphere to form sunspots. In this sense the tachocline acts as a reservoir of intense east-west field, and it is natural to suppose that the solar dynamo is of the $\alpha - \Omega$ type, located in the vicinity of the tachocline. Such dynamos are sometimes called thin-shell dynamos.

In some of the early thin-shell $\alpha - \Omega$ cartoons, the α-effect was associated with magnetostrophic waves generated at the top of the tachocline and supported by the intense east-west field that accumulates there. It turns out that the corresponding values of α are antisymmetric about the equator, as required for dynamo action, and such models predict oscillatory dynamos in which *dynamo waves* migrate from mid-latitudes towards the equator, which is indeed rather sun-like. However, a central difficulty with the early thin-shell models is the dual role played by the east-west field, B_θ, in the α-effect. Of course such a field is required to generate an azimuthal EMF and hence a dipole field, yet too strong an azimuthal field may suppress the helical motion required by an α-effect, a phenomenon known as α-quenching. In order to circumvent this difficulty some dynamo cartoons follow the lead of Parker (1993) and propose a so-called *interface dynamo*, located at the interface between the tachocline and convection zone. The idea is that the dynamo operates within a narrow shell at the outer edge of the tachocline, with the Ω– and α-effects spatially separated. Thus the Ω –effect is active in the tachocline, driven by the shear, while the α-effect operates just outside the tachocline, forced by cyclonic convection and relatively free of α-quenching. Of course the α-effect requires some east-west field to spiral up, and the Ω –effect some poloidal field to shear, and so this kind of interface dynamo relies on the notion that there is a controlled exchange of magnetic flux across the interface. Quite how this is achieved in practice is still a matter of speculation.

In one such interface dynamo it is supposed that many of the buoyant east-west flux tubes disintegrate as they rise up through the convection zone, and this provides the weak, filamentary east-west field on which the α-effect can operate. That is, helical turbulence spirals out poloidal flux loops from this east-west field, which in turn drives the azimuthal EMF needed to support the dipole field. To complete the dynamo cycle, these poloidal flux loops must be transported down to the tachocline where they can be sheared back into east-west field. In this particular cartoon, turbulent convection above the tachocline drives the mix of east-west and poloidal flux back down to the surface of the tachocline, where overshooting flux tubes penetrate the upper regions of the tachocline. Finally, the Ω–effect shears the poloidal flux to regenerate the east-west field and the cycle is complete.

However, this is little more than a speculative cartoon. There are many alternative mechanisms of transporting flux to and from the tachocline, such as the slow mean poloidal recirculation (~20 m/s) which also transports magnetic flux across the convection zone. This mean flow runs from the equator to the poles near the photosphere and returns to the equator deep within of the convection zone. Some competing $\alpha - \Omega$ cartoons take advantage of this mean recirculation to transport flux to and from the tachocline.

15.1.5 Prominences, Flares and Coronal Mass Ejections

The solar atmosphere is anything but passive. It is threaded with vast magnetic flux tubes which arch up from the photosphere into the corona and which are constantly

Figure 15.3 Flux tubes arch up from the photosphere. (Courtesy of the TRACE Project, NASA.)

evolving, being jostled by the convective motions in the photosphere (see Figure 15.3). Some of these flux tubes are associated with sunspots; others are associated with so-called *prominences*. Prominences extend from the chromosphere up into the corona, appearing as arch-like, tubular structures of length $\sim 10^5$ km and thickness $\sim 10^4$ km. They contain cold, chromospheric gas, perhaps 300 times colder than the surrounding coronal gas. This relatively cold plasma is threaded by a helical magnetic field of $\sim 10 \rightarrow 100$ Gauss, which is much weaker than that in a sunspot, but larger than the mean surface field of a few Gauss. A prominence is itself immersed in and surrounded by thinner flux tubes which arch up from the photosphere, criss-crossing the prominence. Some flux tubes lie below the prominence, providing magnetic pressure to its underside, rather like a magnetic hammock or magnetic cushion. Other flux tubes lie above, pushing down on the prominence. The flux tubes which overlie a prominence are sometimes referred to as a *magnetic arcade*.

Quiescent prominences are stable, long-lived structures which survive for many weeks and contain around a billion tons of gas (Figure 15.4b). However, a rearrangement of the magnetic field surrounding a quiescent prominence can turn it into an *eruptive prominence* (Figure 15.4a). This rises up through the corona at ~ 1 km/s, pushing coronal plasma and magnetic flux ahead of it and disappearing into the solar wind in a matter of a few hours. The mass which is propelled from the sun in this way is an example of a *coronal mass ejection* (CME). Often these

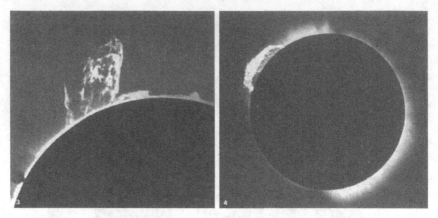

Figure 15.4 (a) An eruptive prominence. (b) A quiescent prominence.

eruptive prominences are accompanied by *solar flares*, particularly if they occur within active regions where the magnetic field is strong.

As yet, there is no self-consistent model for solar flares, although all current theories agree that the power source is stored magnetic energy, triggered by magnetic reconnection as the local magnetic field reconfigures itself into a lower energy state. The largest flares may release around 10^{25}J, while intermediate-sized flares might release $10^{22} \rightarrow 10^{23}$J. The very large flares are rare events, perhaps occurring once or twice a year, while smaller flares are more common, occurring on a daily basis at sunspot maximum. By way of contrast, nano-flares, of strength $\sim 10^{17}$J, occur continuously and, as already noted, may be a major source of coronal heating.

The largest flares are called *two-ribbon flares* and the standard cartoon for such an event is the following. Consider a prominence which is supported by a magnetic cushion and has a magnetic arcade overlying it. These field lines have their roots in the turbulent photosphere and so are continually adjusting to the photospheric jostling of the flux-tube foot-points, evolving in a quiescent manner through a sequence of largely force-free equilibria. Typically the photospheric motion does work on the coronal field lines and so their magnetic energy slowly increases with time as magnetic tension builds up. Moreover, an X-like saddle point usually forms below the prominence in those field lines that separate the helical flux within the prominence from the overlying arcade and underlying cushion.

Now suppose that the prominence starts to rise, perhaps because of a build-up of magnetic pressure in the magnetic cushion. The field lines near the X-point will become increasingly stretched, with flux tubes of opposite polarity pushed together. Eventually, magnetic reconnection occurs. This, in turn, allows the arcade flux tubes to pinch off, releasing magnetic energy (Figure 15.5). Such

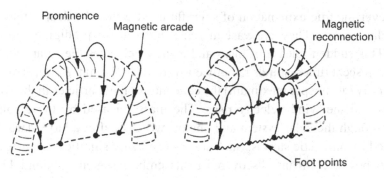

Figure 15.5 A cartoon of a two-ribbon flare.

a reconnection may occur in a matter of minutes and is referred to as the *impulsive phase* of the flare. When this occurs, the downward force associated with the overlying flux tubes is suddenly removed and so the prominence is propelled explosively upward (at ~10 km/s) by the magnetic pressure in the underlying magnetic cushion. The subsequent *main phase* of the flare, which lasts hours, consists of a rising prominence which pushes flux loops and coronal plasma ahead of it.

During reconnection, transient electric fields are generated by $\partial \mathbf{B}/\partial t$ and these electric fields accelerate charged particles to extremely high speeds. Some of the released energy propagates down the arcade field lines to their foot-points in the chromosphere. The footprints of these field lines then appear as two energetic lines of Hα emission, which is the signature of such flares and the origin of their name.

A closely related type of solar eruption is a coronal mass ejection. CMEs are vast, occupying as much as 45° of the edge of the solar disc. In a typical CME, around 5 → 50 billion tons of coronal material is propelled into interplanetary space. Like flares, CME activity follows the 11-year cycle, with one or two per day at sunspot maximum and fewer at minimum. Near sunspot minimum CMEs are restricted mostly to the equatorial regions, but at maximum they occur at most latitudes. Many CMEs are associated with eruptive prominences, with the CME frequently preceding the eruptive prominence, and those CMEs which occur in active regions are frequently accompanied by solar flares.

It has to be said, however, that the precise relationship between flares and CMEs is still poorly understood. Flares can occur before, during or after a CME, and so the association between the two phenomena is rather loose. Indeed, the relationship between flares, eruptive prominences, CMEs and the coronal magnetic field remains a matter of active debate. Clearly, the entire process is much more complicated than suggested above.

Whatever the true explanation of solar flares, it cannot be denied that they are spectacular events. They are vast in scale and release prodigious amounts of energy. This sudden release of mass and energy enhances the solar wind which, even in quiescent times, spirals radially outward from the sun. At times of vigorous solar activity (at sunspot maximum) the concentration of particles in the solar wind increases and the wind velocity rises. The mass released by these solar flares sweeps through the solar system and one or two days after a large flare the earth is buffeted by magnetic storms. Such storms can cause significant damage, as one Canadian power company discovered to its embarrassment. Around 11 March, 1989, a large solar flare burst from the surface of the sun, and as dawn broke on 13 March, six million Canadians found themselves without power.

15.2 The Solar Wind

The plasma in the solar corona is too hot to be gravitationally bound to the sun. Rather, it spirals outward through interplanetary space at supersonic speeds until it is slowed down by a termination shock at the edge of the solar system. As the solar wind sweeps past the earth it has a speed of around 400 km/s.

15.2.1 Why Is There a Solar Wind?

The idea that coronal plasma might be ejected from the sun has been around since the late 1920s as a partial explanation for terrestrial magnetic storms. However, this was originally viewed as an intermittent phenomenon, operating primarily at sunspot maximum. The existence of a continuous solar wind was only finally settled in the late 1950s. Observational evidence in favour of a continuous wind began with the study of comet tails in the early 1950s, but crucially in 1958 E.N. Parker noted that it was impossible for the sun to support a static corona given the high plasma temperatures observed in the corona. Rather, he predicted that the coronal plasma expands outward in a continuous stream. At first, Parker's prediction of a continuous wind was considered controversial, but his prediction was quickly confirmed the following year by the Soviet *Luna* satellites.

The wind consists mostly of a collisionless, ionised hydrogen plasma. It is extremely tenuous, with a staggeringly low number density of around 5 protons per cm^3 at 1 AU. It flows out along open magnetic field lines which have their origin in coronal holes in the sun's atmosphere. At a distance of 1 AU the typical strength of the interplanetary field and the sound speed in the wind are around 10^{-4} G and $50 \rightarrow 100$ km/s, respectively. So, with a characteristic velocity of around 400 km/s, the wind is supersonic by the time it reaches the earth. As it passes around the earth it confines the geomagnetic field to a region called the

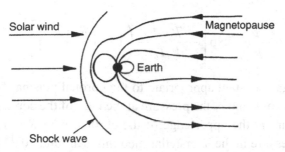

Figure 15.6 The interaction of the solar wind with the earth's magnetic field. The geomagnetic field is confined to a region called the magnetosphere.

magnetosphere. A shockwave forms upstream of the magnetosphere and the earth's magnetic field lines are highly elongated in the downstream direction (Figure 15.6). The boundary between the wind and earth's field is called the magnetopause. After passing the earth, the wind continues to spiral through the solar system at super-sonic speeds, eventually slowing down at the edge of the solar system as it encounters a termination shock at a radius of ~90 AU. The wind finally terminates at the heliopause, at a distance of around 100 → 150 AU. This marks the edge of the heliosphere, beyond which we encounter the interstellar medium.

It is instructive to consider briefly why a static corona is not tenable. For simplicity we take the corona to be spherically symmetric and we ignore the coronal magnetic field. Let us suppose that there is indeed a static corona grav-itationally bound to the sun and follow through the consequences of this assump-tion. Heat is then transferred through the corona by conduction with a thermal conductivity, κ, proportional to the plasma temperature raised to the power of 5/2, $\kappa = \kappa_0 T^{5/2}$. Ignoring coronal heating, the heat flux out through a sphere of radius r within the corona is independent of r and equal to

$$\dot{Q} = -4\pi r^2 \kappa_0 T^{5/2} \frac{dT}{dr}. \tag{15.8}$$

Since $\dot{Q}$ is independent of r, and $T \ll T_c$ for $r \gg R_c$, this integrates to give $T/T_c = (R_c/r)^{2/7}$, where T_c is the temperature at the base of the corona, $r = R_c$. This temperature profile may be combined with the static force balance (in sphe-rical polar coordinates),

$$\frac{dp}{dr} = -\frac{GM_\odot \rho}{r^2}, \tag{15.9}$$

and the ideal gas law, $p = \rho R_g T$, to yield the radial pressure distribution in the corona,

$$\frac{p}{p_c} = \exp\left[-\frac{7GM_\odot}{5R_gT_cR_c}\left(1 - \left(\frac{R_c}{r}\right)^{5/7}\right)\right]. \tag{15.10}$$

(Here R_g is the gas constant appropriate to the coronal plasma, G is the universal gravitational constant, and p_c the pressure at the base of the corona.) Crucially, this predicts a pressure at the upper edge of the corona, $r \gg R_c$, of $p_\infty \approx 10^{-8}\text{N/m}^2$. However, the pressure in the interstellar medium is around 10^{-14} N/m^2, and so we have a contradiction. Evidently a static model of the corona is not tenable, and in 1958 Parker concluded that the corona must continually expand into interplanetary space in the form of a continuous wind.

15.2.2 Parker's Model of the Solar Wind

A simple, steady, spherically symmetric model of the solar wind, which ignores magnetic fields, can be constructed as follows. We start by returning to the radial force balance, which is now modified to incorporate spherically symmetric motion:

$$u_r\frac{du_r}{dr} = -\frac{1}{\rho}\frac{dp}{dr} - \frac{GM_\odot}{r^2}. \tag{15.11}$$

To this we must add mass conservation in the form

$$\dot{m} = 4\pi r^2\rho u_r = \text{constant}. \tag{15.12}$$

We now consider the case of a polytropic equation of state, $p \sim \rho^\gamma$, with a local sound speed of $c_s = \sqrt{\gamma p/\rho}$. (Parker, by contrast, used an isothermal model.) Since the polytropic law yields $p^{-1}dp/dr = (\gamma/\rho)d\rho/dr$, the radial momentum equation can be rewritten in the form

$$u_r\frac{du_r}{dr} = -\frac{d}{dr}\left(\frac{c_s^2}{\gamma-1}\right) + \frac{d}{dr}\left(\frac{GM_\odot}{r}\right), \tag{15.13}$$

which possess the conserved energy integral $e = u_r^2/2 + c_s^2/(\gamma-1) - GM_\odot/r$. In principle, the variation of u_r with radius can be determined from this energy integral, with c_s^2 evaluated using $c_s^2 \sim \rho^{\gamma-1}$ and ρ from $\rho = \dot{m}/(4\pi r^2 u_r)$. However, it turns out to be more enlightening to rewrite the radial momentum equation in a different form. Invoking the continuity equation (15.12), along with $p^{-1}dp/dr = (\gamma/\rho)d\rho/dr$, we have

$$\frac{1}{\rho}\frac{dp}{dr} = -c_s^2\left[\frac{1}{u_r}\frac{du_r}{dr} + \frac{2}{r}\right], \tag{15.14}$$

and our radial momentum equation becomes

$$u_r\left[1 - \frac{c_s^2}{u_r^2}\right]\frac{du_r}{dr} = \frac{2c_s^2}{r} - \frac{GM_\odot}{r^2}. \tag{15.15}$$

This brings out the idea that there are two distinct flow regimes: subsonic and supersonic. The point where the Mach number is unity, $c_s = u_r$, is called the *sonic radius*. Integrating (15.15) shows that the radial velocity increases with r from small subsonic values near the sun to $c_s = u_r$ at the sonic radius,

$$r = r_s = GM_\odot/(2c_s^2). \tag{15.16}$$

It then becomes progressively more supersonic as r is further increased. In short, the behaviour is like that of a convergent-divergent Laval nozzle.

In practice, however, this model is too simplistic, as the wind is far from spherically symmetric, nor is it steady. Moreover, magnetic fields play a crucial role in guiding the plasma up through the corona and into the solar wind. At the base of the corona, u_r is small and **B** large, so inertia is weak and the plasma is constrained to follow magnetic field lines. These field lines have foot-points which are anchored in the photosphere, and so the flow near the photosphere rotates with the sun and is restricted to regions of open field lines, such as coronal holes. By way of contrast, at large radii u_r is large and **B** weak and so we have the opposite situation, with the interplanetary field constrained to follow the plasma flow. Here the magnetic field lines are dragged through interplanetary space by the radial wind while being deformed into a spiral pattern as their foot-points rotate with the sun (Figure 15.7). The surface which divides the regime of strong fields and weak inertia from that of weak fields and strong inertia is called the *source surface*. It has a radius, R_s, which is a few multiples of the solar radius, and it is convenient to classify the source surface as the point where the corona ends and the solar wind starts.

The distinctive structure of the interplanetary field is known as the *Parker spiral*, and it may be determined as follows. We consider the equatorial plane of the sun and look for a steady solution of the induction equation, $\partial \mathbf{B}/\partial t = \nabla \times (\mathbf{u} \times \mathbf{B})$, in a frame of reference which rotates with the sun. In such a frame, steadiness demands $\nabla \times (\mathbf{u} \times \mathbf{B}) = 0$, or equivalently $\mathbf{u} \times \mathbf{B} = \nabla\varphi$ for some φ, which satisfies $\mathbf{u} \cdot \nabla\varphi = 0$. If $\mathbf{u}$ and $\mathbf{B}$ are both radial at the source surface, then $\nabla\varphi = 0$ and $\varphi = \varphi_s = $ constant at that surface. However we also have $\mathbf{u} \cdot \nabla\varphi = 0$, and so

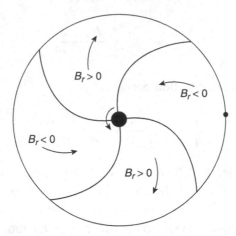

Figure 15.7 The Parker spiral.

integrating out along a streamline we find that φ remains equal to φ_s for all $r > R_s$. We conclude, therefore, that $\nabla \varphi = \mathbf{u} \times \mathbf{B} = 0$ at all radii outside the source surface. In such a situation we have

$$B_\theta/B_r = u_\theta/u_r, \quad r > R_\mathrm{s}, \tag{15.17}$$

in cylindrical polar coordinates, (r, θ, z). (Note that the velocity components are measured in the rotating frame of reference.)

Now in an inertial frame of reference the equatorial solar wind at large radius is approximately constant and radial, though there is some small azimuthal motion which the plasma acquires below the source surface, $r = R_s$. This azimuthal motion is of order $\Omega_\odot R_s$, $\Omega_\odot$ being the solar angular velocity. So for $r > R_s$ we have, in the solar system inertial frame,

$$\mathbf{u} \approx u_{sw}(r)\hat{\mathbf{e}}_r + \Omega_\odot R_s \hat{\mathbf{e}}_\theta, \tag{15.18}$$

with the corresponding velocity in the rotating solar frame being

$$\mathbf{u} \approx u_{sw}(r)\hat{\mathbf{e}}_r - \Omega_\odot (r - R_s)\hat{\mathbf{e}}_\theta. \tag{15.19}$$

It follows from (15.17) that, at large radius,

$$B_\theta/B_r = -\Omega_\odot r/u_{sw}, \quad r \gg R_s. \tag{15.20}$$

This describes the spiralled pattern of $\mathbf{B}$ in the equatorial plane. For $u_{sw} = $ constant, (15.20) represents an Archimedean spiral (Figure 15.7).

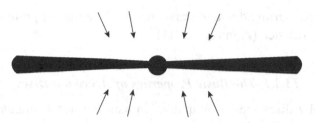

Figure 15.8 When an accretion disc is formed, mass first falls onto the disc and then flows through the disc to the star.

15.3 Accretion Discs

Massive objects, like stars or black holes, often slowly acquire mass from their surroundings through gravitational attraction. This is especially true of the central black holes in active galactic nuclei, of young protostars, which grow by accreting mass from the surrounding protostellar cloud, and of the dying star in a binary star system, which is particularly compact and leaches mass from the adjacent, less-massive star. In all three cases this leads to the formation of a thin, rotating disc of gas which surrounds the growing star or black hole. Such discs, whose axes are usually aligned with the star's rotation axis, are called *accretion discs*. Once they are formed, mass first falls onto the disc, rather than directly onto the star itself, after which it slowly spirals inward through the disc under the influence of gravity (Figure 15.8).

A key question in accretion disc theory is how the mass that spirals inward through the disc manages to shed its angular momentum. Although the magnetic field that threads through a disc can carry off some angular momentum, the dominant mechanism is thought to be turbulent diffusion from the interior of the disc back out towards the rim. Indeed, as we shall see, in the standard disc model the rate at which mass accretes onto the central object is directly proportional to the turbulent diffusivity of the disc. This in turn raises the question of how turbulence is triggered and maintained within an accretion disc. It is now widely believed that, for the more strongly ionised discs, the classical Chandrasekhar–Velikhov instability triggers the turbulence. Moreover, it is thought that essentially the same dynamical mechanism subsequently maintains the turbulence (see, for example, Balbus and Hawley, 1998). This classical instability, now more commonly known as the magnetorotational instability (or MRI for short), is discussed in §7.7. Its essential feature is that a surprisingly weak transverse magnetic field can destabilise a rotating, conducting fluid which is otherwise centrifugally stable. Such a mechanism is thought to be crucial in hot discs such as those in binary star systems. However, it is less clear that MRI can operate in the cooler, more weakly ionised discs that surround young stars, and the dynamics of protoplanetary discs are still widely debated (Armitage, 2011).

15.3.1 The Basic Properties of Accretion Discs

The reason why discs are so ubiquitous in astrophysics is angular momentum conservation. Consider, by way of an example, a protostar of mass M slowly accreting mass from a rotating protostellar gas cloud. The gas is pulled inward by gravity, conserving angular momentum as it falls. At early times the inflow might be approximately spherically symmetric, but it cannot stay that way because conservation of angular momentum distinguishes between mass that spirals inward along the equatorial plane and that which approaches the star along the rotation axis.

To see why this is so, consider the simplest case, where the flow is axisymmetric and we neglect magnetic fields (which is not particularly realistic). We also neglect pressure gradients, which may be plausible during the initial free fall (but not once the disc starts to form), and assume that the gravitational field is dominated by the central star, so that the gravitational potential energy per unit mass is $V = -GM/|\mathbf{x}|$, where $\mathbf{x}$ is measured from the star. We adopt cylindrical polar coordinates, (r, θ, z), centred on the star and with the z-axis aligned with its angular velocity, $\mathbf{\Omega}$. The mechanical energy per unit mass of a particle in free fall is then

$$e = -\frac{GM}{|\mathbf{x}|} + \frac{\Gamma^2}{2r^2} + \frac{u_r^2 + u_z^2}{2}, \tag{15.21}$$

where $\Gamma = u_\theta r$ is the specific angular momentum, which is materially conserved. Now consider material which is spiralling radially inward towards the star along the equatorial plane. The energy per unit mass is now

$$e = -\frac{GM}{r} + \frac{\Gamma^2}{2r^2} + \frac{u_r^2}{2}.$$

As mass flows inward, potential energy is released at a rate of $\sim r^{-1}$, but the azimuthal contribution to the kinetic energy grows much faster, in fact as $\sim r^{-2}$. Clearly the drop in potential energy cannot maintain the growth in kinetic energy forever, and the radial inflow must eventually halt. By way of contrast, the gas flowing in along the rotation axis is subject to no such constraint and can fall directly onto the star. Evidently it is more difficult for material to approach the star via the equatorial plane than along the rotation axis. We conclude that accretion from a rotating cloud cannot be spherically symmetric. Rather, a disc of gas

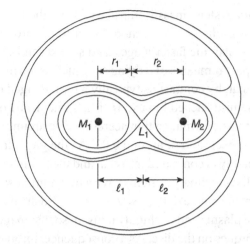

Figure 15.9 Equipotential surfaces in the orbital plane for a typical binary star system.

gradually builds up around the growing protostar, with its axis aligned with the rotation axis of the star. Once the disc is formed, mass is accreted onto the star by first falling onto the disc.

Perhaps the two most studied classes of fully developed accretion discs are those which lie within binary star systems, which is a common configuration, and those that develop around protostars. We shall briefly describe both, starting with the former. Consider two close stars with masses M_1 and M_2, $M_1 > M_2$, with both stars orbiting about their combined centre of mass. There will be a strong gravitational coupling between the stars and the nature of that coupling is best understood by considering equipotential surfaces in the orbital plane (Figure 15.9). Very close to M_1 and M_2, the equipotential surfaces are approximately spherical and centred on the two masses, while somewhat further out the equipotential surfaces are dumbbell-shaped and engulf both stars. A figure-of-eight equipotential surface marks the transition from one class of surface to the other, and the two halves of the figure-of-eight are known as the *Roche lobes* of M_1 and M_2. The two Roche lobes are connected via a saddle point, which is known as the inner Lagrangian point and is labelled L_1 in Figure 15.9.

When in equilibrium, a star is bounded by an equipotential surface, and so stars much smaller than their Roche lobes are approximately spherical. However, if one of the stars swells to fill its Roche lobe (a so-called *semi-detached binary system*), mass can escape from one lobe to the other through the Lagrangian point and fall onto the companion star. The more massive star, M_1, has the larger of the two Roche lobes, and so it is the less massive star that is most likely to fill its lobe. Consequently, in a semi-detached binary system, mass is transferred from the less massive to the more massive star, and this results in an accretion disc forming around M_1.

The more massive a star, the faster it ages, and so in a semi-detached binary M_1 is often the highly compact remnants of a dying star, such as a white dwarf or a neutron star. The companion star, by contrast, is typically a middle-aged (i.e. main-sequence) star. So in a typical semi-detached binary, M_1 is a white dwarf or a neutron star and mass flows from M_2 to M_1, forming a thin accretion disc around M_1. Such discs have a thickness to radius aspect ratio of around 0.03. As the gas spirals through the disc, half of its potential energy is converted into heat and then radiated away, as discussed below. So accretion discs in binary systems have a high luminosity. They are also hot, with temperatures in the range 10^3 K $< T < 10^7$K. Such discs are composed of relatively dense H/He plasma and are highly ionised, so that magnetic fields can exert significant Lorentz torques on the disc. As a consequence, turbulence is thought to be triggered and maintained by MRI. The magnetic field tends to be particularly intense close to the star, where a dipole field often halts the radial inflow within the accretion disc, diverting the mass up along field lines and towards the polar regions of the star. Accretion rates are typically around $\dot{m} \sim 10^{-8}M_\odot$/year, where $M_\odot$ is the mass of the sun.

Many protostars also possess accretion discs. These are composed of relatively cool ($T \le 10^3$K) H_2 gas with the addition of some dust. They have an aspect ratio of around ~0.1, a mass of the order of 1 per cent of the mass of the protostar, and an accretion rate of $\dot{m} \sim 10^{-8}M_\odot$/year. Such discs often provide an environment conducive to planet formation as dust agglomerates into rocks which can then act as seeds for protoplanets. Indeed, in their later stages of development, protostellar discs are often referred to as protoplanetary discs. Because of their modest temperature, protoplanetary discs have a low ionisation and interact only weakly with magnetic fields. While the ionisation in protoplanetary discs may be sufficiently high in the surface layers for MRI to trigger turbulence, there are also extended dead zones where MRI cannot operate (Armitage, 2011). As a consequence, the mechanisms by which angular momentum is transported across such discs are not well understood.

It is instructive to consider the distribution of angular momentum and energy across a fully developed accretion disc. As we shall see, unlike gravitational collapse during the pre-disc phase, turbulent diffusion and dissipation cannot be neglected, and indeed they are central to the very existence of the disc. For simplicity, we shall continue to neglect magnetic fields (a poor assumption) and assume that the disc is axisymmetric and steady-on-average. As before, we adopt cylindrical polar coordinates, (r, θ, z), centred on the star and aligned with the rotation axis. Let $\Omega(r)$ be the angular velocity distribution in the disc, $\Gamma = \Omega r^2$ the specific angular momentum, and M the mass of the star. A simple analysis is possible because

(i) the star is usually considerably more massive than the disc, and so dominates the gravitational field,
(ii) the radial velocity is much smaller than the azimuthal motion, $|u_r| \ll \Omega r$,
(iii) the radial pressure gradient is negligible because the disc is too thin to support significant radial variations in pressure.

The radial force balance is then, simply,

$$\Omega^2 r = \frac{GM}{r^2}, \qquad (15.22)$$

which leads to the Keplerian orbit

$$\Omega(r) = \sqrt{GM/r^3}, \qquad (15.23)$$

or, equivalently,

$$\Gamma(r) = r^2 \Omega(r) = \sqrt{GMr}. \qquad (15.24)$$

The gravitational potential energy in the disc is, of course, $V = -GM/r$, and so the total mechanical energy per unit mass is simply

$$e = \frac{1}{2} \Omega^2 r^2 - \frac{GM}{r} = -\frac{GM}{2r}. \qquad (15.25)$$

We conclude that, as the flow spirals inward, it loses mechanical energy to heat, which is then radiated away from the surfaces of the disc. Let r_i be the inner radius of the accretion disc, which is roughly the radius of the star, and let us assume that the outer radius of the disc is much larger than r_i. Then the total loss of mechanical energy per unit mass within the disc is $GM/2r_i$, and so the net radiation of energy from the disc (its luminosity) is given by

$$L_{disc} = \frac{\dot{m} GM}{2r_i}, \qquad (15.26)$$

where $\dot{m}$ is the radial mass flow rate through the disc. Note that, by measuring L_{disc}, we may estimate $\dot{m}$. The kinetic energy at $r = r_i$ is likewise $GM/2r_i$, and so the rate at which kinetic energy is deposited onto the star is also

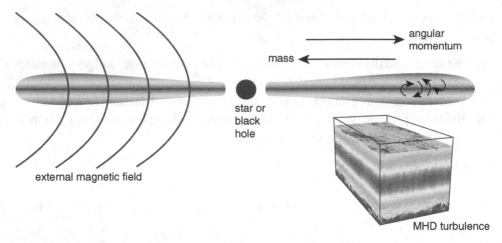

Figure 15.10 Schematic of an accretion disc. The inward advection of angular momentum is balanced by outward diffusion. (Courtesy of Phil Armitage.)

$$\dot{T} = \frac{\dot{m}\,GM}{2r_i}.$$

This kinetic energy is converted into heat near the surface of the star, which eventually manifests itself as radiation. Thus, half of the liberated potential energy is converted (via turbulence) into heat and then radiated from the surfaces of the disc, while the other half is converted into kinetic energy which is then deposited onto the star.

Note that, unlike the pre-disc collapse phase, Equation (15.24) tells us that Γ is not materially conserved. Rather, it is clear from (15.24) and (15.25) that the turbulent diffusion of angular momentum, and turbulent dissipation of mechanical energy, are both essential if we are to maintain a steady radial inflow through the disc (Figure 15.10). We shall address both of these issues in the next section, where we outline the so-called standard disc model.

Note also that the velocity distribution (15.24) is centrifugally stable by Rayleigh's linear criterion,

$$\Phi(r) = \frac{1}{r^3}\frac{d}{dr}\left(r^2\Omega\right)^2 > 0.$$

Indeed, numerical experiments show that it is also non-linearly stable to centrifugal modes of significant amplitude (Balbus and Hawley, 1998). So identifying the origin and nature of disc turbulence is central to interpreting accretion dynamics, and this has proved especially difficult in the weakly ionised protoplanetary discs.

15.3.2 The Standard Model of Accretion Discs

We now describe the standard model for accretion discs. It is somewhat of a minimalist model in the sense that it parameterises the turbulence in highly simplistic way and ignores the potentially important role played by magnetic fields in transporting angular momentum away from the disc. Nevertheless, it is a useful starting point. As before, we adopt cylindrical polar coordinates, (r, θ, z), centred on the star and aligned with the rotation axis. For simplicity, we take the disc to be axisymmetric and steady-on-average, assumptions that are readily relaxed.

Let us start with angular momentum transport within the disc. From (15.24) we have $\Gamma \sim \sqrt{r}$, so as the gas spirals inward it must somehow shed its excess angular momentum, and this is achieved by turbulent diffusion. The simplest way to parameterise the effects of turbulence is to use an eddy viscosity, ν_t, which we take to be a function of radius, r. We can then employ the laminar equations of motion with ν replaced by ν_t, on the understanding that all quantities, such as Γ, have been time-averaged. Neglecting magnetic torques, the azimuthal equation of motion is

$$\rho \frac{D\Gamma}{Dt} = \rho(\mathbf{u} \cdot \nabla)\Gamma = \frac{1}{r}\frac{\partial}{\partial r}\left(r^2 \tau_{r\theta}\right) , \qquad (15.27)$$

where

$$\tau_{r\theta} = \rho \nu_t r \frac{d\Omega}{dr} , \qquad (15.28)$$

is a Reynolds stress. In a steady-on-average flow, the mass conservation equation is

$$\nabla \cdot [\rho \mathbf{u}] = 0, \qquad (15.29)$$

and so (15.27) can be rewritten as

$$\nabla \cdot (\rho \Gamma \mathbf{u}) = \frac{1}{r}\frac{\partial}{\partial r}\left[\rho \nu_t r^3 \frac{d\Omega}{dr}\right] . \qquad (15.30)$$

It is instructive to integrate this over an annular ring within the disc, from r to $r + \delta r$. This yields

$$\oint \rho \Gamma \mathbf{u} \cdot d\mathbf{S} = -\frac{d}{dr}(\dot{m}\Gamma)\delta r = 2\pi \frac{d}{dr}\left[\nu_t r^3 \frac{d\Omega}{dr}\int_{-\infty}^{\infty}\rho dz\right]\delta r, \qquad (15.31)$$

which is more commonly written as

$$\frac{d}{dr}(\dot{m}\Gamma) = -2\pi \frac{d}{dr}\left[\nu_t r^3 \frac{d\Omega}{dr}\Sigma \right],$$ (15.32)

where

$$\Sigma = \int_{-\infty}^{\infty} \rho \, dz$$

is usually called the surface density. Expression (15.32) tells us that any radial inflow of angular momentum must be counterbalanced by an equivalent outward diffusion in order to keep the angular momentum profile constant. Integrating (15.32) with respect to r then yields

$$\dot{m}(\Gamma - \Gamma_0) = -2\pi \, \nu_t r^3 \frac{d\Omega}{dr}\Sigma,$$ (15.33)

for some constant Γ_0. Moreover, for a Keplerian orbit we have $r^3 d\Omega/dr = -(3/2)\Gamma$, and so our angular momentum transport equation simplifies to

$$\dot{m}(\Gamma - \Gamma_0) = 3\pi \nu_t \, \Gamma\Sigma.$$ (15.34)

Expression (15.34) emphasises the fact that the accretion rate, $\dot{m}$, is controlled by the eddy viscosity, ν_t.

The constant Γ_0 is determined by the boundary conditions near the inner radius of the disc, $r = r_i$. Typically, the star rotates more slowly than the inner portions of the disc, and so there is a transition region near $r = r_i$ where (15.33) holds but (15.34) does not. The details of the transition near $r = r_i$ are complex and depend on the strength of the solar dipole field, but it is natural to assume that there is a maximum in Ω as the Keplerian rotation profile, $\Omega \sim r^{-3/2}$, merges into the slower stellar rotation rate. It is therefore common to assume that $d\Omega/dr = 0$ at a point where $\Gamma \approx \Gamma_i = \sqrt{GMr_i}$. We shall adopt this boundary condition, so that (15.33) yields, in the transition region,

$$\dot{m}(\Gamma - \Gamma_i) = -2\pi \, \nu_t r^3 \frac{d\Omega}{dr}\Sigma.$$ (15.35)

Away from the transition region, where the Keplerian orbit holds, (15.35) simplifies to

$$\dot{m}[1 - \sqrt{r_i/r}] = 3\pi \nu_t \Sigma.$$ (15.36)

We now return to the question of the energy balance within the disc, and to Equations (15.25) and (15.26). We have already noted that, as mass spirals radially inward, half of the liberated potential energy is converted into kinetic energy, and the other half into heat which is subsequently radiated away from the surfaces of the disc. We shall treat the disc as an optically thick black body and assume that the thermal energy is radiated away as soon as it is generated. Moreover, in the interests of simplicity, we shall at first ignore the diffusive flux of mechanical energy by turbulent Reynolds stresses within the disc. (We shall correct for this shortly.) Given these assumptions, the energy balance for an annular section of the disc, δr, is

$$2[\sigma T_s^4]2\pi r(\delta r) = \frac{d}{dr}(\dot{m}e)\delta r, \qquad (15.37)$$

where σ is the Stefan–Boltzmann constant, T_s is the surface temperature of the disc and σT_s^4 the radiated energy per unit surface area and per unit time. Substituting for e using (15.25), we find

$$\sigma T_s^4 = \frac{\dot{m}GM}{8\pi r^3}. \qquad (15.38)$$

This gives a surface temperature of

$$T_s(r) = \left[\frac{\dot{m}GM}{8\pi\sigma r^3}\right]^{1/4}, \qquad (15.39)$$

and the total luminosity of the disc as

$$L_{disc} = 2\int_{r_i}^{\infty}[\sigma T_s^4]2\pi r dr = \frac{\dot{m}GM}{2r_i}, \qquad (15.40)$$

in accordance with (15.26). (We have taken the outer radius of the disc to be much larger than r_i.) Of course, the kinetic energy held in the fluid at $r = r_i$ is deposited onto the surface of the star.

While this analysis is reassuringly simple and yields the correct value of luminosity, it is flawed. In particular, (15.37) does not allow for the turbulent redistribution of kinetic energy within the disc by virtue of turbulent diffusion. So let us now repeat the calculation, allowing for turbulent diffusion. The mechanical energy dissipation rate per unit volume is

$$\tau_{r\theta} \, r \frac{d\Omega}{dr} = \rho \nu_t \left[r \frac{d\Omega}{dr} \right]^2, \tag{15.41}$$

which tells us that the rate of generation of heat, integrated through the thickness of the disc, is

$$\nu_t \left[r \frac{d\Omega}{dr} \right]^2 \Sigma. \tag{15.42}$$

This differs from the right-hand side of (15.37) by an amount equal to the turbulent diffusion of mechanical energy, as discussed in, for example, Davidson, 2013. Substituting for the eddy viscosity using (15.36), and assuming that this heat is radiated away as black-body radiation, we obtain

$$2\left(\sigma T_s^4\right) = \frac{\dot{m}}{3\pi} \left[r \frac{d\Omega}{dr} \right]^2 \left(1 - \sqrt{r_i/r} \right), \tag{15.43}$$

where T_s is the disc surface temperature. For a Keplerian orbit this yields the classical result

$$\sigma T_s^4 = \frac{3\dot{m} GM}{8\pi \, r^3} \left(1 - \sqrt{r_i/r} \right), \tag{15.44}$$

which also integrates to give $L_{disc} = \dot{m} GM/2r_i$. This differs from estimate (15.38) because the latter does not allow for the turbulent diffusion of energy within the disc. Expression (15.44) yields a surface temperature profile of

$$T_s(r) = \left[\frac{3\dot{m} GM}{8\pi\sigma r^3} \right]^{1/4} \left[1 - \sqrt{r_i/r} \right]^{1/4}. \tag{15.45}$$

Let us now consider the vertical structure of the disc. This is determined from a balance between the axial pressure gradient and the z-component of the gravitational force,

$$\frac{\partial p}{\partial z} = -\frac{GM\rho z}{r^3} = -\rho\Omega^2 z. \tag{15.46}$$

Let H be the half-width of the disc, whose precise definition is not important, and let the subscript c indicate a property measured at the mid-plane of the disc. Then (15.46) yields $p_c/H \sim \rho_c \Omega^2 H$, from which we conclude $c_s \sim \Omega H$, where c_s is the isothermal sound speed, $c_s = \sqrt{p_c/\rho_c}$. This provides the simple but useful estimate $H \sim c_s/\Omega$, to which we shall return.

In order to use the standard disc model to predict quantities like $T_s(r)$, we need to estimate $\dot{m}$, which is itself controlled by ν_t through (15.36). In short, we need a turbulence closure estimate of ν_t. Typically this is achieved using the popular *alpha-viscosity* proposal of Shakura and Sunyaev (1973). This is essentially a mixing-length estimate of the form

$$\nu_t = \alpha\, c_s H, \quad \alpha \sim 1, \tag{15.47}$$

which amounts to an assertion that the turbulent velocity fluctuations scale as c_s. Since the only constraint on the free parameter α is that it be of order unity, there is, of course, considerable freedom when it comes to a comparison with observations. The value of c_s in (15.47) is based on the mid-plane pressure and density. If we ignore contributions to p from the radiation pressure, then we have $p/\rho = R_g T$ for a gas constant R_g, and hence the sound speed is $c_s = \sqrt{R_g T_c}$, where T_c is the mid-plane temperature. Our estimate of the eddy viscosity now becomes $\nu_t = \alpha\sqrt{R_g T_c}H$, and expression (15.36) predicts

$$\dot{m} = 3\pi\alpha\sqrt{R_g T_c}H\,\Sigma, \qquad r \gg r_i. \tag{15.48}$$

We now combine this with (15.44) and the estimate $H \sim c_s/\Omega$. Ignoring vertical gradients in temperature, so that $T_s \approx T_c$, we find, for $r \gg r_i$,

$$T_c \sim \frac{\dot{m}^{1/4}}{\sigma^{1/4}}\left[\frac{GM}{r^3}\right]^{1/4}, \tag{15.49}$$

$$H \sim \frac{R_g^{1/2}\dot{m}^{1/8}}{\sigma^{1/8}}\left[\frac{GM}{r^3}\right]^{-3/8}, \tag{15.50}$$

$$\rho_c \sim \frac{\sigma^{3/8}\dot{m}^{5/8}}{\alpha\, R_g^{3/2}}\left[\frac{GM}{r^3}\right]^{5/8}. \tag{15.51}$$

In practice, taking $T_s = T_c$ is not always a good approximation and often a radiation opacity model is invoked to relate the two temperatures. Typically one introduces a dimensionless optical depth, $\tau(\rho, T)$, satisfying $\sigma T_c^4 = \tau\sigma\, T_s^4$, and this alters (15.49) – (15.51) to read

$$T_c \sim \frac{\tau^{1/4} \dot{m}^{1/4}}{\sigma^{1/4}} \left[\frac{GM}{r^3}\right]^{1/4},$$

$$H \sim \frac{\tau^{1/8} R_g^{1/2} \dot{m}^{1/8}}{\sigma^{1/8}} \left[\frac{GM}{r^3}\right]^{-3/8},$$

$$\rho_c \sim \frac{\sigma^{3/8} \dot{m}^{5/8}}{\tau^{3/8} \alpha \, R_g^{3/2}} \left[\frac{GM}{r^3}\right]^{5/8}.$$

A common disc opacity model (the free-free-absorption model) has $\tau = K\rho_c T_c^{-7/2}\Sigma$, where K is a dimensional constant. This results in the particularly simple scaling $\tau \sim \hat{K}\dot{m}^{1/5}\alpha^{-4/5}$, where $\hat{K}$ is second-dimensional constant. The power-law dependence of T_c, H and ρ_c on GM/r^3 then remains unchanged from (15.49) to (15.51), but their scaling on $\dot{m}$ is slightly altered. These scaling laws are discussed in, for example, Frank, King and Raine (2002).

As noted earlier, this disc model is highly simplified, requiring axial symmetry, steadiness and weak magnetic forces. The parameterisation of the turbulence is also somewhat naïve. The first two of the restrictions above are readily removed, but the third is not. Magnetic fields typically thread through the disc, exerting torques on the gas and allowing excess angular momentum and mass to be carried away. This can have important consequences in the case of highly ionised discs. The simplistic turbulence modelling, with its free parameter, also leaves much to be desired. Nevertheless, the model is a useful starting point for further discussion.

15.3.3 The Chandrasekhar–Velikhov Instability Revisited

The discussion above highlights the crucial role played by turbulent diffusion in setting the rate of accretion. In highly ionised accretion discs, the primary source of turbulence is thought to be MRI. Recall from §7.7 that a rotating fluid which is stable according to Rayleigh's centrifugal criterion can be destabilised by an axial magnetic field, even a very weak field. In particular, an ideal fluid threaded by a mean axial field, B_0, is stable to axisymmetric disturbances, provided

$$r\frac{d\Omega^2}{dr} > 0, \tag{15.52}$$

which is a more exacting precondition than Rayleigh's centrifugal criterion,

$$\frac{1}{r^3}\frac{d}{dr}\left(r^2\Omega\right)^2 > 0. \tag{15.53}$$

A stronger result can be obtained by considering the short wavelength limit, where it may be shown that a necessary and sufficient condition for instability is $d\Omega^2/dr < 0$ and

$$r\left|\frac{d\Omega^2}{dr}\right| > k^2\frac{B_0^2}{\rho\mu} = k^2 v_a^2, \tag{15.54}$$

where v_a is the Alfvén velocity and k is the magnitude of the wavenumber (Balbus and Hawley, 1998). For a thin disc, k will be dominated by k_z, and so for a highly ionised Keplerian disc the flow is unstable whenever $v_a < \sqrt{3}\,\Omega/k_z$. Given that the smallest value of k_z in a disc is $\pi/2H$, our instability criterion for a highly conducting disc becomes

$$v_a < \frac{2\sqrt{3}}{\pi}\Omega H. \tag{15.55}$$

In practice, a finite magnetic diffusivity, λ, sets a lower limit on the value of v_a that can cause an instability, and order-of-magnitude arguments suggest that this limits the unstable range to

$$\frac{\lambda}{H} < v_a < \frac{2\sqrt{3}}{\pi}\Omega H. \tag{15.56}$$

The lower limit in (15.56) explains why MRI cannot occur in the weakly ionised regions of protoplanetary discs.

16

Plasma Containment in Fusion Reactors

It is easier to write ten volumes on theoretical principles than put one into practice.

Tolstoy

16.1 The Quest for Controlled Fusion Power

Some projects in physics and engineering proceed at an astonishing pace, whilst others hit unforeseen and formidable problems, making progress slow and incremental. Compare, for example, the development of space flight with the quest for power generation through controlled nuclear fusion. Sputnik was launched in the fall of 1957, and a mere twelve years later the Apollo programme had put a man on the moon. The subsequent exploration of the universe through spacecraft-based instruments has progressed at a remarkable pace, heralding a renaissance in solar physics and planetary science. By way of contrast, the possibility of commercial power generation from controlled nuclear fusion has been viewed as tantalisingly just beyond our reach since the early pioneers, such as Andrei Sakharov and Lyman Spitzer, first promoted the idea over half a century ago.

The first patent for a fusion reactor was registered by the UK Atomic Energy Authority in 1946, but at that time all such studies were classified. Research into controlled fusion was declassified in 1955 and the following year, the year before Sputnik was launched, experimental studies began in Moscow into controlled fusion using a tokamak to magnetically confine the plasma. The design was, in part, inspired by a proposal of Andrei Sakharov some five years earlier. (The term 'tokamak' is a contraction of the Russian for *toroidal magnetic chamber*.) Of course, there were simultaneous attempts in the West at plasma confinement, most notably in the USA and the UK. These used various alternative magnetic confinement schemes, such as the *z*-pinch and stellarator, but such schemes encountered severe problems with MHD instabilities and associated plasma break-up. In 1968, Soviet scientists finally announced to their somewhat sceptical western

514

counterparts that reactor T-4 in Novosibirsk had reached the high temperature (10^8K) needed for a fusion reaction. The sceptics soon became converts and the tokamak was rapidly adopted by the west as the preferred means of plasma containment. Several large tokamaks have since been built.

The first controlled release of deuterium-tritium fusion power finally came in 1991 in the JET (Joint European Torus) reactor at Culham in the UK. However, despite this significant achievement, confinement times in tokamaks have remained persistently low (less than a second) with the plasma plagued by a zoo of MHD instabilities. As a result, fusion reactors have yet to produce more energy from the reaction than consumed during its initiation. Moreover, in the absence of a self-sustained fusion reaction, those complex engineering systems, which are so essential for heat extraction and tritium breeding in any commercial fusion reactor, have yet to be tested in anger. So those early dreams of a commercial fusion reactor have remained stubbornly elusive.

The fusion community has, nevertheless, remained resilient. Undaunted by past frustrations, plasma physicists and nuclear engineers are currently constructing the most ambitious tokamak to date. Called ITER (International Thermonuclear Experimental Reactor), it is being constructed at the nuclear facility in Cadarache, France. It had an initial budget of 10 billion Euros, although projected costs are much larger. Its supporters hope that ITER will yield a self-sustained fusion reaction by around 2027, with ten times more power out than in. If successful, the quest for a commercial fusion reactor will progressively shift from one of plasma physics and magnetic containment to one of nuclear engineering and reactor design (first wall design, reactor cooling and tritium breeding). However, it is far from clear that the battle to control MHD instabilities has been won, and the spectre of so-called 'edge-localised' modes at the outer edge of the plasma still casts a shadow over ITER, as discussed below.

This chapter provides a brief introduction to controlled fusion, with an emphasis on the MHD aspects of plasma containment. Those interested in more details could do worse than consult Boyd and Sanderson (2003), Freidberg (2014) or Biskamp (1993) for a thorough discussion of stability, or Bühler (2007) for an overview of reactor cooling and tritium breeding. We start, however, by recalling the basic requirements of controlled fusion.

16.2 The Requirements for Controlled Nuclear Fusion

In fusion, the reacting nuclei are electrically charged, so a large kinetic energy is needed to overcome the Coulomb repulsion, thus allowing particles to get sufficiently close for fusion to occur. Consequently, the traditional problem of controlled fusion has been the need to maintain a high temperature (in excess of 10^8K) and for long enough (greater than several seconds) to establish a self-sustaining fusion reaction. This, in turn, requires stable confinement of the fusion plasma in a 'magnetic bottle' for extended periods of time, eliminating all contact with solid surfaces.

The Lawson criterion provides a minimum condition that must be met in order to get more energy out of the plasma than is put in. It takes the form

$$n\tau > F(T), \tag{16.1}$$

where T is the plasma temperature, n is the number of nuclei per unit volume, τ is the confinement time, and the form of the function F depends on the reactants. The most promising reaction for laboratory plasmas, that which requires the lowest temperature, is between the two isotopes of hydrogen: deuterium and tritium. It is

$$_1D^2 + {}_1T^3 \rightarrow {}_2He^4 + {}_0n^1 + 17.6 \text{ MeV},$$

and the Lawson criterion applied to this reaction at a temperature of 10^8K demands $n\tau > 10^{20} \rightarrow 10^{21}\text{m}^{-3}\text{s}$. Tokamaks use magnetic fields to confine the plasma away from solid surfaces and can maintain plasma densities of up to 10^{20} ions m^{-3} (Boyd and Sanderson, 2003), so confinement times in excess of several second are required to initiate a self-sustaining fusion reaction. The high plasma temperature in a tokamak is maintained partly through Ohmic heating, partly by neutral beam injection, and partly by radio frequency waves. The energy released by the reaction manifests itself mostly in the form of the kinetic energy of the neutrons, with around 20 per cent associated with the kinetic energy of the He nuclei.

Deuterium is abundant, being one part in 6500 in naturally occurring hydrogen. On the other hand, tritium has a half-life of only 12 years, and so is hard to find, produce and store. Consequently, any commercial fusion reactor would have to breed tritium from lithium, or lithium-lead eutectic, using the neutrons released from the fusion plasma. Two possible reactions are

$$_0n^1 + {}_3Li^6 \rightarrow {}_1T^3 + {}_2He^4 + 4.8 \text{ MeV},$$

and

$$_0n^1 + {}_3Li^7 \rightarrow {}_1T^3 + {}_0n^1 + {}_2He^4 - 2.8 \text{ MeV}.$$

The Li^6 reaction is exothermic, but consumes a neutron, while the Li^7 reaction is endothermic, but does not consume a neutron. Natural lithium is 92.6 per cent Li^7, and so most proposed breeder units contain a mixture of the two isotopes.

In any fusion reactor the plasma must be enveloped by a moderator blanket which slows down the neutrons and absorbs their energy. Between the blanket and the plasma sits the so-called *first wall*, which provides a seal around the plasma and is subject to particularly intense radiation levels. In tokamak reactors it is proposed that the moderator blanket double as the tritium breeder, so it comprises liquid lithium or, in more recent proposals, liquid lead-lithium eutectic. This blanket then serves the three distinct roles of (i) breeding tritium, (ii) absorbing and then removing the neutron fusion energy and (iii) shielding the superconducting magnet coils from a high neutron flux.

Of course, the heat absorbed in the moderator blanket must be carried out of the reactor so that it can be used to produce electrical power. Consequently, a cooling system is also required, with helium being the preferred coolant. Some proposals invoke a dual cooling system, with helium cooling all the walls, particularly the first wall, and liquid lead-lithium eutectic flowing through ducts within the blanket, extracting the neutron-induced volumetric heating. One difficulty here is that pumping liquid metal through an intense magnetic field requires a very large pressure gradient in order to overcome the opposing Lorentz forces, so the design of such a cooling system is non-trivial. When coupled with the need to breed and extract tritium, the blanket design becomes a major engineering challenge in its own right (see, for example, Bühler, 2007).

16.3 Magnetic Confinement and the Instability of Fusion Plasmas

16.3.1 The Topology of Confinement

The first step in producing controlled fusion power is to use magnetic fields to confine and pressurise the plasma in a stable configuration. So let us return to §7.5.1 and the issue of magnetostatic stability. The magnetostatic equation,

$$\mathbf{J} \times \mathbf{B} = \nabla p, \tag{16.2}$$

yields

$$\mathbf{J} \cdot \nabla p = \mathbf{B} \cdot \nabla p = 0, \tag{16.3}$$

and so the $\mathbf{J}$ and $\mathbf{B}$ lines lie on surfaces of constant pressure. If the plasma is to be confined away from walls, then each constant pressure surface must be closed within the plasma, and since the surfaces cannot intersect they must be nested, as shown in Figure 16.1. Moreover, these surfaces cannot be simply connected, like the surface of a sphere, since any closed curve on such a surface around which

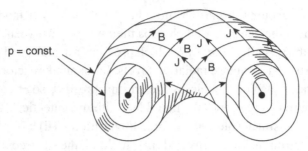

Figure 16.1 In confined plasmas, the **B** and **J** lines sit on constant pressure surfaces which take the form of nested tori.

$\oint \mathbf{B} \cdot d\mathbf{r}$ is non-zero would demand current flowing through that surface in accordance with Ampère's law,

$$\oint \mathbf{B} \cdot d\mathbf{r} = \mu \int \mathbf{J} \cdot d\mathbf{S}.$$

However, this is forbidden by (16.3). So the constant pressure surfaces must have a more complicated topology, and the simplest option is a set of closed, nested tori, as shown in Figure 16.1. This is the topology adopted in tokamaks. The surface of vanishing area which lies at the centre of these nested tori is called the *magnetic axis*.

One measure of the success of a particular magnetic confinement scheme is the β-ratio of the plasma, defined as

$$\beta = \frac{p}{B^2/2\mu},$$

where B is, by convention, taken as the magnetic field external to the plasma. A high β is desirable, in part because the Lawson criterion demands a high plasma density, and in part because the fusion power density scales approximately with the square of the plasma pressure. In practice, however, the achievable value of β is severely limited by the stability characteristics of the particular confinement scheme, as discussed below. It rarely rises above 0.3 and is more typically around $0.03 \rightarrow 0.1$ in tokamaks.

16.3.2 Sausage-Mode and Kink Instabilities Revisited

The magnetic confinement of a fusion plasma is an exercise in high-R_m MHD. In principle, the field containing the plasma could be purely *poloidal*, wrapped

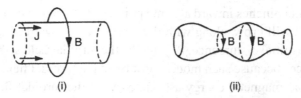

Figure 16.2 The z-pinch, showing (i) the confinement principle, and (ii) instability of the pinch.

around the plasma in the r–z plane and pinching radially inward towards the magnetic axis, as shown in Figure 16.2. Such a field is supported by *toroidal* (i.e. azimuthal) currents within the plasma, running parallel to the magnetic axis.[1] This current may be induced in the surface of the plasma by transformer action, at least for a limited period of time. However, we saw in §7.5.1 that such a configuration is unstable to sausage-mode instabilities, as indicated in Figure 16.2. That is, if I is the net toroidal current flowing around the plasma loop, then Ampère's law tells us that the mean poloidal field at the surface of the plasma is $B_P = \mu I / 2\pi a$, where a is the minor radius of the torus. If the plasma pinches in at some point, conserving I in the process, then the external poloidal field increases, along with the associated magnetic pressure, accentuating the initial disturbance.

In §7.5.1 we showed that such an instability may be stabilised by trapping a toroidal magnetic field within the plasma, running parallel to the magnetic axis. Such a field may be produced by wrapping a superconducting solenoid in the shape of a torus around the plasma, which generates a toroidal field both inside and outside the plasma. The toroidal field within the plasma, B_T, pushes radially outward from the magnetic axis towards the plasma surface, a force that may or may not be counterbalanced by the magnetic pressure associated with an external toroidal field.[2] Such a toroidal field trapped within the plasma is readily shown to stabilise sausage-mode disturbances. Consider the simplest case where the toroidal field is small outside the plasma and relatively uniform within it. Then $B_T = \Phi / \pi a^2$ within the plasma, where Φ is the net magnetic flux circulating around the plasma

---

[1] When discussing axisymmetric fields in cylindrical polar (r, θ, z) coordinates, it is usual to refer to the r-z components as *poloidal* and to the θ component as either *azimuthal* or *toroidal*. When describing tokamaks in particular, it is conventional to adopt the term toroidal in preference to azimuthal, so we shall break with our convention in this chapter and switch terminology from azimuthal to toroidal for the θ component. Note that confusion can easily develop when the plasma torus is 'unwrapped' to become a long cylinder, as in Figure 16.2. In such cases, what was the poloidal field in the context of a torus now apparently becomes an azimuthal field in the cylindrical context! In order to avoid confusion we shall always have in mind a torus, albeit one that may look locally like a cylinder, and so the field in Figure 16.2 is labelled as poloidal.

[2] The radially inward pinch force (per unit volume) associated with the net toroidal field, integrated from the magnetic axis, $s = 0$, to the plasma surface, $s = a$, is $\left(B_T^2 / 2\mu\right)_{s=a} - \left(B_T^2 / 2\mu\right)_{s=0}$, where B_T is the toroidal field and s is the distance from the magnetic axis.

loop. If the plasma pinches inward at some point, conserving Φ in the process, then the toroidal field will grow, along with the associated outward Lorentz force, and this counters the initial disturbance. Physically, the toroidal field stabilises sausage-mode disturbances because such modes must bend the toroidal field lines, which in turn increases the magnetic energy associated with the toroidal field. Thus work must be done against the Faraday tension within the field lines, a tension which tends to push the **B**-lines back towards a parallel configuration.

In §7.5.1 we did a simple, pseudo-static stability calculation where we assumed that there was both a poloidal field surrounding the plasma, pushing inward, and a toroidal field trapped within the plasma, pushing outward. In the stability calculation we ignored any external toroidal field, as well as curvature effects associated with the major axis of the torus (i.e. we considered the undisturbed plasma to be cylindrical). This simple calculation showed that such a configuration is stable to sausage-mode disturbances, provided that $B_T^2 > B_P^2/2$. Of course, there are still many potentially dangerous three-dimensional instabilities in such cases, such as the kink instability.

We may use standard normal-mode analysis to generalise this simple calculation to include an external toroidal field and to allow for more complex disturbances. As above, we take the major axis of the torus to be very large and so we may consider the plasma to be confined to a cylinder of radius a, and we use s to represent the perpendicular distance from the magnetic axis. Suppose that the field inside the plasma is uniform and purely toroidal, B_{in}, while that outside has both a poloidal component, B_P, and a toroidal component, B_{ex}, the external toroidal field again being assumed uniform (Table 16.1). Such a configuration can be supported, at least in part, by surface currents at the edge of the plasma that generate both the external poloidal field, $B_p = \mu I/2\pi s$, and the jump in toroidal field across the interface. (I is the net current flowing along the plasma cylinder.)

We now look for disturbances of the form $f(s)\exp[j(m\varphi + kx - \varpi t)]$, where φ is the poloidal angle and x is measured along the magnetic axis. Taking the plasma to be incompressible, which is usually a conservative assumption when it comes to plasma stability (Freidberg, 2014), the linearised perturbation equations of ideal MHD yield the dispersion relationship

Table 16.1 *Magnetostatic configuration in the model problem of (16.4).*

	Internal field, $0 < s < a$.	External field, $s > a$.
Toroidal field component	$B_T = B_{in}$ = constant	$B_T = B_{ex}$ = constant
Poloidal field component	0	$B_p = \mu I/2\pi s$

$$\frac{\varpi^2}{k^2} = \frac{B_{in}^2}{\rho\mu} - \frac{\left(B_{ex} + mB_p(a)/ka\right)^2}{\rho\mu} \frac{I'_m(ka)K_m(ka)}{I_m(ka)K'_m(ka)} - \frac{B_p^2(a)}{\rho\mu} \frac{I'_m(ka)}{kaI_m(ka)}, \quad (16.4)$$

I_m and K_m being the usual modified Bessel functions. (See, for example, the derivation in Boyd and Sanderson, 2003.) Evidently ϖ is either real (stable solutions) or pure imaginary (exponential growth). Since I_m, I'_m and K_m are all positive, and K'_m negative, the first two terms on the right are strictly positive and so instability arises when the third term on the right dominates the other two.

Consider first the case of the sausage-mode instability, $m = 0$. Here both the internal and external toroidal magnetic fields are stabilising, and for precisely the same reason: a sausage-mode instability needs to bend the toroidal field lines and this requires a source of energy, with the Faraday tension tending to pull the field lines back to a parallel configuration. The poloidal field, on the other hand, is destabilising for the reason discussed above and illustrated in Figure 16.2. When the external toroidal field is zero we have

$$\frac{\varpi^2}{k^2} = \frac{B_{in}^2}{\rho\mu} - \frac{B_p^2(a)}{\rho\mu} \frac{I'_0(ka)}{kaI_0(ka)}, \quad (16.5)$$

and since the maximum value of the destabilising term $I'_0(ka)/kaI_0(ka)$ is $\frac{1}{2}$, we conclude that the plasma is stable, provided $B_{in}^2 > B_P^2(a)/2$. Of course, we arrived at exactly the same conclusion in §7.5.1, though in that case through a simple consideration of quasi-static changes in magnetic pressure.

When considering modes in which $m > 0$ it is convenient to focus on the long wavelength limit of $ka \to 0$, so that the modified Bessel functions may be replaced by power laws. In that case (16.4) simplifies to

$$\frac{\varpi^2}{k^2} = \frac{B_{in}^2}{\rho\mu} + \frac{\left(B_{ex} + mB_p(a)/ka\right)^2}{\rho\mu} - \frac{B_p^2(a)}{\rho\mu} \frac{m}{(ka)^2}. \quad (16.6)$$

This may be rewritten using the equilibrium condition

$$p_0 + \frac{B_{in}^2}{2\mu} = \frac{B_{ex}^2}{2\mu} + \frac{B_P^2(a)}{2\mu}, \quad (16.7)$$

in the form

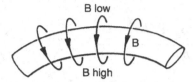

Figure 16.3 The kink instability.

$$\frac{\varpi^2/k^2}{B_{ex}^2/\rho\mu} = 1 + \frac{\left(B_{ex} + mB_p(a)/ka\right)^2}{B_{ex}^2} - \frac{B_p^2(a)}{B_{ex}^2}\left[\frac{m}{(ka)^2} - 1\right] - \frac{p_0}{B_{ex}^2/2\mu}. \quad (16.8)$$

Consider the case of $m = 1$, which is the kink instability shown in Figure 16.3, and for which (16.8) reduces to

$$\frac{\varpi^2}{k^2} = \frac{B_{ex}^2}{\rho\mu}\left[2 + \frac{B_p(a)}{B_{ex}}\frac{2}{ka} + \frac{B_p^2(a)}{B_{ex}^2} - \frac{p_0}{B_{ex}^2/2\mu}\right]. \quad (16.9)$$

Evidently, the poloidal field is destabilising through the second term in the square bracket, while the toroidal field is, once again, stabilising. The instability mechanism is clear from Figure 16.3. As the plasma column begins to buckle, the magnetic pressure associated with the poloidal field increases on the concave side of the column and decreases on the convex side. These changes in pressure exacerbate the initial disturbance and so an instability grows. The toroidal field, on the other hand, is stabilising because the Faraday tension in the field lines resists bending.

As noted above, β is small in most magnetically confined fusion plasmas, and so the last term on the right of (16.9) may often be neglected. Also, in tokamaks, the toroidal field is dominant. In such cases we obtain a particularly simple criterion for the avoidance of the kink instability:

$$|B_p(a)/B_{ex}| < ka = 2\pi a/L, \quad (16.10)$$

where L is the wavelength of the instability. We may think of the toroidal field as stiffening the plasma column, and if it is too weak the column buckles under the pressure of the poloidal field.

The idea of using combined poloidal and toroidal fields is the basis of the magnetic configuration used in tokamaks, as shown schematically in Figure 16.4. The toroidal plasma current, which supports the external poloidal field, is induced by transformer action (at least, during the start-up phase). The toroidal field within the plasma, on the other hand, is induced by

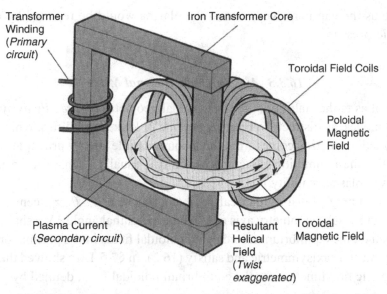

Figure 16.4 Schematic of the magnetic configuration in a tokamak, which uses combined poloidal and toroidal fields.

superconducting coils arranged around the plasma, rather like a torus-shaped solenoid. Loosely speaking, we may think of the poloidal magnetic field as pressurising the plasma, with the toroidal field providing stability, although it turns out that in many configurations (particularly those with a relatively high β) the toroidal field provides most of the pressurisation by maintaining a higher value outside the plasma than inside. Of course, within the plasma individual field lines are helical, as shown in Figure 16.4.

In such a geometry, the wavelength of the kink instability, L in (16.10), cannot be greater than $2\pi R$, where R is the major radius of the torus. So, for a tokamak, the criterion for avoiding kink instabilities becomes

$$|B_p(a)/B_{ex}| < a/R.$$

This greatly restricts the magnitude of the poloidal field (and associated toroidal current) which may be sustained. For this reason the toroidal field is dominant (typically ten times the poloidal field strength) and the pitch of the helical field lines is small.

The tokamak geometry provides additional stability to certain modes by virtue of the close proximity of the superconducting coils used to support the toroidal field. The idea is that, if the plasma were to drift outward towards the coils, the poloidal field trapped between the plasma and the coils would increase as the gap narrowed, and so the plasma would be repelled by a rise in magnetic pressure.

16.3.3 *Axisymmetric Internal Modes*

Some hint as to the stability of such a toroidal geometry to *globally axisymmetric* internal modes, as distinct from sausage-mode or kink instabilities, which are not globally axisymmetric, comes from the incompressible energy principle of §7.5.2. (Recall that the assumption of incompressibility is usually a conservative one when it comes to plasma stability.)

Let us employ global cylindrical polar coordinates, (r, θ, z), centred on the torus, where z is the symmetry axis and θ is the azimuthal angle.[3] We shall place no restriction on the equilibrium toroidal and poloidal fields in the plasma, other than that they must be axisymmetric and satisfy (16.2). In §7.5.1 we showed that, if Φ is the magnetic flux function for the equilibrium poloidal field, defined by

$$\mathbf{B}_p = \nabla \times [(\Phi/r)\hat{\mathbf{e}}_\theta] = \left(-\frac{1}{r}\frac{\partial \Phi}{\partial z}, \ 0, \ \frac{1}{r}\frac{\partial \Phi}{\partial r} \right), \qquad (16.11)$$

then the equilibrium pressure distribution and equilibrium toroidal field take the form

$$p = p(\Phi), \quad rB_\theta = \Gamma(\Phi), \qquad (16.12)$$

for some function Γ. Moreover, the poloidal force balance yields an expression for the equilibrium toroidal current

$$\mu\mathbf{J}_\theta = \nabla \times \mathbf{B}_p = r^{-1}\left(\Gamma\Gamma'(\Phi) + r^2\mu p'(\Phi) \right)\hat{\mathbf{e}}_\theta,$$

which comes directly from (7.37). This, in turn, yields

$$\mu\mathbf{J} = \Gamma'(\Phi)\mathbf{B} + r\mu p'(\Phi)\hat{\mathbf{e}}_\theta. \qquad (16.13)$$

Note that the component of $\mathbf{J}$ perpendicular to $\mathbf{B}$ is associated with $p'(\Phi)$, as it must, since $\mathbf{J} \times \mathbf{B} = \nabla p$.

Turning to stability, in §7.5.2 we showed that the second-order change in magnetic energy resulting from the application of a solenoidal displacement field, $\mathbf{\eta}$, to the equilibrium magnetic field $\mathbf{B}_0$ is, in an ideal plasma,

---

[3] Note that many authors use θ for the poloidal angle, rather than the azimuthal angle. However, in line with the rest of the book, we shall reserve (r, θ, z) for cylindrical polar coordinates.

$$\delta^2 E_B(\boldsymbol{\eta}) = \frac{1}{2\mu} \int [b^2 + \mathbf{B}_0 \cdot \nabla \times (\boldsymbol{\eta} \times \mathbf{b})] \, dV, \quad \mathbf{b} = \delta^1 \mathbf{B} = \nabla \times (\boldsymbol{\eta} \times \mathbf{B}_0),$$

$$(16.14)$$

or equivalently,

$$\delta^2 E_B(\boldsymbol{\eta}) = \frac{1}{2\mu} \int [b^2 - \mu \mathbf{b} \cdot (\boldsymbol{\eta} \times \mathbf{J}_0)] \, dV . \tag{16.15}$$

Stability is ensured whenever $\delta^2 E_B(\boldsymbol{\eta}) > 0$ for all admissible solenoidal fields $\boldsymbol{\eta}$, while instability is guaranteed if any solenoidal trial function, $\boldsymbol{\eta}$, can be found that makes $\delta^2 E_B(\boldsymbol{\eta})$ negative. If we restrict ourselves to globally axisymmetric disturbances, then it is readily confirmed that

$$\mu \boldsymbol{\eta} \times \mathbf{J}_0 = (\eta_\theta / r)\nabla(rB_\theta) + \mu J_\theta(-\eta_z, 0, \eta_r) - r^{-1}\boldsymbol{\eta}_p \cdot \nabla(rB_\theta)\hat{\mathbf{e}}_\theta, \tag{16.16}$$

while $\delta^1 \Phi = -\boldsymbol{\eta}_p \cdot \nabla \Phi$, which we label φ, and the first-order change in $\mathbf{B}$ is

$$\mathbf{b} = \mathbf{b}_p + \mathbf{b}_\theta = \nabla \times \left((\varphi / r)\hat{\mathbf{e}}_\theta \right) + r\left[\mathbf{B}_p \cdot \nabla(\eta_\theta / r) - \boldsymbol{\eta}_p \cdot \nabla(B_\theta / r) \right] \hat{\mathbf{e}}_\theta. \tag{16.17}$$

After some vector manipulation, and integration by parts, Equations (16.12) through (16.17) eventually yield, for globally axisymmetric disturbances in an ideal plasma,

$$\delta^2 E_B(\boldsymbol{\eta}) = \frac{1}{2\mu} \int \left[\left(\delta^1 \Gamma + \boldsymbol{\eta}_p \cdot \nabla \Gamma \right)^2 + (\nabla \varphi)^2 - \Psi \varphi^2 \right] r^{-2} dV, \tag{16.18}$$

where $\delta^1 \Gamma = r b_\theta$ and

$$\Psi(r, \Phi) = \mu r^2 p''(\Phi) + \left(\Gamma \Gamma'' + \Gamma'^2 \right),$$

(Davidson, 1994). Note that we have used $\nabla \cdot \boldsymbol{\eta} = 0$ and the boundary conditions $\mathbf{B}_0 \cdot d\mathbf{S} = \mathbf{b} \cdot d\mathbf{S} = \boldsymbol{\eta} \cdot d\mathbf{S} = 0$, as is appropriate for (16.14). (The energy principle in the form of (16.14) assumes the plasma to be incompressible and that it, and its associated magnetic field, are enclosed by a rigid surface, S.)

Evidently η_θ contributes to $\delta^2 E_B$ through $\delta^1 \Gamma$ alone, and this is a result of toroidal field generation by differential rotation in the η_θ field. In those cases where it is possible to choose η_θ such that $\delta^1 \Gamma = -\boldsymbol{\eta}_p \cdot \nabla \Gamma$, the only stabilising term in (16.18) is $(\nabla \varphi)^2$, which corresponds to b_p^2 in (16.15). In any event, the sole source of instability in (16.18) is $-\Psi\varphi^2$, and so a sufficient (but not necessary) condition for stability to internal axisymmetric disturbances is [4]

$$\Psi = \mu r^2 p''(\Phi) + \left(\Gamma\Gamma'' + \Gamma'^2\right) < 0. \qquad (16.19)$$

However, the equilibrium toroidal current is

$$\mu r J_\theta = \Gamma\Gamma'(\Phi) + r^2 \mu p'(\Phi),$$

from which we obtain

$$\mu \frac{\partial J_\theta}{\partial z} = -\left[\mu r^2 p''(\Phi) + \left(\Gamma\Gamma'' + \Gamma'^2\right)\right] B_r = -\Psi B_r. \qquad (16.20)$$

We conclude that axisymmetric stability to internal modes is ensured provided that B_r and $\partial J_\theta / \partial z$ have the same sign, a condition which is readily satisfied. Suppose, for example, that the toroidal current, J_θ, is either positive for all s or else negative for all s. Then according to (16.19) and (16.20) we have stability to axisymmetric disturbances, provided that $|J_\theta|$ is a minimum on the magnetic axis and increases monotonically towards the edge of the plasma.

The physical origin of this criterion, which is partly an artefact of assuming that the plasma (and its associated magnetic field) is bounded by a solid surface, is discussed in Davidson (1994). The key point is that, in this kind of configuration, the magnetic energy associated with the toroidal current can be minimised by placing the current as close as possible to the edge of the plasma.

It turns out, however, that the most dangerous large-scale modes in tokamaks are non-axisymmetric, rather than axisymmetric. This includes the sausage-mode and kink instabilities discussed in §16.3.2. There are also many small-scale instabilities which tend to manifest themselves at the outer edge of a tokamak plasma, as we now discuss.

16.3.4 Interchange and Ballooning Modes

So far we have focused on large-scale instabilities. However, small-scale local modes are also pervasive in confined plasmas. The central idea is that magnetic fields provide stable support because the Faraday tension tends to resist bending of the field lines. (This is how the toroidal field stabilises both the sausage-mode and

---

[4] More generally, the calculus of variations tells us that $\delta^2 E_B > 0$, and stability is ensured, even when Ψ is everywhere positive, provided that the minimum eigenvalue, λ, of $\nabla_*^2 \varphi + \lambda \Psi \varphi = 0$, $\varphi = 0$ on the boundary, is greater or equal to unity (Davidson, 1994).

kink instabilities in §16.3.2.) However, a small-scale, local mode can often adopt an orientation such that $\mathbf{B} \cdot \mathbf{k} \approx 0$, $\mathbf{k}$ being the wave vector. The field lines are then displaced relative to each other but are not bent, and since there is no bending of the field lines there is little resulting stabilisation. Rather, the plasma is free to move, carrying the field lines with it. So, if there is some local source of instability, the field lines will offer little resistance when $\mathbf{B} \cdot \mathbf{k} \approx 0$ and an unstable mode can readily grow. In some configurations this leads to a disturbance, referred to as an *interchange instability*, which is characterised by $\mathbf{B} \cdot \mathbf{k} = 0$ and by adjacent flux tubes exchanging position in a manner that releases stored energy.

In particularly simple geometries, interchange instabilities can also manifest themselves as large-scale modes. Perhaps the simplest example of this is the sausage-mode instability described in §16.3.2. Consider the two extremes of the incompressible model problem discussed in §16.3.2: (a) that shown in Figure 16.2, in which $B_T = 0$ and the magnetic field consists of circular field lines surrounding the plasma column (only now we consider the circular field lines to have partially diffused into the plasma) and (b) the opposite extreme in which $B_p = 0$, so that the field lines are straight, aligned with the axis of plasma column. The first of these is known as the z-pinch, and the second as the θ-pinch. In the θ-pinch, the interchange modes correspond to $k = 0$ in Equation (16.4), so that the axial field lines can exchange position but do not bend in the process. In this case (16.4) tells us that $\varpi = 0$, so that the interchange modes are neutrally stable. This is because, although there is no resistance to the growth of disturbances, neither is there any energy source to drive such an instability. By way of contrast, the interchange modes in the z-pinch correspond to $m = 0$, and in this case there is an instability, the sausage-mode instability, driven by the stored magnetic energy in the curved field lines.

The point is that, like an elastic band, a circular flux loop can reduce its stored energy by contracting. In particular, we can write the stored magnetic energy of a thin, circular flux loop of radius s as $E_B = \left(B^2/2\mu \right) V = \left(\Phi^2/2\mu V \right) (2\pi s)^2$, where Φ is the flux in the loop and V is its volume. If the loop contracts while preserving its flux and volume (we continue to restrict ourselves to incompressible motion), then its magnetic energy must fall. So in the z-pinch, though not the θ-pinch, there is a reservoir of stored magnetic energy which can be released. Of course, as some circular flux loops contract, reducing their energy, incompressibility demands that others expand radially outward, increasing their energy, and so the requirement for a net release of magnetic energy is a little more subtle. In interchange modes we must consider what happens when two adjacent loops at different radii and with different fluxes exchange position. If the inner flux tube has radius s_1 and flux Φ_1 and the outer one has radius $s_2 = s_1 + \delta s$ and flux Φ_2, then the net change in magnetic energy following an exchange of the two loops is readily shown to be

$$\delta^2 E_B = -\frac{4\pi^2 s_1 (\delta s)^2}{\mu V} \frac{\Phi_2^2 - \Phi_1^2}{\delta s} = -\frac{sV}{\mu} \frac{d}{ds} \left(\frac{B}{s}\right)^2 (\delta s)^2. \qquad (16.21)$$

We conclude, therefore, that in the z-pinch magnetic energy is released provided that B/s within the plasma increases with radius at the point where the instability occurs. This condition is satisfied when the magnetic field is localised at the outer surface of the plasma, as in Figure 16.2.

The general idea, then, is that interchange modes, defined by $\mathbf{B} \cdot \mathbf{k} = 0$, will grow whenever there is a source of energy to drive the instability, and in an incompressible motion that source of energy is necessarily magnetic, in which field lines try to contract to reduce their energy. In the simple case of a cylindrical column with a purely toroidal or poloidal field, the interchange modes can be large scale, because of the high degree of symmetry in the base state. However, in more complex toroidal geometries, in which the field lines are helical, the requirement that $\mathbf{B} \cdot \mathbf{k} = 0$ tends to restrict interchange perturbations to spatially localised disturbances, as discussed below.

To help suppress these local instabilities it is often advantageous to vary the pitch, or winding number, of the helical field lines from one constant pressure surface to the next (see Figure 16.1), so that if the plasma locally bulges outward, carrying its field lines with it, the plasma finds its way barred by surrounding field lines of a different inclination. One measure of the pitch of the field lines is the so-called *safety factor*,

$$q(s) = \frac{sB_T(s)}{RB_p(s)},$$

where R is the major radius and, as before, s is the distance from the magnetic axis in the poloidal plane, $0 \leq s \leq a$. Note that, according to (16.10), we require $q(a) > 1$ in order to avoid kink instabilities. The interchange instability is then characterised by a competition between whatever drives the instability and the local shear in the magnetic field, as measured by the normalised rate of change of $q(s)$ across the constant pressure surfaces, $S = d\ln q / d\ln s$.

When discussing non-axisymmetric disturbances, such as interchange modes, it is conventional to rewrite the incompressible energy principle (16.15) in a slightly different form:

$$\delta^2 E_B(\boldsymbol{\eta}) = \frac{1}{2\mu} \int \left[\mathbf{b}_\perp^2 + B_0^2 [\nabla \cdot \boldsymbol{\eta}_\perp + 2\boldsymbol{\eta} \cdot \boldsymbol{\kappa}]^2 - \mu \mathbf{b}_\perp \cdot (\boldsymbol{\eta}_\perp \times \mathbf{J}_{//}) \right.$$

$$\left. -2\mu(\boldsymbol{\eta} \cdot \nabla p)(\boldsymbol{\eta} \cdot \boldsymbol{\kappa}) \right] dV, \qquad (16.22)$$

where // and ⊥ mean parallel and perpendicular to $\mathbf{B}_0$, $\mathbf{J}_{//}$ is the component of $\mathbf{J}_0$ parallel to $\mathbf{B}_0$, and $\nabla p = \mathbf{J}_0 \times \mathbf{B}_0$. (See Exercises 16.1 to 16.5 for the derivation of (16.22).) The vector $\mathbf{\kappa}$ is the curvature of the $\mathbf{B}_0$-lines, $\mathbf{\kappa} = -\hat{\mathbf{e}}_n/R_c$, where $\hat{\mathbf{e}}_n$ is the local unit normal to $\mathbf{B}_0$ (pointing away from the centre of curvature) and R_c is the local radius of curvature. Evidently, there are two potential sources of instability in (16.22), one associated with $\mathbf{J}_{//}$ and another with ∇p (or equivalently $\mathbf{J}_\perp$). Traditionally, these have been referred to as *current-driven* and *pressure-driven* modes, respectively, although both sources of instability are ultimately dictated by the equilibrium magnetic field distribution. It turns out (see, for example, Freidberg, 2014) that the current-driven modes are usually associated with kink instabilities and the pressure-driven modes with localised interchange modes ($\mathbf{B} \cdot \mathbf{k} = 0$), or interchange-like modes ($\mathbf{B} \cdot \mathbf{k}$ small but finite).

If we accept the traditional interpretation of (16.22), then the driving force for local instabilities in a confined plasma is often a radial pressure gradient, and in particular a radial pressure gradient in regions where $(\mathbf{\eta} \cdot \nabla p)(\mathbf{\eta} \cdot \hat{\mathbf{e}}_n) < 0$, or equivalently $\hat{\mathbf{e}}_n \cdot \nabla p < 0$. Since ∇p points inward towards the magnetic axis, this occurs when $\hat{\mathbf{e}}_n$ is directed outward, towards the edge of the plasma. This is termed a region of *unfavourable curvature*, and it is this which drives the interchange modes in a confined plasma. The simplest configuration in which this instability manifests itself is a pinched cylindrical plasma where a balance between $S(s)$ and the destabilising radial pressure gradient leads to a local stability criterion known as Suydam's criterion. This necessary (but not sufficient) criterion requires (see Freidberg, 2014)

$$S^2 = \left(\frac{d\ln q}{d\ln s}\right)^2 > \frac{8s}{B_T^2/\mu}\left|\frac{dp}{ds}\right|.$$

Perhaps we need to add a caveat at this point. When considering incompressible motion, the idea that ∇p can drive an instability seems slightly odd, as there is no source of free energy associated with ∇p which might fuel such an instability. Rather, ∇p is simply dictated by the magnetostatic field $\mathbf{B}_0$, and the ultimate source of any stored energy is magnetic. So, in the incompressible limit, ∇p in (16.22) must be acting as a proxy for some feature of the magnetostatic field. In this regard it is useful to note that ∇p is related to the curvature, $\mathbf{\kappa}$, and magnetic pressure gradient, $\nabla(B^2/2)$, by

$$\mu\nabla p = \mu\mathbf{J} \times \mathbf{B} = \mathbf{B} \cdot \nabla\mathbf{B} - \nabla\left(\mathbf{B}^2/2\right) = B^2\mathbf{\kappa} - \nabla_\perp\left(\mathbf{B}^2/2\right),$$

which follows directly from (4.23) and the idea of a Faraday tension in the field lines. (See also Exercise 16.1 for a derivation of this expression.) Thus the destabilising term, $(\boldsymbol{\eta} \cdot \nabla p)(\boldsymbol{\eta} \cdot \boldsymbol{\kappa}) \sim \eta_\perp^2 (\nabla p \cdot \boldsymbol{\kappa})$, in (16.22) can be rewritten in terms of $\mathbf{B}_0$. For example,

$$\eta_\perp^2 (\nabla p \cdot \boldsymbol{\kappa}) = \frac{\eta_\perp^2}{R_c} \left[\frac{B^2}{\mu R_c} + \frac{\partial}{\partial n} \frac{B^2}{2\mu} \right] = \eta_\perp^2 R_c \frac{\partial}{\partial n} \left(\frac{B^2}{2\mu R_c^2} \right) + \frac{B^2}{2\mu} (2\boldsymbol{\eta} \cdot \boldsymbol{\kappa})^2, \quad (16.23)$$

where, as above, R_c is the local radius of curvature of the field lines and n is the local normal coordinate, increasing away from the centre of curvature. The expression on the right of (16.23) tells us that when we combine $(\boldsymbol{\eta} \cdot \nabla p)(\boldsymbol{\eta} \cdot \boldsymbol{\kappa})$ in (16.22) with the curvature term $(2\boldsymbol{\eta} \cdot \boldsymbol{\kappa})^2 B^2/2\mu$, we obtain $\eta_\perp^2 R_c \partial (B^2/2\mu R_c^2)/\partial n$, which is a local version of the perturbed energy expression (16.21).

Expression (16.23) allows us to interpret the instability mechanism in terms of changes in magnetic energy. Recall that curved field lines, which are subject to a Faraday tension, can reduce their energy by contracting towards the centre of curvature. However, for every flux tube that contracts in an incompressible fluid, another must expand and, as with the z-pinch, energy is released only if the contracting flux tubes release their energy more rapidly than the expanding ones acquire energy. This, in turn, requires that the initial magnetic energy in the contracting tubes be greater than that in expanding ones, as in (16.21). This is the origin of the term $\eta_\perp^2 R_c \partial (B^2/2\mu R_c^2)/\partial n$ in (16.23), which tells us that the equilibrium can be stabilised, despite the curved field lines, provided that the magnetic energy density (normalised by R_c^2) decreases with distance from the centre of curvature. On the other hand, if the magnetic energy increases with distance from the centre of curvature, then the equilibrium is potentially unstable to curvature-driven modes.

The caveat above refers to incompressible motion. It should be said, however, that compressibility cannot always be neglected, although marginally unstable interchange perturbations do tend to be incompressible. In any event, it is conventional to refer to instabilities driven by the last term in (16.22) as pressure-driven modes, and to interpret the dynamics of those instabilities in terms of ∇p. We shall conform to this convention.

In a tokamak, the issue of stability to pressure-driven perturbations is complicated by the fact that there are two contributions to the curvature vector: one from the poloidal field and another from the toroidal field. In the large aspect ratio approximation, we have

$$\boldsymbol{\kappa} \approx - \left(\frac{B_p}{B_0} \right)^2 \frac{\hat{\mathbf{e}}_s}{s} - \frac{\hat{\mathbf{e}}_r}{r},$$

where, as usual, we employ global cylindrical polar coordinates, (r, θ, z), and s is the distance from the magnetic axis in the poloidal plane. While the curvature of the poloidal field lines is always unfavourable, that associated with the toroidal field alternates between good and bad. There is now the possibility that the net curvature is favourable at the inside surface of the torus, where $\hat{\mathbf{e}}_r \cdot \nabla p > 0$, and unfavourable at the outer edge of the torus, where $\hat{\mathbf{e}}_r \cdot \nabla p < 0$, and indeed this is typically the case in a tokamak. In such cases individual field lines pass through both favourable and unfavourable regions as they circulate around the torus. It comes as no surprise, therefore, that the most dangerous pressure-driven modes in a tokamak are concentrated at the outer edge of the torus, where the radial pressure gradient coincides with an unfavourable direction of $\hat{\mathbf{e}}_n$ for the toroidal field. These are similar to interchange modes in that $\mathbf{B} \cdot \mathbf{k}$ is small (though finite), with the instability localised along a flux tube, with a short wavelength perpendicular to the flux tube and a long wavelength parallel to it. So, as with interchange modes, there is little field line bending and so the potentially stabilising Faraday tension is ineffective. This instability is characterised by a ballooning outward of filamentary flux tubes in the region of unfavourable curvature (see Figure 16.5), and so is referred to as a *ballooning instability*.

Loosely speaking, the difference between an interchange perturbation and a ballooning instability is that in the former we need to keep $\mathbf{B} \cdot \mathbf{k} = 0$, so as to

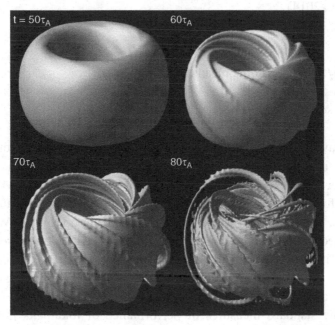

Figure 16.5 The numerical simulation of the non-linear growth of ballooning modes in a spherical tokamak. (From Khan et al., 2007, with permission.)

minimise field line bending, and this means that the instability must grow equally in regions of favourable and unfavourable curvature. In the ballooning mode, however, the instability is concentrated in the region of unfavourable curvature, and this involves a small but finite bending of the field lines, $\mathbf{B} \cdot \mathbf{k} \neq 0$, since the unstable mode grows more in the unstable region. It turns out that concentrating the plasma displacement in the most unstable region is an effective strategy, despite the penalty of a finite field line bending, so ballooning modes are more unstable than interchange modes in most tokamaks.

Crucially, the need to avoid ballooning modes in a tokamak limits the allowable radial pressure gradient in the plasma and this, in turn, demands a low value of β. The resulting limit on β can be estimated by equating the stabilising term in (16.22),

$$b_\perp^2 \sim (\mathbf{B} \cdot \nabla \boldsymbol{\eta})^2 \sim \left(B_p^2/a^2\right)\eta^2 \sim B_T^2\eta^2/q^2R^2,$$

to the final term,

$$\mu(\boldsymbol{\eta} \cdot \nabla p)\boldsymbol{\eta} \cdot \boldsymbol{\kappa} \sim \mu(\eta p/a)\eta/R,$$

which drives the instability. This balance suggests a critical β of the order of $\beta \sim a/q^2R$, where R and a are the major and minor axes of the torus. (See, for example, Biskamp, 1993.) A more detailed calculation gives the limiting value of β as (Freidberg, 2014)

$$\beta \approx 0.2\varepsilon/\bar{q}^2, \tag{16.24}$$

where $\varepsilon = a/R$ and $\bar{q}$ is some mean measure of $q(s)$.

16.4 The Development of Tokamak Reactors

Tokamaks have grown in scale and power ever since the first successful trials of the T-4 tokamak in Novosibirsk in 1968 (top image in Figure 16.6). In the early 1990s, fusion power from experiments was measured in fractions of a MegaWatt and maintained for only extremely short periods of time, but by the late 1990s JET (Figure 16.6, bottom image) was capable of producing up to 16 MW for slightly less than a second. By way of comparison, the proposed fusion power from ITER (Figure 16.7) is 500MW and the design burn time is several hundred seconds. The corresponding growth in reactor size can be judged by comparing the images in Figures 16.6 and 16.7, which show T-4, JET and ITER.

One of the standard measures of performance of tokamaks is the quantity $(n\tau)T$, which is called the fusion triple product. Lawson's criterion tells us that,

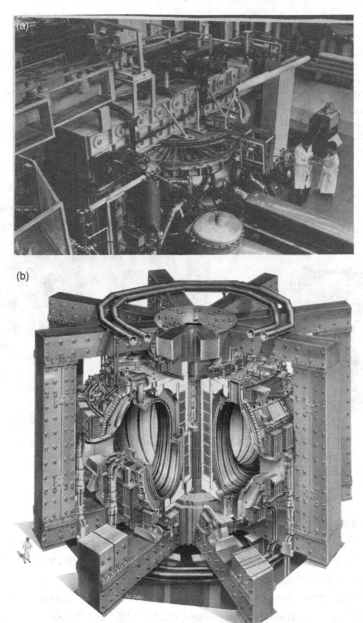

Figure 16.6 (a) The pioneering T-4 tokamak in Russia (https://alltheworldstoka maks.wordpress.com). (b) The JET tokamak in the UK. Notice the difference in scale, as indicated by the size of the people in the figures.

for a given T, this should be as high as possible and in fact $(n\tau)T$ has grown more or less exponentially over the last fifty years. In part, it is this relentless rise in $(n\tau)T$ that leads to optimism that ITER might finally produce a self-

Table 16.2 *A comparison of the design parameters of JET and ITER.*

	Major radius of torus, R (m)	Minor radius of torus, a (m)	Safety factor, $q(a)$	Toroidal current (MA)	Toroidal field on axis (T)	Beta based on toroidal field	Peak power (MW)
JET	3	1.2	~3	4	3.5	0.03	16
ITER	6.2	2.0	~2	15	5.3	0.03	500

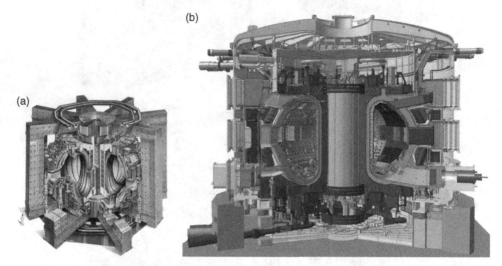

Figure 16.7 A comparison of the JET tokamak in the UK (left) and the ITER tokamak under construction in Cadarache, France (right). The two images are to scale. (ITER organisation, 2013.)

sustaining fusion reaction. ITER itself is much larger than any of its predecessors. Some sense of its size may be gained by comparing the images of JET and ITER in Figure 16.7. A comparison of the design parameters of JET and ITER is given in Table 16.2.

Perhaps the most dangerous MHD instabilities in the current generation of tokamaks are the so-called edge-localised modes (ELMs). These are local instabilities which take the form of filamentary flux tubes ballooning out from the plasma. They occur most violently in and around the so-called *pedestal*, which is a thin layer of plasma at the outside edge of the torus where there are particularly large radial gradients in temperature and pressure. Such a pedestal tends to form at high rates of plasma heating, as the plasma relaxes into the so-called *H-mode configuration*.

There are many types of ELMs, but the most prominent is closely related to the classical ballooning instability. Such instabilities, if unchecked, can cause severe damage to the first wall, and so they constitute a major source of concern for the ITER tokamak. In particular, in order to maintain a self-sustained fusion reaction for several hundred seconds, ITER will need to control all such edge modes, and whether or not this can be achieved has been the subject of some discussion.

In any event, if successful, ITER will demonstrate that a self-sustained (if brief) release of fusion power in a tokamak is feasible. While this would represent an extraordinary tribute to the ingenuity and patience of plasma physicists, spanning more than half a century, it is important to keep in mind that there is a large gap between this kind of 'proof-of-principle experiment' and the reality of an economically viable commercial fusion reactor. Looking beyond ITER, there are, perhaps, three major obstacles to overcome before commercial power generation could be realistically considered. Let us take these in order of difficulty, leaving the most daunting till last.

The first point to note is that tokamaks generate their toroidal current inductively through transformer action (Figure 16.4). This can maintain a transient current for only a short period of time, whereas a commercial fusion reactor must operate in a quasi-steady manner. Clearly, an alternative means of maintaining the toroidal current must be found. While external current drives, such as neutral beam injection, can be deployed in proof-of-principle experiments, their low efficiency makes external current drives problematic when it comes to a commercial reactor. An alternative strategy is to take advantage of a fortuitous phenomenon in fusion plasmas, known as the *bootstrap current*. However, for this to replace the inductively generated current requires a plasma pressure somewhat above that allowed by conventional stability considerations (Freidberg, 2014). So the move from a short-lived experiment to one of a sustained fusion reaction is non-trivial.

The second point concerns the development of the engineering systems required to support a commercial tokamak reactor in a robust and safe manner. In the eyes of many, the development of these systems will be every bit as challenging as confining the plasma in the first place. In addition to the first wall, which must sustain high radiation levels for prolonged periods, there are the superconducting magnets whose size and power are at the very limits of what is achievable, and the reactor blanket, which must breed tritium and extract the neutron-induced heat in a controlled fashion. All of these systems pose formidable challenges, and it is not clear that they can be resolved.

The third major hurdle is that of economics. Recall that back in 1956, as the fission reactor at Calder Hall came online, the press announced that nuclear power would be 'too cheap to meter', echoing a phrase used two years earlier by Lewis Strauss in New York. Of course, the reality proved very different. A glance at Figure 16.7 suggests that a tokamak capable of maintaining even just a short-lived release of fusion power is a highly complex engineering structure, considerably more intricate than, say, a fission reactor. In addition to the cost of the vast superconducting magnets and the reactor chamber, whose first wall will sustain continual radiation damage, a commercial reactor will require a blanket to breed tritium and extract heat. Both the reactor chamber and the blanket will need regular maintenance and, as parts of these structures will become radioactive, this suggests the need for some sort of vast (and expensive) remote handling system. So, even if the engineering turns out to be feasible, which is far from certain, a tokamak reactor may well turn out to be prohibitively expensive to build and maintain. Currently it looks as if the economics some decades from now will favour some combination of solar and wind power, fossil fuels and fission. But then, none of us have a crystal ball!

16.5 Tritium Breeding and Heat Extraction: MHD Channel Flow Revisited

We close this chapter with a brief discussion of proposed blanket designs. Various schemes have been proposed, with the earliest designs using liquid lithium to fulfil the dual roles of coolant and tritium breeder. Later proposals replaced lithium by lithium-lead eutectic and use helium to cool all surfaces, particularly the first wall. In some recent schemes the helium also cools the moderator, carrying the fusion energy out of the blanket. In such proposals the liquid lithium-lead eutectic flows only very slowly, perhaps at 1 mm/s, just fast enough so that tritium can be extracted and processed. In yet other schemes, a dual cooling system is envisaged, with helium cooling all surfaces and the liquid lead-lithium eutectic flowing through poloidal ducts and extracting the neutron-induced volumetric heating from the blanket. In this second case, the liquid metal must be pumped at speeds of around 10 cm/s, which presents a major challenge, given the intensity of the ambient magnetic fields, which directly oppose such pumping. While plasma containment is an exercise in high-R_m MHD, the pumping of liquid-metal coolants belongs to the realm of low-R_m MHD.

Lithium-lead eutectic has a density of $\rho = 9.2 \times 10^3$ kg/m^3, a viscosity of $\nu = 1.4 \times 10^{-7}$ m^2/s and a magnetic diffusivity of $\lambda = 1.06$ m^2/s. If we take a characteristic speed, duct size and field strength of $u = 0.1$ m/s, $d = 0.05$ m and $B = 10$T, we obtain a typical Hartmann number and interaction parameter of

$$\mathrm{Ha} = Bd\left(\frac{\sigma}{\rho\nu}\right)^{1/2} = 1.2 \times 10^4, \quad N = \frac{\sigma B^2 d}{\rho u} = 4.1 \times 10^3.$$

However, $(\mathrm{Ha})^2$ represents the ratio of Lorentz to viscous forces at low R_m, while N is indicative of the ratio of Lorentz to inertial forces, also at low R_m. (See §4.1 for

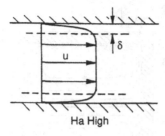

Figure 16.8 Duct flow at large Hartmann number. The magnetic field is upward and the induced core current is out of the page. Hartmann layers of thickness $\delta = d/\text{Ha}$ form on boundaries normal to the magnetic field.

a discussion of the significance of Ha and N.) We conclude that the viscous and inertial forces are negligible outside singular layers, such as boundary layers. Thus we are in the classical regime of high-Hartmann-number duct flow where, outside boundary layers and internal shear layers, the viscous and inertial forces may be neglected. Indeed, the liquid-metal cooling of fusion blankets represents the most extreme case of high-Hartmann-number duct flow encountered in engineering. Of course, such inviscid core flows are surrounded by thin boundary layers in which the viscous and Lorentz forces compete (Figure 16.8).

These kinds of duct flow are discussed briefly in §6.5. Within the core of the duct there is a balance between Lorentz forces and pressure gradients, and if the induced current is recycled through highly conducting duct walls, we find

$$\left|\frac{dp}{dx}\right| \sim \sigma B^2 u \sim N \frac{\rho u^2}{d} \gg \frac{\rho u^2}{d}. \tag{16.25}$$

Evidently, for high values of N the pressure drop becomes extremely large when the duct walls are electrically conducting. Consequently, to minimise the current flowing within the core of the duct, and hence the resulting pressure drop, the duct walls are usually electrically insulated from the coolant. In such cases the induced core current is recycled through thin, low-speed boundary layers which form on the boundaries (Figure 16.9). The channelling of the induced current through these narrow layers increases the effective resistance of the current paths and so reduces the core current, as discussed below.

Perhaps it is worth taking a moment to summarise the different kinds of shear layers that can form is such duct flows. Consider fully developed, high-Ha flow through a square duct of cross-sectional width d and subject to an imposed, uniform field **B**. As noted in §6.5, the duct walls which are normal to the magnetic field develop Hartmann layers of thickness $\delta \sim d\,\text{Ha}^{-1}$, where Lorentz and viscous forces compete. These are the top and bottom layers shown in Figure 16.8, where the

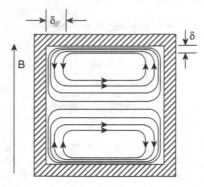

Figure 16.9 Current paths in MHD duct flow. When the duct walls are electrically insulating, the core current is recycled through side layers of thickness $\delta_{//} = d/\text{Ha}^{1/2}$. The flow is into the page.

magnetic field points upward. By contrast, insulated boundaries parallel to the magnetic field develop side layers (Figure 16.9) whose primary function is to recycle current. These are thicker and scale as $\delta_{//} \sim d\text{Ha}^{-1/2}$. Thus the core current is recycled via the side layers, flowing vertically upward or downward within the side layers until it meets the top or bottom boundary of the duct (Figure 16.9). Finally, the current returns through the Hartmann layers, with the current in the Hartmann layers opposite in direction to the core current. Since the role of the core current is to oppose the duct flow, the reversed current in the Hartmann layers is associated with a stream-wise Lorentz force which, in turn, creates large viscous stresses in order to maintain local equilibrium.

The benefit of electrically insulating the duct walls can be verified by integrating the low-R_m form of Ohm's law around one of the closed current paths shown in Figure 16.9:

$$\oint \mathbf{J} \cdot d\mathbf{r} = \sigma \oint \mathbf{u} \times \mathbf{B} \cdot d\mathbf{r} \approx \sigma u B d. \tag{16.26}$$

The integral on the left is dominated by the intense current density in the Hartmann layer which, since $\mathbf{J}$ is solenoidal, is of order $J_c(d/\delta) \sim J_c\text{Ha}$, J_c being the core current. Thus we obtain the estimate $J_c \sim \sigma u B/\text{Ha}$ for insulated boundaries, which may be contrasted with $J_c \sim \sigma u B$ for highly conducting walls. There is a corresponding drop in the Lorentz force and associated pressure drop, with (16.25) replaced by

$$\left|\frac{dp}{dx}\right| \sim \frac{\sigma B^2 u}{\text{Ha}} \sim \frac{N}{\text{Ha}} \frac{\rho u^2}{d}. \tag{16.27}$$

As noted above, N and Ha are both large and often they have similar magnitudes, so that insulating the duct walls can reduce the pressure drop to a more acceptable value, on the order of $\rho u^2/d$.

In addition to thin wall layers, internal shear layers, often called Ludford layers, can also form due to geometrical discontinuities at the walls (e.g. bends and corners). Typically these also have a thickness that scales as $\text{Ha}^{-1/2}$. Within these internal shear layers there is a balance between Lorentz, viscous and inertial forces, and so their internal structure is often much more complicated than that of a simple Hartmann layer.

There is a rich history of characterising these kind of MHD duct flows using classical asymptotic methods, and indeed the anticipation of liquid-metal cooling of fusion blankets has been one of the main motivations for studying MHD duct flows for over half a century. Many of these MHD flows, and their associated shear layers, are described in some detail in Müller and Bühler (2001). From the point of view of blanket cooling, the recurring problem is that the intense magnetic fields induce prohibitive pressure drops in the poloidal coolant ducts, and a variety of schemes have been proposed to minimise these pressure drops (Bühler, 2007).

It is intended that ITER act as a test bed for a number of different blanket designs, although attempts at self-sufficient tritium breeding, as required by a commercial reactor, will be left to later generations of experimental tokamaks. So it looks like, as with controlled fusion in general, this is a story that has a long way to run.

Exercises

(These exercises all relate to energy principle 16.22.)

16.1 The curvature vector for magnetic field lines is $\boldsymbol{\kappa} = \mathbf{t} \cdot \nabla \mathbf{t} = -\hat{\mathbf{e}}_n/R_c$, where $\mathbf{t} = \mathbf{B}/B$ is the unit tangential vector. Expand $(\mathbf{B}/B) \cdot \nabla(\mathbf{B}/B)$ to show that, for magnetostatic fields,

$$B^2\boldsymbol{\kappa} = \mu\nabla p + \nabla_\perp(B^2/2). \qquad (16.28)$$

16.2 Confirm that $\mathbf{B}_0 \cdot [\mathbf{J}_0 \times \mathbf{b} + \nabla(\boldsymbol{\eta}.\nabla p)] = 0$ in a magnetostatic field, where $\mathbf{b} = \nabla \times (\boldsymbol{\eta} \times \mathbf{B}_0)$. Hence, show that $\boldsymbol{\eta} \cdot [\mathbf{J}_0 \times \mathbf{b}]$ in (16.15) may be replaced by

$$\boldsymbol{\eta}_\perp \cdot [\mathbf{J}_0 \times \mathbf{b} + \nabla(\boldsymbol{\eta}.\nabla p)],$$

or equivalently by

$$\boldsymbol{\eta}_\perp \cdot [\mathbf{J}_0 \times \mathbf{b}] - (\boldsymbol{\eta}.\nabla p)\nabla \cdot \boldsymbol{\eta}_\perp. \qquad (16.29)$$

16.3 Confirm that $\boldsymbol{\eta}_\perp \cdot [\mathbf{J}_0 \times \mathbf{b}]$ in (16.29) can be expanded into the form

$$\boldsymbol{\eta}_\perp \cdot [\mathbf{J}_0 \times \mathbf{b}] = \boldsymbol{\eta}_\perp \cdot [\mathbf{J}_{//} \times \mathbf{b}_\perp] + \boldsymbol{\eta}_\perp \cdot [\mathbf{J}_0 \times \mathbf{b}_{//}]$$
$$= \mathbf{b}_\perp \cdot [\boldsymbol{\eta}_\perp \times \mathbf{J}_{//}] + (b_{//}/B_0)\boldsymbol{\eta} \cdot \nabla p$$

and hence show that $\boldsymbol{\eta} \cdot [\mathbf{J}_0 \times \mathbf{b}]$ in (16.15) may be replaced by

$$\mathbf{b}_\perp \cdot [\boldsymbol{\eta}_\perp \times \mathbf{J}_{//}] + (\boldsymbol{\eta} \cdot \nabla p)[(b_{//}/B_0) - \nabla \cdot \boldsymbol{\eta}_\perp]. \tag{16.30}$$

16.4 Expand $\nabla \cdot [\mathbf{B}_0 \times (\boldsymbol{\eta} \times \mathbf{B}_0)] = \nabla \cdot [B_0^2 \boldsymbol{\eta}_\perp]$ to show that

$$B_0 b_{//} = -\nabla \cdot [B_0^2 \boldsymbol{\eta}_\perp] - \mu \boldsymbol{\eta} \cdot (\mathbf{J}_0 \times \mathbf{B}_0) = -B_0^2 \nabla \cdot \boldsymbol{\eta}_\perp$$
$$- \boldsymbol{\eta} \cdot \nabla_\perp (B_0^2) - \mu \boldsymbol{\eta} \cdot \nabla p,$$

and use (16.28) to rewrite this in the form

$$B_0 b_{//} = \mu \boldsymbol{\eta} \cdot \nabla p - B_0^2 \nabla \cdot \boldsymbol{\eta}_\perp - 2B_0^2 \boldsymbol{\eta} \cdot \boldsymbol{\kappa}. \tag{16.31}$$

16.5 Combine (16.30) and (16.31) to show that $b_{//}^2 - \mu \boldsymbol{\eta} \cdot [\mathbf{J}_0 \times \mathbf{b}]$ in (16.15) may be replaced by

$$B_0^2 (\nabla \cdot \boldsymbol{\eta}_\perp + 2\boldsymbol{\eta} \cdot \boldsymbol{\kappa})^2 - \mu \mathbf{b}_\perp \cdot [\boldsymbol{\eta}_\perp \times \mathbf{J}_{//}] - 2\mu (\boldsymbol{\eta} \cdot \nabla p)\boldsymbol{\eta} \cdot \boldsymbol{\kappa}, \tag{16.32}$$

and hence deduce the energy principle in the form

$$\delta^2 E_B(\boldsymbol{\eta}) = \frac{1}{2\mu} \int \left[\mathbf{b}_\perp^2 + B_0^2 (\nabla \cdot \boldsymbol{\eta}_\perp + 2\boldsymbol{\eta} \cdot \boldsymbol{\kappa})^2 - \mu \mathbf{b}_\perp \cdot (\boldsymbol{\eta}_\perp \times \mathbf{J}_{//}) \right.$$
$$\left. - 2\mu (\boldsymbol{\eta} \cdot \nabla p)(\boldsymbol{\eta} \cdot \boldsymbol{\kappa}) \right] dV. \tag{16.33}$$

Appendix A

Vector Identities and Theorems

Grad, div and curl in Cartesian coordinates:

$$\nabla \varphi = \frac{\partial \varphi}{\partial x} \mathbf{i} + \frac{\partial \varphi}{\partial y} \mathbf{j} + \frac{\partial \varphi}{\partial z} \mathbf{k}$$

$$\nabla \cdot \mathbf{F} = \frac{\partial F_x}{\partial x} + \frac{\partial F_y}{\partial y} + \frac{\partial F_z}{\partial z}$$

$$\nabla \times \mathbf{F} = \left(\frac{\partial F_z}{\partial y} - \frac{\partial F_y}{\partial z} \right) \mathbf{i} + \left(\frac{\partial F_x}{\partial z} - \frac{\partial F_z}{\partial x} \right) \mathbf{j} + \left(\frac{\partial F_y}{\partial x} - \frac{\partial F_x}{\partial y} \right) \mathbf{k}$$

Grad, div and curl in cylindrical polar coordinates:

$$\nabla \varphi = \frac{\partial \varphi}{\partial r} \hat{\mathbf{e}}_r + \frac{1}{r} \frac{\partial \varphi}{\partial \theta} \hat{\mathbf{e}}_\theta + \frac{\partial \varphi}{\partial z} \hat{\mathbf{e}}_z$$

$$\nabla \cdot \mathbf{F} = \frac{1}{r} \frac{\partial}{\partial r} (r F_r) + \frac{1}{r} \frac{\partial F_\theta}{\partial \theta} + \frac{\partial F_z}{\partial z}$$

$$\nabla \times \mathbf{F} = \left(\frac{1}{r} \frac{\partial F_z}{\partial \theta} - \frac{\partial F_\theta}{\partial z} \right) \hat{\mathbf{e}}_r + \left(\frac{\partial F_r}{\partial z} - \frac{\partial F_z}{\partial r} \right) \hat{\mathbf{e}}_\theta + \left(\frac{1}{r} \frac{\partial}{\partial r} (r F_\theta) - \frac{1}{r} \frac{\partial F_r}{\partial \theta} \right) \hat{\mathbf{e}}_z$$

$$\nabla^2 \varphi = \frac{1}{r} \frac{\partial}{\partial r} \left(r \frac{\partial \varphi}{\partial r} \right) + \frac{1}{r^2} \frac{\partial^2 \varphi}{\partial \theta^2} + \frac{\partial^2 \varphi}{\partial z^2}$$

Vector identities:

$$\nabla \cdot (\varphi \mathbf{u}) = \varphi \nabla \cdot \mathbf{u} + \mathbf{u} \cdot \nabla \varphi$$

$$\nabla \times (\varphi \mathbf{u}) = \varphi \nabla \times \mathbf{u} + \nabla \varphi \times \mathbf{u}$$

$$\nabla (\mathbf{u} \cdot \mathbf{v}) = \mathbf{u} \times \nabla \times \mathbf{v} + \mathbf{v} \times \nabla \times \mathbf{u} + (\mathbf{u} \cdot \nabla) \mathbf{v} + (\mathbf{v} \cdot \nabla) \mathbf{u}$$

$$\nabla \cdot (\mathbf{u} \times \mathbf{v}) = \mathbf{v} \cdot \nabla \times \mathbf{u} - \mathbf{u} \cdot \nabla \times \mathbf{v}$$

$$\nabla \times (\mathbf{u} \times \mathbf{v}) = \mathbf{u}(\nabla \cdot \mathbf{v}) - \mathbf{v}(\nabla \cdot \mathbf{u}) + (\mathbf{v} \cdot \nabla)\mathbf{u} - (\mathbf{u} \cdot \nabla)\mathbf{v}$$

$$\nabla \times (\nabla \times \mathbf{u}) = \nabla(\nabla \cdot \mathbf{u}) - \nabla^2 \mathbf{u}$$

$$\nabla \times \nabla \varphi = 0$$

$$\nabla \cdot \nabla \times \mathbf{u} = 0$$

Integral theorems:

$$\int_V \nabla \cdot \mathbf{F} dV = \oint_S \mathbf{F} \cdot \mathbf{n} da$$

$$\int_V \nabla \varphi dV = \oint_S \varphi \mathbf{n} da$$

$$\int_V (\nabla \times \mathbf{F}) dV = \oint_S (\mathbf{n} \times \mathbf{F}) da$$

$$\int_S (\nabla \times \mathbf{F}) \cdot \mathbf{n} da = \oint_C \mathbf{F} \cdot d\mathbf{l}$$

$$\int_S (\mathbf{n} \times \nabla \varphi) da = \oint_C \varphi d\mathbf{l}$$

Navier–Stokes equations in cylindrical polar coordinates:

$$\frac{\partial u_r}{\partial t} + (\mathbf{u} \cdot \nabla) u_r - \frac{u_\theta^2}{r} = -\frac{1}{\rho} \frac{\partial p}{\partial r} + \nu \left(\nabla^2 u_r - \frac{u_r}{r^2} - \frac{2}{r^2} \frac{\partial u_\theta}{\partial \theta} \right)$$

$$\frac{\partial u_\theta}{\partial t} + (\mathbf{u} \cdot \nabla) u_\theta + \frac{u_r u_\theta}{r} = -\frac{1}{\rho r} \frac{\partial p}{\partial \theta} + \nu \left(\nabla^2 u_\theta + \frac{2}{r^2} \frac{\partial u_r}{\partial \theta} - \frac{u_\theta}{r^2} \right)$$

$$\frac{\partial u_z}{\partial t} + (\mathbf{u} \cdot \nabla) u_z = -\frac{1}{\rho} \frac{\partial p}{\partial z} + \nu \nabla^2 u_z$$

Appendix B

Physical Properties of Liquid Metals

Metal -	Melting point °C	Reference temperature °C	Density 10^3kg/m^3	Kinematic viscosity 10^{-6}m^2/s	Electrical conductivity $10^6\Omega^{-1}$m^{-1}	Thermal conductivity Wm^{-1}C^{-1}
Titanium	1685	1700	4.1	1.3	0.58	–
Steel*	1495	1600	7.0	0.88	0.71	26
Iron	1535	1600	7.0	0.80	0.72	41
Nickel	1454	1500	7.9	0.62	1.2	–
Copper	1083	1100	7.9	0.51	4.8	160
Aluminium	660	700	2.4	0.60	4.1	95
Magnesium	650	700	1.6	0.80	3.6	81
Tin	232	280	6.9	0.28	2.1	31
Lithium	181	200	0.51	1.2	4.0	47
Sodium	98	100	0.92	0.68	10	89
Woods metal	70	100	9.7	0.29	0.98	8.0
Potassium	64	70	0.82	0.58	7.0	52
Galium	30	70	6.1	0.31	3.8	30
NaK†	−12	40	0.87	0.86	2.6	22
Mercury	−38	30	13.5	0.12	1.0	8.0

* Approximate values for steel with 0.2 per cent carbon
† Sodium-potassium eutectic

References

Part I: From Maxwell's Equations to Magnetohydrodynamics

Chapter 1 A Qualitative Overview of MHD

Davidson, P.A., 1999, Magnetohydrodynamics in material processing. *Annual Reviews Fluid Mech.* **31**, 273–300.

Chapter 2 The Governing Equations of Electrodynamics

Feynman, R.P., Leighton, R.B. and Sands, M., 1964, *The Feynman lectures on physics*, Addison-Wesley.

Jackson J.D., 1999, *Classical electrodynamics, 3rd ed.*, Wiley.

Lorrain, P. and Corson, D., 1970, *Electromagnetic fields and waves*, 2nd ed., W.H. Freeman & Co.

Chapter 3 A First Course in Fluid Dynamics

Acheson, D.J., 1990, *Elementary fluid dynamics*, Clarendon Press.

Batchelor, G.K., 1967, *An introduction to fluid mechanics*, Cambridge University Press.

Davidson, P.A., 2004, *Turbulence: an introduction for scientists and engineers*, Oxford University Press.

Davidson, P.A., 2013, *Turbulence in rotating, stratified and electrically conducting fluids*, Cambridge University Press.

Davidson P.A., Staplehurst P.J. and Dalziel S.B., 2006, On the evolution of eddies in a rapidly rotating system. *J. Fluid Mech.*, **557**, 135–144.

Feynman, R.P., Leighton, R.B. and Sands, M., 1964, *The Feynman lectures on physics*, Vol. II. Addison-Wesley.

Ranjan, A. and Davidson, P.A., 2014, Evolution of a turbulent cloud under rotation. *J. Fluid Mech.*, **756**, 488–509.

Tennekes, H. and Lumley, J.L., 1972, *A first course in turbulence*, The MIT Press.

Part II: The Fundamentals of Incompressible MHD

Chapter 5 Kinematics: Advection, Diffusion, and Intensification of Magnetic Fields

Biskamp, D., 1993, *Nonlinear magnetohydrodynamics*, Cambridge University Press.
Galloway, D.J. and Weiss, N.O., 1981, Convection and magnetic fields is stars., *Astrophys. J.*, **243**, 309–316.
Moffatt, H.K., 1978, *Magnetic field generation in electrically conducting fluids*, Cambridge University Press.
Priest, E., 2014, *Magnetohydrodynamics of the sun*, Cambridge University Press.
Shercliff, J.A., 1965, *A textbook of magnetohydrodynamics*, Pergamon Press.

Chapter 6 Dynamics at Low Magnetic Reynolds Numbers

Chandrasekhar, S., 1961, *Hydrodynamic stability*, Dover.
Davidson, P.A., 1997, The role of angular momentum in the magnetic damping of turbulence, *J. Fluid Mech.*, **336**, 123–150.
Moreau, R., 1990, *Magnetohydrodynamics*, Kluwer Acad. Pub.
Müller, U. and Bühler, L., 2001, *Magnetofluiddynamics in channels and containers*, Springer.
Shercliff, J.A., 1965, *A textbook of magnetohydrodynamics*, Pergamon Press.

Chapter 7 Dynamics at High Magnetic Reynolds Numbers

Arnold, V.I., 1966, Sur un principe variationnel pour les écoulements stationaires des liquides parfaits et ses applications aux problèmes de stabilité non-linéaires. *J. Méc.*, **5**, 9–15.
Balbus, S.A. and Hawley, J. F., 1998, Instability, turbulence and enhanced transport in accretion disks. *Rev. Modern Phys.*, **70** (1), 1–53.
Bernstein, I.B., *et al.*, 1958, An energy principle for hydromagnetic stability problems. *Proc. Roy. Soc. Lond. A.*, **244**.
Biskamp, D., 1993, *Non-linear magnetohydrodynamics*, Cambridge University Press.
Chandrasekhar, S., 1960, The stability of non-dissipative Couette flow in hydromagnetics. *Proc. Nat. Acad. Sci.*, **46**, 253–257.
Chandrasekhar, S., 1961, *Hydrodynamic and hydromagnetic stability*, Oxford University Press.
Davidson, P.A., 2000, An energy criterion for the linear stability of conservative flows. *J. Fluid Mech.*, **402**, 329–348.
Davidson, P.A., 2013, *Turbulence in rotating, stratified and electrically conducting fluids*, Cambridge University Press.
Frieman, E. and Rotenberg, M., 1960, On hydromagnetic stability of stationary equilibria. *Rev. Mod. Phys.*, **32**(4), 898–939.
Kelvin, Lord, 1887, On the stability of steady and of periodic fluid motion. – Maximum and minimum energy in vortex motion. *Phil. Mag.*, **23**, 529.
Moffatt, H.K., 1978, *Magnetic field generation in electrically conducting fluids*, Cambridge University Press.
Moffatt, H.K., 1986, Magnetostatic equilibria and analogous Euler flows of arbitrarily complex topology. Part 2. Stability considerations. *J. Fluid Mech.*, **166**, 359–378.
Velikhov, E.P., 1959, Stability of an ideally conducting liquid flowing between cylinders rotating in a magnetic field. *Soviet Physics JETP*, **36**, 1398–1404.

Chapter 8 An Introduction to Turbulence

Batchelor, G.K., 1953, *The theory of homogeneous turbulence*, Cambridge University Press.

Batchelor, G.K. and Proudman, I., 1956, The large-scale structure of homogeneous turbulence. *Phil. Trans. R. Soc. Lond. A* 248, 369–405.

Biskamp, D., 2003, *Magnetohydrodynamic turbulence*, Cambridge University Press.

Davidson, P.A., 2009, The role of angular momentum conservation in homogeneous turbulence. *J. Fluid Mech.*, 632, 329–358.

Davidson, P.A., 2010, On the decay of saffman turbulence subject to rotation, stratification or an imposed magnetic field. *J. Fluid Mech.*, 663, 268–292.

Davidson, P.A., 2013, *Turbulence in rotating, stratified and electrically conducting fluids*, Cambridge University Press.

Davidson, P.A., 2015, *Turbulence: an introduction for scientists and engineers, 2nd ed.*, Oxford University Press.

Davidson, P.A., Okamoto, N. and Kaneda, Y., 2012, On freely decaying, anisotropic, axisymmetric, Saffman turbulence. *J. Fluid Mech.*, 706, 150–172.

Gödecke, K., 1935, Messungen der atmospharischen Turbulenz in Bodennähe mit einer Hitzdrahtmethode. *Ann. Hydrogr.*, 10, 400–410.

Ishida, T., Davidson, P.A. and Kaneda, Y., 2006, On the decay of isotropic turbulence. *J. Fluid Mech.*, 564, 455–475.

Kolmogorov, A.N., 1941a, Local structure of turbulence in an incompressible viscous fluid at very large Reynolds numbers. *Dokl. Akad. Nauk SSSR*, 30(4), 299–303.

Kolmogorov, A.N., 1941b, Dissipation of energy in locally isotropic turbulence. *Dokl. Akad. Nauk SSSR*, 32(1), 19–21.

Kolmogorov, A.N., 1941c, On the degeneration of isotropic turbulence in an incompressible viscous fluid. *Dokl. Akad. Nauk. SSSR*, 31(6), 538–541.

Kolmogorov, A.N., 1962, A refinement of the concept of the local structure of turbulence in an incompressible viscous fluid at large Reynolds number. *J. Fluid Mech.*, 13 (1), 82.

Krogstad, P.-A. and Davidson, P.A., 2010, Is grid turbulence Saffman turbulence? *J. Fluid Mech.* 642, 373–394.

Landau, L.D. and Lifshitz, E.M., 1959, *Fluid mechanics, 1st ed.*, Pergamon.

Proudman, I. and Reid, W.H., 1954, On the decay of a normally distributed and homogeneous turbulent velocity field. *Phil. Trans. R. Soc. Lond. A*, 247, 163–189.

Saffman, P.G., 1967, The large-scale structure of homogeneous turbulence. *J. Fluid Mech.* 27(3), 581–593.

Tennekes, H. and Lumley, J.L., 1972, *A first course in turbulence*, MIT Press.

Chapter 9 MHD Turbulence at Low and High Magnetic Reynolds Numbers

Batchelor, G.K., 1950, On the spontaneous magnetic field in a conducting liquid in turbulent motion. *Proc. Roy. Soc. London*, A201, 405–416.

Batchelor, G.K., 1953. *The theory of homogeneous turbulence*, Cambridge University Press.

Davidson, P.A., 1997, The role of angular momentum in the magnetic damping of turbulence. *J. Fluid Mech.*, 336, 123–150.

Davidson, P.A., 2009, The role of angular momentum conservation in homogeneous turbulence. *J. Fluid Mech.*, 632, 329–358.

Davidson, P.A., 2010, On the decay of Saffman turbulence subject to rotation, stratification or an imposed magnetic field. *J. Fluid Mech.*, 663, 268–292.

Davidson, P.A., 2013, *Turbulence in rotating, stratified and electrically conducting fluids*, Cambridge University Press.

Davidson, P.A., 2015, *Turbulence: an introduction for scientists and engineers, 2nd ed.*, Oxford University Press.

Federrath, C., Schober, J., Bovino, S. and Schleicher, D.R.G., 2014, The turbulent dynamo in highly compressible supersonic plasmas. *Astro. Phys. Lett.*, 797, L19.

Ishida, T., Davidson, P.A. and Kaneda, Y., 2006, On the decay of isotropic turbulence. *J. Fluid Mech.*, 564, 455–475.

Ohkitani, K., 2002, Numerical study of comparison of vorticity and passive vectors in turbulence and inviscid flows. *Phys. Rev. E.*, 65, 046304.

Okamoto, N., Davidson, P.A. and Kaneda, Y., 2010, On the decay of low magnetic Reynolds number turbulence in an imposed magnetic field. *J. Fluid Mech.*, 651, 295–318.

Saffman, P.G., 1967, The large-scale structure of homogeneous turbulence. *J. Fluid Mech.* 27(3),581–593.

Schekochihin, A.A., Iskakov, A.B., Cowley, S.C., McWilliams, J.C., Proctor, M.R.E. and Yousef, T.A., 2007, Fluctuation dynamo and turbulent induction at low magnetic Prandtl numbers. *New J. Phys.*, 9, 300.

Stribling, T. and Matthaeus, W.H., 1991, Relaxation processes in a low-order three-dimensional magnetohydrodynamic model. *Phys. Fluids B* 3, 1848–1864.

Taylor, J.B., 1974, Relaxation of toroidal plasma and generation of reversed magnetic fields. *Phys. Rev. Lett.*, 33, 1139–1141.

Tobias, S.M., Cattaneo, F. and Boldyrev, S., 2013, MHD turbulence: Field guided, dynamo driven and magneto-rotational. In *Ten chapters in turbulence*, Davidson, P.A., Kaneda, Y. and Sreenivasan, K.R., eds, Cambridge University Press.

Zhdankin, V., Boldyrev, S., Perez, J. C. and Tobias, S.M., 2014, Energy dissipation in MHD turbulence: Coherent structures or nanoflares? *Astrophys. J.*, 795, 127–135.

Part III: Applications in Engineering and Materials

Chapter 10 The World of Metallurgical MHD

Moffatt, H.K. and Proctor, M.R.E., 1984, *Proceedings of the 1982 IUTAM Symposium on Metallurgical Applications of Magnetohydodynamics*, The Metals Society, London.

Chapter 11 The Generation and Suppression of Motion in Castings

Birat, J. and Chone, J., 1982, *4th International Iron & Steel Congress*, London.

Davidson, P.A., 1992, Swirling flow in an axisymmetric cavity or arbitrary profile driven by a rotating magnetic field. *J. Fluid Mech.* **245**: 660–99.

Davidson, P.A.,1995, Magnetic damping of jets and vortices. *J. Fluid Mech.*, **299**: 153.

Davidson, P.A., 1997, The role of angular momentum in MHD turbulence. *J. Fluid Mech.*, **336**: 123–50.

Davidson, P.A. and Hunt, J.C.R., 1987, Swirling, recirculating flow in a liquid metal column generated by a rotating magnetic field. *J. Fluid Mech.* **185**: 67–106.

Marr, H.S., 1982, Electromagnetic stirring in continuous casting of steel. In Moffatt, H.K. and Proctor, M.R.E., *Proc. metallurgical applications of MHD*, The Metals Society.

Takeuchi, E., Masafumi, Z., Takehiko, T. and Mizoguchi, S., 1992, Applied MHD in the process of continuous casting. In *Magnetohydrodynamics in process metallurgy*, Szekely, J., Evans, J.W., Blazek, K. and El-Kaddah, N., The Minerals, Metals and Materials Soc. of USA.

Chapter 12 Axisymmetric Flows Driven by the Injection of Current

Bojarevics, V., Freidbergs, Y., Shilova, E. I., and Shcherbinin, E.V., 1989, *Electrically induced vortical flows*. Kluwer.

Davidson, P.A. and Flood, S.C., 1994, Natural convection in an aluminium ingot: a mathematical model *Metallurgical and materials trans. B.*, **25**B, 293.

Davidson, P.A., Kinnear, D., Lingwood, R.J., Short, D.J. and He, X., 1999, The role of Ekman pumping and the dominance of swirl in confined flows driven by Lorentz forces. *European J. Mech. B*, **18**, 693–711.

Kinnear, D. and Davidson, P.A. 1998. Forced recirculating flow *J. Fluid Mech.*, **375**, 319–344.

Shercliff, J.A., 1970, Fluid motion due to an electric current., *J. Fluid Mech.*, **40**, 241–249.

Chapter 13 MHD Instabilities in Aluminium Reduction Cells

Bojarevics, V. and Romerio, M.V., 1994, Long wave instability of liquid metal-electrolyte interface in an aluminium electrolysis cells: A generalisation of Sele's criterion. *Eur. J. Mech. B*, **13**: 33–56.

Davidson, P.A., 2000, Overcoming instabilities in aluminium reduction cells: A route to cheaper aluminium. *Materials Science and Technology*, **16**, 475–479.

Davidson, P.A and Lindsay, R.I., 1998, Stability of interfacial waves in aluminium reduction cells. *J. Fluid Mech.* **362**, 273–295.

Sneyd, A.D. and Wang, A., 1994, Interfacial instability due to MHD mode coupling in aluminium reduction cells. *J. Fluid Mech.* **263**, 343–359.

Part IV: Applications in Physics

Chapter 14 The Geodynamo

Bin Baqui, Y. and Davidson, P.A., 2015, A phenomenological theory of rapidly rotating turbulence. *Phys. Fluids*, 27(2), 025107.

Christensen, U.R., 2010, Dynamo scaling laws and application to the planets. *Space Sci. Rev.* 152, 565–590.

Christensen, U.R., 2011, Geodynamo models: Tools for understanding properties of Earth's magnetic field. *Phys. of Earth and Planetary Interiors*, 187, 157–169.

Christensen, U.R. and Wicht, J., 2007, Numerical dynamo simulations, In *Treatise on geophysics*, P. Olson, ed., Elsevier.

Cowling, T.G., 1934, The magnetic field of sunspots. *Mon. Not. Roy. Astro. Soc.*, 94, 39–48.

Davidson, P.A., 2013, *Turbulence in rotating, stratified and electrically conducting fluids*, Cambridge University Press.

Davidson, P.A., 2014, The dynamics and scaling laws of planetary dynamos driven by inertial waves. *Geophys. J. Int.*, 198(3), 1832–1847.

Davidson, P.A. and Ranjan, A., 2015, Planetary dynamos driven by helical waves: Part 2. *Geophys. J. Int.*, 202, 1646–1662.

Gailitis, A., Lielausis, O., Platacis, E., Gerbeth, G. and Stefani, F., 2002, Laboratory experiments on hydromagnetic dynamos. *Rev. Mod. Phys.*, 74, 973–990.

Jackson, J.D., 1998, *Classical electrodynamics, 3rd ed.*, Wiley.

Jones, C.A., 2011, Planetary magnetic fields and fluid dynamos. *Ann. Rev. Fluid Mech.*, 43, 583.

Lowes, F.J. and Wilkinson, I., 1963, Geomagnetic dynamo: A laboratory model. *Nature*, 198.

Moffatt, H.K., 1978, *Magnetic field generation in electrically conducting fluids*, Cambridge University Press.

Olson, P., Christensen, U.R. and Glatzmaier, G.A., 1999, Numerical modelling of the geodynamo: Mechanisms of field generation and equilibration. *J. Geophys. Res.*, 104 (B5), 10383.

Parker, E.N., 1955, Hydromagnetic dynamo models. *Astrophys. J.*, 122, 293–314.

Roberts, P.H. and King, E.M., 2013, On the genesis of the Earth's magnetism. *Rep. Prog. Phys.*, 76(9).

Sakuraba, A. and Roberts, P.H., 2009, Generation of a strong magnetic field using uniform heat flux at the surface of the core. *Nature Geoscience*, 2, 802–805.

Sreenivasan, B., 2010, Modelling the geodynamo: Progress and challenges. *Perspectives in Earth Sciences*, 99(12), 1739–1750.

Taylor, J.B., 1963, The magnetohydrodynamics of a rotating fluid and the Earth's dynamo problem. *Proc. Roy. Soc.* A274, 274–283.

Veronis, G., 1959, Cellular convection with finite amplitude in a rotating fluid. *J Fluid Mech.*, 5, 401–435.

Chapter 15 Stellar Magnetism

Armitage, P.J., 2011, Dynamics of protoplanetary disks. *Ann. Rev. Astron. Astrophys.*, 49, 195–236.

Balbus, S.A. and Hawley, J.F., 1998, Instability, turbulence and enhanced transport in accretion disks. *Rev. Modern Phys.*, 70, 1–53.

Carroll, B.W. and Ostlie, D.A., 1996, *Introduction to modern astrophysics*, Addison Wesley.

Cravens, T.E., 1997, *Physics of solar system plasmas*, Cambridge University Press.

Davidson, P.A., 2013, *Turbulence in rotating, stratified and electrically conducting fluids*, Cambridge University Press.

Frank, J., King, A. and Raine, D., 2002, *Accretion power in astrophysics, 3rd ed.*, Cambridge University Press.

Hughes, D.W., Rosner, W. and Weiss, N.O., eds, 2007, *The solar tachocline*, Cambridge University Press.

Mestel, L., 1999, *Stellar magnetism*, Oxford University Press.

Parker, E.N., 1993, A solar dynamo surface wave at the interface between convection and nonuniform rotation. *Astrophys. J.*, 408, 707–719.

Priest, E., 2014, *Magnetohydrodynamics of the Sun*, Cambridge University Press.

Pringle, J.E. and King, A.R., 2007, *Astrophysical flows*, Cambridge University Press.

Shakura, N.I. and Sunyaev, R.A., 1973, Black holes in binary systems. Observational appearance. *Astron. Astrophys.*, 24, 337–355.

Chapter 16 Plasma Containment in Fusion Reactors

Biskamp, D., 1993, *Nonlinear magnetohydrodynamics*, Cambridge University Press.

Boyd, T.J.M. and Sanderson, J.J., 2003, *The physics of plasmas*, Cambridge University Press.

Bühler, L., 2007, Liquid metal magnetohydrodynamics for fusion blankets. In *Magnetohydrodynamics*, Molokov S., Moreau R., Moffatt K., Springer.

Davidson, P.A., 1994, Global stability of two-dimensional and axisymmetric Euler flows. *J. Fluid Mech.*, **276**, 273–305.

Freidberg, J.P., 2014, *Ideal MHD*, Cambridge University Press.

Khan, R., Mizuguchi, N., Nakajima, N., Hayashi, T., 2007, Dynamics of the ballooning mode and the relationship to edge-localised modes in a spherical tokamak. *Phys. Plasmas*, **14**, 062302.

Müller, U. and Bühler, L., 2001, *Magnetofluiddynamics in channels and containers*, Springer.

Index

Printed in the United States
by Baker & Taylor Publisher Services